WORKS ISSUED BY
THE HAKLUYT SOCIETY

WILLIAM ROBERT BROUGHTON'S VOYAGE OF DISCOVERY TO THE NORTH PACIFIC 1795–1798

THIRD SERIES
NO. 22

INTERNATIONAL REPRESENTATIVES OF THE HAKLUYT SOCIETY

Australia	Dr Martin Woods, Curator of Maps, National Library of Australia, Canberra, ACT 2601
Canada	Professor Joyce Lorimer, Department of History, Wilfred Laurier University, Waterloo, Ontario, N2L 3C5
France	Contre-amiral François Bellec, 1 place Henri Barbusse, F92300 Levallois
Germany	Thomas Tack, Ziegelbergstr. 21, D-63739 Aschaffenburg
Japan	Dr Derek Massarella, Faculty of Economics, Chuo University, Higashinakano 742-1, Hachioji-shi, Tokyo 192-03
Netherlands	Dr Leo M. Akveld, Hammerfeststraat 48, 3067 DC Rotterdam
New Zealand	John C. Robson, Map Librarian, University of Waikato Library, Private Bag 3105, Hamilton
Portugal	Dr Manuel Ramos, Av. Elias Garcia 187, 3Dt, 1050 Lisbon
Spain	Carlos Novi, Calle de Vera 26, E-17255 Begur (Girona) and 39 Hazelmere Road, Petts Wood, Kent BR5 1PA, England
Russia	Professor Alexey V. Postnikov, Institute of the History of Science and Technology, Russian Academy of Sciences, 1/5 Staropanskii per., Moscow 103012
USA	Edward L. Widmer, The John Carter Brown Library, Box 1894, Providence, Rhode Island 02912, USA

DONATIONS

The Hakluyt Society and the American Friends of the Hakluyt Society gratefully acknowledge donors in 2009. The Society thinks it appropriate that the generosity of donors should be recorded not only in the Annual Report but also in the books which these donations have helped to make possible. Donations and legacies, and, for British members, Gift Aid, are a valuable supplement to the Society's subscription income, which rarely matches the cost of producing our books to the high standard which we seek to maintain. The officers of the Society, and the editors and series editors of its books, receive no remuneration, and are proud to continue the endeavour which Richard Hakluyt himself inspired, and which we have maintained for 163 years – the endeavour to record, to understand and to interpret the means by which, for better or worse, the various regions, peoples and cultures of the world have become connected with one another.

Mr Edward Alsip
Mr Robert C. Baron
Mr Greg and Mrs Astrid Bear
Mr Matthew Schermer Blum
Dr John R. Bockstoce
Mr Bruce P. Bogert
Mr Charles C. Elk
Dr Norman Fiering
Mr Paul W. Garber
Dr Martin L. Greene
Mr Thomas F. Haas
Mr Todd Hansen
Professor William L. Harris
Mr Warren Heckrotte
Mr Paul Herrup
Dr Ross E. MacPhee and Mrs Clare Flemming MacPhee
Mr William McKinstry
Mr and Mrs Robert M. Norris
Ms Maureen O'Donnell
Mr R. David Parsons
Mr Thomas E. Peckham
Mr Werner Scheule
Professor Paul S. Seaver
Mr Kenneth J. Siple
Mr Edgar L. Weber
Dr Edward L. Widmer
Mr Thomas A. Winter

The Society is especially grateful to Dr Thomas Vaughan CBE and Mrs Elizabeth Crownhart-Vaughan for their donation towards the publication of this particular volume as a mark of friendship for its editor Andrew David.

For Stephen Easton, died 31 December 2009.
For many years printing adviser and devoted friend of the Hakluyt Society,
and its Vice-President for the years 2002–2007.
In affectionate remembrance.

William Robert Broughton by an unidentified artist.
Photograph courtesy of the National Maritime Museum, Greenwich.

WILLIAM ROBERT BROUGHTON'S VOYAGE OF DISCOVERY TO THE NORTH PACIFIC 1795–1798

Edited by
ANDREW DAVID

With an Introduction by
BARRY GOUGH

Published by
Ashgate
for
THE HAKLUYT SOCIETY
LONDON
2010

Published for The Hakluyt Society by

Ashgate Publishing Limited
Wey Court East
Union Road
Farnham
Surrey GU9 7PT
England

Ashgate Publishing Company
Suite 420
101 Cherry Street
Burlington
VT 05401-4405
USA

Ashgate website: http://www.ashgate.com

British Library Cataloguing in Publication Data
Broughton, William Robert, 1762–1821.
William Robert Broughton's Voyage of Discovery to the North Pacific 1795–1798. – (Hakluyt Society. Third series) 1. Broughton, William Robert, 1762–1821 – Travel – North Pacific Region. 2. North Pacific Region – Description and travel – Early works to 1800. 3. Scientific expeditions – North Pacific Region – History – 18th century.
I. Title II. Series III. Voyage of discovery to the North Pacific 1795–1798 IV. David, Andrew, 1924–
915.2'0425–dc22

Library of Congress Cataloging-in-Publication Data
LCCN: 2009944061

ISBN 9780904180978 (hbk)

Typeset by Waveney Typesetters, Wymondham, Norfolk

FSC Mixed Sources
Product group from well-managed forests and other controlled sources
www.fsc.org Cert no. SA-COC-1565
© 1996 Forest Stewardship Council

Printed and bound in Great Britain by
MPG Books Group, UK

Ye Southern, stormy and deserted seas,
Where scarce a vessel tries th' uncertain breeze,
Far from his home my wand'ring Brother braves
Your hidden rocks, and unfrequented waves; ...

And does, unmov'd, your lonely depths explore
Where never human voice was heard before.

Where, more expos'd to storms, the billows roll,
Nor genial climes their boist'rous rage control,
Where hardy firs their gloomy shadows cast
O'er the dark waves, and brave the northern blast,
Mid shelter'd coves, and far retiring bays,
the bartering sailor Nootka's land delays.

From *A Poem on a Voyage of Discovery undertaken by a Brother of the Author's, with other Sonnets, &c*, London, 1792. The anonymous author was Brian Broughton, Fellow of New College, Oxford, and the date of publication shows that it relates to his brother's voyage with Vancouver and not his subsequent one in command of the *Providence*.

CONTENTS

ILLUSTRATIONS AND MAPS

Illustrations

Maps

PREFACE AND ACKNOWLEDGEMENTS

Deciphering the Text

Broughton's writing and those of his amanuenses is usually easy to read, although in a number of places his exact spelling has been difficult to determine. Where uncertainties exist, modern spelling has been adopted, otherwise his spelling, however idiosyncratic, has been followed. The principal difficulties have been to determine whether certain words start with a capital letter or not and whether past participles of verbs in his text end with '…ed', '…'d' or simply '…d' as was common practice at the time of the voyage. The spelling of proper names has also posed a number of problems as Broughton's spelling is inconsistent and often differs from other sources. Where it has been difficult to decipher his actual spelling, that given in the published *Voyage* has often been helpful. Deciphering the exact spelling of some of the other words has been a minor problem; for instance it has been unclear whether Broughton spelled enclosed with an initial 'e' or as inclosed with an initial 'i' or whether he used both spellings indiscriminately. For consistency the modern spelling has been adopted.

There are numerous abbreviations used throughout the journal. Except for inter-cardinal bearings such as NW and standard forms such as Mr, these have been expanded in accordance with Hakluyt Society rules, the missing words being given in italics. Exceptionally words which are routinely abbreviated, such as till for until or bro't for brought, have not been expanded. Punctuation is very inconsistent and often missing. Where necessary this has been supplied in order for the text to make sense. This has been particularly the case with Crosley's letters to Nevil Maskelyne reproduced in Appendix I. The use of capital letters is also inconsistent and sometimes it is difficult to decide which letters are in capitals and which are not. Long sentences have been broken down into more manageable lengths, usually by substituting full stops for semicolons, starting each new sentence with a capital letter. Proper names are also given initial capital letters where necessary, but other words which Broughton has started with initial capital letters have not been amended in accordance with Hakluyt Society editorial policy, except prepositions, pronouns, and definite and indefinite articles. Only words whose nautical meanings are not given in the *Concise Oxford Dictionary* are explained in appropriate footnotes.

Toponymy

The spelling of modern place names is taken from Admiralty Sailing Directions and charts since they contain many more names than are found in modern atlases such as *The Times Atlas*. Thus Russian place names have been transliterated according to the system approved by the Permanent Committee on Geographical Names for British Official Use and the US Board on Geographic Names. Japanese place names, however, pose a problem since no single system has been established by the Japanese for national use. Their Post Office uses the

Modified Hepburn System, which is the system used by British and American Government agencies. On the other hand, the Japanese Hydrographic Office uses a system known as Kokuteisiki. In consequence the spelling of Japanese place names used in this volume differs from names used in *The Times Atlas,* which is based on a different system of transliteration. In addition, the UK Hydrographic Office has now been forced to modify its spelling of Japanese names both on its charts and in Admiralty Sailing Directions since, through international agreements, it has started to publish facsimiles of Japanese charts in its Admiralty series, on which Japanese names have been transcribed into English spelling, omitting all diacritical marks. Thus Kūyshū is now rendered as Kyushu and Tōkyō as Tokyo. The spelling of Chinese names also poses a problem as the former Modified Wade-Giles System of transliteration has been replaced by the Pinyin System in which a name such as Peking has become Beijing and Macao has become Macau. However, certain place names in the former Portuguese colony of Macao, as in the former British colony of Hong Kong, have not yet been changed. Another problem arises in the names of the south-westernmost of the Kurile Islands, since these islands are administered by Russia but claimed by Japan. Where appropriate the Russian names of features in this area are given in footnotes, followed by their names in Japanese when known.

Days and Dates

Broughton has been inconsistent in setting out the day of the week and the dates, sometimes setting them above the day concerned and sometimes in the margin. The actual day of the week is sometimes indicated by the astronomical symbol for the sun, moon and planets, which are also used for the day of the week. Thus ⊙ = Sunday and ☽= Monday, and so on. For consistency the day, date and month are given in bold at the head of each entry ranged hard left.

Marginalia

Entries mainly for variation and occasionally for astronomical observations, entered either in the margin or on the otherwise blank verso pages have been transferred to the main text between square brackets. Where these are given in tabular form this has entailed slight adjustments to accommodate them.

Maps

The track of the *Providence* shown on the various sketch maps is approximate as sufficient reliance could not be placed on their depiction on the charts in the published *Voyage*, nor on the geographical positions given in the journal.

Acknowledgements

We are grateful to the help given by Eric Groves; Alan Perkins, Curator of Scientific Manuscripts, Department of Manuscripts and University Archives, Cambridge University Library; Herbert K. Beals; James E. Hoare; Andrew Cook; Hugh Bett of Maggs Bros Ltd;

Mrs Jennifer Thorp, Archivist, New College, Oxford; Rolf Du Rietz of Uppsala, Sweden; Norman Collingwood; Robert King; the late Anne Yandle, former head of Special Collections, University of British Columbia, who drew our attention to Brian Broughton's 'A Poem on a Voyage of Discoveries'; Professor Glyndwr Williams; Richard Campbell; the late Derek Howse, who supplied some details of Broughton's chronometers, Dr W. R. P. Bourne; Dr Anita McConnell; Dr Gloria C. Clifton, Head of Royal Observatory, Greenwich; Jonathan Betts, Royal Observatory, Greenwich; Roger Quarm, Curator of Paintings, National Maritime Museum, Greenwich; Dr Nigel Rigby, National Maritime Museum; Andrew Davis, Curator of Manuscripts, National Maritime Museum, Greenwich; John Page, Philip Clayton-Gore, Adrian Webb, Guy Hannaford and Liz Hill of the UK Hydrographic Office; Wendy Thirkettle, Manx National Heritage; D. Howlett, Foreign and Commonwealth Office; Anne Wheeldon, Archivist, Archives & Local History, London Borough of Hammersmith & Fulham; Ian Criddle, Assistant Local & Naval Studies Librarian, Plymouth City Council, Central Library; Isabel Simó Rodríguez, Directora, Archivo General de Indias, Seville; Eleanor Gawne, librarian, and the staff of the Caird Library at the National Maritime Museum at Greenwich, who also assisted Barry Gough in his Broughton research when a Caird Fellow of the museum; the staff of The National Archives at Kew; the late Dr John Naish; Manuel Álvarez Casado of the Archivo General de Indias, Seville; Carlos Novi; Raymond John Howgego, series editor for this volume; Barrie Fairhead who prepared the page layout; Sarah Charters of Ashgate Publishing, who saw this volume through the press; and Professor Will Ryan.

GLOSSARY

Chinese (C), Japanese (J), Korean (K) and Russian (R) geographical terms used in this volume. Japanese geographical terms used in *The Times Atlas*, which differ slightly from those used in this volume due to a different system of transliteration, are given in brackets. Chinese geographical terms include both Pinyin and Wade-Giles systems of transliteration.

Foreign term	Language	English meaning
a, ak	K	mountain
am	K	rock
bando	K	peninsula
banka	R	shoal, bank
bong	K	mountain peak
bukhta	R	bay, inlet
chau	C	island
chia	C	cape, bluff
chiao	C	creek, rock, reef, shoal, islet, cape, point
chon	K	river
chou	C	island
dan	K	cape, point
dao	C	island
do	K	island, islands
gawa	J	river
gora	R	mountain, hill
guba	R	gulf, bay, inlet
gunto	J	island group
hae	K	gulf
hana	J	cape, point
hang	K	harbour, point, cape
hanto	J	peninsula
hsi	C	west, mountain, stream
iwa	J	rock
jiang	C	river
jiao	C	point, cape, reef
kaikyo	J	strait
kawa	J	river
kang	C	hill
ko	J	harbour, port
kok	C	headland
kundo	K	archipelago

liedao	C	group of islands
liman	R	estuary
mal	K	point, cape
man	K	bay
men	C	channel, strait
misaki	J	cape, point
more	R	sea
mori	J	forest, wood
mys	R	cape, point, headland
nos	R	headland
ostrov(a)	R	island(s)
pando	K	peninsula
pi	C	cape
port	R	port
proliv	R	strait
reka	R	river
retto	J	chain of islands
reyd	R	roadstead
saki (zaki)	J	cape, point
san	J	mountain
san	K	mountain, hill
se	J	bank, islet, reef, shoal
shan	C	mountain, hill
shar	R	strait, channel
shima (jima)	J	island
shio	J	current
sho	J	bank, reef, shoal
shuidao	C	channel
shoto	J	archipelago
su	J	bank
sudo	K	strait
suido	J	strait, channel
tai	C	big, great, large
take	J	hill, mountain
tao	C	island
to	J	island, islet
tou	C	cape, headland, point
vulkan	R	volcano
wan	C	bay, gulf
wan	J	bay
yama	J	mountain
yo	K	rock
yolto	K	island chain
yu/yü	C	island, islet
zaliv	R	gulf, bay, inlet
zhou	C	shoal, islet

ABBREVIATIONS

ADM	Admiralty papers either in The National Archives or in the National Maritime Museum
AGI	Archivo General de Indias, Seville
AGN	Archivo General de la Nación, Mexico City
AHN	Archivo Histórico Nacional, Madrid
CUL	Cambridge University Library
NMM	National Maritime Museum, Greenwich
RGO	Royal Greenwich Observatory, Board of Longitude Papers
RUSI	Royal United Services Institute Papers, held in the National Maritime Museum
TNA	The National Archives, Kew
UKHO	The United Kingdom Hydrographic Office, Taunton

INTRODUCTION

Barry Gough

1. Background to the Voyage

The voyage of discovery that is the subject of this book was the last of the Royal Navy's exploring voyages to the Pacific Ocean in the eighteenth century. As such it was the culmination of a British national effort that had included the expeditions of John Byron, Samuel Wallis, Philip Carteret, James Cook and George Vancouver. Each of these expeditions had significant objectives. Initially the Admiralty merely intended Broughton to rejoin Vancouver and, on 31 December 1793, he was ordered by the Admiralty to receive on board the astronomer John Crosley for passage to Nootka Sound to join Vancouver's *Discovery,* as a replacement for William Gooch,[1] who had been murdered in Oahu on his way to join Vancouver. Meanwhile final negotiations were taking place in Madrid to settle the Nootka Sound dispute,[2] culminating in the signing on 11 January 1794 of the Third Nootka Sound Convention. As a result Broughton was ordered to carry dispatches for Vancouver from the Duke of Portland, one of His Majesty's Secretaries of State, dated 11 September 1794.[3] These no doubt included copies of the Third Nootka Sound Convention, as well as informing Vancouver that Lieutenant Thomas Pearce of the Marines had been appointed as the British commissioner to implement this final convention and that Pearce had already sailed to carry out his assignment. On 3 October 1794 the Admiralty also furnished Broughton with copies for himself and Vancouver of a treaty of alliance signed on 25 May 1793 between Spain and Great Britain.[4] The Admiralty clearly expected Broughton to rejoin Vancouver and to assist him to complete the survey of the north-west coast of America. However, his instructions – which he is known to have received on 4 October 1793 – do not appear to have survived.[5] According to Broughton, these directed him that, should he fail to meet Vancouver, he 'should survey the southern coast of the south-west part of South America, upon the idea that Captain Vancouver, who had similar orders, would not be able to fulfill them'.[6] In the event, having heard in Hawaii that

[1] TNA, ADM 2/125, pp. 247–8.

[2] In 1789 the Spanish established a garrison on Nootka Island in Nootka Sound and stationed there the frigate *Princesa,* under the command of Estéban José Martínez, in order to forestall a possible Russian settlement. Unknown to the Spanish, British vessels were already using Nootka Sound as a base for trading for sea otter pelts. Shortly after his arrival Martínez seized several British ships, including the *Argonaut,* commanded by James Colnett, and the *Princess Royal.* When news of their seizure reached London, the British fleet was mobilized and war with Spain was only averted by the eventual release of the two ships and the signing of the First Nootka Sound Convention in 1790. A Second Nootka Convention was signed in 1793, which still did not resolve the dispute.

[3] TNA, ADM 12/62, cut 38.

[4] Ibid., ADM 3/114, pp. 78–78v.

[5] The editors, over a period of two decades, have ransacked every possible file of naval and state papers, and of those who conducted or implemented policy, in search of Broughton's instruction but, sadly, without success.

[6] W. R. Broughton, *Voyage of Discovery to the North Pacific Ocean,* London, 1804, p. 64.

Vancouver had sailed for England and that he had called at Valparaíso on his way home, Broughton assumed that Vancouver had been able to comply with this instruction. The field was therefore left to Broughton, in the unique circumstances that he found himself, to determine what his next assignment ought to be.

After consulting his officers, he decided to explore the northern part of Japan and the waters of north-east Asia. Just as one explorer benefited from the work of his predecessors so did Broughton from those who had gone before. Thus Broughton's voyage also had political and scientific objectives. Each succeeding expedition added new details to the chart and increased European knowledge of this vast oceanic space and littoral, and brought or sent home new appreciations of places previously little known in the islands and shores of the Pacific. It was in regard to the north-west Pacific, and to the coasts of Tartary, Japan, Korea and China, that Broughton made his contribution to geographical knowledge. In so doing he established for himself a notable place in the annals of British naval hydrography and more generally in the history of exploration.[1] By determining and verifying by astronomical observations the main features of the Asiatic shore and islands, especially with regard to longitude, he delineated the size of the North Pacific Ocean and established the distances between North America and north-east Asia.

Broughton's voyage therefore signified a shift in British hydrographic inquiry from one side of the North Pacific to the other. Here, on the coasts of north-east Asia, a great European power, Russia, had long established a foothold. In 1689, by the Treaty of Nerchinsk, Russia and Imperial China had agreed on the former's most southerly extension or limit, along the Amur River.[2] Russia also had long had an interest in the Japanese islands, and the question of the insularity of Sakhalin was still undecided in Europe at the time of Broughton's voyage. Similarly, and a decade before Broughton, the French naval officer Jean-François de Galaup de la Pérouse had made his own examination of this puzzling matter. La Pérouse never returned to France, his ships having been wrecked on Vanikoro, one of the Santa Cruz Islands in the South Pacific, and literally lost to history until the early nineteenth century. Details of his geographical discoveries, which had been sent back to Paris in September 1787 from Petropavlovsk, were apparently not known to Broughton, though if they were (and it seems unlikely) Broughton gives no hints.[3]

[1] G. S. Ritchie, *The Admiralty Chart: British Naval Hydrography in the Nineteenth Century* (new edn, with an introductory essay by Andrew David), Edinburgh, 1995. See also James E. Hoare, 'Captain Broughton, HMS *Providence* (and her Tender) and his Voyage to the Pacific 1794–8', *Asian Affairs*, 31, 3 (October 2000), pp. 303–12; another with the same title, *Papers of the British Association for Korean Studies*, 7 (2000), pp. 49–60, and 'Captain Broughton, HMS *Providence* (and her Tender) and Japan 1794–98' in J. E. Hoare, ed., *Britain and Japan: Biographical Portraits*, Vol. 3, Folkestone, Japan Library, 1999, pp. 1–12, 353.

[2] This treaty was signed after a military altercation. It is the opinion of one scholar that neither China nor Russia had well thought out policies as to the future of the Amur River. The driving force seems to have been China's demands for commercial control. The treaty is printed in Frank A. Golder, *Russian Expansion on the Pacific, 1641–1850: An Account of the Earliest and Later Expeditions made by the Russians along the Pacific Coast of Asia and North America; Including some Related Expeditions to the Arctic Regions*, Cleveland, 1914, pp. 290–93. A subsequent treaty, Kiatka 1727, designed to renew the existing peace arrangements, laid down the rules for Russia–China trade at that location and that location only. This is printed in Basil Dmytryshyn, E. A. P. Crownhart-Vaughan, and Thomas Vaughan, eds and trans., *Russian Penetration of the North Pacific Ocean to Siberia and Russian America, 1700–1797: Three Centuries of Russian Eastward Expansion*, Vol. 2: *A Documentary Record*, Portland, Oreg., 1988, pp. 70–78.

[3] The official account of the voyage was published in Paris in 1797 by M. L. A Milet-Mureau, with an English translation published as *A Voyage Round the World, Performed in the Years 1785 ... 1788, by the Boussole and Astrolabe, under the Command of J. F. G. de la Pérouse*, 2 vols and atlas, London, 1799. For a general description of the voyage, see John Dunmore, *The Journal of Jean-François de Galaup de la Pérouse 1785–1788*, 2 vols, Hakluyt Society, 2nd ser. 179–80, London, 1994, I, pp. cli–cxcii.

Broughton faced a similar circumstance to that of La Pérouse – the perennial occupational hazard of hydrographers – of losing his ship to an unknown danger. But great foresight, and the experience of British discoverers, notably Cook, had pointed to the wisdom of having two surveying vessels working together. Broughton lost his command, the *Providence*, but providentially, he lived to tell the tale through his foresight in purchasing a small schooner in Macau to act as a tender. This vessel, which has its own interesting history (see below), became the means of the survival of the *Providence*'s ship's company – itself a harrowing tale of sailing in distant and hitherto uncharted seas – which one day will be properly recognized as one of the classics of sailing heroics in the face of near disaster. The American maritime fur-trader turned historian, William Sturgis, reflecting on the hazards of sailing the Pacific on discovery, wrote in 1846, 'Nearly all the early and distinguished navigators, who discovered and explored the northern regions of the Pacific, met the fate that too often awaits the pioneers in bold and hazardous undertakings, and found a premature death, by violence or disaster, or disease brought on by incessant toil and exposure.'[1] The experience of La Pérouse has already been recounted. What of some of the others? Bering died of scurvy on Ostrov Beringa or Bering Island; Cook was killed at Kealakekua Bay; Charles Clerke succumbed to tuberculosis in the Bering Sea; the American fur-trader John Kendrick was accidentally killed at Honolulu; and Kendrick's second son, Solomon, was killed after the piracy of his vessel by Haida in the Queen Charlotte Islands.[2] Against these examples of misfortune stand William Broughton and George Vancouver, who both completed their long and hazardous voyages, though not without incident. Their lives are inextricably related in the historical record and the annals of North Pacific discoveries.

As in many respects Broughton's voyage can be regarded as an extension of Vancouver's voyage, it seems appropriate to briefly relate the reasons that led to Vancouver sending Broughton, then the commanding officer of the *Chatham*, back to London on 12 January 1793, from Monterey. In September 1792, Vancouver, who had been appointed the British commissioner to implement the First Nootka Sound Convention signed in 1790, had reached an impasse in his negotiations in Nootka Sound with the Spanish commissioner, Juan Francisco de la Bodega y Quadra, over the restoration of British rights there, in accordance with the terms of this convention. The relations between the two were most cordial but difficult with regard to substance and interpretation. In all his negotiations with Bodega, Vancouver was aided by Broughton, who had an excellent command of the French language in which most of the negotiations appear to have taken place. Vancouver could not agree with the Spanish interpretation of the terms of the convention. In particular, he refused to accept the Spanish position regarding the extent of land the Spaniards had seized from the John Meares syndicate in 1789. He wisely wrote to London requesting further instructions. For this purpose, in September 1792, he sent Lieutenant Zachary Mudge of the *Discovery* to London in the Macao-registered brig *Fenis and St Joseph*, by way of Canton and the Cape of Good Hope. Meanwhile, in August of that year, the storeship *Daedalus* arrived at Nootka, fresh from the Hawaiian Islands, bringing supplies to Vancouver but also carrying the sad news of the tragic and senseless killing at Waimea Bay, in Oahu, of Lieutenant Richard Hergest,

[1] 'The Northwest Fur Trade by the Hon. William Sturgis', *Hunt's Merchants' Magazine*, 14 (1846); reprinted in Briton C. Busch and Barry M. Gough, eds, *Fur Traders from New England: The Boston Men in the North Pacific, 1787–1800: The Narratives of William Dane Phelps, William Sturgis and James Gilchrist Swan*, Spokane, Wash., 1997, p. 91.

[2] For other examples and additional details, see ibid., pp. 91–2 and p. 91n.

Vancouver's close friend, and the astronomer William Gooch. The latter had been sent out under instructions from the Board of Longitude to cooperate with Vancouver in scientific discoveries.[1] Gooch's death deprived Vancouver of an astronomer, and a new one therefore had to be sent out in the next possible ship. (This was John Crosley, who subsequently sailed with Broughton.) However, when Vancouver reached Monterey in November 1792, he was informed by Bodega, who had preceded him there, that the latter had received intelligence from Madrid, which suggested that Spain and Great Britain had reached agreement with respect to the territories at Nootka.[2]

As a result, Vancouver decided to send Broughton, who was privy to all the transactions in Nootka, to London to clarify matters, taking with him dispatches and copies of the Spanish charts, which he had been given by Bodega, so that the new negotiations would not again founder for lack of information. Vancouver also suggested indirectly that the Admiralty should send out another vessel to assist him – or to complete his mission – and he also suggested, strongly, that if one were to be sent, Lieutenant Broughton should command her.[3]

Bodega offered to take Broughton with him to the Spanish naval base at San Blas and to arrange for him to travel across New Spain (Mexico) to Vera Cruz on the Gulf of Mexico. Bodega therefore wrote from Monterey to Viceroy Juan Vicente de Güemes Pacheco y Horcacitas, the second Conde de Revillagigedo, in Mexico City, requesting the necessary passport for Broughton to cross New Spain.[4] Broughton and Bodega left Monterey on 12 January 1793 in the Spanish brigantine *Activa*, reaching San Blas on 1 February. Revillagigedo, however, was unhappy to be revealing the weaknesses of New Spain to one of Spain's traditional enemies, but was unwilling to repudiate the actions of his commissioner. His solution was to instruct Bodega to appoint a trustworthy person to accompany Broughton.[5] Bodega therefore instructed the naturalist José Mariano Moziño to accompany Broughton as far as Mexico City.[6] The passport arrived at San Blas on 14 February and the following day Broughton set off for Mexico City accompanied by Moziño, whom he had met at Nootka.[7] Broughton and Moziño reached Mexico City on 8 March, where Broughton met Revillagigedo. Broughton continued his journey on 23 March, accompanied this time by Teniente de Fragata Secundino Salamanca y Humara, who was also known to Broughton as Salamanca and had served in the schooner *Sutil* during the joint Anglo-Spanish survey in Doubtful Sound in 1792. On reaching Vera Cruz, they had to wait for over a fortnight for Capitán de Fragata Cayetano Valdés y Flores, who was also well known to Broughton, as he was the commanding officer of the schooner *Mexicana* during the joint Anglo-Spanish survey referred to above. They sailed for Havana on 17 April in the Spanish frigate *Minerva*, where

[1] The voyage of the *Daedalus* has yet to be fully examined as to its historical importance. Aspects of it are dealt with in Greg Dening, *The Death of William Gooch: A History's Anthropology*, Melbourne, 1995.

[2] W. Kaye Lamb, ed., *George Vancouver, A Voyage of Discovery to the North Pacific Ocean and Round the World*, Hakluyt Society, 2nd ser. 163–6, 4 vols, London, 1984, 2, pp. 728–9.

[3] Vancouver to Philip Stephens, 13 January 1793: TNA, ADM 1/2629, ff. 49–50v; quoted in full in Lamb, *Vancouver's Voyage*, IV, pp. 1582–3.

[4] Bodega to Revillagigedo, 30 December 1792; AHN, Estado 4290 and AGN, Historia 70, cuaderno 25, f. 3.

[5] Revillagigedo to Bodega, 7 February 1793; AGN Historia 70, cuaderno 25, f. 17.

[6] Freeman Tovell, *At the Far Reaches of Empire: The Life of Juan Francisco de la Bodega y Quadra*, Vancouver, 2008, pp. 277–81.

[7] For an account of Moziño's activities and observations at Nootka see Iris H. Wilson Engstrand, trans. and ed., *Noticias de Nutka: An Account of Nootka Sound in 1792 by José Mariano Moziño*, Seattle, 1970.

the three transferred to a Spanish brig, reaching Cadiz about six weeks later.[1] From there Broughton made his way first to Madrid[2] and then to London, where he was able to deliver Vancouver's dispatches to the Admiralty on 16 July 1793[3] only two months later than Mudge. Broughton was thus able to give the British ministry further details of what had transpired at Nootka Sound and on the Pacific Coast of North America.[4] There is no proof, as has been claimed,[5] that he later returned to Madrid as part of the British delegation to give advice to the British ambassador and plenipotentiary in the preliminaries leading to the negotiation of the Third Nootka Sound Convention that finally ended the Nootka Sound controversy and specified the respective rights and limitations of the two powers in that vicinity. But it can be safely said that he gave advice, for it was eagerly awaited by senior members of the British cabinet, W. W. Grenville and Henry Dundas, that helped shape such demands and responses that that administration pressed on the Spanish government through the British ambassador to Madrid. However, it seems as though news of Vancouver's difficulties had already reached London, possibly via Madrid, as a Second Nootka Convention had been signed on 12 February 1793. During the time these negotiations were taking place a treaty of alliance and trade between Spain and Great Britain was also being discussed. The execution of Louis XVI and its accompanying excesses brought the traditional enemies towards an adjustment of their differences and a treaty was signed on 25 May 1793, which was ratified by both governments a few weeks later.[6] This treaty facilitated a final settlement with Spain over the Nootka Sound dispute and Lord St Helens, British Ambassador in Madrid, was again charged with the task of negotiating with Spain. No doubt Broughton, on passing through Madrid, had been able to give the Ambassador valuable advice of someone who had personal knowledge of what had taken place between Vancouver and Bodega and the difficulties they had encountered in implementing the 1790 convention.[7] Eventually, after difficult negotiations, the Third Nootka Convention was signed in Madrid on 11 January 1794.[8]

On 3 October 1794, following a meeting of the Board of Admiralty, attended by the Earl of Chatham, Lord Arden, Admiral Affleck and Sir Charles Middleton, the following minute was recorded:

> The Board having early in the month of May 1793 receiv'd by Lieutenant Mudge, a Letter from Captain Vancouver of the Discovery Sloop of the 21st September, 1792, accompanied with a Journal of his Proceedings and Copy of the Correspondence between Señor Don Juan Francisco de la Bodega

[1] Broughton does not note the exact date of his arrival at Cadiz in his journal, but it was probably during the last week in May. For a full account of his journey across New Spain and his voyage to Cadiz, see Appendix VI.

[2] In a letter to Vancouver, Conde Revillagigedo reported Broughton's safe arrival in Madrid; Lamb, *Vancouver's Voyage*, IV, p. 1417.

[3] Ibid, 4, p. 1582 and TNA, ADM 1/2629.

[4] Grenville to H. Dundas, 10 June 1793, TNA, CO 5/187, copy in BC Archives, GR 329, p. 120. This document makes clear that a company of Marines was being raised, with final orders being given in that same month, for an occupation of Nootka if Spain did not accede to British demands. Broughton's arrival, and news, allowed the more pacific course, and by January 1794 the British finally got all that they had demanded in principle.

[5] William Ray Manning, 'The Nootka Sound Controversy', *Annual Report of the American Historical Association for the Year 1904*, Washington, 1905, pp. 469–71.

[6] Ibid., p. 468. See also p. 39 below.

[7] It seems unlikely that Broughton was sent back to Madrid as suggested in Lamb, *Vancouver's Voyage*, I, p. 109.

[8] For full Spanish texts of the three conventions see Warren L. Cook, *Flood Tide of Empire: Spain and the Pacific Northwest, 1543–1819*, New Haven and London, 1973, pp. 544–8, and for the English texts see Manning, 'The Nootka Sound Controversy', pp. 454–6 and 467–70.

> y Quadra and himself; and on the 16th of July following and another Letter from him by Lieutenant (now Captain) Broughton of the 13th of January, with the Journals, Charts and Drawings which accompanied it, giving a more detailed account of his Proceedings in execution of the Service on which he had been employed, Resolved that he be acquainted that their Lordships entirely approve of his Proceedings, and transmitted to Mr Secretary Dundas copies of the said Letters with their Inclosures for His Majesty's Information and Commands; the result of which will be communicated to him in the Pacquet from His Grace the Duke of Portland, one of His Majesty's Principal Secretaries of State, with which Captain Broughton is charged and by whom he will receive their Lordships order for his further Proceedings ...
>
> A Treaty between His Britannic Majesty and the King of Spain having been signed the 25th of May 1793; Resolved that Printed Copies thereof be sent to Captain Vancouver of the Discovery Sloop at Nootka Sound, and to Captain Broughton of the Providence Sloop at Spithead for their Information and Guidance.[1]

Meanwhile, Lieutenant Thomas Pearce of the Marines, who had previously served under William Bligh in the *Providence* during the latter's second breadfruit voyage, 1791–3, had been appointed British commissioner to implement the Third Nootka Convention in place of Vancouver. He left England in the late summer of 1794 and travelled to Nootka on board Spanish vessels by way of La Coruna, Cadiz, Havana, and Vera Cruz, from where he travelled overland to Mexico City, which he reached on 30 November. He sailed from San Blas on 13 January 1795 on board the Spanish frigate *Activa* to Monterey, where he was joined by Brigadier General José Manuel de Alava, the Spanish commissioner, reaching Nootka Sound on 16 March, where a brief ceremony took place.[2] On 25 April 1795, Pearce wrote to Philip Stephens, Secretary to the Admiralty, from Tepic informing him of his proceedings at Nootka.[3] Pearce also wrote to Secretary of State, the Duke of Portland, on the same date, which was published in the *London Gazette* No. 13813 (12–15 September 1795). Since Pearce had left England before Broughton had sailed, he was able to inform Alava that the *Providence* was bound for Nootka, enabling them both to leave letters with Maquinna for Broughton.[4]

2. Broughton and Vancouver

William Robert Broughton was born on 22 March 1762, son of Charles Broughton, a gentleman of Hammersmith in the county of Middlesex. Sadly, little is known of his childhood or family. His later connections were with the church, the navy and army, and with the gentry of Cheshire. His brother, the Revd Brian Broughton, five years his junior, was for a time

[1] TNA, ADM 3/114, ff. 78–78v.

[2] Manning, 'Nootka Sound Controversy', pp. 281–478. Pearce's correspondence, along with almost all other lieutenants' correspondence, appears to have been destroyed in the reorganization of the Admiralty Record Office in 1808–9. However, incoming letters from Pearce dated between 6 July and 27 December 1794 are noted in the *Admiralty Digest* for that year; TNA, ADM 12/61. A copy of his letter of 27 December 1794, reporting his arrival in San Blas and his intention of sailing to Monterey to meet Brigadier General José Manuel de Alava is held in TNA, HO 28/19, f.248, and a copy of his full report to Philip Stephens from Tepic, dated 25 April 1795, reporting their proceedings at Nootka is held in TNA, HO 28/20, ff. 192–3. A copy of his report to the Duke of Portland of the same date is reprinted in the *London Gazette*, No. 13813, 12–15 September, 1795.

[3] Copy in TNA, HO 28/20, ff. 192–3.

[4] Cook, *Flood Tide*, pp. 421–3, 426 and Lamb, *Vancouver's Voyage*, I, pp. 186–7. Little is known of Pearce's subsequent career. He was promoted to captain on 9 May 1795 and placed on half-pay in 1801.

a Fellow of New College, Oxford.[1] Broughton probably went to sea as a volunteer first class, and thus was destined for the quarterdeck. As a midshipman he served on the North American Station during the American War of Independence and was taken prisoner during an action at Boston. As a lieutenant he served in the *Burford* and fought in actions between Vice-Admiral Sir Edward Hughes and the French Commodore Chevalier Pierre-André de Suffren in the Indian Ocean.

On 28 December 1790, aged 28, Broughton was appointed in command of the brig *Chatham* to take part in Vancouver's expedition. His appointment to the *Chatham* was not as a result of any request that Vancouver might have made. It may, however, have been owed to Captain John Knight, under whom Broughton had served in the *Harlem* during the American War of Independence and again in the *Victory* during the 'Spanish Armament', when the British fleet was mobilized on 30 April 1790 due to deteriorating relations with Spain following the seizure of two British merchantmen in Nootka Sound. Knight was a prolific 'amateur' surveyor and Broughton may have learned his surveying skills from him. In any event, once on the north-west coast of North America this was soon to change as he made an impressive entry into this line of work. On 23 November 1791 Broughton discovered a cluster of small islets and rocks, which he named Knight's Islands after his former captain. They were sighted again the following day by Vancouver, who named them The Snares, by which they are now known. A few days later, on 29 November, Broughton was the European discoverer of an island to the east of New Zealand's South Island, which he named Chatham Island in honour of the First Lord of the Admiralty, the Earl of Chatham. Broughton and several of his officers landed in an effort to establish friendly relations with the local inhabitants, the Moriori, but a clash occurred in which one of the Moriori was killed. Broughton made some useful ethnographic observations and produced a sketch survey of the south side of the island.[2] In May 1792, on the north-west coast, Broughton examined a cluster of islands, now known as the San Juan Archipelago, lying east of Haro Strait and west of Rosario Strait, and bounded on the south by the Strait of Juan de Fuca and on the north by the Strait of Georgia.[3] In October of that same year he explored the Columbia River for about one hundred miles upstream from the river bar and entrance, making an outstanding contribution to the exploration and charting of that river, as Vancouver attested.[4] Near the upriver extremity of his navigation, at

[1] Brian Broughton was baptized on 29 June 1766 at Hammersmith, the son of Charles Broughton of that borough. After attending Winchester School he matriculated at New College, Oxford, on 29 April 1785 at the age of 18, becoming a fellow in 1787, which at the time was automatic after two years' study. He resigned his fellowship in 1812, by which time he had gained his MA, having been presented with the rectory of Long Ditton, Surrey. He died on 8 January 1838. He was the author of a poem titled 'Copse Hill Grove' and also *Poetical Reflections in North Wales* in 1801 with six aquatint views as well as his anonymous lengthy elegiac, *A Poem on a Voyage of Discovery, undertaken by a Brother of the Author, with other Sonnets, &c*, London, 1792.

[2] Lamb, *Vancouver's Voyage*, I, pp. 63, 380–83, 387–8. See also Anne Salmond, *Between Worlds: Early Exchanges between Maori and Europeans 1773–1815*, Auckland, 1997, pp. 195–204.

[3] J. Neilson Barry, ed., 'Broughton's Log of a Reconnaissance of the San Juan Islands in 1792', *Washington Historical Quarterly*, 21, 1 (January 1930), pp. 55–60. See also Lamb, *Vancouver's Voyage*, IV, 1558–63, and Richard W. Blumenthal, ed., *With Vancouver In Inland Washington Waters: Journals of 12 Crewmen, April–June 1792*, Jefferson, N.C., 2007, pp. 109–39. This last contains his rough journal, 28 April–25 June 1792, and the San Juan examination, 18–25 May 1792.

[4] Broughton's own account of his Columbia River examination has not survived. But Vancouver is known to have used it when compiling the account of his voyage, which is printed in Lamb, *Vancouver's Voyage*, II, pp. 747–70. See also, Andrew David, 'John Sherriff on the Columbia, 1792: an Account of William Broughton's Exploration of the Columbia River', *Pacific Northwest Quarterly*, 83, 2 (1992), pp. 53–9.

Possession Point, he claimed the river and surrounding territory for His Majesty King George III. Broughton's 'zeal and exertions' in the Columbia River exploration won Vancouver's approval,[1] for this was difficult and hazardous work. Broughton had acquitted himself well in all his naval service to date and Vancouver also came to trust Broughton's diplomatic skills. During the voyage, Broughton had become familiar with the Spanish empire and officials in the Western Hemisphere and, from his experiences afloat and ashore with the Spanish, may have had, in addition to his fluency in French and, in the circumstances, a basic and most useful command of the Spanish language.

3. Broughton and Bodega

During the various social and diplomatic encounters in Nootka Sound Broughton was on excellent terms with Bodega and later, on the voyage from Monterey to San Blas, they became firm friends. Before the two parted in San Blas, Bodega commissioned Broughton, should he return to the north-west, to buy for him in London four dozen shirts 'of the finest trim' and a number of nautical instruments, including an Arnold chronometer (No. 56), a small pocket chronometer (Earnshaw No. 2), a Ramsden sextant 'with three telescopes', a thermometer, a glass 'to determine distances at sea', and an artificial horizon.[2] On his return to London, Broughton purchased these items for his Spanish friend. They cost Broughton £250 and, in enclosing the bill, he assured Bodega that 'the instruments were the finest in England'. The Arnold box chronometer cost £84, the pocket chronometer £47:15:6, the Ramsden sextant £19:9:0, the artificial horizon £2:12:0 and the shirts £78:4:6. As had been arranged, a cousin of Bodega, Pedro de la Quadra, living in Cadiz, paid the bill with 22,640 pesos.[3] Broughton took these items with him in the *Providence*, but, on leaning that Bodega had died on 26 March 1794, aged forty-nine,[4] on reaching Monterey in June 1796, he handed them to the governor Don Diego Borica. The governor then sent them by the first available vessel to Francisco Hijosa, the comisario in San Blas, where the instruments were stored in the royal warehouse on the orders of Viceroy Revillagigedo. As Freeman Tovell points out, 'As the file ends there, it will never be known what happened to either these elegant shirts or the nautical instruments, but it takes no stretch of the imagination to assume that the Spanish navy appropriated them for its own use'.[5]

4. Broughton and the *Providence*

Broughton was appointed in command of the *Providence* on 3 October 1793 and promoted to Commander. His immediate duty was to prepare her for sea and to see that all necessary repairs were satisfactorily completed, that provisions and stores were received and stowed correctly, and that her guns and ammunition were embarked – in short, to see that all the necessary details were attended to for a long voyage to distant seas and extensive sailing.

[1] Lamb, *Vancouver's Voyage*, II, p. 747.
[2] Tovell, *Far Reaches of Empire*, pp. 333–4, quoting AGN, Californias 37, ff. 179–185v.
[3] Broughton to Bodega, in French from London, 28 July 1793, AGI, Estado, legajo 39.
[4] Tovell, *Far Reaches of Empire*, p. 329.
[5] Ibid.

Figure 1. The *Providence* (detail), from a watercolour by George Tobin, c. 1791. Photograph courtesy of the Mitchell Library, State Library of New South Wales.

The *Providence*, mounting 12 guns, was the ninth Royal Navy vessel of the name. Having originally been intended for the West India trade, she had been bought on the slip by the Navy Board from Mr Perry's at Blackwall in 1791 to replace the *Bounty*, lost in the mutiny, for a further attempt to carry breadfruit from Tahiti to the West Indies, again under William Bligh.[1] She proved to be, as Bligh had discovered, highly suitable for distant voyaging. She was not, as some authorities state, a small vessel. Measuring 406 tons, she was a little larger than Vancouver's *Discovery*. Her measurements were: 107 feet 10¼ inches overall (89 feet 6 inches on the waterline), 29 inches 2½ inches beam, and 12 feet 3¼ inches keel to rail. Her draft was 16 feet 6 inches.[2] Bligh described her as a superior ship of her class, with three decks. The *Providence* under Bligh had reached St Vincent on 23 January 1793, where over 500 plants were landed; the rest were unloaded at Jamaica. She then sailed for England. The *Providence* arrived at Deptford in the River Thames on 6 August in company with the *Assistance* under Nathaniel Portlock, another veteran of Cook's voyages and of the north-west coast of America. A month later Bligh made a final entry in his log: 'This voyage has terminated with success, without accident, or a moment's separation of the two

[1] For details of Bligh's successful voyage see Douglas Oliver, *Return to Tahiti: Bligh's Second Breadfruit Voyage*, Melbourne, 1988.

[2] David Lyon, *The Sailing Navy List: All the Ships of the Royal Navy – Built, Purchased and Captured – 1688–1860*, London, 1993, p. 234.

ships.'[1] On paying off, the *Providence* was turned over to the Dockyard officers and was soon taken in hand for another mission, unconnected with the war with France that had commenced in February 1793. The war had, in fact, delayed the return of the *Providence* and *Assistance* to England. These same circumstances obliged merchant vessels and warships bound from United Kingdom ports to sail in convoy, and later this also delayed Broughton, making a possible rendezvous with Vancouver all the more unlikely.

On 3 October 1793 Broughton was appointed in command of the *Providence*,[2] then being refitted in Woolwich, with Zachary Mudge as his First Lieutenant. The following day Broughton was instructed that the dockyard had been ordered that the *Providence* was to be fitted and stored for a voyage to remote parts, manned with 100 men, and victualled for twelve months, with all species of provisions. Broughton was to cause the utmost dispatch to be used in getting her ready for the above service and, having done so, he was to fall down to Galleon's Reach to take in guns and gunner's stores, after which he was to repair to the Nore for further orders.[3] Broughton received further instructions on 29 November, which amended the *Providence*'s complement to 115.[4] On 23 September 1794 Broughton was also ordered to take out a number of seamen for the *Discovery* and the *Chatham* to complete their complements,[5] but as he only gives the number of officers and crew embarked on board the *Providence* as 115,[6] these additional seamen may have already been embarked. On 31 December 1793 Broughton received orders to receive on board Mr John Crosley, as well as his servant and baggage. Crosley, 'being appointed by the Board of Longitude to make Astronomical Observations in a Voyage to Nootka Sound and elsewhere,'[7] joined the *Providence* the following day.[8] As Broughton's tasks would have included taking out provisions and supplies to Captain Vancouver and further carrying on the extensive survey, the *Providence* had to be fully stowed. On 9 January 1794 the Lords Commissioners of the Admiralty wrote to the Commissioners for taking Care of Sick and Wounded Seamen that

> Captain Broughton, Commander of His Majesty's Sloop Providence, who is going on a voyage to remote Parts, having requested that she may be supplied with the undermentioned Articles for the use of her Company during the said voyage, viz.

Portable Soup	1500 pounds	
Pulvis Antimonialis	30	bottles
Elixir of Vitriol	36	bottles
Rob of Oranges & Lemons	42	bottles
Bark Rendered [*word indistinct but likely Rendered*][9]	One hundred weight	

> You are hereby required and directed to cause the said sloop, which is now at Woolwich, to be supplied therewith accordingly, with all the dispatch ...[10]

[1] Oliver, *Return to Tahiti*, p. 257.

[2] TNA, ADM 6/24, f. 270v.

[3] Ibid, ADM 2/124, pp. 506–7.

[4] Ibid, ADM 2/125, pp. 167–8.

[5] Ibid, ADM 2/126, minute dated 23 September 1794.

[6] Broughton, *Voyage*, p. xvi.

[7] Ibid, ADM 2/125, pp. 247–8.

[8] Ibid, ADM 36/13497.

[9] Bark of the Cinchona tree, which when rendered produces quinine; used as a febrifuge and tonic. Also known as Peruvian or Jesuits' bark.

[10] Admiralty to Sick and Hurt Board, 9 January 1794, NMM, ADM E/45. Fresh lemons were acknowledged to be the best antiscorbutic, but the demand for chests of lemon or of lemon juice exceeded available supply. For discussion of the issues of the day, see Brian Vale, 'The Conquest of Scurvy in the Royal Navy 1793–1800: A Challenge to Current Orthodoxy', *The Mariner's Mirror*, 94, 2 (May 2008), pp. 160–75.

These were medicines, including antiscorbutics, supplied for the voyage as a counter to the ravages of scurvy. (In any event, no mention of scurvy during this voyage appears in Broughton's narrative nor in any report by John Floud, the surgeon.) It may be pointed out, however, that even though James Lind, in 1745, had determined, by experiment, that the use of oranges and lemons was the most effective antiscorbutic – he tried all sorts of items and concoctions including elixir of vitriol three times a day on an empty stomach, vinegar ditto, assorted gruels and potions, the bigness of a nutmeg, and, not least, an electuary of garlic, mustard seed, radish and balsam of Peru and gum myrrh in barley water acidulated with tamarinds – it was not until Sir Gilbert Blane became a Commissioner of Sick and Wounded in 1795, that a standard issue of citrus fruits to all ships became general.[1]

5. The Voyage from England to Macau

If Broughton kept a journal for the early part of the voyage as far as Tahiti, it has not survived. Nor has the *Providence*'s log for the same period. The published account of the voyage, which contains a number of errors and omissions, must therefore have been based on various missing documents. Consequently, it is the principal source for this part of the voyage. It has, however, been possible to supplement this, partly from two logs covering the period from 24 January 1794 to 24 January 1795 kept by William Chapman, the ship's Master and, of less value, a log kept by Lieutenant James Giles Vashon between 14 January and 6 October 1795. Letters from John Crosley, the astronomer, to Nevil Maskelyne, the Astronomer Royal, contain additional information for this part of the voyage.[2]

On 18 February 1794 the *Providence* dropped down river to Galleon's Reach to embark her guns and gunner's stores, where she remained until the 28th when she moved to Gravesend, as being a better place than Woolwich to obtain seamen. Here she remained between 1 and 30 March, when, still short of seamen, she proceeded to the Nore, where volunteers were received from the *Sandwich*, flagship of Vice-Admiral John Dalrymple, to complete her complement. Thus the crew were all volunteers, except one who joined later in Plymouth, having been sent on board the flagship there by the civil power and from thence to the *Providence*.[3] On 16 April Broughton was instructed to proceed to Spithead without loss of time and to remain there until further orders.[4] Sailing from the Nore, as escort to a convoy of merchant ships, the voyage started inauspiciously when on 23 April the pilot ran the *Providence* aground on the Brake Sand.[5] Broughton laid out a kedge anchor and hove her off. After briefly anchoring in the Downs, Broughton continued on to Spithead, where he anchored on 28 April. On 8 May the *Providence* moored alongside Dockyard Wharf in Portsmouth and two days later she was taken into dock where five copper sheets were replaced on her starboard side. While in Portsmouth, William Bayly, formerly one of the astronomers on Cook's second voyage and

[1] J. Lind, *A Treatise on the Scurvy*, 3rd edn, London, 1772, pp. 144–50. Lind's account is also printed in John Hattendorf, R. J. B. Knight, A. W. H. Pearsall, N. A. M. Rodger and Geoffrey Till, eds, *British Naval Documents 1204–1960*, The Navy Records Society, 1993, pp. 523–4.

[2] See Appendix I.

[3] This was Patrick Sherry who died on 6 June 1796; Broughton, *Voyage*, p. 60.

[4] TNA, ADM 2/125, p. 518.

[5] A small drying bank on the western side of the Gull Stream, the channel between the Kentish coast and the Goodwin Sands.

the astronomer on his third, but now headmaster of the Royal Naval Academy, rated the *Providence*'s chronometers. It seems likely that they were sent to him previously by the Board of Longitude for that purpose. The following day the *Providence* undocked and anchored once more in Spithead, where she remained until 10 September when she moved to St Helens, the fleet anchorage in the SE approaches to Portsmouth, NNE of Bembridge Point, the eastern extremity of the Isle of Wight.

It was not until 4 October 1794 that Broughton received his orders, which were secret, with an additional one to put himself under the command of Captain Drury in HMS *Trusty* and to proceed to sea with his convoy, bound for the Mediterranean. Broughton was to keep in company with Drury as long as their routes were the same.[1] The convoy, escorted by the *Polyphemus*, *Providence*, *Saturn*, *Fury* and *Trusty*, left St Helens on 21 October but when the wind changed to south the following morning the convoy and escort bore up for Tor Bay, where they remained at anchor until 25 October, when they were able to resume their voyage down channel.[2] Off the Lizard they spoke with the frigate *Artois* with the captured French frigate *La Revolutionaire* in tow.[3] On 28 October, when well to the west of the Isles of Scilly, in 7°20′W, a severe gale was encountered and the convoy became dispersed. Drury then made the signal to make for the nearest port. Broughton looked into Falmouth, but with no signs of the *Trusty*, he continued on to Plymouth, where he anchored in Cawsand Bay on 30 October.

Broughton was detained in Plymouth until 29 January 1795 when Admiral Parker[4] flying his flag in the *Raisonable* made signal to sail. On 15 February a favourable north wind enabled more than 400 sail to get under way and proceed to sea, where the fleet under the command of Admiral Lord Howe was waiting to escort them down channel. On 19 February, well clear of the French coast, Admiral Howe's fleet parted company and the convoy separated into various components, the *Providence* continuing for the time being with Admiral Parker's West India fleet until 1 March, when that too parted company. This left the *Providence*, *Reliance*, *Supply* and the frigate *Iris*, with her convoy, to continue in company. On 6 March the four ships anchored off Santa Cruz in Tenerife to obtain fresh provisions and water ship. On 8 March the *Iris* and her convoy sailed for the Cape Coast and on the 13th the *Providence*, *Reliance* and *Supply* sailed in company bound for Rio de Janeiro. On 20 March Ilha de Santo Antão, the NW island of Ilhas do Cabo Verde, was sighted and on 22 April Ilha da Trindade, enabling the position of this isolated island to be verified. On 2 May they passed Cabo Frio and on the 5th Broughton anchored in the entrance to Rio de Janeiro, moving the following day to an anchorage within the harbour, north of lhas da Cobras, where he was joined by the *Reliance* and *Supply*.

[1] Broughton, *Voyage*, p. 4. As stated earlier, Broughton's secret instructions do not appear to have survived.

[2] TNA, ADM 52/3292/9 and 10 and Crosley to Maskelyne, 18 November 1794, in Appendix I, pp. 228–30. There is no mention of Tor Bay in Broughton, *Voyage*, p. 4.

[3] *La Révolutionaire*, a 44-gun French frigate, commanded by Henri-Alexandre Thévenard, was intercepted by a British squadron on 21 October 1794, about 30 miles west of Ushant. The superior sailing of the 38-gun frigate *Artois*, Captain Edmund Nagle, enabled her singly to bring the French frigate to action; she defended herself with great gallantry, but struck her colours to the *Artois* when the rest of the British squadron closed in; William James, *The Naval History of Great Britain*, 6 vols, London, 1837, 1, p. 211. *La Révolutionaire* was taken into the Royal Navy with the same name and was eventually broken up in 1822.

[4] Admiral Sir Hyde Parker (1739–1807), who was going to the West Indies to take over as commander-in-chief, Jamaica, 1796–1800. He was commander-in-chief of the British fleet at the first Battle of Copenhagen in 1801 and is chiefly remembered for the signal he sent to Nelson to leave off the action, which Nelson is said to have ignored by lifting his telescope to his blind eye.

Broughton spent eighteen days in Rio de Janeiro, during which the rigging was overhauled and the decks recaulked. The ship's company were constantly supplied with fresh provisions, and wine was purchased at a reasonable rate. The Portuguese authorities were, however, very suspicious and the three ships were watched by guard-boats day and night. Broughton took the opportunity of calling on Captain John Hunter on board the *Reliance*, who was returning to New South Wales to take up his appointment as Governor in succession to Arthur Phillip. Together they called on the Portuguese Governor, the only time that Broughton went ashore. Crosley, however, was allowed to land on Ilha das Enxadas to rate the chronometers and to fix its position. Matthew Flinders, who was senior master's mate on board the *Reliance*, also landed on the island to take similar astronomical observations.[1]

On 24 May Broughton set sail alone as the *Reliance* and the *Supply* were not ready to proceed. As he intended to round Van Diemen's Land and then cross the Pacific from west to east due to the lateness of the season,[2] Broughton took with him Hunter's orders for Port Jackson.[3] Broughton first set a course for Gough Island, named after Captain Charles Gough of the East India Company ship *Richmond*, who had sighted it on 3 March 1732,[4] in order to fix its position. The island had been discovered by Gonçalo Alvarez in 1505 and laid down on a Portuguese chart of 1519.[5] The island was one of several in the South Atlantic which were to have been examined by Captain Henry Roberts in the *Discovery* for their suitability as a base for future operations.[6] Broughton's visit to the island may have been a belated fulfilment of this objective. Gough Island was sighted on 10 June in poor visibility. Broughton remained off the lee of the island overnight in the hope of landing, but the weather, particularly a very heavy swell, made this impossible. Some indifferent observations placed the island in 40°19′S, 9°27′W, considerably different from the position of 40°3′S, 2°30′W, as it was laid down in all the charts and books they had on board, as noted by Crosley.[7]

From Gough Island Broughton set a course to pass south of the Cape of Good Hope and then south of Van Diemen's Land (Tasmania), which was sighted on 2 August. Broughton had hoped to examine the unknown area between Furneaux Islands and Cape Howe, but contrary winds prevented this and carried the *Providence* north of Sydney. In consequence, Broughton entered Port Stephens, just north of the present day city of Newcastle, which he surveyed with Crosley's assistance (see Fig. 3 on p. 12). Providence Bay, extending seven miles north of Yacaaba Head, the NE entrance point of Port Stephens, and Broughton Islands situated eight miles NE of Yacaaba Head, commemorate his visit. Here he encountered four convicts who had escaped from Sydney in 1790 in a four-oared boat, but had got

[1] Geoffrey Ingleton, *Matthew Flinders: Navigator and Chartmaker*, Guildford, Surrey, and Alphington, Victoria, 1986, p. 23. In 1801–3 Flinders, in command of the *Investigator*, went on to carry out an important survey of Australia.

[2] Broughton may have been given discretional powers in his secret orders.

[3] Broughton, *Voyage*, p. 12. The contents of these orders is not known.

[4] James Horsburgh, *Directions for Sailing to the East Indies*, 2 parts, London, 1809, I, p. 62.

[5] Robert Headland, *Chronological List of Antarctic Expeditions & Related Historical Events*, Cambridge, 1989, pp. 53 & 69.

[6] In 1789 Secretary of State William Wyndham Grenville proposed sending an expedition to examine certain islands in the South Atlantic, including Gough Island. The *Discovery*, commanded by Henry Roberts, was prepared for this expedition with Archibald Menzies to be appointed as surgeon and naturalist. However, when war seemed possible over the Nootka Sound crisis the expedition was abandoned and the *Discovery* was sent to the north-east Pacific, commanded by George Vancouver instead; TNA, HO 28/6, ff. 304–12, and ADM 1/4153, f. 43; David Mackay, *In the Wake of Cook*, Beckenham, Kent, 1985, pp. 42–5.

[7] Crosley to Maskelyne, 6 October 1795, CUL, RGO, 14/68, ff. 98–104.

no further than Port Stephens. They came willingly on board and were taken back to Sydney to complete their sentences.

Broughton finally reached Sydney on 27 August where he handed over Captain Hunter's orders to Major Paterson, the acting governor. During their stay there, due to the constant bad weather encountered during the run from Rio de Janeiro, it was found necessary to recaulk, refit the ship and overhaul the rigging. Crosley took the opportunity to set up his observatory tent on Bennelong Point, to rate the chronometers, in the same spot where Malaspina and his officers had set up their observatory two years earlier.[1] The seine was hauled daily to provide fresh fish for the ship's company and, by permission of Major Paterson, a boat was sent daily to Garden Island for fresh vegetables, which were grown there for the town, free from the unwanted attentions of the local fauna. On 8 September the *Reliance* and the *Supply* arrived, having been 97 days from Rio de Janeiro.

On the morning of 13 October Broughton got under way, bound for Tahiti. After an uneventful passage, taking them north of New Zealand, Broughton entered Matavai Bay on 29 November. The following morning, when warping the ship into a better anchorage, an iron stocked anchor was swept up, which had formerly belonged to the *Bounty* and had been abandoned when the mutineers had cut her cables when making their final departure from Tahiti. The observatory tent was set up on Point Venus to enable Crosley to rate the chronometers, but the surf breaking on the shore caused the ground to tremble so much that observations were impossible. It is to be regretted that the only surviving account of occurrences ashore is the statement that 'The natives afforded us every assistance in our various pursuits, and amply supplied us with provisions and vegetables'.[2] On 10 December Broughton weighed anchor and at noon he hove to in order to hoist the boats and to take leave of their friends, which included Pomare, his wife 'Itia, his two brothers Teri'irere and Vaetua and the high priest Ha'amanemane. On their departure the Tahitians were saluted with a four-gun salute. A course was then set for Hawaii, possibly to obtain information about Vancouver. On 16 December a small uncharted island was sighted, which Broughton thought was a new discovery. He therefore 'named it Carolina Island in compliment to the daughter of Sir P. Stephens of the Admiralty'.[3] It was, however, not a new discovery, having been sighted by Quiros in 1606, who named it San Bernardo, and it had possibly been discovered by Magellan in 1521.[4] The daughter's name was in fact Caroline and the island is now named Caroline Island on charts and maps.

The island of Hawaii was sighted on 1 January 1796, but on approaching land the trade wind was lost. Nevertheless several canoes came alongside, with pigs and vegetables. On the 2nd a vessel was sighted under sail to the NW, which the natives informed Broughton was an English brig. It was in fact the schooner *Prince William Henry*, Captain William Wake.[5] Broughton subsequently purchased this schooner in Macau on 29 December 1796; see p. 107 below. Light and variable winds at first prevented Broughton reaching Kealakekua Bay. He

[1] Andrew David, Felipe Fernández-Armesto, Carlos Novi and Glyndwr Williams, eds, *The Malaspina Expedition, 1789–1794: The Journal of the Voyage by Alejandro Malaspina*, 3 vols, The Hakluyt Society, ser. 3, nos 8, 11, 13, London, 2001–2004, III, pp. 78–9.

[2] Broughton, *Voyage*, p. 26. Broughton's surviving journal starts on their departure from Tahiti.

[3] Ibid., p. 29.

[4] Andrew Sharp, *The Discovery of the Pacific Islands*, Oxford, 1960, pp. 5–6.

[5] Michael Roe, ed., *The Journal and Letters of Captain Charles Bishop on the North-West Coast of America, in the Pacific and in New South Wales*, Hakluyt Society, 2nd ser. 131, Cambridge, 1967, p. 135.

therefore sent an officer inshore in the pinnace to obtain information about Vancouver from an American vessel at anchor in the bay. This was the *Lady Washington*, Captain Roger (or Robert) Simpson. Broughton had encountered her a few years earlier when he was in command of the *Chatham*, and the *Lady Washington* was commanded by John Kendrick. Simpson was able to pass on information he learned from the English brig that the *Discovery* and *Chatham* had sailed for England from Valparaíso in July or August 1795.

Broughton spent seven weeks in the Hawaiian Islands, presumably to avoid arriving at Nootka during the depth of winter. The observatory was landed and set up in a field adjacent to the morai, where Vancouver had set up his observatory, for Crosley to obtain fresh rates for the chronometers. No difficulties were encountered such as Cook had faced. John Young, principal agent of Kamehameha and a seasoned friend of the English, offered assistance with canoes and food. Kamehameha himself, the most powerful chief in the island of Hawaii, was absent on a raiding party to an adjacent island in his quest to bring all the Hawaiian Islands under his control. The ship's company was permitted shore leave, something which the cautious Vancouver would not have allowed. In the circumstances, no objection to shore leave could be raised. Of immediate concern, however, was that since crossing the equator the *Providence* had made from two to four inches of water per hour and the opportunity was taken to unstow the holds and heel the ship in an attempt to find the cause, but without success. Many fine things were said of Vancouver by the Hawaiians, and the cattle that he had left were sighted. Broughton gave a pair of goats, some geese and ducks. Mudge gave the Hawaiians his pigeons.

On 1 February 1796 Broughton made his departure from the bay, leaving the *Lady Washington* at anchor. Two days later he anchored off the island of Maui, opposite the village of Lahaina, but few provisions could be obtained there. So on the 6th Broughton got under way again and on the 7th anchored in Waikiki Bay, now called Mamala Bay, on the south coast of Oahu, just to the west of Diamond Head. Here Kamehameha came on board. He seemed to be engrossed in completing the sailing vessel *Britannia* which Vancouver had put in hand for him, and in forcibly extending his rule to the neighbouring islands, contrary to Vancouver's advice. Broughton learned from him that the Hawaiians executed for the murder of Hergest and Gooch in Oahu during Vancouver's visit were not in fact the culprits but had merely been nominated for that part to appease Vancouver. Broughton anchored briefly off a harbour called Fair Haven, now identified as Honolulu, in order to survey it, but his survey has not survived. On 16 February Broughton anchored off Waimea, on the south coast of Kauai, finding the island at war. In spite of this, they received an abundance of vegetables and forty hogs. Broughton attempted to mediate between the warring parties, but without success and so he set sail for the adjacent island of Niihau, anchoring there on 19 February. The following day some fresh provisions were obtained, but, with the weather deteriorating, Broughton got under way and on 22 February set a course for Nootka Sound. At noon on 25 February Broughton altered course to the west to run down the longitude of an island called Donna Maria Lajara, laid down on Arrowsmith's 1794 'Map of the World' and said to have been discovered in 1781. With no sight of land in its supposed position and after running down 5° of longitude, Broughton abandoned his search and altered course once more to the north. A large swell from the SE quarter obliged them to pump out the ship every two hours.

On 15 March land was sighted and the following evening Broughton anchored in the entrance to Nootka Sound amidst showers of hail, sleet and snow. The land lay under a blanket of snow, presenting to Broughton's eye a desolate appearance. Next morning Broughton sent Mudge in a well-armed boat into Friendly Cove for intelligence. He returned at noon with the news that there were no ships in the cove and that the spot where the Spanish settlement had been was now occupied by an extensive native village. Since Lieutenant Thomas Pearce, who had been appointed as the British Commissioner to implement the Third Nootka Convention, had left England several months before Broughton received his secret instructions, they could well have included orders for him to confirm that the Spanish had not reoccupied the sound. On the 17th, with a favourable sea breeze, Broughton entered the sound and anchored two miles NNE of Friendly Cove. Here the principal chief Maquinna and another chief, Clupananutch, brought off to the ship several letters dated March 1795 at Nootka, almost a year earlier. These were from Lieutenant Pearce and Brigadier General Alava informing Broughton of the abandonment of Nootka Sound in accordance with the agreement reached between the two Courts. From the letters Broughton learned that Vancouver had sailed from Monterey for England in December 1794, and that the Nootka controversy had been settled at last. Any fears that the natives, known to be numerous, might pose a difficulty proved false.[1]

The leak having increased on the way north, it had now become imperative to find and repair it, for which an examination of the hull was essential. As this was the season of heavy weather a snug spot, well protected from high winds and heavy seas, was required. Broughton therefore decided to proceed to Marvinas Bay, situated on the west side of Cook Channel, about four or five miles north of Friendly Cove, where a suitable site was found between an islet and Nootka Island, where the ship was made steady by hawsers. Nearby the beach was found to be convenient for laying the ship on shore and, accordingly, the *Providence* was moved there. The sails were unbent and the yards and topmasts struck. Next, tents were erected alongside and the holds were cleared of provisions and landed, together with the anchors, cables and guns. The *Providence* was then hauled on shore, but it was not possible to find the leak. On 10 April a vessel was seen in the offing, which proved to be the *Lady Washington*, which had put into Nootka Sound to repair her leaks. A few days later she moored alongside the *Providence*, enabling all the spirits, slops and perishable stores to be put on board her. Next a wharf was built adjacent to the *Providence*. She was then heaved down to reveal the port side as far as the garboard strake[2] when it was discovered that 14 feet of false keel had been carried away and that a copper bolt which was meant to secure the strake to the floor timbers was missing. Only the copper sheathing had prevented the uninterrupted inflow of water. Meanwhile the carpenters were employed sheathing the *Lady Washington* and making other repairs. On completion of the repairs, the sloop had to be cleaned throughout and then dried with fires before reloading could be carried out. As pointed out by Admiral Ritchie, the episode seemed to have been regarded by Broughton as a minor inconvenience and to have been a remarkable example of the efforts expended by seamen in those days and of the ingenuity that officers and men on detached service had to exercise.[3]

On 16 May the *Providence* was warped out of Marvinas Bay and a light breeze carried her down to Friendly Cove. On 21 May, with a light land breeze, Broughton made his departure. In a letter to the Admiralty Broughton says that they had been supplied by the natives with venison, fish and vegetables, and that they had parted with Maquinna on the friendliest terms,

[1] According to Warren Cook, Broughton 'was disappointed that he had come halfway around the world on a wild goose chase'; Cook, *Flood Tide*, p. 427. However, this is unsubstantiated from anything Broughton has written.

[2] The first range of planks laid upon a ship's bottom, next to the keel.

[3] Ritchie, *Admiralty Chart*, p. 67.

adding that no vessel of any nation had been in the Sound since the Spaniards quitted it.[1] Broughton next anchored briefly in Neah Bay, known to the Spanish as Nuñez Gaona, on the southern shore of the Strait of Juan de Fuca. There for four months in 1792 the Spanish had kept a frigate in the port and had set up a rough establishment ashore, before it was found to be unsuitable as an alternative to Nootka Sound and so was abandoned. In his secret orders Broughton may have been instructed to confirm that it too had been abandoned. Broughton found it afforded good anchorage, with plenty of wood and water.[2] With nothing to detain him, Broughton spent just one day there, before shaping a course south towards Monterey, where he hoped to get further intelligence of Vancouver.

Making his way south, and into warmer temperatures, Broughton sighted Point Reyes, the conspicuous promontory forming the western entrance point of Drakes Bay, on 4 June. Sounding as they passed the point, Broughton entered the bay and anchored in 15 fathoms, 1½ miles offshore. 'With two boats we explored the shore, where I conceive the place is in which Sir Francis Drake anchored 1579.'[3] Vancouver's earlier identification of Drakes Bay, which he visited on 14 November 1792, had been based entirely on Spanish authorities, probably as a result of discussions he had with Bodega, Alcalá-Galiano and Valdés at Nootka. In the face of bad weather and heavy seas, Vancouver had thought it prudent not to enter and examine this renowned anchorage and, accordingly, he continued on to San Francisco.[4] Broughton, by contrast, was able to examine the bay more closely and would have pressed his investigations further inshore, beyond a remarkable sandy patch, except for surf breaking across the bar to Drakes Estero.[5] He noted that the bay afforded good shelter, except in easterly winds. Deer were seen on the hills and in the valleys, and communications were made with the natives on the shore. The surf prevented a landing, but a sailor swam ashore with knives, trinkets and other items as presents.

Broughton then continued to the south, past the Farallon Islands, without sighting either them or the entrance to San Francisco Bay on account of haze. On 6 June he entered Monterey Bay, where he moored ship after saluting the Spanish fort with eleven guns, which was returned with an equal number. Here he met a reception similar to that accorded to Vancouver in Monterey by José Joaquin de Arrillaga in 1793. The governor, Don Diego Borica, a colonel of cavalry, was coldly civil and would only meet absolutely essential needs. There was no social intercourse. Broughton handed over to the Governor, for Bodega's executor, the four dozen shirts and the various nautical instruments, including two chronometers he had bought for 'my late much valued and lamented friend'.[6] The governor, however, would not allow the British to erect the observatory ashore for Crosley to rate the ship's own chronometers. Nor were they allowed to ride or walk into the country. Borica did, however,

[1] Broughton to E. Nepean, 15 June 1796, UKHO, OD 177.

[2] On 2 October 1792, Vancouver, at Nootka Sound, was told by Fidalgo, in command of the *Princesa*, freshly arrived from Neah Bay, that the bay had been abandoned by the Spanish because the location had been found exposed to the weather, and that the anchorage was very bad because of its rocky bottom; Lamb, *Vancouver's Voyage*, II, p. 684. Broughton, however, found safe anchorage here; see p. 47.

[3] Broughton, *Voyage*, p. 59. Point Reyes lighthouse is in 37°59′36″N, 123°1′21″W.

[4] Lamb, *Vancouver's Voyage*, II, p. 701.

[5] An indication of the breakers at the bar of Drakes Estero is shown on the US Coast Guard Chart of 1860, printed in Samuel E. Morison, *The European Discovery of America: The Southern Voyages, A.D. 1492–1616*, New York, 1974, p. 675. Also James F. Imray, *North Pacific Pilot, Part I. North America (West Coast)*, 4th ed., London, 1885, f.p. 234.

[6] For full details of these items and their eventual fate, see p. xxviii above.

agree to forward a letter from Broughton to the Admiralty, giving an account of his proceedings so far, but it does not appear to have reached London, although a duplicate copy was brought back by Mudge when he was sent home from Macau in June 1797. Broughton's original letter may have suffered the same fate as a letter Mudge wrote to his patron John Stuart, First Marquess of Bute, Ambassador to the Court of Spain. Mudge's letter gave details of the voyage so far and an account of Bute's son, Lord George Stuart, who was a midshipman on board the *Providence*. By the time this letter reached Madrid, Bute had been recalled and the letter was deposited in the Archivo General de Indias.[1]

Broughton could not have known that the Spanish, under orders of the Viceroy of New Spain, the Marqués de Branciforte, who had replaced Revillagigedo, had decided to send a warship north to Bucarelli Bay, in present day Alaska, to look for any suspicious foreign activities. This was the *Sutil*, commanded by José Tobar y Tamáriz, but due to bad weather she only got as far as Nootka, where nothing suspicious was encountered. The intention of this voyage is not quite clear, except to pre-empt any attempt that the British might make to reoccupy Nootka. The *Sutil* left San Blas on 16 March 1796, three months before Broughton reached Monterey. Branciforte would have known that Broughton was returning to the north-west coast, but it seems likely that the first news that he was on the coast would have been when he heard of his arrival at Monterey. There were other factors leading to Broughton's frosty reception by Borica. These were the years of renewed Anglo-Spanish tension, despite the British government making all efforts to avoid giving provocation to Spain after the outbreak of war with France. In August 1796 an alliance between Madrid and Paris was arranged and a month later Spain declared war on Britain, one of the justifications being that the British had abused the First Nootka Convention and that British vessels were carrying on contraband trade on the coasts of Peru and Chile.[2]

The time had come for Broughton to decide on his future proceedings. Broughton wrote in his journal at this juncture, 'My orders from the Admiralty were, that I should survey the southern coast of the south-west part of South America, upon the idea that Captain Vancouver, who had similar orders, would not be able to fulfil them. But as I now had certain intelligence that he had left this port eighteen months before, and that both the ships, *Discovery* and *Chatham*, under his command, were in good condition, I had not the smallest doubt of his ability to comply with his instructions.'[3] In view of this, Broughton decided that his future proceedings depended on his own discretion 'in such a manner as might be deemed eligible for the improvement of geography and navigation.'[4] He therefore ordered his officers to give their opinions in writing how this could best be achieved. He was delighted to discover that their opinions coincided with his own, namely to survey the coast of Asia commencing at the island of Sakhalin and ending at the Nanking River (that is, the Yangtze or Chang Jiang). It was also his intention to complete the survey of the adjacent islands, namely the Kuriles, and those of Jesso and Japan (that is, Hokkaido and Honshu), left unfinished in Cook's third voyage. In so doing the limits of both continents of America and Asia would be known and the knowledge of the North Pacific Ocean would be complete. Broughton also gives his reasons in the preface to the published account of his voyage, 'It should be remembered that,

[1] AGI, Estado, 39, N.14.8 (5). For a copy of this letter see Appendix I.

[2] Cook, *Flood Tide*, p. 425.

[3] See pp. 245–50 below. Broughton was of course unaware that the *Discovery*'s damaged mainmast had forced Vancouver to abandon this survey.

[4] See p. 50 below.

in the third volume of Cooke's last work, Captain King observes that the navigation of the sea between Japan and China afforded the largest field for discovery: and the survey of this unknown part of the North Pacific Ocean was particularly recommended by the Honourable Daines Barrington in his *Miscellanies*.'[1] The Hon. Daines Barrington, KC, was an eminent English judge and a fellow of the Royal Society of London, who used his influential position to promote maritime exploration, especially towards finding the long-sought Northwest Passage. On 17 February 1774 he laid before the Royal Society a plan for a voyage to attempt to find a passage from the Pacific, which led to the dispatch in 1776 of James Cook in command of the *Discovery* and *Resolution* on the third of his great voyages. In 1781 Barrington published his influential collection of essays, simply entitled *Miscellanies*,[2] with the aim of promoting the completion of the work of exploration left undone at the conclusion of Cook's third voyage. In the preface to his essay 'The Possibility of Approaching the North Pole Discussed' he wrote that 'The Coast of Corea, the Northern part of Japan, and the Lequieux Islands, should also be explored'.[3] Another reason put forward by Broughton for his decision to cross the Pacific to survey parts of the coasts of Asia was that so far Crosley had not had any opportunities of complying with his instructions from the Board of Longitude in making observations and ascertaining the positions of unknown places,[4] since the *Providence*'s voyage so far had followed Vancouver's tracks.

Broughton was thus bound for an area where British mariners and British merchant interests, including the East India Company, had great expectations and hopes of opening trade.[5] A hint of what would develop in the future occurred a few years earlier, when James Colnett, in the merchant ship *Argonaut*, had sailed from Macau in 1791 to trade with Japan and Korea only to meet Japanese resistance to any open trade. In addition, Korean trade prospects were a chimera. Colnett's opinion was that 'all Nations improve by commerce which I have every reason to believe will soon be extended to Corea and Japon where a great Call for furs will happen ... an Embassy to Japon might open that trade much sooner than is believed'.[6] Colnett's pioneering voyage would presage other attempts, and safe navigation in those seas necessitated the survey work that Broughton was about to undertake and the drafting of plans that would become the charts of the future.[7] As Colnett knew, the existing charts were faulty and unreliable. The London merchant adventurer Richard Cadman Etches, principal in the King George's Sound Company, was correct when he told the presiding genius of British science and discovery, Sir Joseph Banks, on 19 May 1792, 'I cannot but observe that the Island[s] in the Archipelago of Japan are very

[1] Broughton, *Voyage*, p. iv.

[2] Daines Barrington, *Miscellanies*, London, 1781. For details of his life and influence, see entry in *Oxford Dictionary of National Biography*.

[3] Barrington, *Miscellanies*, p. vii.

[4] See p. 251 below.

[5] James Cook, Captain John Blankett, Alexander Dalrymple, Daines Barrington, John Henry Cox, John Meares and Lord Macartney were all proponents of opening trade with Japan, the main hope being to use the outer islands as a means of breaching Japanese officialdom and resistance, Robert J. King, '"The long wish'd for object" – Opening the Trade to Japan, 1785–1795', *The Northern Mariner/le marin du nord*, 20, 1 (January 2010), pp. 1–34.

[6] James Colnett, 'A Voyage for Whaling and Discovery Round C. Horn to the Pacific Oceans made under Protection of the Rt. Hon. Lords Commissioners of the Admiralty, Performed in the Merchant Ship Rattler by J. Colnett, Lieutenant in the R.N., 1793–1794', BL, Add. MS 30,369, f. 251.

[7] TNA, ADM 55/142. See also F. W. Howay, ed., *The Journal of Captain James Colnett Aboard the Argonaut from April 26, 1789 to Nov. 3, 1791*, Toronto, The Champlain Society, 1940.

erroneously laid down.'[1] The British administration of William Pitt hoped for the opening of a Japanese field of commerce, long-denied, and intended that Britain's first ambassador to China, Lord Macartney, should visit Japan after completing his mission to China, but circumstances prevented Macartney from acting on this. But an indication of British intentions had been made, and the advantage of having reliable charts was a harbinger of future commercial and diplomatic prospects.[2]

Broughton took his departure from Monterey on 20 June, bound for Kealakekua Bay in order to rate his chronometers, which he had been unable to do at Monterey, and to replenish his water, before making for the coast of Japan. Due to the unsocial conduct of Colonel Borica, Broughton felt he was justified in not saluting the fort on sailing, as naval etiquette demanded, even though it was evident that they expected this compliment from the preparations the Spaniards had made. On 25 June the *Providence* passed through the charted position of an island called Paxaros[3] or Bird Island on various charts on board. Although they could see six or eight leagues around the horizon there was no indication of any land in the vicinity. In addition they had hardly sighted any birds since leaving the coast of California. On 2 July Broughton then ran down the longitude of an island named St Maria la Gorda on Henry Roberts's 'A General Chart exhibiting the Discoveries made by Captn James Cook', again without success. Broughton had already searched for this island on 25 February en route from Niihau to Nootka when using Arrowsmith's 1794 'Map of the World on a Globular Projection'. After a favourable passage Broughton anchored in Kealakekua Bay on 6 July.

As a taboo existed ashore it was not until 10 July that the observatory was erected in the same place as before, for Crosley to obtain fresh rates for the chronometers, which had altered considerably since leaving Nootka Sound and Monterey. Equal altitudes were obtained on the first day, but clouds prevented any observations for a further seven days. Cloudy weather persisted and it was not until the 22nd that the observatory was dismantled and brought on board. The same evening Broughton got under way and set a course to the west. The Hawaiian king's agent, John Young, who had accompanied Broughton to Oahu on his previous visit, had returned, but Kamehameha had not. On 25 July Broughton hove to in Mamala Bay when Kamehameha and some other chiefs came on board from Honolulu. The *Britannia* was in the same state as before and Broughton presented Kamehameha with several useful articles needed to complete her. Kamehameha gave permission for the British to make a settlement in any of the islands. Broughton considered that Oahu would be the first choice on account of Honolulu, adding that there was another harbour in the island with a large basin where the natives collected pearls in considerable numbers – a clear reference to Pearl Harbor.

[1] Quoted in Robert King, '"A Regular and Reciprocal System of Commerce" – Botany Bay, Nootka Sound, and the Isles of Japan', *The Great Circle: Journal of the Australian Association for Maritime History*, 19, 1 (1997), pp. 1–29, p. 21. King comments on an accompanying chart that the inaccuracies depicted on it, especially with respect to Japan, on the eve of Broughton's survey, were indicative of the poor state of knowledge in general concerning that country, adding that on it 'the outline of the Main Island [Honshu, called in the chart, Niphon] is grossly distorted, Great Liqueo [Ryukyu, or Okinawa] is indicated as being much larger than it actually is, Matoosmae [Matsmai, present day Hokkaido] is shown much smaller than it is, Sagaleen [Sakhalin] is shown as a small island occupying only a fraction of the area of the actual island, and the Southern Kuriles are shown far too large.', ibid. King also analyses in his article the proponents of opening trade with Japan at this time, and the ways of doing so.

[2] Helen H. Robbins, *Our First Ambassador to China*, London, 1908, p. 380. In addition to paramount trade considerations, matters of culture and science also motivated British diplomacy. See P. J. Marshall, 'Britain and China in the Late Eighteenth Century', in Marshall, *'A Free Though Conquering People': Eighteenth-Century Britain and its Empire*, Aldershot, 2003, ch. 12.

[3] The archaic Spanish spelling of Pajaros = birds.

That evening they took leave of their friends. Broughton then set a course for Kauai, anchoring there the next day off Waimea River to obtain water, which was only acquired with some difficulty. Provisions could only be bought for guns and powder, which was strictly forbidden. Broughton therefore got under way and set a course for Niihau, where he hoped to obtain some provisions, anchoring in Yam Bay, probably Nonopapa Landing, in the hope of procuring some yams. A few yams were obtained, but unfortunately two marines were killed in an unprovoked attack by the natives. Broughton retaliated by burning a village and destroying a number of canoes.

On 31 July Broughton continued on towards Japan. The following day an island, now known as Nihoa, was sighted, shown as the twin islands of Bird I. and Montagues I. on 'A Chart of the Northern Pacific Ocean ...' in John Meares's account of his voyages.[1] This account led Broughton to conclude that the European discoverer of Nihoa had been Captain Douglas in 1789, whereas in fact James Colnett had sighted the island on 21 March 1788 in the *Prince of Wales* in company with the *Princess Royal*, Captain Charles Duncan.[2] Colnett, however, was not the first person to sight Nihoa as there are the remains of some ancient Polynesian stone buildings on the island.[3] In excellent visibility Broughton could see only a single island, but Colnett noted it 'Shewed like two Isles but I think it was Join'd by a low narrow beach'.[4] On 4 August Broughton was once again in the latitude of the island of Donna Maria Lajara, which he had searched for earlier in the year, but again no birds were sighted and there was no indication of land. On 24 August the *Providence* was on the parallel of Rica de Plata shown on Henry Roberts's 'A General Chart exhibiting the Discoveries made by Captn James Cook', but once again there was no sign of land in the vicinity.

On 7 September, as she approached the north coast of Honshu, the *Providence* was struck by a series of violent squalls. They lasted only for an hour during which most of the sails were blown away and the ship heeled over so much that Broughton feared that the masts would go by the board. For the hour it lasted it was one of the worst gales he had ever encountered. At daylight the following morning the coast of Honshu was sighted and Broughton altered course to the north. On the 9th they passed Port Nambu (modern Miyako Wan) and on the following morning they rounded Shiriya Saki, the NE extremity of Honshu. Broughton then set a course across the eastern part of Tsugaru Kaikyo, the strait separating Honshu and Hokkaido, towards Uchiura Wan, a deep inlet on the SW coast of Hokkaido which Broughton named Volcano Bay on account of the active and dormant volcanoes in the vicinity.

Broughton, who was now entering an area where the charts he held on board the *Providence* were of little help, commented in the margin of one version of his journal, 'It should be remember'd that we had no account whatever relating to these parts excepting Cooke's 3d voyage & general chart, in which Matsmai is laid down as an island North of Japan, which we suppose this to be.'[5] This was clearly a reference to Henry Roberts's general chart of the world, on which

[1] John Meares, *Voyages made in the Years 1788 and 1789, from China to the North West Coast of America*, London, 1790, f.p. i, with a view of the island, f.p. 80, where it is named 'The new discover'd Sandwich Island'. As Montagu is the family name of the Earl of Sandwich, this may account for Bird Island's twin island with that name.

[2] Robert Galois, ed., *A Voyage to the North West Side of America: The Journals of James Colnett, 1786–89*, Vancouver and Toronto, 2004, p. 204.

[3] *Pacific Islands Pilot*, Vol. III, 10th edn, Taunton, 1982, p. 280.

[4] Galois, *Journals of James Colnett*, p. 204.

[5] NMM, RUSI/NM/132 (8).

Hokkaido is depicted as series of six small islands named the Islands of Jeso, the SW island being named Matoosmaee, while Sakhalin is named Sagaleen and is depicted as a small island almost attached to Russian Tartary, 'taken from a Russian MS. Chart, got by us at the Island of Oonalashka'.[1] Broughton may also have held Arrowsmith's 1790 'Chart of the World on Mercator's Projection', which names the island to the north of Tsugaru Kaikyo as Matmay. Broughton was also aware of the discoveries made in 1643 by Maarten Gerritszoom de Vries in the *Castricom* and of Martin Spanberg (Martyn Petrovich Shpanberg) in the *Arkhangel Mikhail* in 1739, brief details of which are given by James King in the published account of Cook's third voyage.[2] De Vries was a captain in the Dutch East India Company (VOC) who sailed on 3 February 1643 from Batavia (Jakarta) in the *Castricom* accompanied by the *Breskens*, which soon lost contact, in search of islands to the NNE of Japan, reputed to be rich in precious metals. Passing to the east of Hokkaido and the southernmost of the Kurile Islands, de Vries was the European discoverer of Ostrov Iturup, which he named Staten Eylandt, and Ostrov Urup, which he named Compagnies Landt and on which he landed. He returned to the Pacific through Proliv Friza, the Russian for Vries Strait, between Ostrov Urup and Ostrov Iturup. Martin Spanberg, a Danish officer in Russian service, took part in the first Kamchatka Expedition in 1725–30 under Vitus Bering, and in 1731–44 he took part in the Second Kamchatka Expedition. In 1738 he charted the northernmost of the Kurile Islands as far south as Ostrov Urup. The following year he reached Honshu where he tried unsuccessfully to open relations with the Japanese. On 3 July 1739, on his return voyage to the north, he anchored off a large island, now known as Ostrov Shpanberga, where he remained until 9 July during which time he made contact with the natives and obtained thirteen casks of good water.

On 12 September, as Broughton entered Uchiura Wan, three fishing boats came alongside and the fishermen came on board. When Broughton enquired from them whether the island was called Matsmai, they invariably pointed to the west. This was in the direction of the principal town of the region, near the SW tip of Hokkaido, called Matsumae, which may be the reason for the misnaming of the island. Broughton understood from the fishermen that the island was in fact called Insu, which was the name he adopted in his journal and on his charts. The fishermen were the Ainu, and Broughton gives a good description of them and their canoes in his journal.[3] Broughton found a safe anchorage in Muroran Ko, on the SE side of Uchiura Wan, which he called Endermo from its Ainu name. The town of that name is set in attractive country with good soil, and with fresh water available in the vicinity. Broughton thought most English plants would grow there. Soon after the *Providence* anchored, some Japanese officials arrived on board and told the Ainu to leave. Some of the Japanese, nevertheless, proved friendly and helpful, and a few days later brought a chart of the world, which appeared to be Russian, on which was a collection of coats of arms of the world, from which they were able to discover that the *Providence* had come from the British Isles. They also allowed Broughton to copy a large chart of the islands north of Japan and the following day brought a chart they had made themselves. In return Broughton presented them with a copy

[1] James Cook and James King, *A Voyage to the Pacific Ocean ... for making Discoveries in the Northern Hemisphere*, 3 vols, London, 1784, I, p. lxxxii.

[2] Ibid, 3, pp. 388–9. See also James Burney, *A Chronological History of North-Eastern Voyages of Discovery*, London, 1819, p. 156. Although he does not mention it, Broughton may have had on board a copy of Gerhard Friedrich Müller, *Voyages from Asia to America for Completing the Discoveries of the north West Coast of America*, London, 1761.

[3] See p. 72 and pp. 79–81 below.

of Henry Roberts's general chart of the world. During the *Providence*'s stay, Crosley made the necessary observations to rate the chronometers and the bay was surveyed (see Fig. 4 on p. 74). On the morning of the 28th Broughton got under way, anchoring that evening in the entrance to the harbour. On the morning of 30 September one of the sailors, Danish-born Hans Oldson, died after falling from a tree. He was buried on Daikoko Shima in Muroran Ko.[1]

Broughton liked the Ainu, who he found gentle, inoffensive and polite. Their speech was slow and timid. He noted that their language contained many Japanese words. He described their food as mainly dried fish boiled with a little oil, but noted that they also made use of wild berries and other plants. They kept bears and eagles in cages. He thought these were for food, since the Ainu would not part with them. He also felt that the local inhabitants were under tight control by the Japanese, who tried to drive them away when they approached the visitors, and he noted that the Japanese seemed to eat better than the Ainu.

On 1 October Broughton took his departure from Uchiura Wan and, after rounding Erimo Misaki, the SE extremity of Hokkaido, he altered course to the north to follow the eastern coast of Hokkaido towards the Kurile Islands. He was now following in the tracks of De Vries in the *Castricom* in 1643, Spanberg in the *Arkhangel Mikail* in 1739 and John Gore commanding the *Resolution* and *Discovery* on their homeward voyage in 1779 at the conclusion of Cook's. This was a time of miserable sailing, with heavy weather and thick fog adding to the gloom and making observations difficult. Broughton described the land, with snow lying in valleys. There were few signs of habitation, but sea birds, whales and porpoises were abundant. On 7 October Broughton brought to north of Ostrov Shpanberga, known to the Japanese as Shikotan To,[2] the island where Spanberg watered in 1739.[3] Broughton sent a boat inshore, the crew finding the island to be more productive than they expected, with springs of water and very good pasturage. Crowberries and a species of cranberry were found in abundance, but no inhabitants were seen. A good observation with three sextants made the latitude 43°50′30″N, which agreed within 30″ with that obtained by Spanberg. Broughton now steered NNE for land he could see in that direction. As he closed the land he could distinguish a passage between the islands. This was Proliv Yekateriny, the strait separating Ostrov Kunashir and Ostrov Iturup, known respectively to the Japanese as Kunashiri Suido, Kunashiri To and Etorofu To.

Once clear of the strait Broughton altered course to the NE and for the next seven days took what observations he could to fix the positions of Ostrov Iturup and Ostrov Urup. On 13 October he sighted Ostrova Chernyye Brat'ya, a group of small islands between Ostrov Simushir and Ostrov Urup, which had been seen by La Pérouse in August 1787. On 29 August the French explorer subsequently passed through the strait separating Ostrova Chernyye Brat'ya and Ostrov Simushir, naming it Passage de la Boussole. It is now known as Proliv Bussol'. On 16 October Broughton reached the small island of Marukan, now known as Ostrov Simushir, where, according to the published account of Cook's third voyage, there was a Russian settlement.[4] An officer was sent inshore to examine a small harbour at the northern end of the island. On his return he reported that the Russian settlement had been abandoned,

[1] The grave was destroyed when a lighthouse was erected in 1891, but the Muroran Rotary Club erected a stone to the memory of this sailor and this early event in Anglo-Japanese relations. This story derives from MS 6681, Papers of Harold Williams, folder 18, National Archives of Australia.

[2] The sovereignty of several islands to the north of Hokkaido is in dispute between Japan and Russia.

[3] Müller, *Voyages from Asia to America*, p. 29.

[4] Cook and King. *Voyages to the Pacific Ocean*, 2, p. 500.

but that the Russian arms carved in wood and painted remained, as well as crosses erected here and there. The natives were similar to those of Muroran Ko in appearance, but spoke a different language. They were dressed in bearskins, with Russian boots and cotton handkerchiefs on their heads. The harbour is now called Bukhta Broutona (Broughton Bay), having been so named by Adam J. von Krusenstern in his *Atlas* (St Petersburg, 1813). (See Fig. 6 on p. 90.)

The onset of bad weather prevented any further contact with the shore or any of the other Kurile islands and, with the approach of winter, Broughton got under way and set a course to the south. However, on 18 October, just south of Ostrov Simushir he fell in a severe gale and fractured his arm, making further exploration difficult. He therefore passed through Ostrova Chernyye Brat'ya and, on reaching open water, set a course to the SW towards Honshu. On 4 November course was altered to close Taito Saki, the SE extremity of Honshu, named White Point on the chart 'Part of Japan or Nipon' in the published account of Cook's third voyage.[1]

On 9 November the coast of Honshu was sighted in the vicinity of Taito Saki and the following day several fishing boats approached the *Providence* and threw some fish on board, which were gratefully received. Broughton described these boats in some detail[2] and included sketches of them in the published account of his voyage.[3] Next day a group of islands were seen to the SSW and the distinctive high mountain of Fuji-san to the NW. The islands were the Izu Shoto, a chain of islands in the seaward approaches to Tokyo or 'the famous Port of Jeddo', as Broughton referred it, using the former European name for this city. Broughton spent three days surveying these islands. During the night of 12 November, when well to the west of the principal islands of Izu Shoto, he passed a short distance north of Zeni Su, a group of islets and reefs, the highest only twelve metres in height, without sighting them, and the following night passed a short distance south of their dangers, again without sighting these islets of which he had no knowledge. Clearly this was a lucky escape.

After completing his examination of Izu Shoto, Broughton continued to the SW, and on 17 November he sighted Shiono Misaki, the southernmost point of Honshu. When Kyushu was sighted in indifferent visibility on 23 November, Broughton altered course to the south. The following day the 1,118-metre-high active volcano in the centre of Sakura Shima, close east of the town of Kagoshima, was sighted. On 28 November, having been set some distance to the SE, Broughton found himself in the vicinity of several islands laid down in Arrowsmith's charts, with nothing to indicate the presence of any such islands. These were the non-existent Todos los Santos charted in 30°45′N, 139°22′E.[4] On 1 December the island of Okinawa was sighted. This is the SW island of Nansei-shoto or the Ryukyu Islands, named in Broughton's journal as Great Liqueo and on his chart as the Great Lieuchieux. It was also known at the time as Loo-Choo[5] and Lewchew,[6] with probably other variations as well. The island was only visible at times and, after sighting Kerama Retto off the SW extremity of Okinawa, named in some charts as the Matchi Islands, Broughton continued to the SW in

[1] Ibid, 3, f.p. 395.

[2] See pp. 95–6.

[3] Broughton, *Voyage*, f.p. 139. See Fig. 7 on p. 95.

[4] Henry Stommel, *Lost Islands*, Vancouver, 1984, pp. 108, 132, 133.

[5] Basil Hall, *Account of a Voyage of Discovery to the West Coast of Corea and the Great Loo-Choo Island*, London, 1818.

[6] John M'Leod, *Voyage of His Majesty's Ship Alceste, along the Coast of Corea to the Island of Lewchew*, 2nd edn, London, 1818.

search of some islands supposed to be situated between Okinawa and Taiwan (formerly known as Formosa).

On 3 December a low island was sighted to the west. This was Miyako Shima, which Broughton called Typinsan, the easternmost island of Sakishima Gunto. The next two days were spent sailing along the southern side of this group of islands, taking the necessary bearings to fix their positions. On 6 December the east coast of Taiwan was sighted, and after rounding its southern extremity Broughton set a course for Macau. On 10 December Broughton passed the rocky islet of Zhenyantou, known in those days as Pedro Blanco because of its white summit. Two days later, after threading his way through the islands in the eastern approaches to Macau, Broughton anchored off Macau, twenty-two months after his departure from Plymouth. Prior to his arrival Broughton mustered the ship's company and ordered both the seamen and officers to deliver up all their journals, remarks and drawings, and also enjoining them to secrecy over their proceedings since leaving England.

On his arrival Broughton sent an officer to wait on the Portuguese governor of Macau, Jose Manuel Pinto, to request a pilot to take the *Providence* into the Typa, the important anchorage off Macau. On getting under way the following morning, Broughton saluted the fort with eleven guns, which was returned with an equal number. It was not, however, until the end of the month that Broughton had sufficiently recovered the use of his arm for him to leave the ship. His first task was to look for a suitable consort. 'During the recent long nights and days at sea in thick and blustery weather,' Admiral Ritchie suggests that Broughton 'had pondered on the risks he was taking in a single vessel in uncharted waters. If *Providence* ran aground there would be no ship to stand by. Cook had learnt this lesson on his first voyage ...'[1] Broughton had heard that a suitable vessel was to be found in Lark's Bay, also known as Dirty Butter Bay, a small and shallow haven on an island about nine miles SW of Macau.[2] Here he found the 87-ton schooner *Prince William Henry*, which he purchased for £1,500[3] and subsequently fitted out as a tender, to assist his surveying.[4] Another important purchase in Macau was box chronometer Arnold No. 45 from the East Indiaman *Henry Addingon*, Captain Richard Atherton Farrington,[5] as the performance of Earnshaw No. 1 had become erratic.

[1] Ritchie, *The Admiralty Chart*, p. 70.

[2] Amasa Delano, *Narrative of Voyages and Travels in the Northern and Southern Hemisphere: Comprising Three Voyages Round the World together with a Voyage of Survey and Discovery in the Pacific Ocean and Oriental Islands*, Boston, 1817, p. 43. John Kendrick and James Colnett were among the north-west traders who used Lark's Bay. Here, too, in about 1780 a depot of opium was established on two British vessels stationed in the bay, and from where they sold it. Chinese officials regarded this place as a smugglers' haven. In 1794, after a time of complaint and harassment by the Chinese, and additional trouble from pirates, the parties concerned in the trade were induced to bring one of their ships, exclusively with opium, to Whampoa, roadstead for Canton. In 1799 the Chinese government enacted special laws against the import of opium but the trading practice continued unabated; Anon, 'The Traffic in Opium with China ...', *The Calcutta Christian Observer*, September 1837, pp. 485–6. This article was reprinted from *Chinese Repository*, April 1837.

[3] £1,169 for the vessel, the balance to pay off the crew; Broughton to the Commissioners of His Majesty's Navy, 29 December 1796, UKHO, OD 177.

[4] Alan Reid, 'Broughton's Schooner', *The Mariner's Mirror*, 64, 3 (1978), pp. 241–4; Andrew C. F. David, 'Broughton's Schooner and the Bounty Mutineers', ibid., 63, 3 (1977), pp. 207–13. Broughton's reasons for the purchase are stated in his letter to Nepean, 1 January 1797, TNA, ADM.1/1517, and Cap B512. Broughton's purchase was not that built by the mutineer Morrison as stated by John Barrow in *The Eventful History of the Mutiny and Piratical Seizure of HMS Bounty*, London, 1831, p. 161.

[5] CUL, RGO 14/68, f. 145.

Broughton lived ashore for most of the period from January to April 1797, working on his survey of the Kuriles and transmitting it together with a copy of his proceedings and his future intentions to the Admiralty. On 30 December the *Suffolk*, wearing the flag of Rear Admiral Peter Rainier, and the sloop *Swift*, commanded by Lieutenant Thomas Hayward, one of the officers set adrift with Bligh after the mutiny of the *Bounty*, joined the *Providence* in the Typa, sailing again on 7 January for Madras with a convoy of four homeward-bound East Indiamen. Broughton had served under Rainier in the 68-gun *Burford* on the North American Station.[1] The local Chinese assisted the *Providence*'s ship's company with making cordage. In February the Portuguese Governor, with several gentlemen, dined on board the *Providence*, and was saluted with eleven guns.

6. Conclusion of the Voyage and the Wreck of the *Providence*

After refitting both vessels and taking on supplies, water and provisions for fifteen months, Broughton was almost ready to depart. First, however, Crosley landed on a nearby island, where on 8 April he obtained the necessary observations to determine the rate of Arnold No. 45 and its error on mean time. Then, on 11 April 1797, Broughton took his departure to continue his survey to the north. Within days the crew went down with dysentery, which Broughton attributed to the hazy and foggy weather encountered, though others, probably correctly, thought it could be attributed to the water taken on at Macau. There was also mortality among their sea stock of pigs, particularly those taken on board at Tahiti and Hawaii. After sighting Taiwan on 7 May, Broughton had hoped to have proceeded to the north through Taiwan Strait, but contrary winds forced him to pass to the south of the island, which was rounded on 9 May. He then headed NNE for Sakishima Gunto to examine the north side of this archipelago. On 12 May the *Providence*'s foremast was found to be rotten, which was repaired by fishes[2] and other means. The top gallant mast was also brought down on deck to lessen the strain on the foremast. The same day Yonakuni Shima, an isolated island about midway between Taiwan and Iriomote Shima, the westernmost island of Sakishima Gunto, was sighted. Two days later Iriomote Shima itself was sighted, followed the next day by Ishigaki Shima, where a boat was sent inshore to obtain some intelligence of the inhabitants. The landing party made their way to a small village, which was deserted except for an elderly man who treated them with great civility. Once the landing party had re-embarked, other inhabitants came down to the beach. They appeared harmless and inoffensive.

The two vessels then continued to the east in apparently deep water, when several small islands were seen well to the south. On the evening of 17 May the mate of the watch reported that there was no appearance of land ahead or to windward. Accordingly, Broughton decided to stand on for another hour before trying for soundings and then plying to windward for the night. At this juncture Broughton went below to plot his position, which he found was to the north of Miyako Shima, which they had passed to the south of the previous year.

Almost immediately breakers were sighted ahead and on each bow, with the vessel making 4½ knots. The officer of the watch, the third Lieutenant, James Giles Vashon, put the helm up, but the *Providence* ran heavily aground on Yaebi Se, a coral reef extending nine miles north

[1] Nicholas Tracy, *Who's Who in Nelson's Navy: 200 Naval Heroes*, London, 2006, p. 63.

[2] Long pieces of hard wood, convex on one side and concave on the other, bound opposite each other to strengthen the mast and secured by bolts and hoops or stout rope called woolding.

of Ikema Shima, a small island off the northern end of Miyako Shima. The signal of danger was made to the schooner and Mr Chapman, the master, was sent to anchor her as near as possible in an attempt to haul the *Providence* off the reef. The sails were braced aback and the anchors cleared away, but after her head paid off to the south she struck again and remained fast. When the carpenter reported that the water was up to the orlop deck and that she was bilged forward, all attempts to haul the ship off the reef were abandoned. At this stage the officers were unanimous that nothing could be done to save the ship. Broughton therefore gave orders to the officers to save their commissions and to prepare the boats for embarking the crew. By now the *Providence* was nearly on her beam ends. The boats were then lowered and the crew ferried across to the schooner, without loss of life, Broughton and the officers being the last to leave in the pinnace. Some papers and navigational equipment were saved, but the personal effects of the officers and crew were all lost, so quickly had the water risen in the vessel. Among the papers saved were Broughton's journal and the ship's log and among the instruments saved was Arnold box chronometer No. 45.

The schooner had come to with both her anchors down, but one was then hove up. At 4 a.m., however, the other anchor cable parted and the schooner was lucky not to have shared the fate of the *Providence*. Broughton remained off the wreck in the schooner during the forenoon while the boats were sent to the wreck to see what could be saved, which included a kedge anchor and cable, some sails and cordage, but no provisions. Broughton therefore decided he had no option but to seek help from the nearest island, which was Miyako Shima, eventually anchoring the schooner off the present day village of Hirara at the SW end of the island. Here they were treated with great civility and all their needs were supplied. At first Broughton contemplated leaving seventy men here, while he sought help in Macau, but decided this would be impracticable without the use of force.

On 23 May, after a stay of five days and having been amply supplied with provisions, Brought set sail for Macau in the schooner, accompanied by the pinnace and cutter. The long boat had been left behind, fully rigged, as a present for the friendly islanders. The voyage back to Macau was hardly a pleasant one, with part of the crew having to be carried in the boats as the schooner was not large enough to accommodate 112 persons. During the day the boats were normally under sail, but at night they were usually being towed. On 26 May, off Iriomote Shima, a canoe brought them off water and potatoes, the inhabitants being apparently aware of the shipwreck. Four days later they rounded the south point of Taiwan and set course for Macau. On 4 June the cutter was sent ahead to Macau to inform the governor of their return and to order provisions, while Broughton, in the schooner, altered course to the north for Zhu Jiang or Canton River. After passing through Hu Men, the famous Boca Tigris, he came to an anchor on 5 June off Huang-pu, known in those days as Whampoa. By now Broughton had decided to continue his survey in the schooner and, having selected thirty-five seamen for this purpose, he needed to make arrangements for passages back to England for those of his crew he did not wish to retain. The greater number were transferred to the sloop *Swift*, Lieutenant Thomas Hayward, and the East Indiamen she was to convoy to the East Indies. Shortly after sailing the *Swift* was unfortunately lost in a typhoon with all hands, and thus it was that some of the survivors of the wreck of the *Providence* subsequently perished at sea. John Crosley was sent home in the packet *Crescent*, which was also damaged in the typhoon and forced to return to Macau for repairs, eventually reaching England on 7 July 1798.[1] Zachary Mudge was also

[1] For his subsequent career see p. lvii.

sent home, presumably in one of the East Indiamen, taking with him duplicates of Broughton's journal and correspondence.[1] The vessel he sailed in was also caught by the typhoon and Mudge had to open the package they were in to dry out its contents.[2] James Vashon was retained in the schooner as Broughton considered that the stranding would have been avoided had Vashon put the helm down, instead of up. In consequence Broughton intended to apply for his court martial, when one could be convened.

On 11 June Broughton dropped down to Macau and between 14 and 25 June observations were obtained to establish the rate of Arnold No. 45, in the same place where Crosley had observed. The observer was probably William Chapman. On sailing Broughton expressed in his journal his hopes and the possible difficulties ahead:

> We now sail'd a second time in the prosecution of the voyage, not indeed with the most flattering hopes of success, as the season was far advanc'd & the vessel in many respects ill-suited to the purpose. But still there was a prospect of acquiring a geographical knowledge of the Tartarean & Corean coasts; nor did I wish to fail in any endeavour that might promote the improvement of science by the discovery of parts of the world as yet but imperfectly known. The officers and men were equally dispos'd with myself to do their respective duties, and we departed in good health.[3]

On 30 June, P'eng-hu Ch'un-tao or the Pescadores Islands, off the west coast of Taiwan, were sighted. Some observations were taken of the western side of these islands; Broughton regretted that he was unable to examine their eastern side or to have some communication with the shore. After rounding the northern part of Taiwan, Broughton passed to the north of the harbour of Kelang, now the site of the major port of Chi-lung Kang. He then altered course to pass north of Sakishima Gunto. On 6 July, Senkaku Shoto, a group of uninhabited islets situated 90 miles ENE of the NE extremity of Taiwan, were passed, followed next day by Kobi Sho, the summit of an extinct volcano, situated 17 miles NE of the principal island of the group. Later the same day he passed about three miles north of the steep-to[4] 83-metre-high islet of Sekibi Sho. On 8 July, Kume Shima, the westernmost island of Okinawa-shoto was sighted. As Broughton closed the island, the schooner was boarded by some of the inhabitants, who enquired if they were from Hollandia. Assisted by a favourable current, Broughton threaded his way through a number of islands and on 10 July anchored in the harbour of Napachan, present day Naha Ko, near the SW end of Okinawa.

Broughton spent two somewhat frustrating days in Naha Ko. Provisions and water were sent off to the schooner, but it was clear that the Okinawans wanted Broughton to depart as soon as possible. Two officers were sent ashore, but they were escorted back on board about an hour later, after making some useful observations. On 12 July Broughton got under way and he spent the next two days working his way round the west coast of Okinawa, taking what observations he could. After rounding the northern end of the island, Broughton altered course to the NNE, passing to the east of Amani Gunto and Osumi Gunto. He next spent eight days attempting to proceed through Osumi Kaikyo, the strait between Tanega Shima and Kyushu, as it had been his intention to cross the Korea Strait and carry out a running

[1] Now held in UKHO, OD 177.

[2] For his subsequent career see p. lv.

[3] See pp. 136–7 below; repeated with variations in Broughton, *Voyage*, p. 222.

[4] 'STEEP-TO ... Said of a bold [steep] shore, admitting of the largest vessels coming very close to the cliffs without touching the bottom': W. H. Smyth, *The Sailor's Word-Book*, London, 1867, p. 654.

survey of the east coasts of Korea and the Gulf of Tartary from south to north. However, light winds and adverse currents frustrated this attempt and, on 24 July, Broughton altered course to the east and then NE to make the south coast of Honshu. He then continued to the north along the east coast of the island, making what observations he could. On 12 August he was once again at anchor in Muroran Ko.

Broughton spent ten days in Muroran Ko, refitting the schooner, taking observations for rating the chronometer and obtaining wood and water. From his Japanese friends Broughton acquired a chart of the Japanese Islands, with a strict injunction not to disclose from whom he had obtained it. On 21 August Broughton got under way and anchored in the entrance to the bay to await a favourable wind. On the 23rd the wind shifted to the east and Broughton weighed anchor and began working his way out of Uchiura Wan and through Tsugaru Kaikyo, the strait separating Hokkaido and Honshu. As the combined tidal stream and current generally runs to the east through the strait, it was not until 30 August that Broughton was clear of the strait and opposite Matsumae, the principal town of the region, situated on the SW extremity of Hokkaido. It seemed as though preparations were being made to receive them, but Broughton considered the schooner's force was so insufficient that it would be unwise to put himself in the power of the Japanese.

Broughton now began working his way up the little-known west coast of Hokkaido. On 7 September he was off the western entrance to La Pérouse Strait, through which the French explorer passed in August 1787. Unknowingly, Broughton was now about to survey where La Pérouse had explored in 1787. From the north coast of Honshu La Pérouse's two frigates had crossed the Sea of Japan, where they made their landfall on the mainland of Asia on 11 June 1787 in the vicinity of Mys Povorotnyy, about fifteen miles SE of present day Nakhodka. La Pérouse then began working his way up the west coast of the Gulf of Tartary, anchoring twice without encountering any inhabitants. He then crossed the gulf to the west coast of Sakhalin, anchoring in two different locations where the inhabitants informed him that there was a navigable strait between Sakhalin and the mainland and another navigable strait to the south between Sakhalin and Hokkaido. On 27 July La Pérouse reached his farthest north in the gulf, where he sent an officer in the longboat to examine the head of the gulf, but the officer's report was inconclusive. Encountering heavy seas and shallow water, La Pérouse now found shelter about fifteen miles south of Bukhta Sushcheva, in what he called Baye de Castries, present day Zaliv Chikhacheva, with the port of De-Kastri on the west side of the bay, where he remained between 27 July and 2 August. Here he was told by natives, who had come from Segalien River (Reka Amur) in canoes, that on leaving the river they had pushed their canoes over a sandbank, covered with seaweed, which joined Sakhalin to the mainland. In consequence La Pérouse abandoned his plan to send his longboat ten or twelve leagues to the head of the gulf to investigate further. In this he acted with necessary caution, for he had lost six officers and fifteen seaman when two of his boats had capsized in Lituya Bay on the coast of Alaska. With the onset of southerly winds, accompanied by some fog and cold weather, La Pérouse prudently decided to make his way south. Southerly wind now carried his two ships south and on 11 August La Pérouse passed through the strait between Sakhalin and Hokkaido that now bears his name, and on 29 August he entered the North Pacific through Proliv Bussol', as mentioned above.

Broughton's running survey of the Gulf of Tartary was not as comprehensive as that of La Pérouse. This is hardly surprising considering the size of both the schooner and her crew. Unlike La Pérouse Broughton did not land on the west coast of Sakhalin, but he did penetrate about eight miles farther north than La Pérouse, as pointed out by the Russian explorer Adam

Krusenstern.[1] In addition, Broughton had reached the head of the gulf almost seven weeks later than La Pérouse, and with the imminent onset of winter he could not linger too long. When the depth decreased to three fathoms Broughton tacked and came to anchor in four fathoms off Bukhta Sushcheva. He then sent the master inshore to look for shelter. When Chapman returned he informed Broughton that he was convinced that a passage to the north was pretty evident. Broughton did not agree, but to settle the matter he got under way at daylight and stood to the north. When the depth decreased to two fathoms Broughton could see the head of the gulf, which appeared to be closed in by low land, while in the distance high land could be seen in a NE direction. There also was no evidence of tide, nor was there any indication of any river flowing into the head of the gulf, which convinced Broughton that a passage did not exist. He therefore tacked and stood to the south. Chapman, however, was correct as described below. Broughton then continued to the south along the west coast of the gulf, taking what observations he could. On 12 October the island of Tsushima, situated about halfway between Kyushu and the south coast of Korea, was sighted. After spending a day in the vicinity of the island, Broughton headed for the coast of Korea and on 14 October he entered the harbour of T'hosan, the present day port of Pusan. Broughton spent a week in Pusan, during which fresh supplies of wood and water were obtained, but little in the way of provisions. The natives were friendly, but were anxious that Broughton should depart as soon as possible. He remained, however, a few more days in order to obtain a fresh rate for the chronometer. Despite some opposition by the Koreans, Broughton was also able to produce a rough survey of this important harbour during his stay (see Fig. 9 on p. 189).

On 21 October Broughton got under way and spent the next five days working his way through the numerous islands forming the Korean Archipelago, off the SW coast of Korea, before passing close west of Quelpart Island, present day Cheju Do. On 29 October, having satisfied himself there were no islands to the south of Cheju Do, Broughton altered course to the west in order to survey the west coast of Korea. However, on 3 November he found himself off the entrance to the Nanking River, now known as the Yangtze or Chang Jiang. From here Broughton steered NE towards the SW extremity of Korea and on 6 November he sighted Soheugsan Do, the southernmost island of Huksan Chedo, a group of islands situated 70 miles NW of Cheju Do. Northerly gales then prevented any further progress towards Korea, forcing Broughton to abandon his proposed examination of its western coast. He therefore altered course to the south and on 10 November, after passing to the west of Cheju Do, he sighted the isolated group of islands known as Danjo Gunto, situated about 100 miles west of Kyushu. Two days later Broughton sighted various islands forming Osumi Gunto and, before noon on the 13th, Tokara Shima was sighted. He now altered course to the SW and then west, passing well to the north of Amami Gunto, Okinawa Gunto and Sakishima Gunto. On 19 November, Uotsuri Shima and Kobi Sho, which he had passed 2 miles to the north of on 7 July that year, were sighted. Keeping well to the north of these islands Broughton continued to the west and two days later sighted the northern part of Taiwan through the haze, before altering course to the south towards Macau. On 23 November the coast of China was sighted and two days later Broughton passed two miles north of Zhenyantou (Pedro Blanco). On 27 November, having taken from the officers and crew all remarks relating to the voyage, the schooner was once again safely at anchor in the Typa off Macau.

[1] A. J. Krusenstern, *Voyage Round the World in the Years 1803, 1804, 1805, & 1806*, 2 vols, London, 1813, II, pp. 181–2.

Broughton's journal ends with his arrival at Macau, where he remained until 21 January 1798, drawing his charts before sailing for Madras to join Admiral Rainier for the expected court martial to enquire into the loss of the *Providence*. Being informed that Rainier was expected in Trincomalee, Broughton proceeded there and on 19 May the court martial was duly assembled. Broughton was acquitted of any blame for the loss of the sloop, but due to the

> neglect of Lieutenant James Giles Vashon in not having caused a proper lookout to be kept at each cathead after Dark, conformable to the written orders of William Robert Broughton Esq of the said Sloop and to his not taking the proper steps, by throwing all aback to extricate her from the Danger, after it was first discovered ... they did therefore adjudge the said Lieutenant James Giles Vashon to be dismissed His Majesty's Service. But in consequence of his having served in the capacity of an Officer upon particular professional service in His Majesty's Schooner *Providence*, with his former Commander, ever since the loss of the said Sloop, and other Circumstances, the Board begged to recommend his Case to the Consideration of the Commander in Chief.[1]

Rainier clearly did not confirm the sentence as Vashon was appointed almost immediately to the 44-gun frigate *Sybille* in the East Indies fleet.[2] Chapman did not return home immediately either as he was appointed to the *Trident* on 28 May 1798.[3] On 24 May Broughton delivered up the schooner to Admiral Rainier and on 15 June he sailed for Madras, finally arriving back in England on 6 February 1799. The schooner *Prince William Henry* was commissioned at Trincomalee as HM Schooner *Providence*, the tenth with that name. She served on the East Indies Station for a time, and then returned to British waters. On 2 October 1804 she met her end as a fireship in an abortive attack on Boulogne by Admiral Keith.

7. The Voyage in Retrospect and Broughton's Subsequent Career

Details of Broughton's further service may be recounted briefly. He was promoted to post captain on 28 January 1797 during the voyage. On his return home he spent two years on half-pay while editing his account of the voyage for publication, in conjunction with his brother, the Revd Brian Broughton, Fellow of New College Oxford, who saw it through the press. It was published in 1804 by T. Cadell and W. Davies in the Strand, London, with the title, *A Voyage of Discovery to the North Pacific Ocean: in which the Coast of Asia, from the Lat. of 35° North to the Lat. of 52° North, the Island of Insu, (Commonly known under the Name of the Land of Jesso,) the North, South, and East Coasts of Japan, the Lieuchieux and the Adjacent Isles, as well as the Coast of Corea, have been Examined and Surveyed. Performed in His Majesty's Sloop Providence, and her Tender, in the years 1795, 1796, 1797, 1798. By William Robert Broughton.*[4]

On 23 June 1801 Broughton was appointed in command of the 50-gun *Batavier* employed as a floating battery in the Queen's Channel in the Thames Estuary. Then on 17 May 1802 he was

[1] TNA, ADM 1/5344; minutes of court martial held on 19, 21 & 22 May 1798.

[2] For Vashon's subsequent career see pp. lv–lvi.

[3] For Chapman's subsequent career see p. lvi.

[4] A German edition appeared in 1805. A French translation was published in Paris, 1807. A reprint of the English edition appeared in 1967, published by N. Israel of Amsterdam and Da Capo Press of New York.

appointed in command of the *Penelope*, 36 guns, under Commodore Sir William Sydney Smith, and employed off the Dutch coast and in the English Channel. In 1806 Broughton was in a squadron that chased the French ship-of-the-line *Le Vétéran*, commanded by Jérôme Bonaparte, the younger brother of the French Emperor, off the French coast, forcing her to take shelter in the little port of Concarneau on the south coast of Britanny.[1] On 30 May 1807 Broughton took command of the *Illustrious*, 74 guns, and in her in 1809, under Admiral Lord Gambier, he saw action against the French Brest squadron in Basque Roads, the anchorage in the mouth of the Charente. In 1810, still in the *Illustrious*, he took part in the capture of Mauritius, and in 1811 he was Commodore in charge of an expedition to Java, during which he was criticized as being unduly cautious. Nevertheless Batavia (Jakarta) surrendered to British forces on 8 August. The following day, however, Rear-Admiral Sir Robert Stopford arrived and, despite the fact that he was outside the limits of his station at the Cape of Good Hope, he superseded Broughton, much to the latter's indignation. On Stopford's departure Broughton resumed his command. On his return to England in 1812, Broughton applied for a court martial on Stopford for unjustly depriving him of his command, with all its attendant benefits including a share of prize money. This was a case he was unlikely to win and was certain to harm him. The Admiralty refused Broughton's request, and approved Stopford's course of action.[2]

On his return to England Broughton was placed on half-pay at a time when Britain was still at war with France and was now at war with the United States. At the peace in 1815 he was made a Companion of the Bath and on 12 August 1819 he was appointed a Colonel of Marines. He returned to active service on 31 May 1815 when he was given command of the 100-gun *Royal Sovereign*, followed on 30 August by being appointed in command of the 74-gun *Spencer*, a guard ship at Plymouth. It was here, in November 1817, that his son William (of whom more below) was entered as a first class volunteer on board his father's ship. On 28 September 1818, Broughton was superceded in command of the *Spencer* and was placed on half-pay. It was his last command.

Broughton retired to Florence, Italy, with his family, where he died on 12 March 1821. He was buried on 14 April in the English Burial Ground at Leghorn (Livorno), and a handsome and elaborate headstone was erected over his grave (see Fig. 2). Its lengthy inscription attests to his abilities and service:

> Sacred to the memory of William Robert Broughton, Esq., Captain in the Royal Navy of England and Colonel of Marines. His professional career was honorable to himself and beneficial to his country, in two voyages of discovery he traversed the Pacific Ocean with the perseverance, intrepidity, and skill of a British seaman. On the intricate coast of Java, as commander in chief of the English squadron he steered the fleet to victory and secured that valuable island to his sovereign. After having braved and overcome danger for forty-seven years in the service of his country, on the 12th of March 1821 in the 59th year of his age, he died suddenly at Florence in the bosom of his family, to whom he was endeared by those qualities which ameliorate the evils and enliven the joys of domestic life. It is now the consolation, as it was the happiness of his afflicted widow and children, that the character of a brave and gallant officer was united in the object of their sorrow, that of a good Christian.[3]

[1] James, *Naval History*, IV, pp. 205–7.

[2] Gerald S. Graham, *Great Britain in the Indian Ocean: A Study of Maritime Enterprise, 1810–1850*, Oxford, 1967, p. 50n. See also C. Northcote Parkinson, *War in the Eastern Seas, 1793–1815*, London, 1954, pp. 413–19.

[3] Edmond S. Meany, *Vancouver's Discovery of Puget Sound*, Portland, Oreg., 1957, pp. 265–6.

Figure 2. Grave of Captain William Robert Broughton, in the English Burial Ground at Leghorn (Livorno), in Italy.
Photograph courtesy of the Howay Collection, University of British Columbia.

He was survived by his wife. The only known likeness of Broughton is a portrait in oil, showing the sitter wearing captain's undress uniform of the 1812–25 pattern (see *Frontispiece*); it is held in the National Maritime Museum, Greenwich.[1]

In October 1802 Broughton married his cousin, Jemima, youngest daughter of the Revd Sir Thomas Delves Broughton, Bart., of Doddington Hall, near Nantwich, Cheshire. They had three daughters and one son, William, who was born 23 October 1804. The son's distinguished career in the Service may be briefly recounted. As mentioned above, he entered the Navy as a first class volunteer in his father's command, the *Spencer*, in 1817. Made lieutenant in 1825, he was promoted to commander in 1830. On 7 September of that year he was badly wounded when in command of the sloop *Primrose*. In her he engaged an exceptionally powerful Spanish slave ship *Velos Passagero* bound from Whydah (Oudah), in the Gulf of Benin, to Havana, and after a desperate action captured her. For his gallant conduct on this occasion he was advanced to post captain in 1831.[2]

Broughton's accomplishments as a naval officer and a hydrographic surveyor are commemorated by the many places named after him, notably: Broughton's Bluff, in the Lewis and Clark Recreation Site, Troutdale, Oregon; Broughton Islands, off the east coast of Australia; Broughton Island, one of the Snares Islands, New Zealand; Broughton Arm, Dusky Sound, New Zealand; Broughton Streets in Victoria and Vancouver, British Columbia; Broughton Island, the

[1] Information from Roger Quarm, Curator of Paintings, National Maritime Museum. The portrait was bought at Christies, 14 January 1972.

[2] Entry 'William Broughton, Esq.', in John Marshall, *Royal Naval Biography*, IV, pt. II, London, 1835, pp. 474–7.

Broughton Archipelago and Broughton Lagoon, Queen Charlotte Strait, all in British Columbia; and Bukhta Broutona in Ostrov Simushir, one of the Kurile Islands. Broughton Bay in Vladivostok did not survive as a place name; it is now Zolotoy Rog, the Golden Horn.

Broughton's achievements, especially as regards his surveys in north-east Asian waters, soon entered the historical record. The timing of his voyage was significant. He had examined the coasts of Japan and Korea at a time when European scientific knowledge of these waters was sparse. At Hokkaido and Sakhalin Broughton had forestalled the Russians, to the disappointment of the latter. 'We now have a whole new field of discoveries before us', complained Herman Ludwig van Löwenstern, the cartographer who sailed with Krusenstern in 1803–6, 'for everything we have done up to now at Matmai [Hokkaido] and Sachalin, Broughton is supposed to have already described in his newly published voyage, so they say.'[1] Krusenstern, who included details of Broughton's discoveries in his own account, admired 'the penetrating eye of Captain Broughton'.[2] British mariners held Broughton in high regard. The historian Admiral James Burney, who had sailed with James Cook on the latter's second and third voyages to the Pacific, compared some of Broughton's observations with those of La Pérouse's discoveries.[3] Broughton's voyage has also become integrated into the general hydrographic history of Japan.[4]

As for British hydrographic goals at this time, it can be said that Broughton met every expectation of the government and, in fact, anticipated its intention as announced in the Order in Council of 12 August 1795, whereby Alexander Dalrymple was appointed Hydrographer to the Admiralty and an office established to carry on the business of preparing charts for the fleet and to the benefit of safe navigation. That order in council acknowledged that there was a 'want of sufficient information respecting the navigation of those parts of the world to which their services may be directed, and with which they are sometimes totally unacquainted'.[5]

The corpus of Broughton's plans and charts made a substantial contribution to British and world empirical data concerning the seas of north-east Asia, and they did so at an important juncture in world affairs.[6] France, Russia and Japan were all engaged in examining these same

[1] Victoria Joan Moessner, trans., *The First Russian Voyage Around the World: The Journal of Hermann Ludwig von Löwenstern, 1803–1806*, Fairbanks, Alaska, 2003, p. 337. This account contains not a little second-hand evidence, including the amusing story that Broughton kept two monkeys in his cabin, and when he, in Port Jackson, went on deck to greet some visitors left them unattended. They subsequently got hold of his papers and ripped them all to pieces. 'That is the report I am supposed to send to the Admiralty', he told his guests when they entered the cabin and saw pieces of paper lying about the cabin; ibid., p. 396. The same source tantalizingly says that Broughton was warned before the *Providence* went aground and was lost but that his 'imperviousness could not stand having anyone tell himself anything. He sailed straight ahead and [was] shipwrecked'; ibid.

[2] Krusenstern, *Voyage*, II, p. 182.

[3] James Burney, *A Chronological History of the Discoveries in the South Sea or Pacific Ocean*, 5 parts, London, 1803–17, III, pp. 430–32.

[4] See, for instance, *Tsuko ichiran Dai* 6 [Summary of Navigations, 6], Osaka, Seibundo Shuppan, 1967, pp. 361–9 and Murorann Shiritus Muroran Toshokan, ed., *Providence-go raiko to Muroran* [HMS *Providence* visits Japan and Muroran], Muroran, 1969. This was the age of feudal clan of the Tokugawa shogunate. Visits of foreign ships are recounted in the literature of this era. The *Providence* received a prominent place in this literature. Information from Azusa Tanaka, Director, Division of Interlibrary Services, the National Diet Library, Tokyo.

[5] Vice-Admiral Sir Archibald Day, *The Admiralty Hydrographic Service, 1795–1919*, London, 1967, p. 13.

[6] Some years passed before other British hydrographers could continue Broughton's work. Basil Hall found Broughton's pioneering labour commendable. See Hall, *Account of a Voyage*, the essential account of the visit of HMS *Alceste* and *Lyra* there in 1816. The disclosure of the details of 'Loochoo' (Ryukyu Retto) may be followed in Barry Gough, ed., *To the Pacific and Arctic with Beechey: The Journal of George Peard of H.M.S. Blossom, 1825–1828*, Hakluyt Society, 2nd ser. 143, Cambridge, 1973, pp. 205–14.

waters, and the British contribution made by Broughton was of significance. Broughton made contributions to marine exploration on both sides of the Pacific, but it is in Asian waters that he was a pioneer. Broughton's contribution in north-east Asian waters largely completed the work of Cook and La Pérouse. Broughton's account of his voyage, published in 1804, appeared long before other works had laid down the essentials of the geography of this part of the world and its offshore islands. This was Broughton's achievement.

In another branch of science Broughton had also made a contribution, namely his ethnological observation based on incidental experiences of cross-cultural encounters. His experiences, though undeniably anecdotal, entered the historical record. As such they became noteworthy in the meeting of East and West in these seas. In 1816, as part of the British embassy to China by Lord Amherst, the frigate *Alceste* and the sloop *Lyra* proceeded north to the west coast of Korea, the eastern boundary of the Yellow Sea, while the Ambassador travelled overland to the Imperial capital. Captain Basil Hall, who commanded the *Lyra*, had a copy of Broughton's *Voyage* with him, and in Sakishima Gunto he found Broughton's account of the natives of Typinsan (Miyako Shima) and his visit to Napachan (Naha Ko) to be of considerable interest. Broughton, he noted, had been received by the inhabitants of Napachan with great kindness. They had supplied his wants but had objected to his landing, and, in fact, had sent back to the vessel some of the officers who had been sent on shore to examine the town.[1] Looked at from a different angle we may say that the investigative ways of Broughton's age had been checked by the resistance to foreigners of these particular Asian peoples. In the circumstances, the surveyor had no option but to withdraw in necessary if reluctant acceptance of reality.

8. Subsequent Careers of Broughton's Principal Officers and his Astronomer

Zachary Mudge. Mudge was sent back to England from Macau on 11 June 1797, following the loss of the *Providence*, taking with him dispatches and a copy of Broughton's journal. He was promoted commander in 1797 and advanced to post captain in 1800. Two years later he took command of the frigate *Blanche* in the West Indies, and in 1805, when carrying dispatches from Jamaica to Barbados, intended for Nelson, his ship was captured by a French squadron, and burnt at the end of the engagement. For the next five years he commanded the frigate *Phoenix* in the Bay of Biscay and off the coast of Portugal. His last command was the 74-gun ship *Valiant*, in 1814–15. In 1830 he was promoted Rear Admiral, in 1841 Vice-Admiral and in 1849 Admiral. He died in Plympton on 26 October 1852.

James Giles Vashon. Following his court martial in Trincomalee, the sentence of which was not confirmed, Vashon was appointed to the frigate *Sybille*, in which he served with distinction during the capture of French frigate *Forte* in the Bay of Bengal on 1 March 1799. He was promoted captain on 28 May1802 and appointed in command of the frigate *Fox* and sent to Macau, to escort the East Indiamen *Dover Castle* and *Asia* back to Madras, taking a route north of Luzon, then through Dampier Strait to Amboyna, north of Timor and south of Java and Sumatra, reaching the vicinity of Madras on 24 September 1802.[2] In 1803, in the same frigate, he took part in the storming of Bate Fort on the west coast of India when he was

[1] Hall, *Account of a Voyage*, p. 183.

[2] An interesting track chart by Vashon of this cruise is held in the UKHO, 595 on Be1.

severely wounded, a musket ball remaining in his body for the rest of his life.[1] Nevertheless Vashon surveyed Bate Harbour,[2] which was published by Alexander Dalrymple on 7 April 1807, although wrongly attributed to J. V. Vashon. In 1811, while in command of the 36-gun frigate *Thalia*, Vashon held the appointment, between 9 October and 29 December, of Commodore commanding HM ships and vessels employed in and around Jamaica and the Bahamas.[3] The following year, still in the *Thalia*, he escorted a homeward bound convoy from the West Indies. He died in January 1814 at Ballavargher in the parish of Braddon, in the Isle of Man, and was buried on 15 January in the Old Kirk Braddan, which was the parish church for Douglas until 1876. He was survived by his wife Louisa. He was the draughtsman of the engraved chart of the NE Coast of Asia and Japanese Isles in the published account of Broughton's voyage.

William Chapman. Chapman did not return home immediately from Trincomalee as he was appointed to HMS *Trident* on 28 May 1798. The following year he surveyed St Mary's Isles and the Permeira (Mūlki) Rocks off the West Cost of India in the newly renamed schooner *Providence*, while still Master of the *Trident*.[4] On 4 December 1805 Chapman was appointed Master of the 74-gun ship *Illustrious*, once more under Captain Broughton, and sailed in her to join Admiral Collingwood's fleet based in Gibraltar, remaining on board until 21 October 1807 when Broughton annotated his master's log stating that Chapman was too unwell to sign it himself. During this commission Chapman surveyed the river Tagus and the Strait of Gibraltar, which were published by William Faden in 1807 and 1810. A more comprehensive version of Chapman's survey of the river Tagus was also published by Faden in 1810, which was republished by the Hydrographic Office in 1823 after the Admiralty had purchased the copper plate from Faden.[5] Faden also published a smaller version of this survey in 1812 in his *España Maritima, or Spanish Coasting Pilot*, together with Chapman's sailing directions. This too eventually became an Admiralty chart. After a period on half-pay Chapman was next appointed master of the 74-gun *Spencer*, Captain J. Quilliam and flagship of Rear Admiral R. Stopford, based on Plymouth, remaining in her until she was paid off on 10 January 1809. This enabled Chapman to survey Hands Deep in the approaches to Plymouth in 1808 and partly to survey Plymouth Sound, which was published by William Heather on 1 March 1809. Chapman was next appointed as Admiralty surveyor in command of the surveying sloop *Sorlings* on 11 July 1809 to carry out surveys in the approaches to the River Thames, but he died on 2 April 1810. In 1800 Chapman applied for remuneration from the Board of Longitude for astronomical observations taken in the *Providence*[6] and was awarded £20 on 6 March 1806.[7]

[1] Louisa Vashon petitioned the Admiralty for a special pension on account of her husband's death being brought about by the musket ball remaining in his body; TNA, ADM 1/5071, letter Pro V50, dated 2 Sept 1814, written from Douglas, Isle of Man.

[2] UKHO, u95 on Ba3.

[3] His journal for this appointment survives; TNA, ADM 50/77.

[4] UKHO, B636 on Ba1. When commanded by Lieutenant John Rogers she was also used by Peter Heywood in 1802 to survey the shoals off the NE tip of Ceylon (Sri Lanka), ibid., u85 on Ba3.

[5] Andrew David, 'Admiralty Charts and William Faden's Copper Plates', *IMCoS Journal*, Winter 2008, No. 115, pp. 7–11.

[6] CUL, RGO 14/11.

[7] CUL, RGO 14/17/176. See also Derek Howse, 'Britain's Board of Longitude: The Finances, 1714–1828', *The Mariner's Mirror*, 84, 4 (1998), pp. 400–417.

John Crosley. Prior to this appointment to the *Providence*, Crosley had spent just under three years as assistant to the Astronomer Royal, Nevil Maskelyne.[1] After surviving the wreck of the *Providence*, Crosley was sent home from Macau in the packet *Crescent*, but after being struck by a typhoon in the South China Seas, she was forced to return to Macau for repairs, Crosley eventually reaching England on 7 July 1798. On 24 August 1798 Crosley was awarded £400 by the Board of Longitude for observations taken during the voyage of the *Providence* in addition to his salary.[2] After the voyage Crosley complained to Nevil Maskelyne, the Astronomer Royal, that during the voyage his sleep had been disturbed and his health injured because his hammock was continually knocking against a partition.[3] In 1801 he was appointed astronomer to the *Investigator* under Matthew Flinders to complete the survey of Terra Australis, but ill health forced him to leave the expedition when it reached the Cape of Good Hope and to return home. Between 1811 and 1814 he was employed by the Board of Longitude in conjunction with Samuel Flinders recalculating the longitudes obtained during the *Investigator*'s voyage using lunar distances observed at Greenwich rather than the predicted distances given in the *Nautical Almanac*, which Flinders had used during the voyage. For this Crosley was awarded £500.[4] At the same time he appears to have recalculated the longitudes obtained during the *Providence*'s voyage, which he inserted in two copies of the published account of the voyage.[5] He was a member, and for many years president, of the Spitalfields Mathematical Society. He died in Brighton on 22 September 1817.[6]

9. Broughton and the Insularity of Sakhalin

The failure of Broughton to establish the insularity of Sakhalin was to have repercussions many years later. Even today it is not clear whether it was the Russians or the Japanese who first established that Sakhalin was indeed an island. In Russian eyes the Bering voyages had left two geographical puzzles unfinished. First, a myth persisted that an isthmus connected Mys Shelagskiy in northern Siberia to North America, in spite of the voyages through Bering Strait of Semen Dezhnev in 1648, Vitus Bering in 1728, James Cook in 1778 and Charles Clerke in 1779. The second puzzle was that the survey of the coast about the river Amur and the island of Sakhalin was incomplete. From the perspective of Russian explorations the former was finally disproved in 1823 by Lieutenant Ferdinand von Wrangell who had examined the north coast of Siberia by dog team.[7] From the perspective, too, of the same government, an official attempt to explore and chart the waters and harbours of the River Amur and Sakhalin had still to be carried out, in spite of the incomplete examinations of three prominent surveyors, La Pérouse, Broughton and Krusenstern. The Russian Government, alarmed

[1] Derek Howse, *Nevil Maskelyne: the Seaman's Astronomer*, Cambridge University Press, 1989, p. 156.

[2] CUL, RGO 14/17. See also Howse, 'Britain's Board of Longitude', p. 407.

[3] Howse, *Maskelyne*, p. 187, quoting Maskelyne to ?Banks, 23 January 1801, NMM, PST/76/173–4.

[4] CUL, RGO 14/7, minute dated 3 March 1818.

[5] One copy was sold by Maggs Bros of London in 1988; the other copy is held in the Caird Library, National Maritime Museum; see p. lxiv.

[6] *The Gentleman's Magazine*, Vol. 87 for 1817, Part 2, p. 472, which states incorrectly that Crosley was a member of the Philosophical Society near Spital-Square. No record of such a society has been found.

[7] See Ferdinand Wrangell, *Narrative of an Expedition to the Polar Sea in the Years 1820, 1821, 1822, & 1823*, Fairfield, Wash., 1981.

at the rise of British commercial interests in the North Pacific that followed James Cook and James King's advertisement of the lucrative prospects of the sea otter trade, sent a scientific and cartographic expedition under Joseph Billings to the North Pacific in 1785 and intended to send a small squadron of five ships designed to visit Kamchatka, open up trade with Japan and explore the north-west coast of America, claiming it for Catherine II. This exploring expedition was to be under the command of Captain Grigory I. Mulovsky, who received his instructions in April 1787. He and the ships under his command, however, never left Kronstadt in the Baltic, for war with Sweden broke out in November 1787, and the project was postponed indefinitely after Mulovsky was killed on 17 July 1789 at a battle near the island of Bornholm.[1]

As related earlier, La Pérouse set out in 1787 to chart the east coast of Asia, the only shore that James Cook had never seen, ten years before Broughton's similar reconnaissance. La Pérouse had hoped to discover whether Sakhalin was an island or was joined by an isthmus to the mainland of Asia, but with the onset of southerly winds, accompanied by some fog and cold weather, he prudently decided to make his way south. Presumably he concluded that if such a channel existed then it could only be for shallow-draught vessels.[2] Broughton in 1797 faced the same geographical conundrum. He had reached about eight miles farther north of where La Pérouse had concluded his survey. But although William Chapman, master of the schooner, was of the opinion that a passage to the north was evident, Broughton too concluded that it was a cul-de-sac. He was wrong, as was La Pérouse before him. The Japanese explorer Mamiya Rinzo later, in 1809, proved the insularity of Sakhalin.[3] The Russians may, however, have had prior knowledge that Sakhalin was an island. In 1731, the French astronomer and geographer Joseph-Nicolas Delisle, then resident in St Petersburg and a member of the St Petersburg Academy of Sciences, was asked by the Academy to produce a chart for Bering's projected Second Kamchatka Expedition to show what was then known of the relation of eastern Asia to America. A copy of this manuscript chart, titled 'Carte Dressée en 1731 Pour Servir a la recherche des Terres et des Mers situées au Nord de la Mer du Sud', was found in 1925 by the American geographer Frank A. Golder in the archives of the Service Hydrographique de la Marine in Paris.[4] This chart clearly shows Sakhalin as a small island off the mouth of 'R. D'Amour'. Also, in October 1778, in Unalaska, Grigory Izmaylov allowed Cook to copy a chart made by Captain of the Fleet Wawseelee Irkeechoff, who had 'explored in 1758, the coast of Tartary between Okhotsk and the River Amur to Japon'.[5] Henry Roberts later used Izmaylov's map for his 'A General Chart: Exhibiting the Discoveries made by Captn James Cook'.[6] Although the Izmaylov map does not appear to have survived, a portion of it seems to have been copied later as an inset

[1] Robert J. King, 'The Molovsky Expedition and Catherine II's North Pacific Empire', *Australian Slavonic and East European Studies* (University of Queensland), 21, 1–2 (2007), pp. 97–122.

[2] Dunmore, *La Pérouse*, II, pp. 312–14.

[3] John J. Stephan, *Sakhalin: A History*, Oxford, 1971, pp. 37–41, quoting a variety of sources.

[4] Portefeuille 72, Division 2, as Pièce1; F. A. Golder, *Bering's Voyages*, 2 vols, New York, 1925, I, p. 70, n 148, with the chart reproduced, f.p. 72. This chart is also reproduced in Sven Waxell, *The American Expedition*, London, Edinburgh and Glasgow, 1952, also f.p. 72.

[5] J. C. Beaglehole, ed., *The Journals of Captain James Cook: III: The Voyage of the Resolution and the Discovery, 1776–1780*, 2 parts, Hakluyt Society, extra ser. 36a, Cambridge, 1967, pp. 452–3. Beaglehole was unable to find out anything about this captain, either from English or Russian sources and no map of his explorations has come to light.

[6] Cook and King, *Voyage to the Pacific Ocean*, I, p. lxxxii.

onto one of Broughton's manuscript charts to show the insularity of Sakhalin, with the legend, 'This Coast is taken from a Russian MS Chart got at Kamchatka by Captn King'.[1] On this inset Sakhalin is depicted as an island, albeit a small one, and the lower reaches of the Amur are shown with reasonable accuracy. Russian knowledge of the insularity of Sakhalin prior to 1782 is also supported by an atlas compiled by Vasily Fedorovich Lovtsov at Bol'sheretsk in that year, in which Chart 1 in the atlas clearly shows Sakhalin as an island.[2] Sagalin I. is also shown detached from the Asian mainland on Thomas Jeffrey's 'A Map of the Discoveries made by the Russians on the North West Coast of America. Published by the Royal Academy of Sciences at Petersburg'.[3]

The Russian circumnavigator Adam Krusenstern reached the northern tip of Sakhalin in August 1805, but made no attempt to proceed further south, being 'convinced that no passage exists to the south of the Amur between the coast of Tartary and this [Sakhalin] island.'[4] Almost three decades passed before the Russians renewed their explorations, when another Russian naval officer, Gavrilov, reached the mouth of the Amur from the north, but did not extend his search to the south.[5] Then in 1849 Captain-Lieutenant Gennady I. Nevelskoi was ordered to carry out a survey in the transport *Baikal* of the estuary and the mouth of the Amur and to determine whether a strait separated Sakhalin from the mainland.[6] The following year he discovered a navigable channel, four miles wide, which was named Proliv Nevel'skogo after its discoverer. The greatest secrecy was kept by the Russian authorities so as not to alarm China or to arouse the hostility of other powers, notably Britain. Russia's subsequent move down the Amur was proposed by General Count Nikolai Murav'ev, holding the dual position of Commander-in-Chief and governor general of Eastern Siberia, whose headquarters were at the Russian America Company post at Irkutsk on Lake Baikal. He believed that Britain had designs on the Amur.[7] In keeping their discovery or rediscovery of the insularity of Sakhalin secret, including where the Amur flowed into the Pacific, was to the disadvantage of British and French naval forces during the Crimean War in the Pacific, 1854–6.[8]

Thus for many years after Broughton's voyage a veil of secrecy existed over the location of the mouth of the Amur and the insularity of Sakhalin and, above all, the existence or non-existence of a navigable passage between the island and the mainland. At his farthest north Broughton was over 60 miles from the mouth of the river, and he had concluded to his own

[1] UKHO, 514/2 on Se. For a reproduction of this portion of Izmaylov's map see Derek Hayes, *Historical Atlas of the North Pacific Ocean*, London, 2001, p. 126. Unfortunately the caption omits mention of the legend and so it appears as though this portion of Izmaylov's map was part of Broughton's survey.

[2] The Lovtsov atlas is held in The Provincial Archives of British Columbia, Victoria, Vancouver Island. See Lydia T. Black, *The Lovtsov Atlas of the North Pacific Ocean*, Kingston, Ont., and Fairbanks, Alaska, 1991.

[3] Müller, *Voyages from Asia to America*, facing title page.

[4] Krusenstern, *Voyage*, II, p.178.

[5] Basil Dmytryshyn, E. A. P. Crownhart-Vaughan and Thomas Vaughan, eds and trans., *The Russian American Colonies to Siberia and Russian America, 1798–1867*, Vol. 3: *A Documentary Record*, Portland, Oreg., 1989, p. 486.

[6] Dmytryshyn, Crownhart-Vaughan and Vaughan, *Russian and American Colonies*, pp. 475–8.

[7] Ibid., p. 482.

[8] An introduction to Broughton's examination is Ian R. Stone, 'W. R. Broughton and the Insularity of Sakhalin', *The Mariner's Mirror*, 82, 1 (February 1996), pp. 76–80. On La Pérouse, Broughton, et al., see Stephan, *Sakhalin: A History*, pp. 33–41. On later matters of military geography and operations, see Stephan, 'The Crimean War in the Far East', *Modern Asian Studies*, 3, 1969, pp. 268–74 and Barry Gough, *The Royal Navy and the Northwest Coast of North America, 1810–1914: A Study of British Maritime Ascendancy*, Vancouver, 1971, pp. 108–30.

satisfaction that no navigable passage existed. Even so, history presents odd twists. The following may be briefly related to show how 'After the event even a fool is wise', as Homer put it. Evidence from La Pérouse, Broughton and Krusenstern about the apparent non-existence of a navigable channel was sufficient for the geographer Alexander G. Findlay to note in 1851 that 'it is not absolutely determined whether Saghalin be an island or a peninsula, but ... all evidence certainly tends towards the latter opinion ... It must therefore be considered that Saghalin is joined to the continent by a flat sandy neck of land.'[1] Thus at the outbreak of war with Russia in 1854 neither Findlay, the Commanders-in-Chief of the Royal Navy's China and Pacific Stations or their French counterparts were aware that the Russians had solved this mystery.

In 1854 a combined British-French squadron, commanded by Rear-Admiral David Price, appeared before the heavily fortified port of Petropavlovsk in Kamchatka. They found an inferior Russian naval force there, protected by the formidable shore defences. Dismayed by what he saw, Admiral Price took his own life just before a combined amphibious attack was to have been made. The allies pressed on nevertheless and landed their force of bluejackets and marines, which was repulsed with great losses, forcing the allied squadron to abandon the attack. The British squadron then recrossed the North Pacific to Esquimalt in Vancouver Island to prepare for a fresh attempt the following year. A sidelight to this was the construction of hospital facilities at Esquimalt as a possible receiving depot for wounded sailors and marines from a further attack. But when the British and French returned in force to Petropavlovsk under Rear Admiral H. W. Bruce the following year, they found the place deserted, and all the Russian warships had vanished. They had apparently escaped into the Gulf of Tartary, where they were found by an allied squadron from China commanded by Commodore Charles Elliot, in Zaliv Chikhacheva (La Pérouse's Baye de Castries) at the head of the gulf. In the belief that he would need reinforcements, Elliot withdrew and returned with more ships, only to discover the Russian ships had escaped into the Amur through Proliv Nevel'skogo.[2] No matter how embarrassing this was to British arms and allied attempts at cooperation, hindsight should not detract from Broughton's achievements. There are many examples of missed identifications. Cook, for example, failed to identify the Strait of Juan de Fuca in 1778, and Vancouver failed to find the mouth of the Fraser River in 1792. None of these mariners should be the butt of armchair geographers of a later time; they did their jobs to the best of their abilities and with the materials and resources they had at their disposal. Their work was not finite, and it was left to others to complete their charts.

[1] Alexander G. Findlay, *A Directory for the Navigation of the Pacific Ocean*, 2 parts, London, 1851, pt 1, p. 618.

[2] Gough, *The Royal Navy and the Northwest Coast*, ch. 5; Stephan, 'The Crimean War', pp. 257–77; and see, especially, Jon D. Grainger, *The First Pacific War: Britain and Russia, 1854–1856*, London, 2008, pp. 20–27.

TEXTUAL INTRODUCTION TO THE JOURNAL

1. Broughton's Journal

There are various copies of Broughton's journal held in the UK Hydrographic Office in Taunton and in the National Maritime Museum in Greenwich.

The UK Hydrographic Office
The UK Hydrographic Office holds two parts of Broughton's journal and a collection of letters.[1]

1. 'A Journal of the proceedings of His Majesty's Sloop Providence from making the Coast of Japan Sept. 7th 1796 and Her Navigation to & from the Kurile Islands to Her arrival at Formosa on the 7th December 1796, William Broughton Esqe Commander'. Annotated in faded pencil 'Opend and Examined this journal July 14, 1797 after the Tyffoon the Ship having lost her rudder [and her] cargo shifted ... Z Mudge'.

The text on recto side only of 22 unnumbered folios + 10 blank sheets on laid paper, 250 × 400 mm, w/m fleur-de-lys = W, bound in original marbled boards titled 'HMS Providence – Journal by Com*mande*r W. Broughton, 1796 – Kuril Islands'; rebound in modern blue buckram.

2. 'A Journal of the proceedings of his Majesty's Sloop Providence and her Tender from Macao in China to the Coast of Tartary in 52° N Lat & 141° E Longitude, commencing the 1st of April 1797, ending the 27th of November 1797 By William Robert Broughton Esq^{e} Commander'

The text on recto and verso of 39 unnumbered folios + 1 blank sheet on laid paper, 225 × 365 mm, w/m bugle in a shield, surmounted by a crown, over a circle containing the date 1795 with another circle below containing the initials SF, bound in original marbled boards, titled, 'HMS Providence – Journal by Com*mande*r William Broughton 1797 – Russian Tartary'; rebound in modern blue buckram.

3. A collection of 11 letters titled on a half sheet 'Letters to the Admiralty from Capt*ai*n Broughton of his Maj*esty*'s Sloop *Providence* and her Tender in 1796, 1797 & 1798'. 240 × 370 mm, w/m EDMEADS & PINE, bound in modern blue buckram. Possibly Broughton's letter book. This collection of letters partly duplicates those in the Admiralty papers in The National Archives, but some of these letters are the only extant copies, the originals probably having been lost in transit or, in the case of those sent from Monterey, were not forwarded from Madrid.

[1] UKHO, OD 177.

The National Maritime Museum

The National Maritime Museum holds a collection of papers relating to Broughton's voyages in the *Chatham* and *Providence*, originally presented to the Royal United Services Institute in the nineteenth century by Broughton's only son, Captain William Broughton. The papers were transferred to the National Maritime Museum in 1968 and are held in RUSI/NM/132 (1) to (11) with (4) not allocated. The items relevant to this work are:

(2) 'The Journal of a Tour across the Continent of New Spain from S[t] Blas in the Northern Pacific Ocean to La Vera Cruz in the Gulph of Mexico by Lieut. W. R. Broughton, in the year 1793, Commander of H.M. Brig *Chatham*': the text from 12 January to 30 May 1793, written mostly on recto and verso on 32 folios on laid paper, 186 × 233 mm, no discernable watermark, bound in sheepskin.

(3) A collection of unbound papers written on folded foolscap paper, passed between Broughton and his brother Brian, during the editing of Broughton's voyage. They include four incomplete pages of a draft titled 'Preface to Broughton's Voyage', a draft notice 15 lines long 'Shortly to be published by Capt[n] Broughton ...', followed on the same and succeeding pages by 'Memoranda from Perouse's Voyage' and 'Passages from Payne's *Epitome*' (John Payne, *An Epitome of History*, London, 1794). Extracts from *History of the Voyages and Discoveries made in the North,* translated from the German by Johann Reinhold Forster (London, 1786), are partly reproduced with numerous errors as a footnote to pages 89–90 of the published *Voyage*, followed by an incomplete letter from Broughton to 'My dear Brian', drawing attention to the extracts from Forster which Broughton had received from William Coxe, the author of *Account of the Russian Discoveries between Asia and America,* and also drawing attention to Sir George Staunton, *An Authentic Account of an Embassy from the King of Great Britain to the Emperor of China* (3 vols, London, 1797), referred to on page 243 of the published *Voyage*. A draft description of the island and inhabitants of Lieuchieux (Okinawa Shima) is partly reproduced on pages 241–4 of the published *Voyage*, and a draft for the description of the natives of Volcano Bay (Uchiura Wan), is partly reproduced on pages 105–9 of the published *Voyage*.

(5) The text from 11 December 1795 to 17 October 1796 on recto side of 90 unnumbered folios, with some marginalia on verso, mainly observations for longitude and variation, on laid paper 248 × 400 mm, watermark fleur-de-lys = W, bound in original marbled boards with the title 'Capt[n] Broughton 1795, 1796 HMS *Providence*' with a later annotation, 'Presented by Captain W. Broughton, RN, son of ...'

(6) The text from 18 October to 12 December 1796 on recto side only of 30 unnumbered folios on laid paper, 190 × 317 mm, signed by Broughton. Unbound.

(7) The text from 20 June to 10 September 1796 on recto side only with some marginalia on verso, 19 folios on laid paper, 200 × 310 mm, no discernable watermark, bound in original marbled boards. This partly duplicates RUSI/NM/132 (5).

(8) The text from 11 September 1796 to 15 June 1797 mainly on recto but some text on verso, 46 folios on laid paper, 200 × 320 mm, w/m Britannia in shield = monogram of C. Hale + 1800, bound in original marbled boards; annotated on front 'transcribed by B. Broughton'. This partly duplicates RUSI/NM/132 (5) and (6).

(9) The text from 27 June to 21 October 1797 mainly on recto but some text on verso on 40 unnumbered folios on laid paper, 200 × 320 mm, w/m Britannia in shield = monogram of C. Hale + 1800; annotated on front 'transcribed by B. Broughton'. This partly duplicates parts of RUSI/NM/132 (10) and (11).

(10) The text from 1 April to 30 September 1797, 46 folios, 245 × 380 mm on laid paper, w/m fleur-de-lys in shield, crown above, GR below, bound in original marbled boards without title.

(11) The text from 1 October to 27 November 1797, 20 folios on laid paper, 240 × 370 mm, w/m EDMEADS & PINE = fleur-de-lys in shield, crown above, monogram E & P below, bound in paper.

2. Surviving Logs

The National Archives holds four logs relating to the voyage.

The Ship's Log
ADM 55/147. Partly log and partly journal, from 20 June 1796 to 27 November 1797. It consists of 128 folios; (ff. 127v to 128 being 'Remarks on the Japanese & Corean Vessels')[1] on laid paper, 275 × 405 mm, w/m PORTAL & BRIDGES = fleur-de-lys in shield, surmounted by a crown over GR. It is titled:

A Log of the proceedings on board his Majesty's Sloop *Providence* from the 20th of June 1796 in her passage from California to the Sandwich Islands, and from thence to the Coast of Japan, the Land of Insu and Kurile Islands, with her return to China in December of the same Year. The proceedings continued from China in 1797 to the loss of the said Sloop off the Island of Taypingshan on the 17th of May 1797: And the prosecution of the Voyage in the Tender from July 1797 to the end of the Year in which time the Islands of Lieuchieux, the Eastern Coast of Japan, the Land of Insu with the Tartarean and Corean Coasts were examined

by William Robert Broughton Esqre Commander
of the above Vessels

and signed at the bottom of f. 127, 'NB This Log Kept & written by me, as witness my hand William Robert Broughton'.

Vashon
ADM 51/4490. Log.14 January to 6 October 1795. It is very brief and contains little of interest.

Chapman
ADM 52/3292/9 and 10. Logs. Both from 24 January 1794 to 24 January 1795 They are not duplicates but slightly different versions of one another. They are useful in establishing the exact dates of events in the early part of the commission before the *Providence*'s final departure from Plymouth in February 1795.

3. The Published Account

The published account of the voyage was edited by Broughton with, according to an entry in Francis Edwards Catalogue 525 (1930), item 68, assistance from his brother, the Reverend

[1] Reproduced in Broughton, *Voyage*, pp. 378–9, with minor changes in punctuation and capitalization, and in Appendix II.

Brian Broughton, who also saw it through the press. Item 68 in the catalogue offered for sale a copy of the *Voyage*, together with 15 pages of correspondence between the Revd Brian Broughton and the publishers Cadell and Davies between 23 April 1803 and 23 August 1804. Unfortunately the whereabouts of this copy and the correspondence is not known.[1] Brian Broughton's involvement is clearly more extensive than just seeing the book through the press since his brother was sending him extracts from various publications for incorporating in the text (see p. lxii above). In addition the *Voyage* contains a number of errors and misprints, which a naval officer would surely have spotted (see p. lxix below).

4. Annotated Copies of the Published Account

Two annotated copies of Broughton's *Voyage* have been sighted. One is in the National Maritime Museum, which carries the bookplates of Charles Broughton and W. E. Delves Broughton. It has a number of annotations in ink and pencil, some of which have generated footnotes in this edition, which are quoted as 'in annotated *Voyage*'. This copy was sold by Sotheby's under 'Eccles' on 27 April 1953 when it was bought by the London book dealer Maggs Bros, probably acting as agent for the museum.[2]

The other annotated copy was sold by the London book dealer Maggs Bros in 1988; but its present whereabouts is not known. Photocopies, however, were taken at the time of some of the annotations, which are to all intents the same as those in the copy now held in the National Maritime Museum. The title page of this copy carries the hand-stamp of the Spitalfields Mathematical Society, of which Crosley was a member and for many years president, and also the hand-stamp of the Royal Astronomical Society. When the Mathematical Society was dissolved in 1845 its library was made over to the Royal Astronomical Society. It seems highly likely therefore that this was Crosley's personal copy.

5. The Graphic Records

The surviving graphic records are all held in the UK Hydrographic Office, Taunton.

1. Sketch of Port Stephens on the Eastern Coast of New Holland by W. R. Broughton Esqr Commander of H.M. Sloop *Providence* 1795. (C642/2, Australia folio 1). See Fig. 3 on p. 12.

2. Plan of Indermo Harbor in the Isl[d] of Matsmay. By Captain W. R. Broughton1797. (B46, Japan folio 1). See Fig. 4 on p. 74.

3. Sketch of T'hosan Harbor S.E. Coast of Corea by Capt[n] W. R. Broughton 1797. (B43, Korea folio 1). See Fig. 9 on p. 189.

4. A Sketch of the Liquieux Islands with the track of his Majesty's Sloop *Providence* in 1796 & 1797. (515 on Pb).

[1] We are grateful to Rolf E. Du Rietz for this reference. See also Rolf E. Du Rietz, *Bibliotheca Polynesiana*, Oslo, 1969, pp. L and 50.

[2] See also Du Reitz, *Bibliotheca Polynesiana*, pp. L and 50.

5. A Chart of the N.E. Coast of Asia & Japanese Isles with the Track of his Majesty's Sloop *Providence* and her Tender in 1796 & 97 under the Command of W. R. Broughton. (514/1 on Se).

6. Chart of the East Coast of Japan with the Kurile and Liquieux Is explored by Captⁿ Broughton in His Majesty's Sloop *Providence* in 1796. (514/2 on Se).[1]

7. A Chart of the N.E. Coast of Asia, Japanese, Lieuchieux and Madjicosemah Islands and the Track of His Majesty's Sloop *Providence* and her Tender in 1796 and 1797 under the Command of W. R. Broughton Esqr. (513, Press 10c).

8. A continuation of the former CHART with the track of the *Providence* & Tender in 1796 & 97 under the Command of W. R. Broughton. (513 on Be1).

6. The Chronometers

Broughton sailed with Earnshaw No. 1[2] and Earnshaw box chronometer No 248 and Earnshaw pocket chronometer No. 1522, the latter two belonging to John Crosley. Broughton also took with him Earnshaw pocket chronometer No. 2 and Arnold box chronometer No. 56, which he had bought in London for Bodega y Quadra (see p. xxviii), but on his arrival in Monterey in June 1796, Broughton learned that Bodega had died and so handed them over to the Governor to be delivered to Bodega's executor. On 24 September 1796, Earnshaw No. 248 stopped with a broken mainspring, and by 11 December 1796 Earnshaw No. 1 was going so indifferently that in February 1797 Broughton brought Arnold Box chronometer No. 45 from the East Indiaman *Henry Addingon*.[3]

Arnold No. 45 at least was saved when the *Providence* was wrecked and may have been in the schooner at the time. In any event it was used for the subsequent surveys carried out in the schooner. The fate of Earnshaw No. 1 is not known. According to Crosley it fell into one of the boats. In answer to my query, Jonathan Betts, the chronometer expert at the National Maritime Museum, could find no references to Earnshaw No. 1. However, he suggested that it was probably also a pocket chronometer since it fell out of Crosley's pocket as he was getting into the boat. He furthermore suggests that Nos 1 and 2 were probably merely numbers allocated by the observatory and not by Earnshaw, and that it was just possible that they were Earnshaw pocket chronometers Nos 1501 and 1502, a series started in 1791 at the Royal Observatory trials for Bligh's watch for his second bread-fruit voyage in the *Providence* (he chose No. 1503[4]). The probability that they were pocket chronometers is also borne out by a letter from Crosley to Earnshaw on 7 November 1804 in which he wrote regarding the *Providence*'s voyage, 'your three timekeepers decidedly

[1] For a later addition to this chart which depicts the insularity of Sakhalin, see p. lix, n. 1 above.

[2] Probably a pocket chronometer; see below.

[3] CUL, RGO 14/68, f. 145. The *Henry Addington*, Captain Richard Atherton Farington, was at Whampoa on 17 February 1797; Anthony Farrington, *Catalogue of East India Company Ships' Journals and Logs 1600–1834*, London, 1999, p. 312.

[4] British Horological Institute, *Earnshaw's Appeal*, facsimile edn, 1986, p. 37.

surpassed Mr Arnold's box timekeeper', thus suggesting that the three were not box chronometers.[1]

7. The Instruments and Navigational Books

Instruments

On 20 January 1794 the Navy Board were instructed to supply Broughton with the following instruments:[2]

2 Sextants (Ramsden)[3]
1 Achromatic Telescope
1 Case of Telescopes Night and Day
2 Surveying Compasses with Sights
1 Variation Compass of Walkers[4]
1 Station Pointer
2 Copying Glasses of Zires[5]
2 Parallel Rulers do
1 Chain & Lines
2 Brass edged Rulers of Zires
1 Case of Instruments
1 Pair of Beam Compasses
1 Marine Barometer
1 Plain [*sic*] Table complete
2 Thermometers
1 Do with a Case for sinking
1 Pocket do
1 Artificial Horizon (Complete)
1 Best Azimuth Compass
1 Good Stop Watch with a Second Hand for Observations.

A pocket thermometer and the stop watch do not appear to have been supplied, but Broughton was also supplied with:

1 Boats water thermometer
3 Compass cards
1 Bottle of quicksilver
1 large packing case
A portable observatory supplied by a Mr Smith of Knightsbridge at £20, presumably the marquee they were ordered to supply on 23 October 1793 as mentioned below.

[1] Ibid., p. 226. For confirmation that the Earnshaw chronometer Broughton bought for Bodega was 'a small pocket chronometer', see Tovell, *Far Reaches of Empire*, p. 333.
[2] TNA, ADM 2/271, pp. 430–31, Admiralty to Navy Board, and NMM, ADM A/2852.
[3] Jesse Ramsden (1731–1800), the well-known instrument maker whose sextants were specially prized.
[4] Ralph Walker (1749–1824).
[5] An unrecorded maker.

Crosley was not supplied with instruments by The Board of Longitude, having been instructed to take charge of those supplied to William Gooch on meeting with the *Discovery* (see p. 251). However, he equipped himself with a number of instruments and chronometers, most of which were lost in the wreck of the *Providence*. For a list of these see his claim for their loss or damage in Appendix I.

Navigational Books

It is not known what navigational books were supplied to the *Providence*, but these would certainly have included copies of the *Nautical Almanac* for the years that she was expected to be in the North Pacific and also copies of the second edition of the *Tables Requisite*, published in 1781. In addition William Chapman, as Master, was expected to provide himself 'with the proper instruments, maps and books of navigation' in accordence with *Regulations and Instructions Relating to His Majesty's Service at Sea*.[1] John Crosley, however, provided himself with numerous books on navigation and astronomy. For a list of these see his claim for their loss in Appendix I.

8. Other Equipment Supplied by the Navy Board

To assist in the survey work the Navy Board was instructed on 23 October 1793 to supply the *Providence* with a marquee and tents as supplied to the *Discovery*, with the addition of four small boat tents to contain the crew when on a surveying service, and two six-oared yawls in lieu of cutters and two more to be carried in frame.[2] On 20 February 1794 the Navy Board was further instructed to supply the *Providence* with a 'canteen' for the marquee, which was duly supplied by John Trotter & Co, who had supplied the marquee; the canteen consisting of chairs, beds, blankets, cutlery, and so on.[3] On 28 February 1794 the Navy Board was also instructed to supply the *Providence* with a variety of hardware and other articles as presents for the natives of the islands Broughton might visit.[4]

9. Broughton's Survey Methods

Broughton gives very little indication in his journal of the methods he used to carry out his surveys. These would, however, have been very similar to those used by Cook and Vancouver. In surveying harbours and sheltered anchorages, such as Muroran Ko in Hokkaido, Broughton would first have measured a base on a suitable stretch of coastline by means of Gunter's chain. Next he would have built up a rudimentary scheme of triangulation, orienting the survey correctly by measuring at one end of his base the angle between one of his triangulation stations and the sun to obtain a true bearing, while observing magnetic variation so that magnetic bearings could be converted into true bearings. Meanwhile the observatory would have been set up ashore at one of his stations to obtain its geographical position so that the

[1] Brian Lavery, ed., *Shipboard Life and Organisation, 1731–1815*, Publications of the Navy Record Society, Vol. 138, 1998, pp. 25–7.

[2] TNA, ADM 2/271, pp. 135–6.

[3] NMM, ADM A/2852.

[4] TNA, ADM 2/272, p. 71.

survey could be located correctly. Next the coastline and topography would be surveyed. The positions of rocks and shoals may have been fixed by stationing a boat over them and fixing their positions by bearings taken by compass or possibly by horizontal sextant angles to and from various triangulations stations, probably marked by flags. Soundings would be obtained from a boat by lead-line, fixed by compass bearings to various triangulation stations. The height of the tide would also have been observed during the survey so that soundings could be reduced to a low water datum.

On open stretches of coastline, Broughton would have carried out a running survey, which had been brought to near perfection by Cook and Vancouver. Such a survey is described by John Robertson, whose manual[1] formed part of John Crosley's library and was probably also held by Broughton and William Chapman, the *Providence*'s master. In such a survey Broughton would have set up a series of geographical positions on shore, some distance apart, whose positions were established for both latitude and longitude by means of astronomical observations and his chronometers. The *Providence* would then proceed along the coast, between these positions, at a suitable distance offshore so that all prominent features could be easily identified. 'Ship stations' would be established from time to time from which compass bearings could be taken to prominent coastal features. Soundings would be taken at and between each ship station. The courses steered between the ship stations and the distance between them would be measured by log and entered in a survey book. A rough sketch would be made of the coastline as the ship proceeded and entered in the survey book, in which all the bearings and other relevant hydrographic information would also be noted, including the *Providence*'s latitude obtained each day at noon, when observing conditions were favourable. Using all this data, adjusted as necessary, the position of the various ship stations could then be plotted on the survey sheet and the positions of the various coastal features could be established and the coastline sketched in from the survey book.[2]

[1]John Robertson, *The Elements of Navigation; Containing the Theory and Practice. With all the Necessary Tables*, 2 vols, London, 1754, with further editions in 1764, 1772, 1780, 1786 and 1796.

[2] For a full description of this method see Andrew David, 'Vancouver's Survey Methods and Surveys', in Robin Fisher and Hugh Johnston, eds, *From Maps to Metaphors: The Pacific World of George Vancouver*, Vancouver, 1993, pp. 51–69.

ESTABLISHING THE TEXT

Establishing the text has not been straightforward. There is no surviving journal or ship's log covering the voyage from the appointment of Broughton in command of the *Providence* on 3 October 1793 to her departure from Tahiti on 20 June 1796. Two versions of the master's log, however, survive for the period from 24 January 1794 to 24 January 1795, three weeks before the *Providence* made her final departure from Plymouth. A log kept by Lieutenant James Giles Vashon also survives, covering the early part of the voyage between 14 January and 6 October 1795, the day that the *Providence* left Sydney Cove, but it contains only the most basic information. Since the published account of the voyage covering the voyage as far as Tahiti is given in some detail, it must have been compiled either from a copy of the ship's log or from a journal kept by Broughton for this part of the voyage, which clearly have not survived. Further details of events that occurred in this part of the voyage are also given in a series of letters from John Crosley, the astronomer, to Nevil Maskelyne, the Astronomer Royal, and also, to a lesser extent, in letters written by Broughton to the Admiralty. A selection of this correspondence is given in Appendix I. The early part of the voyage in this edition has therefore had to taken from the published account of the voyage, with suitable footnotes where appropriate from the master's and Vashon's logs and from Crosley's and Broughton's correspondence. These additional sources are also used in describing the early part of voyage in the introduction.

The published *Voyage*, however, contains several errors and omissions, among which is the failure to mention on page 4 that the *Providence* was forced to take shelter in Tor Bay between 22 and 23 October 1794 and, on page 5, the name of an island in Plymouth Sound is given as Duke's Island instead of Drake's Island. A further error appears on page 11 where it is stated that the *Reliance*, Captain Hunter, on entering Rio de Janeiro saluted the *Providence*, which was returned with an equal number of guns. Since Captain Hunter was considerably senior to Broughton this statement is clearly incorrect. In fact, as described by Crosley and Vashon and supported by the *Reliance*'s log, the *Reliance* was actually saluting the Portuguese Governor with thirteen guns, which was returned by Fort Sancta Cruz with an equal number. These errors, which no seaman would have made, clearly indicate that Broughton had either left the drafting of this part of the voyage to his brother Brian or that the errors occurred when Brian was seeing the *Voyage* through the press.

For editing the remainder of the voyage Broughton and his brother Brian clearly used the copies of Broughton's journals, which he had retained in his possession. These were eventually deposited, as stated above, sometime in the nineteenth century by his son Captain William Broughton in the Royal United Services Institution, and were eventually transferred to the National Maritime Museum in Greenwich in 1968. Broughton and his brother also made use of the ship's log now held in The National Archives in Kew. It seems likely that the two brothers did not see or make use of the copies of Broughton's journal which had been sent home during the voyage and which were probably deposited by the Admiralty on their receipt in the fledgling Hydrographic Office, where they are still held in its present location in Taunton. The Greenwich versions of Broughton's journal have therefore been used in this edition of the voyage, although the dated watermarks of 1800 on two of them indicate they are later copies. However, the Taunton versions of his journal were originally transcribed and then compared with the Greenwich versions and the necessary corrections made to conform with the later. These corrections consisted mainly of capitalizations of initial letters, abbreviations, etc. Any significant differences between the versions of the journal in Greenwich and Taunton are noted, where appropriate, in footnotes.

A difficulty arose on 18 October 1796 when Broughton broke his arm off Ostrov Simushir (Marukan), one of the Kurile Islands, when his journal was continued daily from the following day until 9 November by an amanuensis under his inspection. But when the journal was being edited either Broughton or his brother wrote in the version they were working on, 'The part of the Journal wanting is from the 18th October 1796 to our arrival in China December 1796 & our continuance there till 1st April'. This was later crossed out and replaced by 'This part at Dulwich'.[1] The missing portion of the journal, written by the amanuensis, was eventually located. The Taunton version of the journal, however, covers this part of the voyage from 9 November to 7 December 1796 in more detail than the amanuensis's account, which has been used in this edition. The Taunton version of the journal for this period is therefore given in Appendix III. However, neither the amanuensis's nor the Taunton versions of the journal cover events in Macau from the time of Broughton's arrival in December 1796 to shortly before his departure on 11 April 1797 to continue his voyage and so this is covered by the ship's log. Broughton or his brother continued to use extracts from the log in editing the remainder of the *Voyage*. The make-up of this edition is therefore:

From 3 October 1793 to arrival at Tahiti	The published *Voyage*
From Tahiti to 17 October 1796	NMM, RUSI/NM/132 (5)
From 18 October to 12 December 1796	NMM, RUSI/NM/132 (6)
Stay in Macau	TNA, ADM 55/147, ff. 40–41
From 1 April 1797 to 30 September 1797	NMM, RUSI/NM/132 (10)
From 1 October 1797 to end of voyage	NMM, RUSI/NM/132 (11)
Variant text in Appendix III from 17 October to 7 December 1796	UKHO, OD 177

In order to avoid a continuous text this edition has followed the published *Voyage* and has been split into two books with a corresponding number of chapters, with appropriate titles.

[1] NMM, RUSI/NM/138 (8).

ERRATA

The Publishers regret that the maps as printed contain some errors in name forms, as listed below.

Map 1:
For Sakhalim *read* Sakhalin

Map 4:
For Proliv Yekatering *read* Proliv Yekateriny
For Vulkan Tya tya *read* Vulkan Tyatya

Map 5:
For Hakura Shima *read* Hekura Shima
For STO NAIKAI *read* SETO NAIKAI
For Nagasaki KO *read* Nagasaki Ko
For Sakuri Shima *read* Sakura Shima
For Takara Shima *read* Tokara Shima
For Shiana Misaki *read* Shiono Misaki

Map 6:
For Fuji Sano *read* Fuji San (mountain)
For Bosa Hanto *read* Boso Hanto
For Unaga Suido *read* Iraga Suido
For Suno Si *read* Suno Saki

Map 7:
For Hado Misaki *read* Hedo Misaki
For Kumo Shima *read* Kume Shima
For Le Shima *read* Ie Shima

Map 8:
For Miyara Wan *read* Miyana Wan

Map 10:
For Motsuba Misaki *read* Motsuta Misaki
For Mys Lamahon *read* Mys Lamanon
For Bukhtz Sushcheua *read* Bukhta Sushcheva
For Zoliv Chikhacheua *read* Zaliv Chikhacheva
For Mys Nizmenny *read* Mys Nizmennyy
For Hadi vostok *read* Vladivostok
For Mys Gamouz *read* Mys Gamova
For Chosen Man *read* Chosan Man

Map 12:
For Malincho *read* Malinche

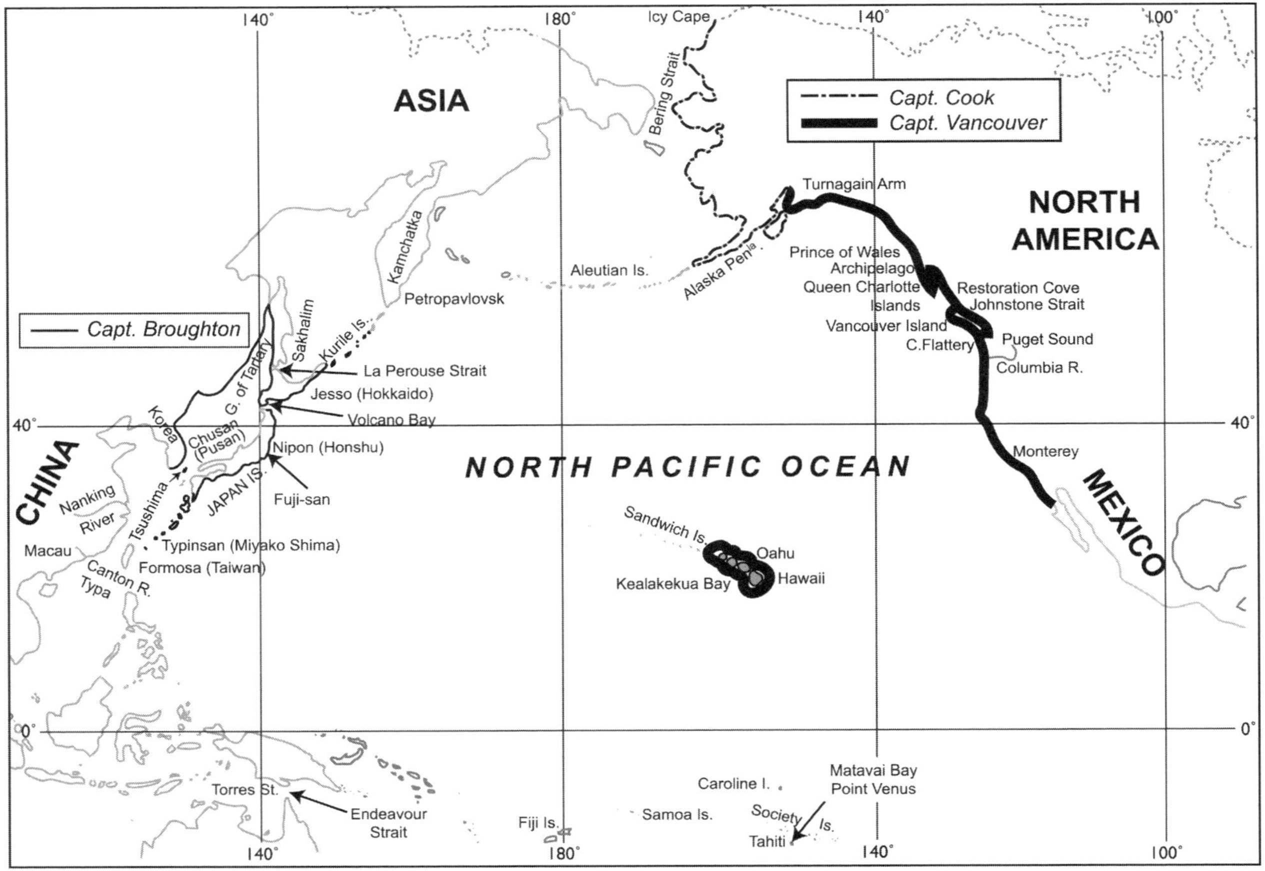

Map 1
The North Pacific in 1797

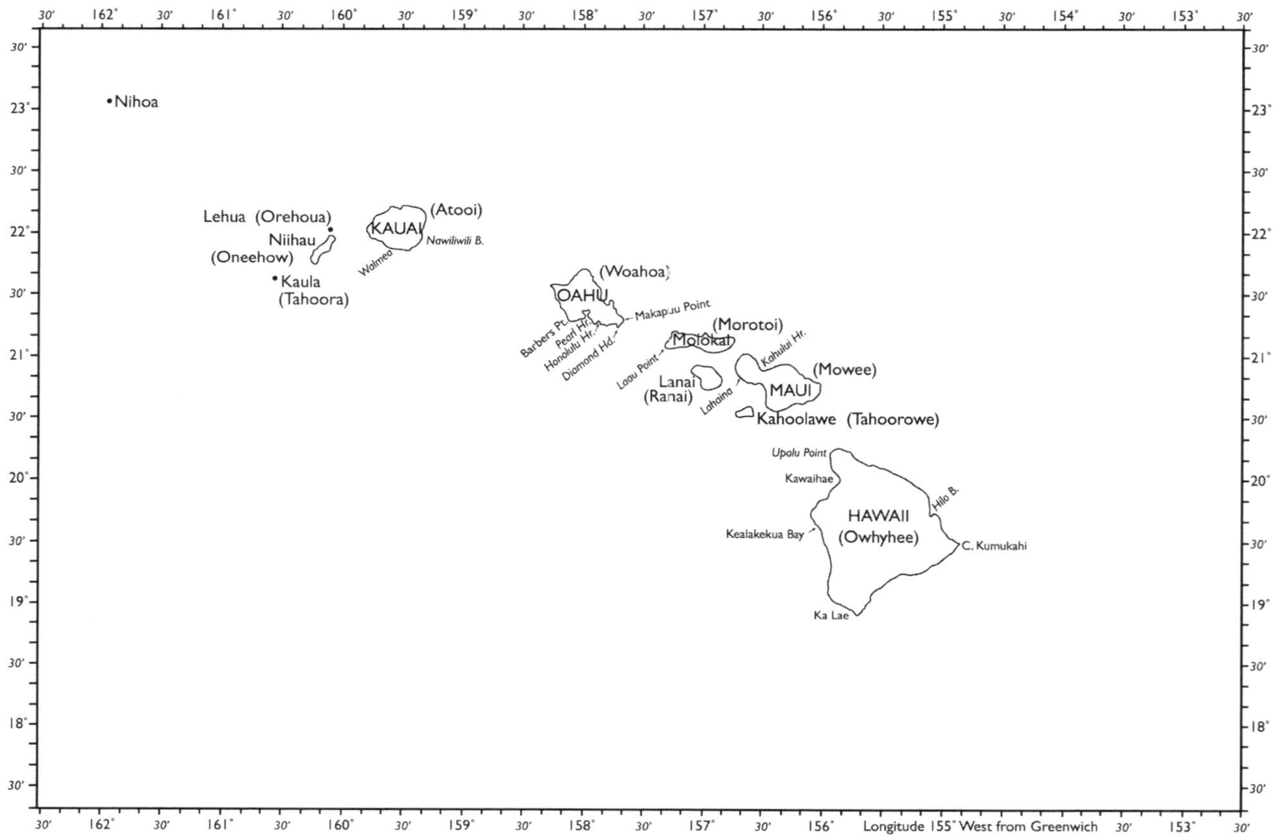

Map 2
The Hawaiian Islands

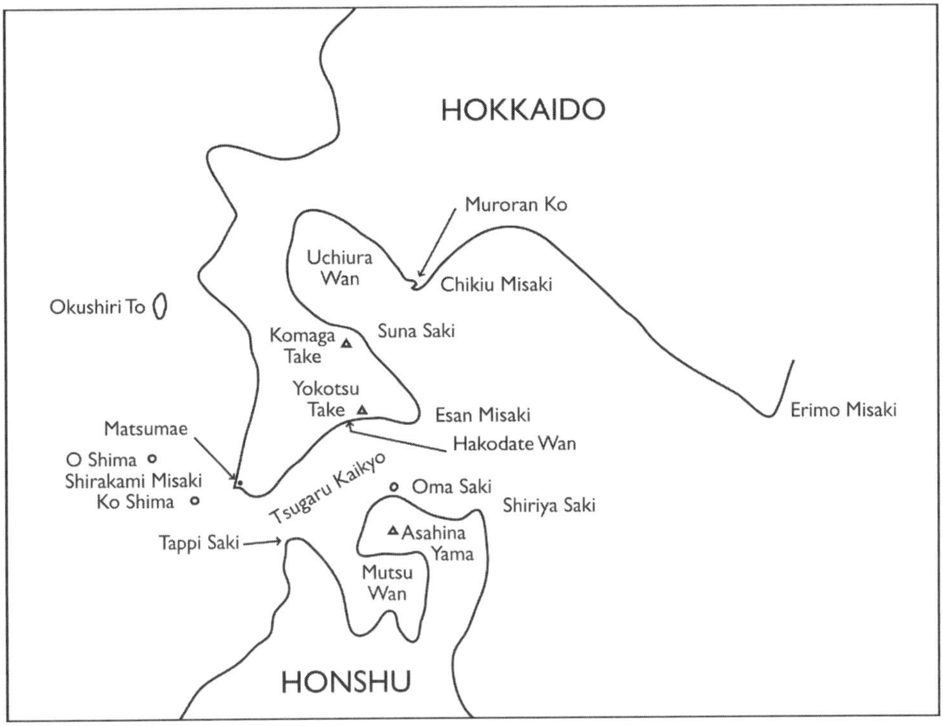

Map 3
Tsugaru Kaikyo and Approaches to Uchiura Wan

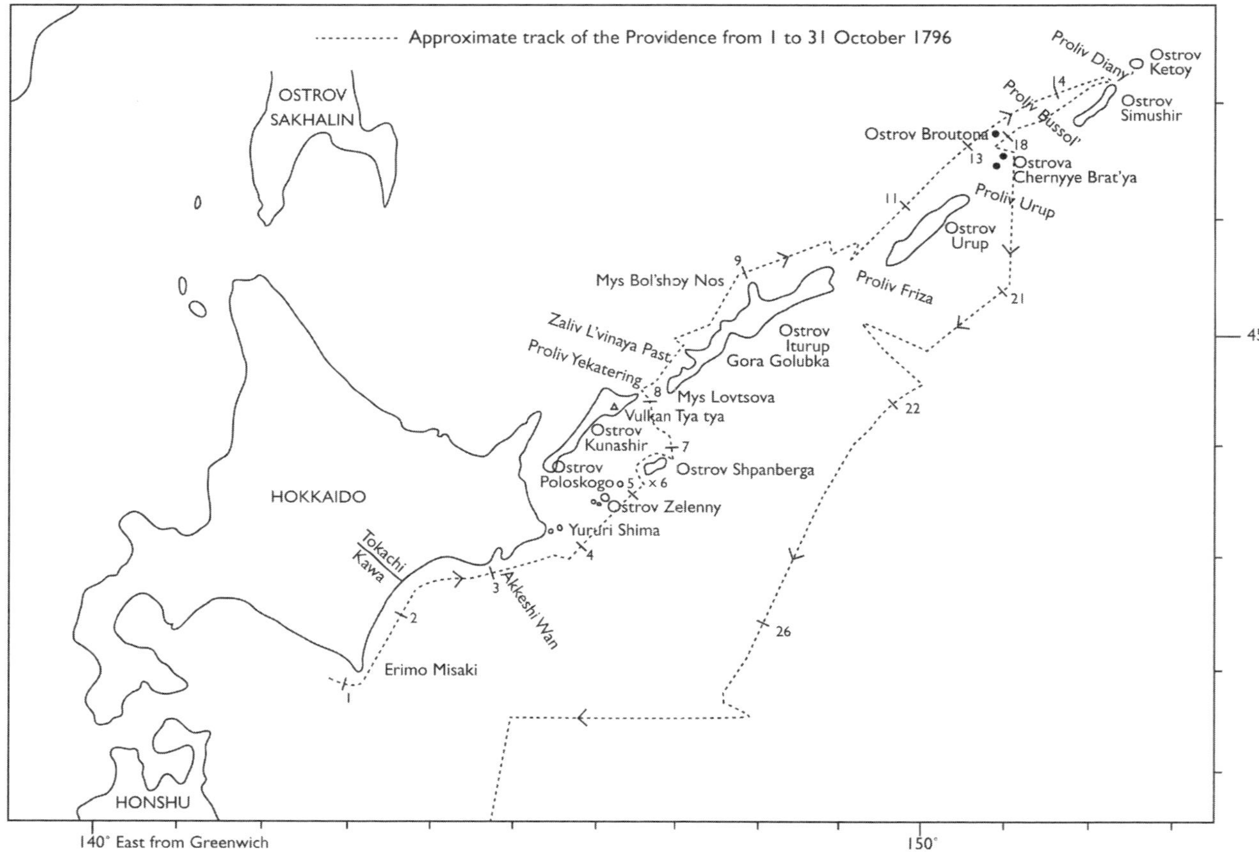

Map 4
The Kurile Islands

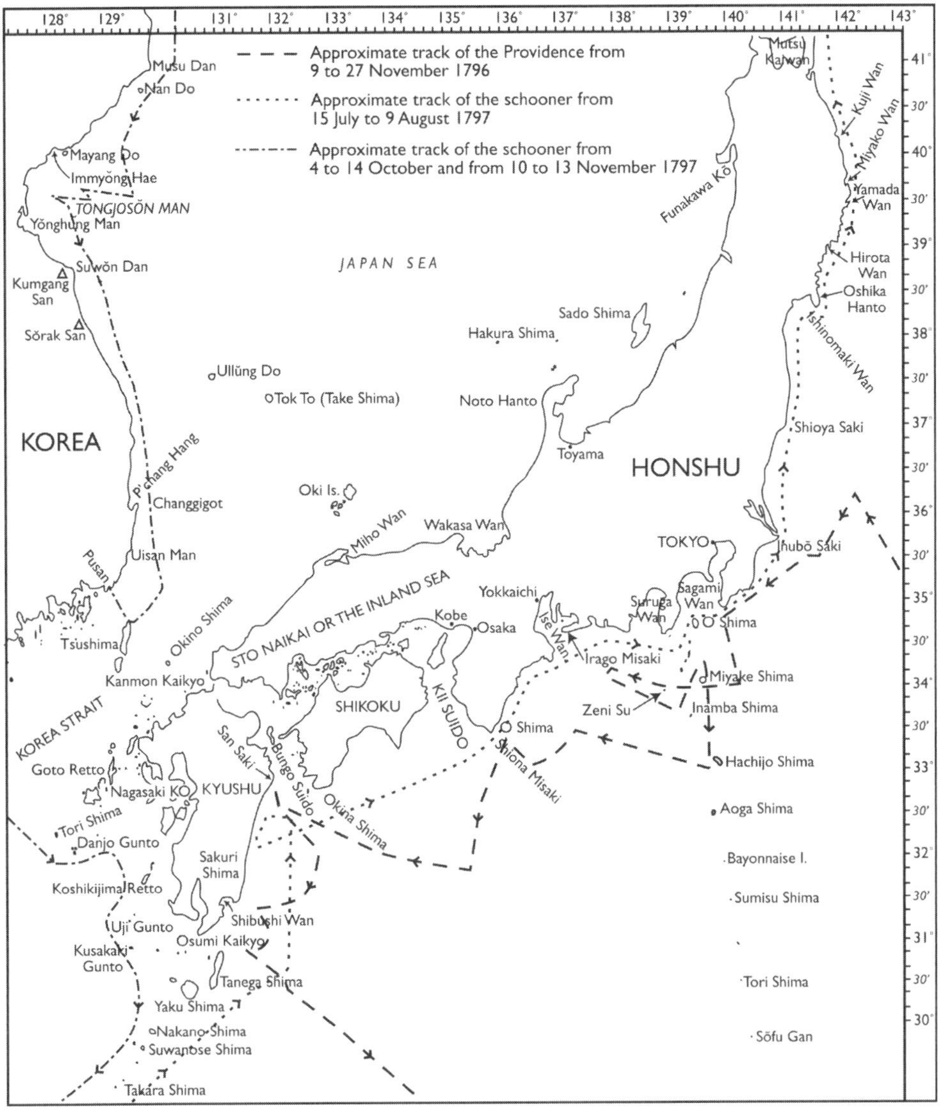

Map 5
Southern Islands of Japan and East Coast of Korea

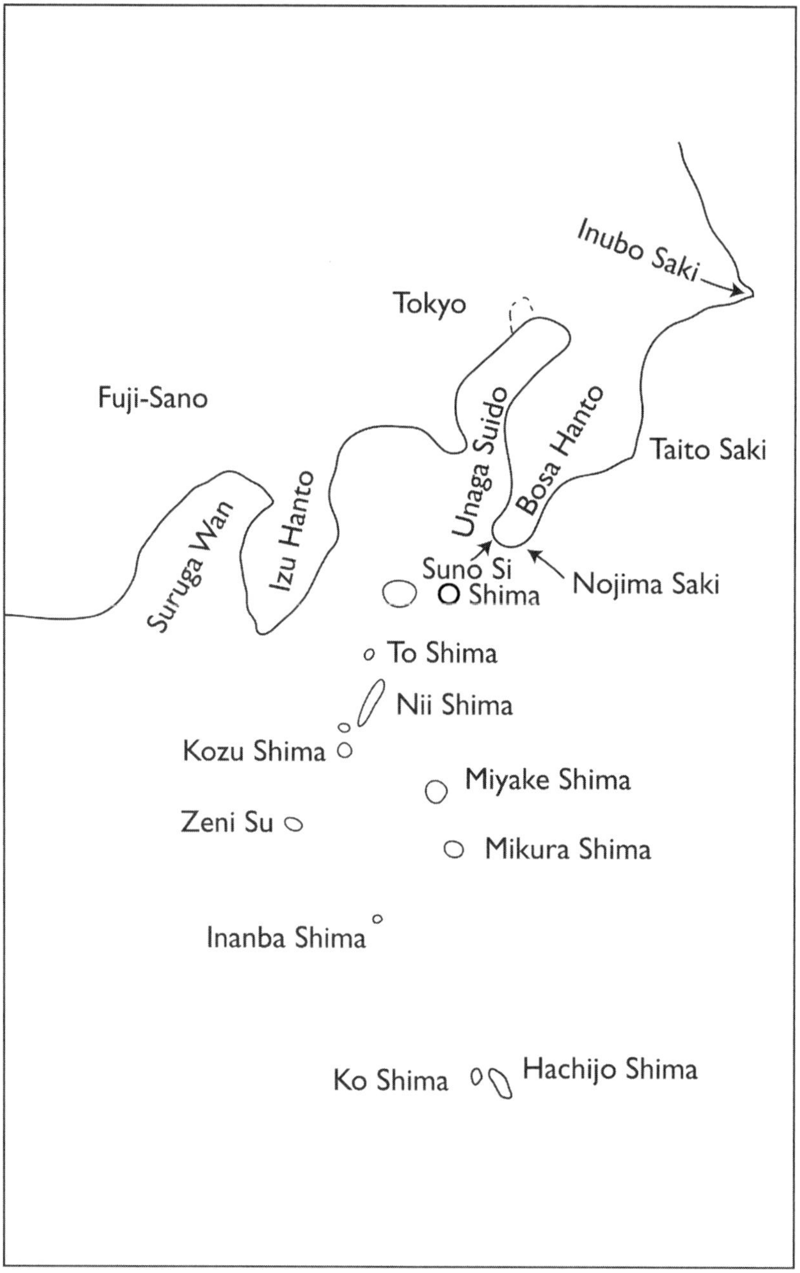

Map 6
Izu Shoto and Approaches to Tokyo

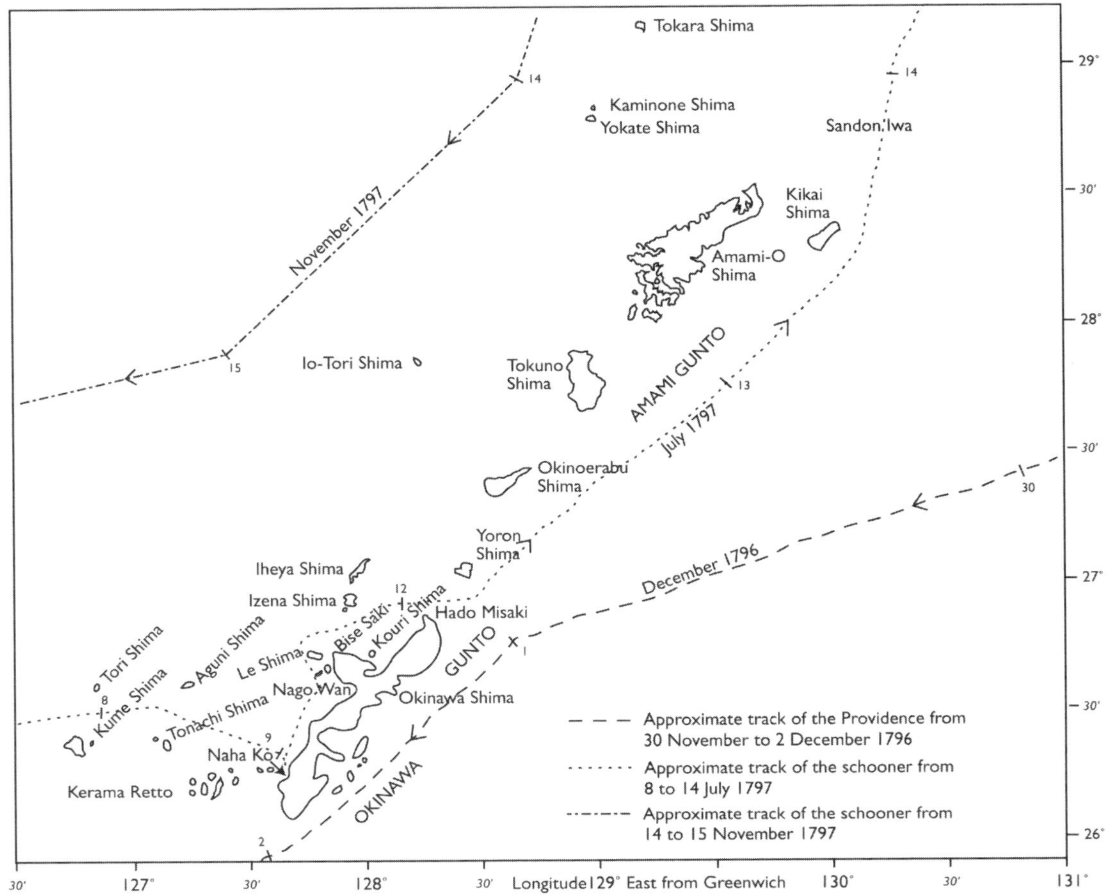

Map 7
Okinawa Gunto and Amami Gunto

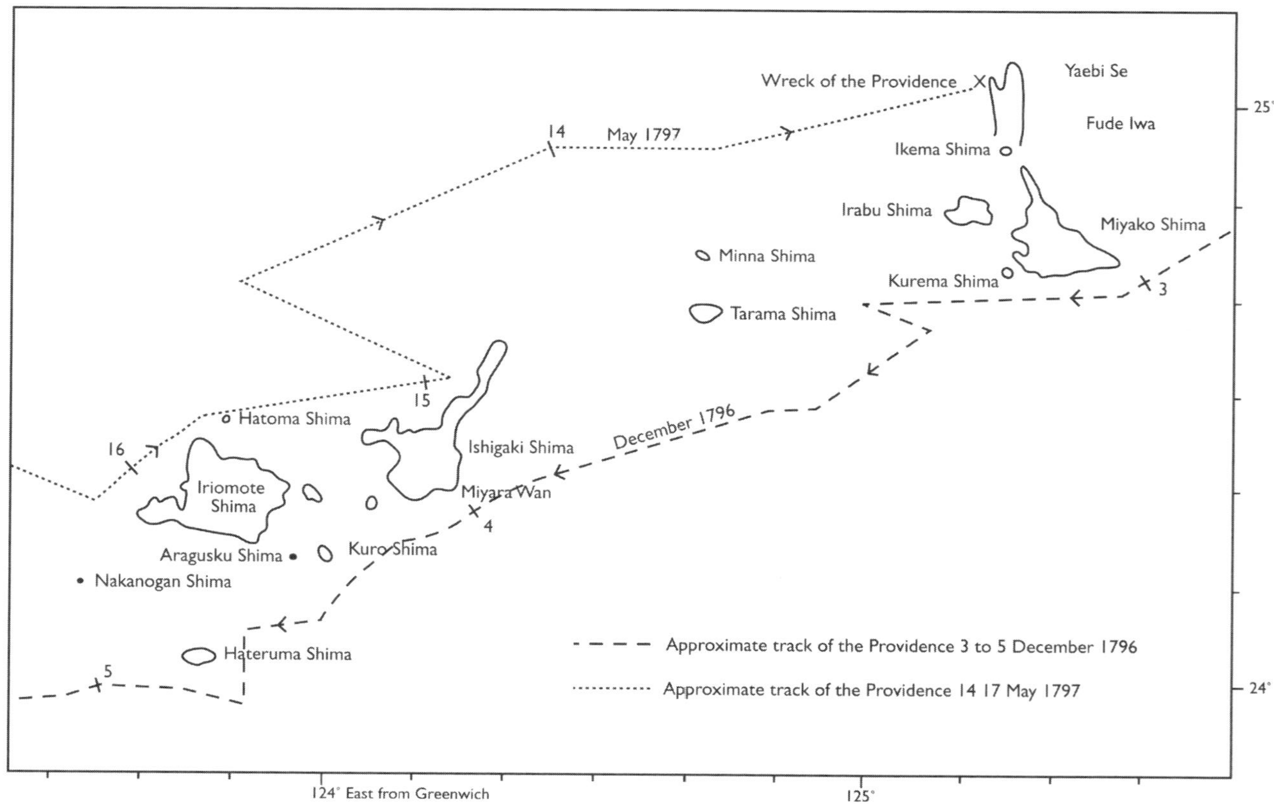

Map 8
Sakishima Gunto

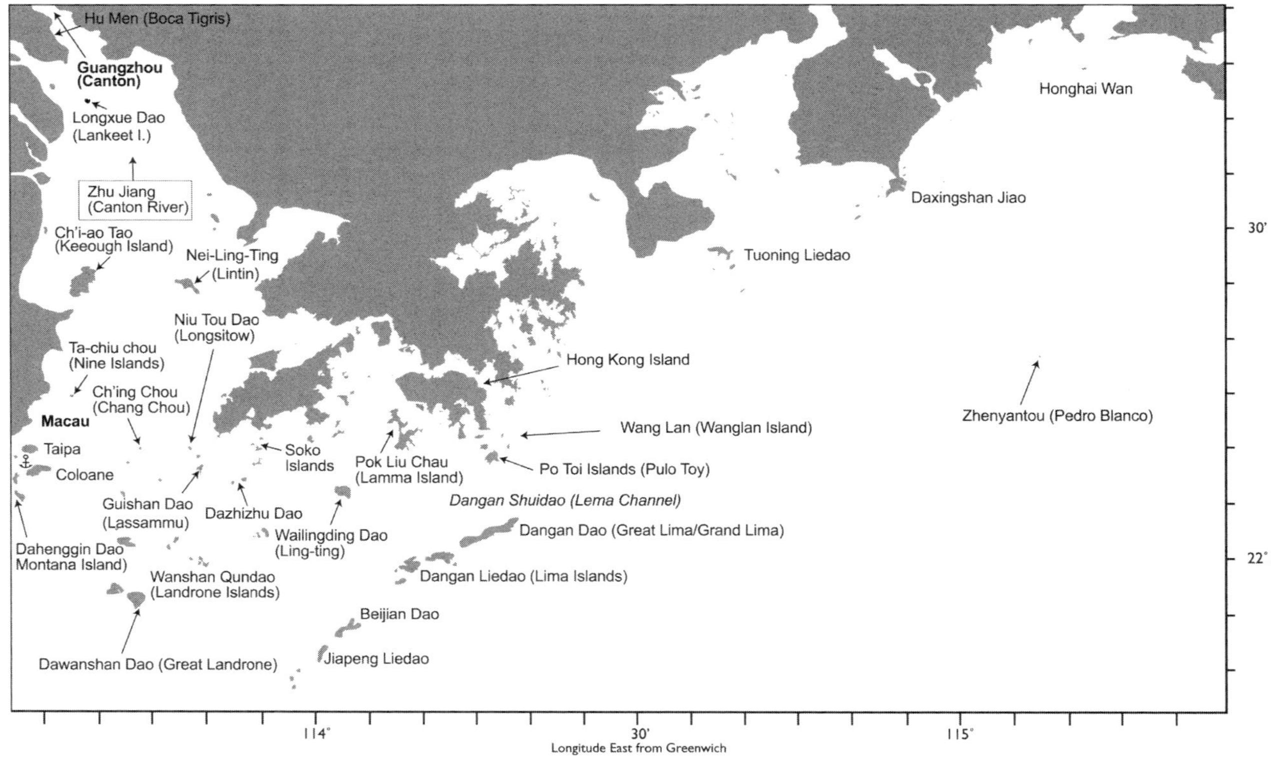

Map 9
The Approaches to Macau

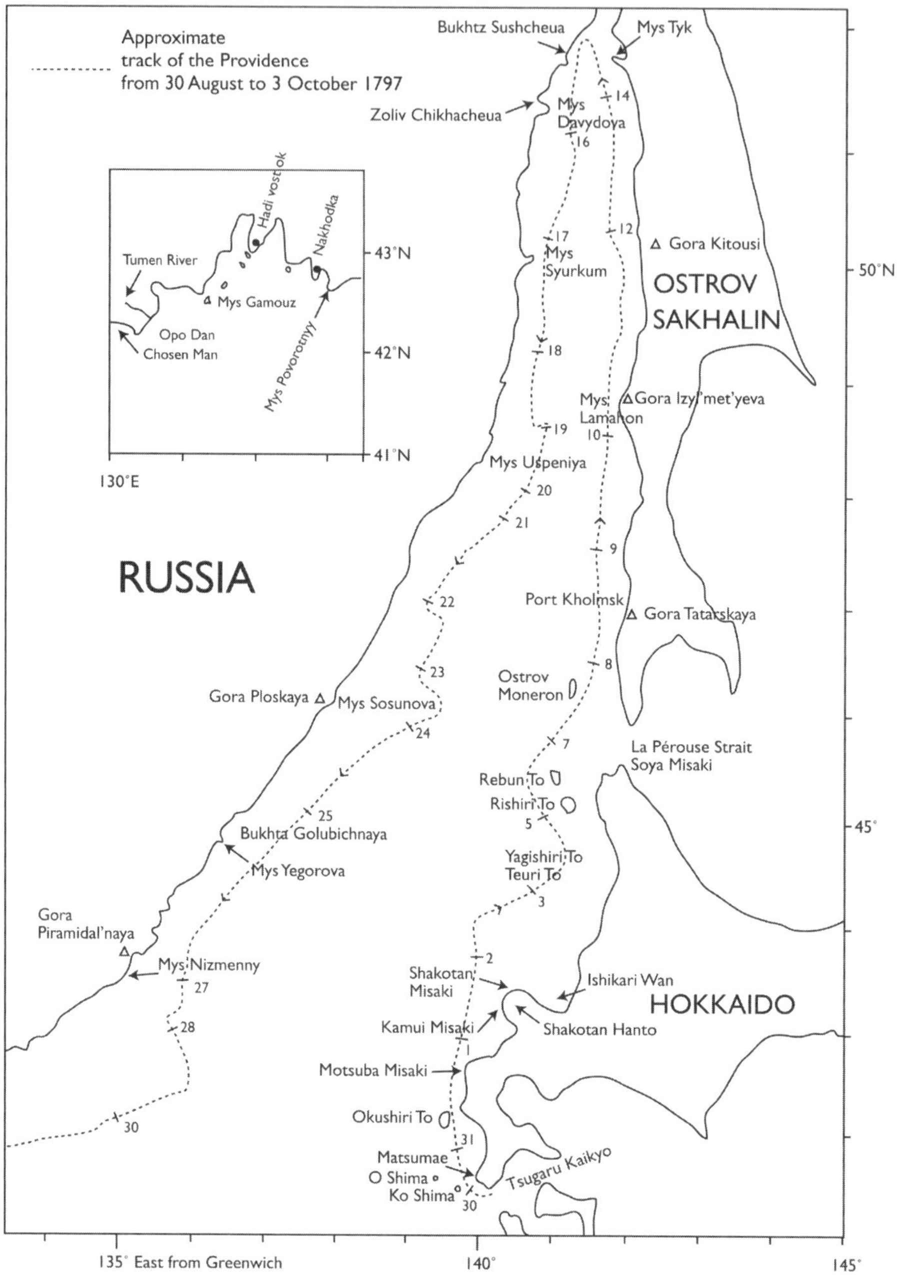

Map 10
The Gulf of Tartary

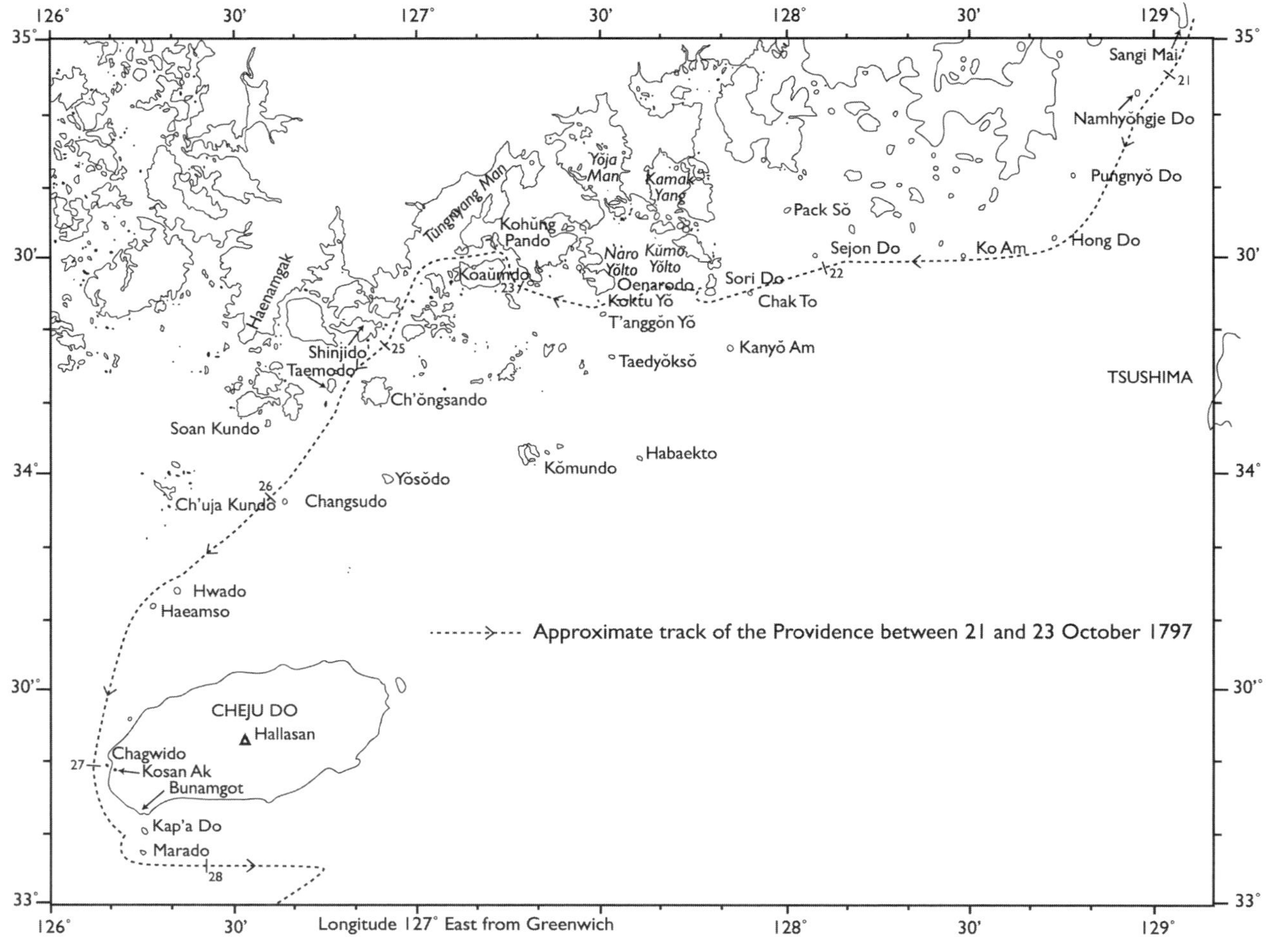

Map 11
The Korean Archipelago

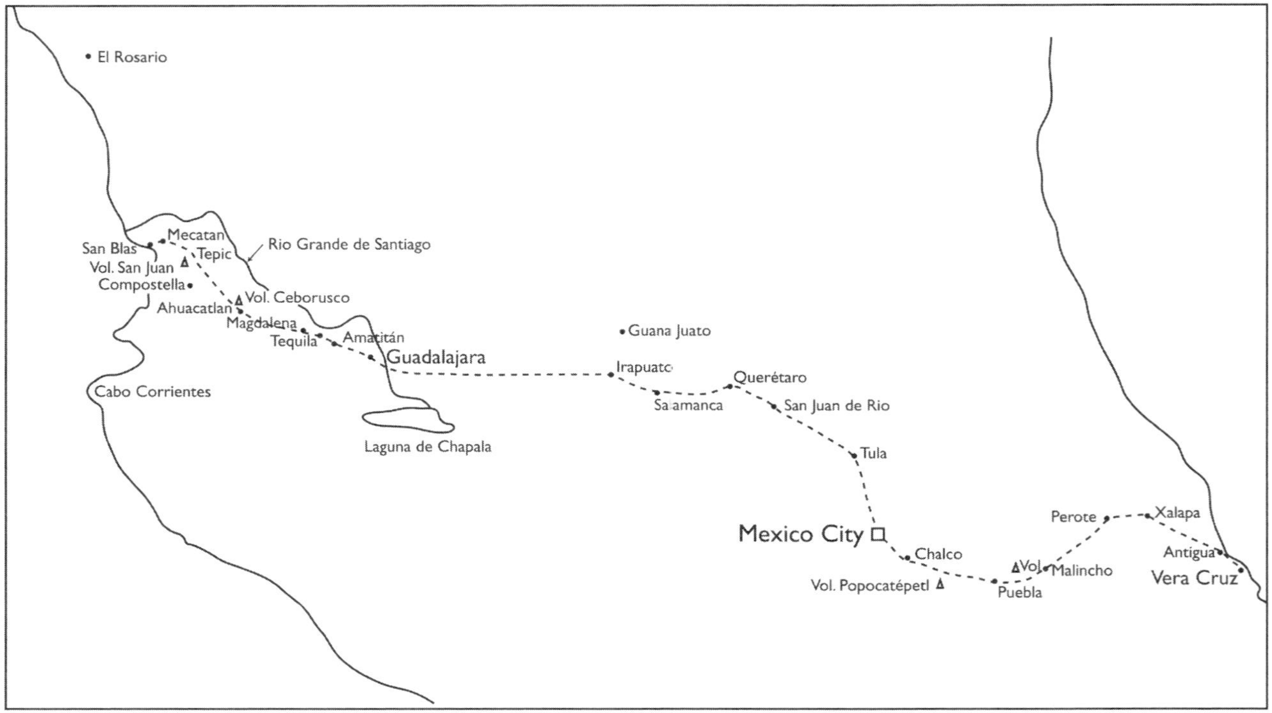

Map 12
Broughton's Journey across New Spain

Book One

Passage to the North-West Pacific and First Expedition to the North

CHAPTER 1

From England to Port Jackson and thence to Tahiti

The Navy-board recommended his Majesty's ship Providence for the voyage on which I was to be employed. She had been commanded by Captain Bligh, and had lately returned from the West Indies after having conveyed the bread-fruit there from Otaheite. This ship had originally been intended for the West-India trade, and was purchased by Government on the stocks, for the express purpose of bringing the bread-fruit trees from the South Seas, in which service she had been engaged for two years. She was river-built,[1] of about 420 tons in burthen, and was singly sheathed with copper, though I think it would be proper, that all ships employed in distant voyages should be sheathed with wood, and coppered over the sheathing. The Providence was taken into dock at Deptford for one tide, and thence transported to Woolwich, at which yard she was ordered to be fitted. On the 3d October 1793, I received my appointment to command her, and the same day commissioned the ship. The fitting her out detained us at Woolwich the remainder of the year; and early in 1794, the ship being ready, we dropped down to Gravesend, as a more convenient place for entering seamen. Here we remained until the end of March, when we proceeded to the Nore, saluting Vice-Admiral Dalrymple, whose flag was flying on board the Sandwich. During the month of April we completed our complement with volunteers from the Sandwich; and having orders to proceed to Spithead, we accordingly sailed with a convoy of merchantmen under our protection. During the night, in passing through the Gull Stream,[2] the pilot run the ship upon the Brake Sand:[3] we were not able to heave off till the morning's tide, when, soon after, we anchored in the Downs,[4] the ship appearing to have received no damage. Vice-Admiral Peyton's flag was flying on board the Leopard, which we saluted; and in the evening, our ship and the convoy worked through the Downs to the westward, with a light westerly wind. In the night the wind drew round to the eastward, and we made all sail at day-light, repeating signals for the convoy to do the same. We passed through the grand fleet lying in Sand-down Bay,[5] under the command of Lord Howe; and anchored at Spithead, having saluted Admiral Sir Peter Parker, whose flag was flying on board the Royal William. During the month of May the ship was docked, but had received no damage in her passage to the Downs:[6] she continued in every respect ready for sea; and the ship's company, as well as the officers, were paid their wages to the end of July. In this month his

[1] I.e. built on the Thames, where she was laid down by Perry at Blackwall.

[2] The channel between the Kentish coast and the Goodwin Sands.

[3] A small drying bank on the western side of the Gull Stream.

[4] The well-known anchorage off the Kentish coast stretching from South Foreland to North Foreland, where outward and inward bound vessels frequently anchored to pick up and land pilots.

[5] Sandown Bay off the SE side of the Isle of Wight.

[6] She was docked on 10 May when five sheets of copper were replaced on her starboard side and then warped out of dock the same day; TNA, ADM 52/3292/10.

Majesty visited Portsmouth, to view the grand fleet after the action of the 1st of June;[1] and every captain had the honour of being presented to him.

[Tuesday] 2 October [1794]

I received my orders, which were secret, with an additional one to put myself under the command of Captain Drury, of his Majesty's ship Trusty, and to proceed to sea with his convoy, then bound for the Mediterranean; nor was I to separate as long as our courses were the same.[2]

[Sunday] 21 October

We sailed from St. Helen's with a fine breeze from the eastward.[3] After clearing the Channel[4] the wind veered to the NW when the Trusty ordered us to proceed to Falmouth. In the night we separated from the men of war, and reached Falmouth;[5] when, not seeing the Trusty, we sailed for Plymouth Sound, where we found her and the convoy at anchor. Vice-Admiral M'Bride's flag was flying on board the Minotaur, and Rear-Admiral Cotton's flag on board the Cambridge in the harbour.

November

During this month, the gales were strong and the weather variable; and as we were anchored in Cawsand Bay, we could not have joined the convoy, had they gone to sea from the Sound with an easterly wind: we therefore endeavoured to work into the Sound; but the ship missing stays off Red Point,[6] we came to with both bowers all standing in foul ground, not the length of two cables from the shore. In this unpleasant situation we were prevented moving by strong easterly winds, till the master-attendant, Mr. Hemmings, brought an anchor lighter to windward of us, and, sending the end of a cable on board, we were enabled to heave off in safety, and run in between Duke's Island[7] and the Main.[8]

[Thursday] 29 January 1795

We continued in this station till the following month, when we run into the Sound. During January 1795, the weather was cold, with frequent gales. Rear-Admiral Parker, in his Majesty's ship Raisonable, made the signal for sailing.

[1] The Battle of the Glorious First of June.

[2] These orders were not entered in any of the relevant volumes of Admiralty correspondence (TNA ADM 2/125–7 and 1347–8) and thus do not appear to have survived.

[3] Broughton fails to mention that, due to a change in the wind direction, the *Providence* and the convoy were forced to anchor in Tor Bay between 22 and 23 October; TNA, ADM 52/3292/10. See also Crosley's letter to Maskelyne dated 18 November 1794 on p. 228.

[4] Off the Lizard the *Providence* spoke with the *Artois*, with the French frigate *La Révolutionnaire* in tow as her prize, which were cheered; TNA, ADM 52/3292/10. See also Crosley's letter to Maskelyne dated 18 November 1794 on p. 228.

[5] On 28 October the convoy was much dispersed in a gale, and on the following day the *Providence* bore up for Falmouth but did not enter it, anchoring in Cawsand Bay on 30 October; TNA, ADM 52/3292/10.

[6] Redding Point situated about ½ mile SW of Drake's Island.

[7] A typographical error for Drake's Island.

[8] For a full account of this incident see Crosley's letter to Maskelyne, dated 20 December 1794, on p. 231.

[Monday] 5 February
In the night we had a strong gale from WNW parted our best bower cable, drove on board a transport, sprung the cathead,[1] and damaged the main channels.[2] Moderate weather, the next day, enabled us to recover the anchor and splice the cable. We again moored the ship, as there was no probability of sailing till the wind came to the NE; when the signal was made for unmooring, and for all officers to repair on board.

[Sunday] 15 February
The whole fleet was under way; nor could any wind have been more favourable than this from the North, as it enabled every ship from Hamoaze[3] and Catwater[4] to proceed to sea, amounting to more than 400 sail, which had been detained equally with ourselves since October. As we stood out to sea, we observed the grand fleet under the command of Lord Howe waiting for the convoy.

[Monday] 16 February
The wind and weather were favourable; the Lizard Point at noon bore from us WbyS; our observed lat*itude* 50°13′30″N. The convoy from Falmouth joined us.

[Thursday] 19 February
In the evening the grand fleet, consisting of 34 sail of the line and seven frigates, parted company with us. The convoy for Portugal separated also the next morning, with the Trusty and Fly sloop. We continued with the West-India convoy, consisting of the Raisonable, Medusa, Iris, Cormorant, the Dromedary store-ship, and 200 sail of merchant vessels.

[Saturday] 21 February
Fresh gales from the SW obliged us to reduce our sails, and dispersed the merchant ships.

[Sunday] 22 February
The wind kept increasing, and veered to the NW.

[Tuesday] 24 February
The gales were strong and variable from the SW: the following day it was calm; but on the next to that, the gale commenced again from the West and NW. This bad weather separated half the convoy.

[1] Strong short beams of timber, projecting almost horizontally over the ships's bow on each side of the bowsprit, which serves to suspend an anchor clear of the bow when ready to let it go.
[2] Broad thick planks of wood projecting horizontally from the side of a ship to spread the shrouds; also known as chain-wales.
[3] The anchorage in the River Tamar upstream from Drake's Island.
[4] The anchorage in the entrance to the River Plym off Plymouth.

[Sunday] 1 March

We parted company with Admiral Parker's fleet, sailing with the Iris frigate, Reliance, and Supply.[1] We generally found our ship sail as well as the merchantmen.

[Monday] 2 March

With a fine breeze from the eastward, we pursued our course to the south.

[Thursday] 5 March

Early on the morning, saw the Canary Islands; at noon, the Peak of Teneriffe[2] bore N88°30′W. Baffling winds prevented our anchoring till the morning of the 6th, when we came to off Santa Cruz in 35 fathoms;[3] the Church[4] bearing West, Punta de Nago[5] EbyN. As our stay was likely to be short, we did not moor. In paying our respects to the Governor, he made some trifling excuse for not inviting us to his table; we however met with that civility, and many others, from Mr. Rooney, an Irishman, who had been settled here some years.[6] The contractor, Mr. Calloghan, supplied us with excellent wine for the ship's company, and beef daily. We also took some live cattle on board. Vegetables were in great plenty: onions and potatoes were the most salutary and useful for sea-store. The soil of the country, as far as I could perceive, is rocky, very scarce, and covered with stones; yet vegetables grow here with a considerable degree of vigour and luxuriancy, principally arising from the fertile quality of the volcanic mould. Fresh water is good here, and may be procured in great quantities for shipping. Landing is often difficult, on account of the surf that breaks on the shore. Some few days before we made this island, the flying-fish first appeared. Owing to the very heavy surf, there was little communication with the shore.

[Sunday] 8 March

The Iris and her convoy sailed.[7]

[Friday] 13 March

We sailed from Teneriffe with light airs from the SW quarter, baffling at times in every direction; nor did we loose sight of the Peak till the 16th, when at noon the wind veered to the northward, and continued a steady breeze. The Reliance and Supply were our inseparable companions. The island of Gomera bore from N20°E to N60°E, and the island of Ferro[8] N50°W distant 10 or 12 leagues.

[1] The *Reliance*, Captain John Hunter, with Commander Henry Waterhouse as second captain, and the *Supply*, Lieutenant William Kent, were part of the convoy which sailed from Plymouth on 15 February. Hunter had been appointed Governor of New South Wales on 5 February 1794. Matthew Flinders was master's mate and George Bass surgeon on board the *Reliance*.

[2] Pico de Teide, 3,715 metres in height.

[3] She was towed in by her boats; TNA, ADM 51/4490.

[4] Identified by Crosley as St Francis's Church; see Crosley's letter to Maskelyne dated 6 October 1795 on p. 233.

[5] Punta de Anaga, the NE extremity of Tenerife.

[6] In May 1791 Vancouver was also entertained by 'Mr Rhoney, an Irish Gentleman, to whose hospitality we were greatly indebted'; Lamb, *Vancouver's Voyage*, I, p. 313.

[7] Bound for the Cape Coast; TNA, ADM 51/1155/2.

[8] Hierro.

[Saturday] 21 March
In the forenoon we made St Anthony, one of the Cape de Verd Islands:[1] at noon the south-west point bore S8°E five or six leagues. Its lat*itude* 17°0′46″N, long*itude* 25°16′26″W.[2] The steady fresh trade-wind soon carried us out of sight of this island, and we pursued a SSE course.

[Wednesday] 25 March
Died Samuel Redriffe, a fine young lad of 18 years of age: a fever of only three days' continuance was the cause of his death.

[Sunday] 29 March
Sultry weather. The wind variable from N*orth* to NW. Frequent calms impeded our progress: at intervals the squalls were heavy, with thunder, lightning, and rain.[3]

[Wednesday] 8 April
We crossed the Equator with gentle breezes from the SE.

[Saturday] 11 April
The trade-wind became fresh and steady; our course SSW, a very good one.

[Wednesday] 15 April
Wind more easterly, and continued ESE when to southward of 10°S latitude.

[Wednesday] 22 April
Made the island of Trinidad;[4] at noon it bore of us S41°E. Observed lat*itude* of this island 21°21′41″S, long*itude* 29°29′W[5] of Greenwich.[6]

[1] Santo Antão, the NW island of Ilhas do Cabo Verde.

[2] In annotated *Voyage*, p. 8, corrected in pencil to:
High Land of St Anthony
Lat 17° 0′48″N
Long 25 13 11 W p*er* Time Keeper No 1
25 24 5 W p*er* Mean of 20 sets of Lunars

[3] The *Providence* was in the Doldrums, characterized by light and variable winds alternating with squalls, heavy rain and thunderstorms.

[4] Ilha da Trindade.

[5] 'E' in Broughton, *Voyage*, p. 9, which is clearly a transcription error.

[6] In annotated *Voyage*, p. 9, corrected in pencil to:
Center of the Island of Trinidad
Lat 20°29′47″S
Long 29 2 29 W p*er* Time keeper No 1
29 19 50 W mean of 38 sets of Lunars

[Thursday] 23 April

We lost sight of the land in the afternoon; next day had light airs from the NE. The wind soon veered to the East, and became steady. As we purposed to reach Rio Janeiro, we steered more to the westward, and in lat*itude* 23°S shaped our course due West, in search of an island said to lie in that parallel.[1]

[Friday] 1 May

Variable winds in all directions. At day-light five strange ships were discovered: they proved to be Brazil merchantmen bound to Lisbon, and had left Rio Janeiro five days before.

[Saturday] 2 May

Early this morning the land was seen, and at noon Cape Frio bore NbyW eight or nine leagues; at the same time, we sounded in 70 fathoms fine sand. By our observations we place this Cape in the lat*itude* of 22°59′41″S, long*itude* 41°53′12″W.[2]

[Thursday] 5 May

The variableness of the weather prevented our reaching the entrance of Rio Janeiro harbour before this day, when we came to an anchor in 28 fathoms sandy bottom. Round Island[3] S88°W; Sugar-loaf Hill[4] N55°W; extremes of Brazil coast from N64°E to S64°E. The weather was now calm.

[Friday] 6 May

In the afternoon a pilot came off, and we got under weigh; but calm weather obliged us to anchor in the entrance of the harbour. The next day we were more fortunate, coming to an anchor within the island of Cobras[5] in 5½ fathoms. There were lying here a Portugueze frigate, and several merchant ships of different nations. The Reliance, Captain Hunter, saluted us coming in, which we returned with an equal number of guns.[6]

[1] A non-existent island with the name Ascençaon shown on one of d'Après de Mannevillette's charts 300 miles west of Ilha de Trindade. It was searched for unsuccessfully by La Pérouse in October 1785 and by Malaspina in September 1789.

[2] 'E' in Broughton, *Voyage*, p. 10, which is clearly a transcription error. In annotated *Voyage*, p. 10, corrected in pencil to:

Cape Frio
Lat 23°1′40″S
Long 41°41′1″W per No 1
41°59′16″W per mean of 38 sets of Lunars

[3] Ilha Redondo.

[4] Pão de Açúcar. In annotated *Voyage*, p. 10, inserted in pencil:

Sugar-loaf Rock at the entrance of the Harbour
Lat 22°57′28″S
Long 42 41 50W per Time k*eepe*r No 1
42 58 51W per mean of 47 sets of Lunars

[5] Ilha das Cobras.

[6] This is undoubtedly an error as Broughton was considerably junior to Captain John Hunter, who was a post captain of several years seniority, whereas Broughton was a commander. It was Broughton who should have saluted

[Sunday] 24 May

We remained at Rio Janeiro till this day, having employed ourselves in overhauling the rigging, caulking the ship, and preparing for sea in every other respect. The ship's crew were constantly supplied with fresh provisions; and we purchased at a reasonable rate wine, rum, and sugar. In the procuring of those articles we had every indulgence we could wish for; but we had also the mortification to be watched by guard-boats day and night; nor could any officer land unless he was attended by a Portugueze officer of equal rank. These restrictions were so unreasonably suspicious, that I went but once on shore, except with Governor Hunter to pay our respects to the Viceroy. The soil about Rio Janeiro is generally good, producing great crops of oranges, pine-apples, melons, sugar-canes, and other tropical plants. There is a great variety of birds; such as parrots, cockatoos, sea-fowl, tropic and men-of-war birds. There are several kinds of the monkey tribe; one remarkably small called the weshtu,[1] scarcely seven inches in length. As Captain Hunter's ships were not ready I proceeded to sea alone, taking his orders for Port Jackson; as I intended, in consequence of the lateness of the season, to proceed to the South Seas by Van Diemen's Land.[2] The situation of Rio Janeiro is in lat*itude* 22°53′17″S and long*itude* 42°51′16″W.[3]

[Monday] 25 May

A breeze springing up from the north-eastward, we soon lost sight of the Brazil coast.

[Tuesday] 26 May

A fine gale carried us to the southward rapidly; and in the latitude of 31°S the wind veered to N*orth* and NW. [29th] We shaped our course SE, reducing our sails as the gale increased.

[Tuesday] 2 June

We had a large following sea, the wind increasing to a strong gale at SW.

Hunter and not the other way round if such a salute was warranted. However, the statement by Crosley that, on entering harbour, the *Reliance* saluted 'the fort of Sancta Cruz with 13 Guns which was returned with an equal number' was almost certainly correct; see Crosley's letter to Maskelyne dated 6 October 1795 on p. 233. This is confirmed by the *Reliance*'s log: 'at 3 Saluted Fort Sancta Cruz with 13 Guns the Fort returned the Same number'; TNA, ADM 51/1169, part 7, f.36 and by the log kept by Lieutenant Vashon, TNA, ADM 51/4490.

[1] Possibly the Tufted-eared Marmoset (*Callithrix jacchus*). The appearance of the only other very small primate to be found near Rio de Janeiro, the Golden Lion Tamarind (*Leontopithecus rosalia rosalia*), is so dramatic, being entirely golden in colour, that Broughton would surely not have confined his description to its being very small.

[2] Broughton was mindful of the difficulties that Bligh had in attempting to round Cape Horn in the *Bounty* in April 1788.

[3] 'E' in Broughton, *Voyage*, p. 12, which is clearly a transcription error. In annotated *Voyage*, p. 12, corrected in pencil to:

Island of Enchados in the H*arbou*r of Rio Janeiro
Lat 22°53′16″S
Long 42°42′58″W per Time k*eepe*r No 1
42°59′59″W per mean of 47 sets of Lunars

[Friday] 5 June
A strong breeze at NW with continual rain. In the latitude of 40°S we steered East, intending to make Gough's Island, and fix its situation. [7th] A heavy gale from the NNW obliged us to bring the ship to the wind in the morning. After laying to under a balanced mizen four hours, we again bore away as the violence of the gale subsided.

The gale from the West returned with redoubled violence;[1] and as the ship had not sufficient velocity through the water to escape the sea, we again brought to under a mizen stay-sail. In the evening the weather became more moderate, and we bore away.

[Wednesday] 10 June
At 1½ past noon, we discovered Gough's Island bearing east five or six miles. Our observation at noon made the lat*itude* 40°19′S, which was indifferent. The gale increased; yet being desirous of making further remarks upon the island, at ½[2] past 3 we brought to the wind under the lee of it, bearing from us NNW. The sea was irregular and confused; during the night the gale subsided; our soundings were at 155 fathoms. No observation at noon.

[Thursday] 11 June
The following day we were equally unlucky, the weather being rainy and hazy; and as there was no chance of it clearing, we resumed our course to the eastward. The breezes were strong at NNW and the rain constant. Gough's Island is high and much broken, not more than two or three miles in circumference or extent. We could not perceive the least sign of vegetation; but as the weather prevented landing, we could not make the remarks we wished, and our idea of its circuit is even doubtful. By our time-pieces we place it in the lat*itude* 40°19′S and long*itude* 9°27′W; which may err two or three miles, as the horizon was very confused.[3]

[Friday] 12 June
Strong breezes from the NNW. We steered to the East, preserving nearly the parallel of lat*itude* 41°30′S although we could not get an observation more than once in three days. The weather became more moderate, with the wind variable to the SW after we got into east longitude.

[Tuesday] 16 June
The wind again returned to its old quarter with increasing violence: in reefing the maintopsail, Hugh MacDonald fell off the yard upon the deck, and was most unfortunately killed

[1] No date is given in Broughton, *Voyage*, p. 12.
[2] '1½' in Broughton, *Voyage*, p. 13, which is presumably a transcription error.
[3] In annotated *Voyage*, p. 14, corrected in pencil to:
Gough's Island
Lat 40°19′0″S
Long 9°37′33″W per Time k*eepe*r No 1
9 36 48 per Time k*eepe*r No 248
The true situation of Gough Island is 40°19′S, 9°57′W.

on the spot. About this time we had the first appearance of the albatross, and the beautiful bird called the pintado or Cape pigeon;[1] also great numbers of sea-gulls, sheerwaters, &c.[2]

[Wednesday] 1 July
The breeze was variable at East and SE, with foggy weather. Several seals came about the ship, and we passed some rock-weed. [7th] A very severe gale from the N*orth* and NW attacked us; and in the forenoon, to avoid the sea, we furled the foresail, and brought the ship to the wind under a storm staysail, it blowing tremendously hard from the NW quarter, and raining violently. In the night we carried away the tiller.

[Tuesday] 14 July
Wind veered to the West: squally and cold.

[Sunday] 19 July
Light breezes and cloudy weather induced us to get up the top-gallant masts and yards; and we set all our small sails, which had been of very little use during this turbulent passage.

[Tuesday] 28 July
Wind at SSE increased to a strong gale; ship was under storm staysails; squalls, always attended with rain.

[Sunday] 2 August
Fresh breezes and cloudy weather; lat*itude* 44°18′S. [3rd] At 1 PM we saw Van Diemen's land generally covered with snow; it bore NbyE, the wind at NNE: at noon the extremes bore from N5°W to 62°W; lat*itude* 44°5′S.

[Tuesday] 4 August
In the evening the land bore from N54°W to 71°W; and the wind remaining at North prevented us seeing more of it, as we had only to keep our wind to the eastward. [6th] We steered NNW, intending to make Furneaux Isle, and explore the unknown space between it and Cape Howe,[3] if the wind permitted.

[Wednesday] 12 August
Strong gales from the East; by which we continued our course to the North, not having been able to make Furneaux's Island. At sunset we saw the land, and altered our course to NNE in consequence; the next morning we had the mortification to find ourselves deceived, as we had

[1] The Cape Petrel (*Daption capense*).
[2] For possible seabirds that might be encountered in this latitude see Gerald Tuck, *A Guide to Seabirds on the Ocean Routes,* London, 1980, p. 70.
[3] The yet undiscovered Bass Strait.

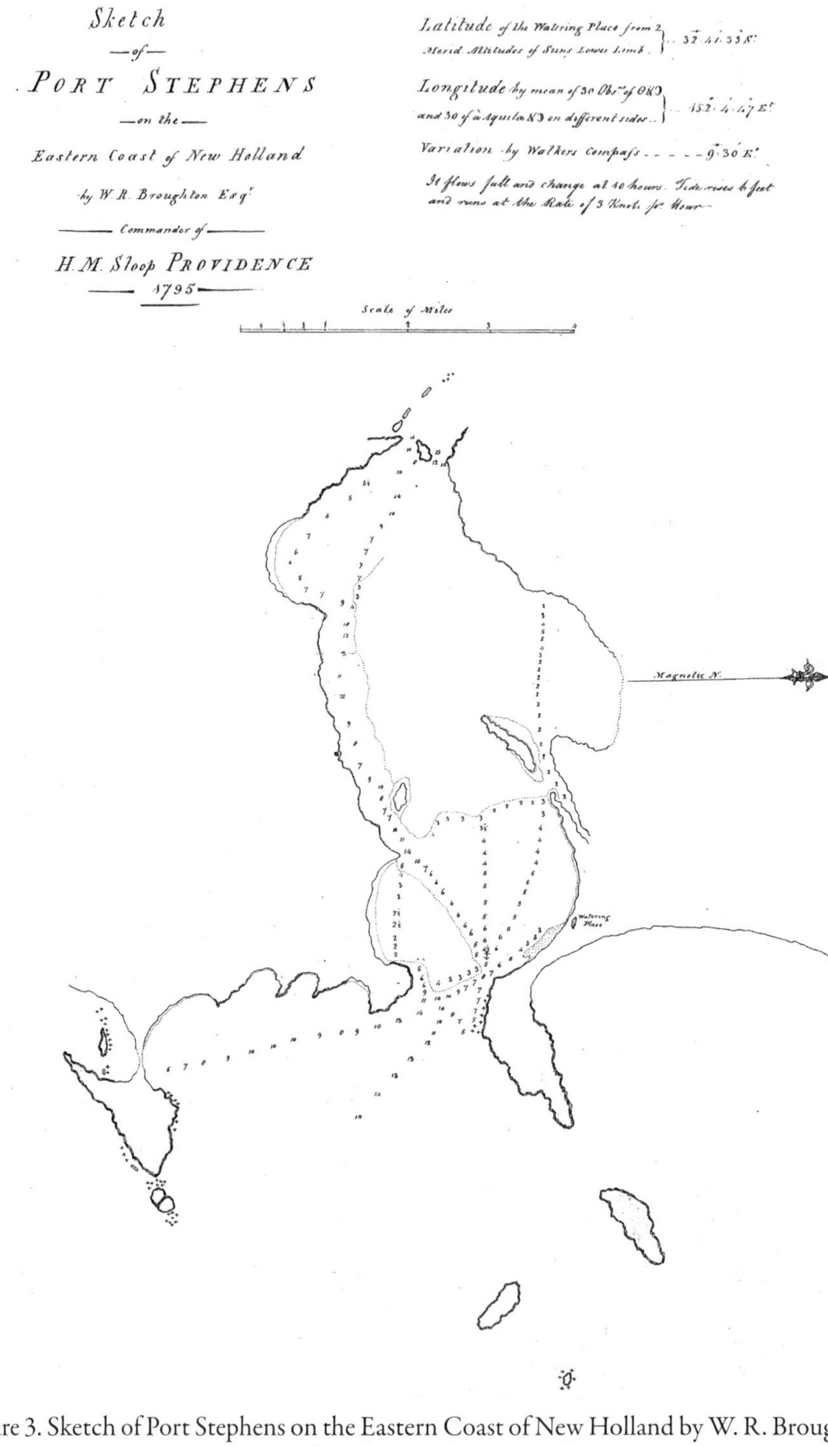

Figure 3. Sketch of Port Stephens on the Eastern Coast of New Holland by W. R. Broughton Esq[r] Commander of H.M. Sloop *Providence* 1795.
Courtesy of the UK Hydrographic Office.

clear weather, and nothing to prove that we were near the coast of New Holland: lat*itude* 34°50′S.

[Thursday] 13 August

The wind from the NW blew direct from the land. We were to the North of Port Jackson.

[Friday] 14 August

At day-light made the land to the NW of us; at noon we had clear fine weather, and Cape Hawke bore N88°W. The extremes of coast from S22°30′W to N2°E. Observed lat*itude* 32°11′S.

[Saturday] 15 August

We kept plying to the windward; at noon Cape Hawke bore N35°W. Islands off Port Stephens S71°W. Lat*itude* 32°38′S.

[Tuesday] 18 August

The wind still remaining fixed in the SW quarter, I found there was no chance of beating to the southward: I therefore determined to enter the port which was in our power. We entered in safety Port Stephens, and anchored in 5½ fathoms sandy bottom. Points of entrance were open from S12°E to 85°E. We steadied the ship with the stream anchor, and sent the boats in search of water. We remained here one week, and completed our watering from a lake on the northern shore with great facility. During our stay, we had frequent intercourse with the natives. They were the same race of people as those described to inhabit Port Jackson and Van Diemen's Land. They were inoffensive, quiet, and docile; and whenever we approached them, removed their women and children. We found here residing with the natives four Englishmen, who had deserted in a boat from Port Jackson five years before. Five came originally, but one had died; those that remained were miserable half-starved objects, depending on the hospitality of the natives for their subsistence, who occasionally supplied them with a part of their provisions, at all times in no great abundance with the inhabitants. Notwithstanding the wretched state in which they existed, the man who had enticed them to desert refused to come on board. We collected some articles to leave with him to make his situation more comfortable; but in the meanwhile, being assured he should be well treated, and probably not suffer for his former conduct, he agreed to come off with the others. One or two of these men were married, and left their wives and children with little regret.[1] The soil round Port Stephens is composed chiefly of sand and decayed vegetables, though in the swamps it is rather of a better kind; but upon the whole the nature of the place is very barren, and unfit for any great degree of cultivation. The sea produces a good variety of fish; such as mullet, toad-fish, a kind of torpedo,[2] flounders, &c. In the woods, were several species of beautiful paroquets;[3] and a small

[1] The four convicts taken back to Port Jackson were John Tarwood, George Lee, George Connoway and John Watson, John Sutton having died; David Collins, *An Account of the English Colony in New South Wales*, 2 vols, London, 1798, 1802, I, p. 356.

[2] An electric ray.

[3] Various species of Lorikeets.

bird of a brown plumage,[1] somewhat resembling the Java sparrow.[2] The quadrupeds that we saw were the kangaroo, dogs,[3] &c. On the beach we observed a variety of curiously marked shells; such as buccinum or whelk, limpets, muscles, oysters, and beautiful specimens of the Venus shell.[4] The natives live chiefly upon fish, fern roots, dog's flesh; and should a dead whale happen to drift upon the shore, it forms a most delicious repast for them. The astronomer made the following observations on shore at the watering place abreast of the ship:

Lat*itude* by means of 4[5] merid*ian* alt*itudes* of the ⊙	32°41′33″S
Long*itude* by mean 4 time-keepers	151°44′44″E
Long*itude* by mean 12 sets of lunar distances	152° 4′47″
Long*itude* by mean diff*erence* of long*itude* between this place and Port Jackson, taking the longitude of Port Jackson at 151°10′3″E that being the mean between Sign*or* Mallespina[7] and Mr. Crosley, and the diff*erence* by the 4[8] watches in long*itude* at 33°46′W[9]	151°43′49″[6]

At noon we hoisted in the boats, and made sail to the southward with the wind easterly. [26th] Port Stephens bore N76°W and the extremes of land N24°E to S75°W: lat*itude* 32°51′S. At midnight we tacked to the northward till daylight, when we again steered to the South and SSW as the wind permitted. We had 70 fathoms water: the forenoon was hazy, the weather rainy, the wind increasing, with every prospect of a gale from the SE which blew directly upon the land. We carried all sail to secure our port before dark. At noon the signal-house on the South head of Port Jackson harbour bore S32°W, and we weathered the North head half a mile by carrying a press of sail; [27th] ½ past noon we were in the entrance of the harbour, and at 1 PM in running up, a pilot came on board. In the afternoon we moored with our bower cables in Sydney Cove, and struck yards and topmasts. In the night, the gale (as predicted) increased to a perfect hurricane, and continued the next day with extreme violence; nor could we send a boat on shore. We could not be too thankful for our safety: for had we remained at sea, most probably the ship never would have cleared the land; as at no time from our run here, were we more than two leagues from the coast. We found Major Paterson, commanding the New South Wales corps, acting as governor;[10] who received us in the most welcome manner,

[1] Not identified.

[2] An exotic cage bird (*Padda oryzivora*).

[3] Presumably dingoes.

[4] A member of the family *Veneridæ*.

[5] In annotated *Voyage*, p. 18, corrected in pencil to '2'.

[6] In annotated *Voyage*, p. 18, corrected in pencil to '152°11′32·5″'.

[7] Alejandro Malaspina. For details of the comparison between Crosley's and Malaspina's observations see Crosley's letter to Maskelyne, dated 6 October 1795 on p. 233.

[8] In annotated *Voyage*, p. 18, corrected in pencil to '2'.

[9] In annotated *Voyage*, p. 18, corrected in pencil to 1°1′29·5″W. Also added in pencil below the table:

Lat	32°41′43″S
Long	151° 46 56E per Time k*eepe*r No 1
	152 6 1E per Time k*eepe*r No 248
	152 6 10E mean of 12 sets of Lunars

[10] Paterson had succeeded Major Francis Grose as Acting Governor upon the latter's departure from the colony in December 1794.

shewing us every civility and attention in his power. Indeed I in particular cannot sufficiently express my gratitude for the hospitality shewn to me both by Major Paterson and his lady during our long stay here. We immediately commenced refitting the ship, caulking her within and without, together with overhauling the rigging &c., and landed tents at Cattle Point[1] for the astronomer to ascertain the rates of the time-pieces. The hauling the seine was continually in use for supplying the ship's company with fish; and by order of Major Paterson, we daily and amply received vegetables for our people by sending a boat to Garden Island.

[Tuesday] 8 September
Governor Hunter arrived with his ships, having been 97 days from Rio de Janeiro.

[Saturday] 12 September
We saluted his excellency Governor Hunter with 13 guns, on reading his commission that appointed him Captain General of New South Wales, &c. &c.

[Tuesday] 6 October
The ship being ready for sea we unmoored and shifted our birth to the Fair Way, cheering the Reliance as we left the Cove.

We took our final leave of Port Jackson in the morning: at noon the south head of Broken Bay bore N34°W, lat*itude* 33°50′S; North head, Port Jackson, S88°W. During our stay we entered several good seamen from merchant ships and the colony to complete our complement: our ship's company was in perfect health. We abstained from following the example of other ships that have touched at this colony, by not taking away any of the convicts: a practice very general in merchant ships, which has tended to corrupt the morals of the South Sea islanders; for in the voyages of the traders to the north-west coast of America, these men have generally deserted by the way, stopping either at the Society or Sandwich Isles. The soil round Port Jackson is light and sandy generally; though, in the more inland parts of the country, it is much better, and produces good crops of corn and other vegetables. Tropical fruits do not answer well here, it being too cold in the winter season; such as have arrived from the Cape grow luxuriantly. The bread-fruit and cocoa-nut have lately been brought here, but did not succeed. Plantains and the sugar-cane have done better, but in all probability will never arrive to any great degree of perfection. There are several kinds of other fruits and vegetables which might deserve a trial here; such as gooseberries and currants, a greater variety of apples, pears, plums, limes, peaches, &c. Experiments might also be made of the tarro root, yams, and sweet potatoes. Annual and perennial flowers are still desiderata here. The birds of this place are of several kinds; such as the black and white cockatoo, variety of paroquets, crows, gulls, shags, &c. The emu is a bird belonging to the genus of the cassowary, grows to a large size, and like the rest of the genus is unable to fly. Quadrupeds are chiefly the kangaroo of two kinds, opossums, flying squirrels, mice, kangaroo rats, and a species of ferret. Reptiles are snakes, some of a large size,

[1] A grassy cliff sloping gently down to a flat rocky ledge on the eastern entrance to Sydney Cove and a favourite observation site with other navigators. The point was later renamed Bennelong Point after an Aborigine of that name much lauded by the early settlers; now the site of Sydney Opera House; Ingleton, *Flinders,* pp. 47–8.

guianoes,[1] lizards, frogs, &c. The bays are well-stored with fish; such as salmon, eels, mullet, the leather jacket, flounders, &c. The shores contain oysters, muscles, conchs, and other shell-fish. Sharks are sometimes caught here of a large size, which produce the only oil the people have to use.

The following astronomical observations were made on shore at Cattle Point; one of those forming Sydney Cove, and where the Spaniards had before fixed their observatory:

Lat*itude* by mean of meri*dian* alt*itude* of the ☉ .... 33°51′47½″ S
Long*itude* from mean of 90 sets 45° to the east and 45° to the West of ☽ .... 151° 9′48″ E[2]
Variations by compass .... 11° 9′ East

The flag-staff on the South head bearing by compass N73°45′E; distant seven miles.

[Wednesday] 14 October

We directed our course to the North of New Zealand, intending to touch at Otaheite in our route to Nootka Sound. At 3 PM the land extended from NW¼West to West, and we soon after lost sight of the coast.

[Monday] 19 October

In lat*itude* 33°10′S and long*itude* 169°E the SW wind changed to a gale from the NE quarter. It was of no continuance; for on the following day it veered to NW, and we again pursued our eastern course.

In 32°51′S and 175°E the winds were northerly, with moderate weather, continuing at NE till this day [24th],[3] when the wind again veered to the NW with a strong breeze.

[28th] Wind now altered to the NNE; our lat*itude* 34°15′S and long*itude* 189°35′E by the time-piece agreeing with our accounted longitude.

[Friday] 30 October

We stood to the Northward, as the wind became more easterly; lat*itude* 35°45′S and long*itude* 192°30′E.

[Monday] 2 November

It blew so hard as to oblige us to lay to under the main-sail. The gale moderated: [3rd] we wore ship at noon and stood to the SE in lat*itude* 32°46′ and 192°25′ long*itude*.

[1] Possibly a misspelling of guiana, a variant spelling of guana or iguana, an arboreal lizard.

[2] In annotated *Voyage*, p. 23, corrected in pencil below the text to:

Lat	33°51′45″S per mean of 14 mer*idian* alt*itude*s of the ☉
Long	150°45′28″E per Time k*eeper* No. 1
	151 4 30 E per Time k*eeper* No 248
	151 10 30 E per mean of 12 sets of Lunars

[3] Some time on this day the *Providence* crossed the 180° meridian when the date should have been put back one day. Thus the 24th should have become the 23rd, etc., until the *Providence* recrossed the 180° meridian on 14 August 1796 when once again the date would be correct.

[Friday] 13 November

In 30°33′S and 197°53′E we stood to the SE, having had no observations since the 9th for the lat*itude*; we found ourselves 1°30′ further north than we expected, in which direction only we had experienced any current.[1] In the night we stood to the Northward; lat*itude* 31°27′S and long*itude* 197°27′E. The wind grew more favourable for our course, and on this day at noon[18th], in the lat*itude* of 28°33′S and long*itude* by watch 201°27′E, [19th] we observed distances ☉ and ☽ making the longitude agree nearly with the watch.

[Wednesday] 25 November

At 2 hours 40 minutes after midnight we made the island of Ohetorea[2] bearing NE6°N, which as it was in our course we stood off for two hours, and again tacking made sail towards it. At noon it bore from S3°E to S3°W. Our latitude was indifferent; it made 22°0′54″S and long*itude* 208°32′48″E.

[Saturday] 28 November

Fresh breezes from the Eastward brought us in sight of the island of Otaheite, bearing to the NNE. At 11 AM we tacked ship, East-point of the land bearing NNE; Low-point, in the centre of the island,[3] N½W off shore; Tiarraboo[4] east point, N74°W.[5] Lat*itude* observed 17°47′32″S, long*itude* 210°44′20″E. Account 208°53′E

[Sunday] 29 November

The wind soon after noon sprung up favourably for our course, and we made all sail. As we passed Ohitepehah Bay[6] we were surrounded by canoes. In the evening at 5 we anchored in 5 fathoms, in Matavai Bay;[7] and the next morning, warped into good anchorage. In so doing, we swept an iron stocked anchor, which had belonged to the Bounty, when she cut her cables, and went to sea from the Bay with the remaining mutineers. In the course of the day we raised our tents in Point Venus, to ascertain the rates of the time-pieces, and compleating our water. The natives afforded us every assistance in our various pursuits, and amply supplied us with provisions and vegetables.[8]

[1] 'bar. 7°49′E' omitted as meaningless; also marked for deletion in annotated Broughton, *Voyage*, p. 25.

[2] Île Rurutu or Hiti-roa, an outlying island of the Society Islands.

[3] Isthme de Taravao, the narrow isthmus joining Tahiti proper with Tahiti-iti.

[4] Presqu'île de Taiarapu or Tahiti-iti.

[5] '10″' in Broughton, *Voyage*, p. 26, which is clearly a transcription error.

[6] Oaitipeha in John Hawkesworth, ed., *An Account of Voyages undertaken ... for making Discoveries in the Southern Hemisphere*, 3 vols, London, 1773, II, p. 158; Baie de Tautira or Cook Anchorage.

[7] 'at 8 fathoms' omitted; it may, however, refer to the next morning's anchorage.

[8] No account has survived covering the *Providence*'s stay in Matavai Bay; the next entry in Broughton, *Voyage*, p. 27 is for 11 December, the day she sailed.

CHAPTER 2

From Tahiti to the Hawaiian Islands

	Feet		
Fore Castle Grating	19	= 4′ 10″	For Dip of the Horizon to correct the Observ'd Altitudes[1]
Fore Castle	16	= 3 49	
Gangway	15	= 3 42	
Quarter Deck	16	= 3 49	
After part d*itt*o	17	= 3 56	
Quarter Gallery	9½	= 2 56	
Poop Rail	20	= 4 16	

NB The Log Line 47⅓ feet to 29″ Glass[2]

Friday 11 December 1795

At ½ past 4 PM No 1 made the Longitude 210°2′03″ when Point Venus bore of us S53°E 12 or 13 miles which will make the Long*itud*e of P*oin*t Venus 210°22′E[3] its situation by Capt*ain* Cook.[4] On the 20th it agreed within 3′ of the Long*itude* of Oahitipeha Bay[5] which was the day after we made the Island. The Elaps'd time being only 12 days. Of course we have no reason to suppose it has alter'd its rate since leaving Port Jackson and makes us less regret that the Astronomer was not able to make any Observations on Point Venus for that purpose. The Point of Land on which the Tents were pitch'd was so much affected by the Surf striking upon it as to give a trembling Motion to the quicksilver. And the pains he took in trying various methods answer'd no purpose. Neither cou'd he take any Altitudes from the Seas Horizon from the Position of the Point & surrounding reef. The other watches made the Long*itude* as follows [observations wanting].

[1] The standard correction for the dip of the horizon for heights of eye from various positions on deck, taken from John Hamilton Moore, *The New Practical Navigator*, various editions, Table XIII, to be subtracted from the observed altitude to obtain the true altitude.

[2] British seaman usually adopted a log-line 50 feet in length in conjunction with a 30-second sand-glass. Having a non-standard sand-glass Broughton has had to shorten his log-line, making it equivalent to a length of 49 feet with a 30-second sand-glass.

[3] '210°32′E' in Broughton, *Voyage*, p. 28, which appears to be a transcription error.

[4] The accepted longitude of Point Venus based on Cook's observations is 149°35′45″W; *Tables Requisite to be used with the Nautical Ephemeris*, London, 2nd edition, 1781, Table XX. The Latitude and Longitude of Places, p. 164.

In annotated *Voyage*, p. 28, inserted in pencil at bottom of page:

Long of Point Venus 210°29′38″E
ditto 210°21′8″E per Time k*eepe*r No 1

[5] A transcription error for Oaitipeha Bay (see p. 17 above); the bay to the west of Pointe Tatatua on the south-eastern part of Tahiti; Baie de Tautira, also known as Cook Anchorage.

From the Island of Otaheite towards the Sandwhich [*sic*] Islands in H.M.S. Providence

Friday 11 December 1795

Moderate Breezes & very pleasant Weather: At Noon Point Venus bore E½N 3 or 4 miles and we hove too to hoist our Boats in and to take leave of our Friends. At ½ past 1 PM Pomarre,[1] Iddeah,[2] his Wife with his two Brothers Orupiah[3] & Whyatua[4] & the High Priest Mannimany[5] left us. I made them severally very handsome Presents and we parted excellent Friends & they assur'd me on my return (which they seem'd anxiously to wish) we shou'd have no more Thefts committed and the Guns before Stolen shou'd be returned.[6] As it gave them pleasure I saluted them with 4 Guns & we made Sail to the Northward. We had not a Man unwell & our Decks were full of Hogs & Vegetables for our further Refreshment. Many of the Hogs weigh'd more than 200 pounds.[7] At 4 PM Point Venus bore S53°E 12 miles. At 5 AM we saw the Island of Tethuroa[8] bearing from S74°E to S60°E 4 or 5 Leagues. The Body of Otaheite S20°E and the Island of Eimeo[9] S4°E. The Wind continued moderate from the NE D*irectio*n and at Noon we had no sight of the Land.

Punish'd 4 Marines for neglect of duty & W*illia*m Hurvart Seaman who deserted from the Tents at Otaheite and was bro't back by the Natives. Course N35°W, 63′, Lat*itude* ⊙ 16°35′45″S, Account 16°41′S, Long*itud*e acc*oun*t. 209°47′E, No 1 209°50′17″E. [*in margin*: Variations[10] per No 3 the 11th PM, 4°19′E; Adams 3°07′ East.]

Saturday 12 December 12

Moderate Breezes with Smooth Water, and very pleasant weather, at ½ past 6 PM we cou'd just discern from the Main Yard the Island of Huahine bearing S56°W 14 Leag*ues*. The Wind kept at NE & we carried all our small Sails. Course N21°W, 78′, Lat*itude* ⊙ 15°24′S, Account 15°24′S, Long*itud*e d*itt*o 209°17′E, No 1 209°14′E. [*in margin*: Variations the 12th PM, No 3 6°10′E; Adams 7°53′E. NB very indifferent, no dependence on them.]

Sunday 13 December

Fresh Breezes and fair Weather with some slight Squalls in the Night which brought the Wind to the East. AM muster'd the Ships company & perform'd Divine Service. Serv'd daily Fresh Pork to the Ships company. Course N13°E, 94′, Lat*itude* ⊙ 13°52′42″S, Account 13°46′S, Long*itude* d*itt*o 209°39′E, No 1 209°30′40″East.

[1] Pomare.
[2] 'Itia.
[3] Teri'irere.
[4] Vaetua.
[5] Ha'amanemane.
[6] This incident not mentioned in Broughton, *Voyage*, p. 27.
[7] 'cwt' (hundredweight) in the MS, which is clearly a transcription error for pounds.
[8] Tetiaroa.
[9] Moorea.
[10] Broughton used two methods for obtaining variation, though he only occasionally indicated which method he used. Variation by amplitude is obtained by observing the magnetic bearing of the sun when on the horizon (eastern amplitude) or setting (western amplitude) and then knowing the sun's declination and the ship's latitude, the sun's true bearing can be calculated, the difference being the variation. Variation by azimuth is obtained by observing the magnetic bearing of the sun when it is above the horizon, when its altitude must also be measured in order to calculate its true bearing.

Monday 14 December

Moderate Breezes and fair weather with slight squalls at intervals & the wind from NEbE to East, we carried all our small Sails. At noon Course N4°E, 86′, Lat*itude* ⊙ 12°26′S, account 12°19′S, Long*itude* d*itto* 209°45′E, No 1 [*blank*].

Tuesday 15 December

Fresh Breezes with variable & squally weather attended with some heavy showers of Rain during the Night. In the afternoon we saw several Tropic Birds.[1] The Wind continued Easterly with some Southing but in the Forenoon back'd round to the NE'ward. Lat*itude* Observ'd 10°51′26″S, Account 10°50′S, Long*itude* d*itto* 209°54′E, No 1 209°36′ East. [*in margin*: Variations the 15th AM, No 3 5°15′E.]

Wednesday 16 December

Moderate Breezes & very warm throughout these 24 Hours with smooth water. The wind continued variable from East to NE. At ½ past 9 AM we saw an Island upon our Weather Beam bearing N78°E. At ½ past 11 we tack'd towards it; and it bore from us at Noon East to E19°S just in sight from the Main Top. I estimated its distance to be 5 or 6 Lea*gues* & the South Extreme was the Highest part cover'd with Trees; most probably Cocoanuts from their appearance, being in detach'd Clumps along the Shore & their Bearings will place it in the Latitude of 9°57′S & Long*itude* 209°35′ East, the Center of it,[2] but it must be recollected the bearings were taken with a small Compass from the Masthead the ∠ subtended by a Quadrant. Course N9°W, 59′, Lat*itude*. ⊙ 9°53′13″S, Long*itude* d*itto* 9°54′E, accounted Long*itude* 209°54′E, No 1 209°18′ East. In my last Run this way we must have pass'd this Island in the Night about 12 or 13 leagues to the Westward of it.[3] I named it [*blank*].[4] [*in margin*: Variations the 16th AM, No 3 4°33′E; Adams 5°48′E; Walker 5°37′E; Inspection 5°0′.]

Thursday 17 December

Light Breezes & variable weather with Squalls throughout the afternoon & Night which prevented our approaching the Island. At 1 PM we tack'd and stood to the NE. At 5 PM the Island bore from E½S to SEbE½E about 5 Leagues. It appear'd to be very low & coverd with Trees & if I am right in its estimated distance its length will be about 5 miles in a N*orth* & S*outh* direction. At Midnight we had a heavy Squall from the Eastward which gradually bro't a steady Breeze from the NE after 2 Hours Calm & we Kept our Wind to the Northward. At Noon we had fresh Breezes. Lat*itude* ⊙ 9°00′S, Account 8°56′S, Longitude d*itto* 209°28′E, Course N18°W, 58′.

[1] Red-tailed (*Phaethon rubricauda*) or White-tailed Tropicbirds (*Phaethon lepturus*).

[2] It is actually in 9°57′S, 150°13″W.

[3] Broughton was in the vicinity of this island in the *Chatham* on 7 February 1792.

[4] In his letter of 15 June 1796 to Evan Nepean from Monterey, Broughton said he named it 'the Island of Carolina in complement to Miss Stephens of the Admiralty (see p. 209 below). She was the illegitimate daughter of Sir Philip Stephens, one of the Lords Commissioners of the Admiralty, whose name was in fact Caroline and not Carolina. It was not, however, a new discovery, having been sighted by Quiros in 1606 and named by him San Bernardo, and possibly discovered by Magellan in 1521; Sharp, *Discovery of Pacific Islands*, pp. 5–6, 61–2. It is one of the Lower Line Group of islands, now part of the Republic of Kiribati, and is correctly named Caroline Island on charts and maps.

Friday 18 December

Moderate breezes & fair weather with a Swell from the NE which we have not experienced before this [which] I attributed to our having passed the Low Islands[1] that extend in a Parallel towards the Marquesas. We had not withstanding many Birds about us particularly Boobies[2] & the Wind in the Forenoon continued at NE & we had some strong Squalls from the same quarter. At Noon we had Fresh Breezes. Course N24°W, 83′, Lat*itude* ⊙ 7°44′44″S, Account 7°40′S, Long*itude* d*itto* 208°53′E, No 1 208°27′20″ East.

Saturday 19 December

Moderate breezes & fair weather throughout these 24 Hours the Wind now veering to the East of NE. Several Birds about us as before. This morning by the carelessness of my Servant our Barometer was renderd useless having the Glass Tube broken, a misfortune we could not Remedy. Lat*itude* ⊙ 6°42′S, account 6°34′S, Long*itude* acc*oun*t 208°7′E, Course N36°W, 78′, No 1 207°36′28″E.

Sunday 20 December

Fresh Breezes & fair weather the Wind continued at NEbE throughout PM. Perform'd Divine Service and read the Articles of War.[3] At Noon Lat*itude* ⊙ 4°59′S, Account 4°53′S, Long*itude* d*itto* 207°52′E, Course N8°W,104′, No 1 207°6′E.

Monday 21 December

Fresh Breezes & very clear pleasant weather. The Wind at EbN in general. A large Swell from the Eastern quarter. At Noon we observ'd in the Lat*itud*e of 3°4′5″S, Account 3°3′S, Long*itude* d*itto* 207°52′E, Course North, 115′, No 1 206°53′50″ East.

Tuesday 22 December

Ditto weather without any Variation, the Wind in the Forenoon came to NE. Lat*itude* ⊙ 1°3′36″S, Account 1°3′S, Long*itud*e d*itto* 207°22′E, Course N13°W, 124′, No 1 206°00′ East.

Wednesday 23 December

Fresh Breezes & fair weather with the wind variable from NE to East. Course N7°W, 117′, Lat*itude* ⊙ [0°]52′2″ North, account [0°] 47′N, Long*itude* Acc*ount* 207°09′E, No 1 made 205°48′27″E. Crossed the Line at midnight in the Long*itud*e of 205°50′ East by No 1.

[1] The Tuamotu Archipelago, shown as the 'Low Islands' on Arrowsmith's 'Map of the World on a Globular Projection', published 1 January 1794.

[2] Masked Boobies (*Sula dactylatra*), Red-footed Boobies (*Sula sula*) and Brown boobies (*Sula leucogaster*) are encountered in this area.

[3] The disciplinary code for the Royal Navy. They were read frequently to the ship's company so no one could plead ignorance of them.

Thursday 24 December
Strong trade Wind & very pleasant weather. The Wind from East to EbS. Course N25°E, 126′, Lat*itude* ⊙ 2°45′26″N, account 2°54′S, Long*itude* acc*oun*t 208°2′E, No 1 206°18′E. [*in margin*: Variations the 24th AM, No 3 5°29′E; Adams 6°28′E; Walker 6°38′E; Inspect*ion* 5°0′.]

Friday 25 December
Ditto Winds and weather. Muster'd the Ships Company & serv'd double allowance of Liquor in complement to the Day. Lat*itude* ⊙ 4°41′26″N, acc*oun*t 4°34′N, Longitude by account 208°44′E, Course N20°E, 122′, No 1 206°56′15″E.

[Saturday][1] 26 December
Fresh Breezes & fine weather with the Wind from East to ENE. Course N9°E, 134′, Lat*itude* ⊙ 6°53′N, 6°37′N account, Long*itu*de ditto 209°04′E, No 1 207°27′50″E, NE Current. [*in margin*: Variation the 26th, No 3 5°16′E.]

[Sunday] 27 December
Fresh Breezes & Squally weather at times with Showers of rain with the wind steady from the NE quarter. AM muster'd the Ships Company and perform'd Divine Service. At Noon Hazey weath*er*. Course N18°W, 133′, Lat*itude* ⊙ 8°59′51″N, Acc*oun*t 8°48′N, Long*itu*de 208°23′E, No 1 206°50′E.

[Monday] 28 December
Moderate Breezes & Cloudy weather. Armourer employ'd repairing the necessary Iron work at the Forge. Exercis'd the Seamen at Small Arms. Course N9°W, 109′, Lat*itude* ⊙ 10°45′32″N, Account 10°38′N, Long*itu*de 207°55′E, No 1 206°24′E.

[Tuesday] 29 December
Fair & pleasant weather. PM we set the Lower Rigging up Fore & aft the wind at East & ENE with very smooth water. Lat*itude* ⊙ 12°37′35″N, Acc*oun*t 12°33′N, Long*itu*de 208°03′E, Course N4°E, 112′, No 1 206°22′18″E.

[Wednesday] 30 December
Fresh Breezes & very fine weather with the wind at EbN. We steer'd North. AM the Breeze increas'd with a large Swell. Course North, 149′, Lat*itude* ⊙ 15°06′13″N, Account 15°04′N, Long*itu*de 208°3′E.

[Thursday] 31 December
Strong Breezes & dark cloudy weather prevented our carrying Top gallant Sails. We steer'd from Noon NbW Wind at EbN. Course N10°W, 151′, Lat*itude* ⊙ 17°35′20″N, Account 17°37′N, Long*itu*de 207°36′E, No 1 205°38′E.

[1] From here symbol for the day of the week omitted.

[Friday] 1 January 1796 – Off the Sandwich Islands
Strong Breezes & Squally weather with a large Eastern Swell. We steer'd NW½N & NW till 1 AM when having run near the parallel of the Lat*itude* of Owhyhee[1] South Point we haul'd our Wind & bro't too for the Night under double reef'd Topsails. At 2 AM we saw the Land distinctly bearing WNW and we had a fine moonlight morning. At 4 we bore away under an easy Sail steering West. At 8 the South point bore N72°E 3 or 4 miles. As we approach'd the Western side of the Island we lost the Trade wind & it became variable with Light airs from every direction. At Noon the Extremes of Owyhee bore from N15°W to S81°E. Course N56°W, 121′, Lat*itude* Acc*oun*t 18°59′N, Long*itude* acc*oun*t 206°04′E. No observations. Offshore 2 Lea*gue*s.

[Saturday] 2 January
Calm & cloudy weather continued throughout the afternoon which enabled several Canoes to visit us with Pigs & Vegetables. At Sun set the Extremes of the Land bore from N12°W to S78°E. At 10 PM a Light Air Sprung up from the Land with which we steer'd along shore. At Daylight we were abreast of Karakakooa Bay[2] having experienc'd a strong current setting to the NW. We soon after Tack'd & made Sail for the Bay with a Light Air from the North. At 7 we saw a Sail in the NW quarter. The Natives inform'd us it was an English Brig that had sail'd from the Bay in the Night.[3] Several Canoes laden with Refreshments came off. At Noon we Observ'd in 19°30′N when the North point of the Bay bore of us S64°E about 3 or 4 miles. At Noon Light Airs and Variable.

[Sunday] 3 January
Light airs & variable prevented our making the Bay. I therefore dispatch'd an officer in the Pinnace to gain information respecting Capt*ai*n Vancouver as we understood there was an American Vessel laying there who cou'd give us some intelligence respecting him. At 5 PM the North point of the Bay bore S48°E 6 or 7 miles, the Current setting us to the NW at the rate of 2 miles p*e*r Hour. At ½ past 9 PM the Pinnace returned confirming what we had before heard of Capt*ain* Vancouver, with the Discovery & Chatham being return'd to England, an English Brig having saild from the Bay yesterday who reported to the Master of their having saild from Valparaiso in July or August 1795 for England by way of Cape Horn. We stood to the SSW all night and at 7 AM with the last of the Land wind we stood in Shore. It soon became calm & we continuing drifting to the NW. At Noon we Obs*erve*d in 19°42′N and the Extremes of the Land bore from N30°E to S50°E. Off shore 3 or 6 Leagues. Wash'd below & muster'd the Ship's Comp*any*.

[Monday] 4 January
We had light airs from the westward till midnight when it became calm & variable without any variation throughout the Forenoon. Notwithstanding we had run in for the Land 5 leagues by

[1] Hawaii; Broughton also uses the spelling Owyhee occasionally.
[2] Kealakekua Bay.
[3] It was the schooner *Prince William Henry*, Captain William Wake; Roe, *Journal of Charles Bishop*, p. 135. Broughton subsequently purchased this schooner in Macau on 29 December 1796; see p. 107 below.

the Log we did not seem to have near'd it, and our distance prevented any communication. At Noon we Observ'd in the Lat*itud*e of 19°43′N and the Extremes of Owhyhee bore from S55°E to N36°E. Off shore 5 or 6 Leagues. It remain'd perfectly Calm with Hazey [weather].

[Tuesday] 5 January
Light Airs & Calms alternately throughout the 24 Hours which prevented our approaching the Land. Neither did we appear to have felt any more the influence of the Current. AM Exercis'd the Seamen at Small Arms. Lat*itude* at Noon made 19°36′48″N and the Extremes of Owhyhee bore from N34°E to S54°E. Off shore 5 or 6 Leagues. Having had no communication with the Shore these 3 days We serv'd Salt Provisions for the 1st time these 5 weeks having expended our Otaheitian Pork exclusive of 3 Hogsheads that were salted for Sea Stores so abundant had been our Supplies during our short residence at that Island.

[Wednesday] 6 January
Throughout the 24 Hours it was Calm and Hazey weather. AM we tried the current which set to the SW 7 f*athom*s p*e*r Hour notwithstanding we had felt a contrary Effect during the Night & had drifted considerably to the North. Lat*itude* at Noon was 19°45′N and the Island of Mowee[1] bore N5°E and the N*orth* Extreme of Owhyhee N45°E. Off shore 6 or 7 Leagues. Found the Gammoning[2] of the Bowsprit Stranded.

[Thursday] 7 January
Light Breezes from the Sea carried us close in with the Land by Dark but not near enough to Anchor. The Current still continued and the Calm weather and at Noon we were 8 or 10 Leagues from the Land. At which time it bore from S46°E to N44°E. No observations.

[Friday] 8 January
Light airs and variable weather till Midnight carried us 8 Leagues in Shore when it became Squally with Rain from the NE quarter & we stood along shore to the Southward. At Daylight the wind veer'd to the South when we tack'd expecting to fetch into the Bay but found the current had affected us more strongly than Ever & we were only off the S*outh* West point of Toe yan-yan Bay[3] when we brought to the wind to procure some Refreshments from the Canoes that came off. Lat*itude* at Noon 19°48′N, Karakakooa Bay bore S41°E 8 Leag*ues*.

In Karakakooa Bay

Light airs and Squally weather during the Night. By 8 pm we had run the Distance of Kowrowa Point[4] & it continued calm till midnight. The wind afterwards came from the land

[1] Maui.
[2] Seven or eight turns of a rope-lashing passed alternately over the bowsprit.
[3] Kawaihae Bay on the NW side of the island of Hawaii, Cook's Toe-yah-yah Bay.
[4] Keawekaheka Point, the NW entrance point of Kealakekua Bay.

& we kept plying to Windward. At 7 am the Point of Ko[w]rowa bore E½S two miles & the Wind veering to the North with the assistance of the Boats we anchor'd in the Bay at 10 AM. The Lady Washington American Brig Saluted us with 7 Guns, we ret*urne*d 5. The Ship was surrounded by Canoes bartering Roots & Fruit with Women in the greatest profusion. In the afternoon we moor'd with the Stream Anchor in 19 f*athom*s the Points forming the Bay bearing from S7°W to W3°N. Off shore 150 fathoms.

We had constant rain & variable weather after our arrival. On the 11th We pitch'd our Tents in the Fields adjoining the Morai[1] for the Astronomer to ascertain the rates of the Watches. L*ieutenan*t Young of the Marines with a Corporal and 7 Privates commanded the Party for their Protection. The Ground being Tabooed no Persons were permitted to come within the Limits exclusive of the Priests belonging to the Morai. They cou'd not possess a more retir'd Situation.

Ever since our crossing the Equator the Ship had made from two to Four Inches Water p*e*r Hour. Our occupations here were chiefly to find out the cause & we unstor'd the Holds &c Heeling Ship for that purpose but all our efforts prov'd unsuccessful. At times we made one Inch p*e*r Hour & often less but in general we had only occasion to pump twice in the 24 Hours. We also clear'd the Bread Room & shifted the whole of the Bread to destroy as many of the We[e]vils by Fire &c as lay in our power. Still we cou'd only destroy a part & many remain'd to our mortification for our daily consumption.

On the 18th of January we Saluted Her Britannic Majesty and the Weather became more favorable for the Astronomer's Observations. On the 20th he got equal Altitudes for the first time. Strong Squalls of Wind from the Land in the Night bro't home our Stream Anchor. On the 22d we now moor'd with the Bower about ¼ of a mile from the Shore in 15 f*atho*ms. From the 26th to the 31st Inst the Weather prevented any Observations. But the Evening of that day he had corresponding altitudes. We immediately struck the Tents and with the assistance of Double Canoes (as our Boats cou'd not land) we got everything off from the Shore & Sail'd at 4 AM from the Bay with the Land Wind. During our Anchorage here for three Weeks we experienced constant Land Winds during Night, which gradually died away by 8 AM and during the Day very light Airs & Breezes from the Sea. The Surf seldom permitted our Boats to Land in which we found no inconveniency from the good Harmony that subsisted between us & the Natives who readily furnished us with Canoes that prov'd a most safe conveyance. I had every reason to be satisfied with our reception & the good behaviour of the Inhabitants. No theft of any kind was committed of any consequence, nor any interruption to our Pursuits. We had nearly the whole of the Ships Company on shore to recreate themselves at different times. They were neither molested nor insulted in their Walks but treated with kindness and good will universally.

We were amply supplied for our daily consumption with Hogs by the power Annex'd to a British Seaman from Ta maah maah.[2] This Man (whose name was J*oh*n Young[3]) had been resident here for Six Years and to his good offices and attentions we were much oblig'd in procuring us what we wanted from the Kings possessions. Ta maah maah with all the Chiefs having been absent with 16,000 Men these 8 Months in an Expedition against the Leeward Islands all

[1] Situated on the eastern side of the bay, near the present day village of Napoopoo; Kakooa on Cook's engraved plan of the bay. Broughton uses the Tahitian *morae* instead of the Hawaiian *heiau*.

[2] Kamehameha, the most powerful chief in Hawaii.

[3] John Young (1744–1835), formerly boatswain of the American brig *Eleanora* of New York, was detained and then befriended by Kamehameha, becoming an important figure in Hawaii, where he spent the rest of his life.

of which he had conquered except Atooi.[1] No Hogs in consequence were to be purchas'd the absent Chiefs having tabooed all their property. A Blind Chief whose name was Mahoa[2] was the executive Person under the control of Young & conducted himself very well with respect to our wants for which civility I repaid him & his Wife's by giving him every thing they ask'd for. The Priests at the Morai paid particular attention to our Gentlemen at the Tents and in no Respect were our wants unsatisfied as far as the general power of the People extended.

They all spoke highly in favor of Capt*ain* Vancouver and I was led to believe by the propriety of his conduct & the good Sentiments incalculated into them by some of the Europeans who have & those who do reside upon the Islands that any Vessel may now touch here with safety for Refreshments & according to their means of purchase be amply supplied As every article of European Manufacture is considerably fallen in Value.

The Cattle that were left here by Capt*ain* Vancouver had bred & were in excellent Order & are likely to stock the Island from their being tabooed for Ten Years. Goats they have in large Quantities. I added a Male and Female to their stock & left under the care of Young a breed of Geese & Ducks. The 1st Lieut*enant*[3] also spar'd them his Pidgeons [*sic*]. I think from the care that are taken of them they are very likely to succeed. Some Grape Vines from Port Jackson with other vegetable Seeds were planted & sown during our Stay. Pumpkins & Melons were not very plenty but we had some excellent Cabbage weighing near 2 lbs they were cultivated at some distance & bro't as a present.

On board the Weather was extremely pleasant the Ther*mometer* varying from 74° to 78°. At the Tents on shore was from 79° to 86°.

On the 20th of January No 1 made the Longitude of Karakakooa 203°46′45″[4] which differs only 11′[5] from the true Long*itude* as settled by Capt*ain* King & Mr Bayly.[6] An error only of 11′ in 108 days being the elaps'd time since her rate was settled by Mr Crosley at Port Jackson.

On 31st of January at Noon No 1 was slow for mean time at Karakakooa 14^{h}31′29″·19 and losing 6″·594 p*er* day on mean time.

The Variation on board the Ship mean of 3 Compasses		8°15′ East
D*itt*o on shore at the Tents by Adams large d*itto*		9°12′ East

No 1	203° 47′ 4″	East by Time keepers. Observations by the Astronomer[7]
Pocket 2	203. 48.19	
Box A 56	199. 36.42	
Box E 248	204. 44.36	

[1] Kauai. [2] Not identified.

[3] Zachary Mudge.

[4] Corrected in pencil in annotated *Voyage*, p. 35, to '203°7′0″'.

[5] Corrected in pencil in annotated *Voyage*, p. 35, to '10′'.

[6] The longitude of Kealakekua Bay as established by the mean of various astronomical observations by Bayly was 203°55′6″E; William Bayly, *The Original Astronomical Observations made in the Course of a Voyage to the Northern Pacific Ocean ... in His Majesty's Ships the Resolution and Discovery*, London, 1782, p. 61. The longitude of the Bay was, however, given as 203°0′E on the inset plan on 'Chart of the Sandwich Islands' in Cook and King, *Voyage to the Pacific Ocean*, III, f.p. 1.

[7] In annotated *Voyage*, p. 35, corrected in pencil to:

No 1	203°47′45″	Long 203°57′45″ per Capt Cooke
Pocket 2	203 49 6	
Box A 56	199 37 26	
Box E 248	204 45 18	

The above account will show what the different Watches made the Long*itude* of the Bay by their Port Jackson Rate. Arnold Box time Piece still keeps so inconsistent a Rate as to render it useless.

Long*itude* of Bay 203°57′55″E by King & Bayly
Latitude of Karakakooa Bay mean of 6 Mer*idian* Altitudes of the Sun 19°28′9″·5N
Mean of 13 distances of Aldebaran West of Moon 204°27′30″E

Monday[1] 1 February
With a Light wind from the Land we ran out of the Bay leaving the American Brig at an Anchor to provide herself with Provisions for her Voyage to the NW Coast. By 8 AM it was Calm & we experienc'd a Current setting us to the NW. At Noon we observ'd in 19°31′N when the North point of Karakakooa Bay[2] bore S72°E 5 or 6 miles. Several Canoes came off.

Tuesday 2 February
In the afternoon we had light Breezes from the Sea and we stood along shore to the NW. At Sunset we were nearly abreast of the NW Point of Toe yan-yan Bay offshore 4 miles. The Extreme of Owyhee bore from S38°E to N10°E. At 10 PM the Wind came off the Land and we steer'd NW for the Island of Mowee. At 3 AM when we open'd the passage between Owyhee & Mowee we had strong Breezes & Squally which oblig'd us to double reef the Topsails. At Daylight the *South*West extreme of Mowee N28°W and Tahoorowa[3] N53 °W. We had Strong Squalls with heavy Rain as we run in for Morokinne.[4] At 6 AM we were abreast of it, when we lost the Strong Wind and we had fair weather with baffling Winds from every direction. At Noon the Island of Morokinne N82°E 3 miles. No Observation.

Wednesday 3 February
Variable winds prevented our reaching the SW part of Mowee before Sunset when we came to anchor off the Village of Rahina[5] in 20 fath*oms* Sandy bottom Distant from the Shore about 1 mile. The Extremes of Tahoorowa bore from S33°E. to S8°E, Ranai[6] from S46°W to N83°W, Mowee N14°W to S51°E and Isthmus of Morotoi[7] N60°W. AM we stay'd the Masts[8] and set the Rigging up Fore & Aft & employ'd the Cutter Watering. Throughout the Forenoon we had light airs & Cloudy weather. Mean of 2 altitudes Obs*erve*d Lat*itud*e 20°52′N at ⚓, Long*itud*e No 1 203°13′45″E.

[1] From here symbol for the day of the week inserted once again.
[2] Keawekaheka Point.
[3] Kahoolawe.
[4] Molokine, a barren crescent-shaped small island, in the centre of Alalakeiki Channel between Maui and Kahoolawe.
[5] Lahaina.
[6] Lanai.
[7] Molokai; the isthmus is presumably the centre of the island between the 1,315-metre-high Kamakou at the eastern end of the island and the 427-metre high Mauna Loa at its western end.
[8] Presumably to set up the main and fore-stays.

At Anchor off the Village of Rahina on the SW part of Mowee

As this place appear'd to me favorable for procuring Fruits & Vegetables we remain'd here till the 6th having moderate & Light Winds from every direction & in general experienc'd a strong current setting to the NW. The Village extends about 2 miles along the Shore. Off the western point of it[1] is a small reef which continues in the direction of the Village about 30 Yards from the Shore & affords landing for Canoes or Boats within it. At each Extreme is an excellent Stream of Water but the Westernmost is the best & most convenient for Ships as you may anchor to the Westward of it in a fine Bay with Clear ground carrying 5 *fatho*m close in. We made frequent excursions on shore, the Natives treating us with every civility. We had great pleasure in observing the excellence of their cultivation & the extent of ground appropriated for that purpose. It reminded us of our Native Country Notwithstanding the different productions of Tarro, Sweet potatoes, Melons, Sugar Canes, Gourds & Pumpkins amidst Groves of Bread Fruit Trees and in different situations Clumps of Cocoanut Trees affording Shady Paths in every directions. This being the Residence of the former King (since dead) the whole Village on the arrival of Ta-maah-maah had been destroy'd & the Ground laid waste. It exhibited at present only miserable Hovels for the Shelter of its Inhabitants who were occasional Residents till the Conqueror had made a distribution of the Island among his followers. No Hogs were to be seen & our Supplies were inferior in every respect to my expectations.

Off the Village of Rahina SW part of the Island Mowee

You may Anchor in the Bay abreast the River from 10 to 7 *fatho*ms clear sandy Bottom & may water any Ship with the greatest convenience. The Ground is so clear you may see it in 20 *fathoms* & several Spots within the Reefs in the Bay wou'd admit of hauling the Seine. I shou'd prefer this Anchorage to any amongst the Islands from its local advantages in every Respect. At Daylight on the 6th we got under weigh with a Light Air from the Land the Current setting us to the NW along Shore. By 10 AM we were well in the passage between Morotoi & Mowee[2] and at ½ past we met the Strong Trade setting in from the Eastward & we steer'd West between Ranai & Morotoi. At Noon we had a Fresh gale & clear weather when the West point of Morotoi[3] bore N74°W 10 or 12 miles. AM wash'd and clean'd below.

Sunday 7 February
Fresh Breezes and hazey weather at 2 PM we pass'd the west point of Morotoi & steer'd WbN for Wohahoo.[4] At 5 PM we were abreast of the East Point of the Island[5] and at ½ past 6 we came too an anchor in Whyteetee Bay[6] in 9½ fathoms, Sandy bottom. The Bluff point[7]

[1] Puunoa Point; a reef, on which the sea usually breaks, extends 2 cables west from the point.
[2] Kalohi Channel.
[3] Laau Point.
[4] Oahu.
[5] Makapuu Point.
[6] I.e. Waikiki Bay, to the west of Diamond Head, now Mamala Bay.
[7] Diamond Head.

bearing S80°E and the West point[1] N83°W off shore about 1¼ miles. The next morning I sent the Boat to sound about the Ship & our situation appearing to be good we moor'd with the Bowers NE & SW. Ta maah maah[2] sent off a message to know if he shou'd fire his great Guns. I advis'd him to save his Powder. Soon after he paid us a Visit attended by all his Chiefs with their Helmet Caps & Cloaks dress'd for the occasion. He was in European Clothes with a Beautiful Yellow Feather'd Cloak on that entirely cover'd him. He made me a present of one of their handsome Dresses & made very liberal offers of Supplies & of furnishing us with water in their Canoes without giving our Boats any trouble. The Wind continued blowing very strong from the ENE with clear & pleasant weather. In the morning of the 10th we parted our Small Bower Cable 25 f*atho*ms from the Clinch[3] & hove into ½ a Cable on the Best Bower both Cables were much rubb'd by the Foul ground which tho' to appearances Clean was a mixture of Sand & Coral. The Natives made a Rope fast to the end of our Cable. In the Evening it became more moderate & we purchas'd our Anchor.[4]

Ta maah maah sent me this day 20 Hogs & some Cocoanuts But Roots & Vegetables we were not able to procure. This Even*ing* the King & all our Visitors left us the Taboo having taken Place & no Canoes cou'd afterwards come off. On the 11th in the morning we got under way & run down abreast of a Small Harbor call'd Fair Haven[5] & came too in 16 f*atho*ms Sandy bottom. The Bluff point of Whyteetee bay S75°E West Ex*treme* of Bay N83°W and the entrance of the Harbor N15°E 2 miles. As my only motive for anchoring here was to make a Sketch of the Harbor the Boats were employ'd on that service for three days. It was discover'd by a Mr Brown Commander of the Merchant Ship Butterworth in 1794. On the 1st of January 1795 he lay at anchor in the Harbor with the Jackall & Prince Le Boo, Merchant Vessels, under his direction. The Butterworth he had before sent home. Having an implicit confidence in the Natives (whom he had before assisted in the Wars) the Vessels were left in a defenceless State, the Crews being chiefly on shore Salting Pork &c of which circumstance the Natives being fully acquainted they boarded them with many Canoes killing the two Commanders John Brown[6] and Rob*er*t Gordon & wounding several Others & took possession of the Vessels. On the 12th Jan*uary* they order'd the Vessels out of the Harbor into Whyteetee Bay & the same Night the English Crews being on board each Vessel they surpriz'd the Natives drove them overboard & retook the Vessels. After touching at Owhyee to acquaint the Englishmen with their proceedings they sail'd for China. Many reasons have been given for this unhappy event But the principals concern'd in the Deed having since lost their Lives there is no ascertaining the Truth at the Islands.

The Harbor tho' of Small Extent is safe & convenient with 5 f*atho*m muddy Bottom when you ride within the Spits. It is form'd by an opening thro' the reefs with a clear Channel in a NNE direction & you may carry in 4 f*athoms*; the Breadth of the Channel is not more than a Cable. It blows generally very Fresh out of the Harbor which wou'd render it necessary to warp in as you have not room for Working. A fine stream of Fresh water empty's itself at the Head. It bears S57°E [N57°W] from Whyteetee Bluff [Diamond Head] 5 or 6 miles, the Entrance.*[7]

[1] Barbers Point. [2] Kamehameha; misspelled here and later, Tai maah maah.

[3] Probably where the inboard end of the cable was attached to a ring by a half hitch with the end stopped back to its own part by seizings, but the outboard end of the cable was similarly attached to the anchor ring by a clinch.

[4] Possibly recovering the small bower anchor by means of a purchase.

[5] Honolulu Harbour. Broughton's survey has not survived.

[6] His correct name is William Brown; Lamb, *Vancouver's Voyage*, III, p. 818n.

[7] Inserted at foot of page '*NB In the Valley above the Harbour was the action that acquir'd the Island. Tiana who came with the King & afterwards joined the Wohahoo People was here kill'd & about 300 of his party.'

Having nearly completed our water on the 14th we got under weigh & parted with our Numerous Visitors who came to us from Whyteetee the day after the Taboo. Indeed I had no inducement to stay as we cou'd not collect any Refreshments whatever. The situation of the People on shore was most miserable. They were nearly starving for want of food & to add to their [dis]comfort they were universally infected with the Itch[1] so much so as to prevent our suffering any Woman on board. No Cultivation was to be seen on Shore & of course no prospect of their future Subsistence. All the Kings attention was engross'd by the Vessel the English Carpenters were building for him. She was just planking in the Harbor as we Sail'd & was about 40 Tons Burthen. In this Vessel with other Boats belonging to him he meant to proceed to Atooi to compleat his conquests. Tytaboree[2] & Korokranee his Brother were the principal Chiefs formerly of Wohahoo; they had made their escape as was suppos'd to Atooi & it was this Tiytaboree who by the assistance of Mr Brown's Men had defeated Tayo[3] the King of Atooi & Brother to the former King of Wohahoo, Titeree[4] who died some time before & was succeeded by his Son Tiytaboree (Korokranee being Chief of Mowee). I tried to persuade Ta maah maah from his expedition but to no Effect. This Folly or Ambition call you what you will is most unfortunate for his Subjects As it is almost impossible they can ever Return again to Windward But will carry Itch & Famine to their New Conquests w[h]ere they must Remain. The European Vessels have furnish'd him with such quantities of Musquets & Ammunition with some 3 or 4 pounders for his Vessels & having 16 Europeans with him there is nothing that he does not think himself equal to perform. At present he has some distant thoughts after taking Atooi of proceeding to Bola Bola[5] one of the Society Islands. An American Vessel having bro't 3 Men from thence left them with him & have I suppose pointed out to him the probability of his Succeeding. During our Stay (excepting the Taboo day) he constantly resided on board. His wants were Numerous & great, wishing his Vessel to be Rigg'd & fitted for Sea &c in every Respect. I did my utmost to comply with his wishes, but I fear I only succeeded in part Notwithstanding my endeavours to do every thing in my power to please him & from the best information I cou'd collect it appeared that Ta maah maah profess'd a great Friendship for English Vessels & seem'd to speak in terms of abhorrence of the different Murders that had been committed at the Islands without his Knowledge and of his determination to prevent it in future. He mention'd also that one of the Men who had been accessory in killing Mr Gooch & Li*eutenan*t Hergest had been put to death by his People & another of them had escap'd to Mowee & explain'd that the Men who were murder'd alongside of the Discovery as an atonement for their Guilt were not the real Perpetrators of the Deed But unhappy Men whom the Chief selected for that purpose to satisfy Capt*ain* Vancouver.[6] Tai maah Motoo[7] the Chief who took an American Vessel some Years ago, came off several times & was refused admittance. This Man openly avows he will take the first Vessel in his power. It has been pointed out to him to beware all American Vessels as they will certainly put him to Death. The other Chiefs frequently came on board & were generally employ'd

[1] Probably scabies.

[2] Kalanikupule; Broughton also uses the spelling Tiytaboree and Trytaboree.

[3] Kaeo.

[4] Kahekili.

[5] Bora-Bora.

[6] Kamehameha was not at Oahu in 1792 when the murders occurred, and so his statement is hearsay evidence.

[7] Kameeiamoku.

either [by] themselves or Deputy's bartering Trinkets & Pearls, many I believe were purchas'd but very few of any quality. The King's People did the same without any Scruple[s]. The principal view seem'd to tend towards getting from us every thing they thought useful And their liberality which Capt*ain* Vancouver speaks so handsomely of particularly the King's seems to have been obliterated by his Conquests.

The entrance of the Harbor is situated in the Lat*itude* of	21°18′N.
Longitude p*er* Watch No 1	202° 0′30″E
Variation of the Compass (mean 3 Compasses)	9.40.40 E*ast*
Flows Full & Change at 3^h Rise of Tide 4½ feet	

Wymea[1] Bay Atooi at ⚓ off Harbor

Monday 15 February

At Noon we observ'd in the Lat*itude* of 21°16′45″N. The mean of 2 meridian altitudes of the ☉ in Wyteetee Bay made the Lat*itude* 21°15′35″[2]N and the Long*itude* p*er* No 1 202°3′30″[3] East. I saluted the King[4] with 4 Guns on his leaving us & we made Sail to the Westward with a fine Breeze from the NE Quarter. It was rather singular we did not discover this Harbor when we anchor'd here in 1792 with Capt*ain* Vancouver but the reason I believe was we did not think of looking for one Tho' I remember our noticing a Break in the Reef as we sail'd thro' the Bay.[5] At 6 PM the east point of Wohahoo bore N9°E 3 or 4 miles and we steer'd NWbW½W for Atooi. After running 58 miles upon that course we saw the Island bearing from NWbW to West 3 or 4 Leagues dist*ant*. We haul'd our Wind to the North till 7 AM when the extremes bore from N27°W to S30°W Offshore 3 miles. We made sail to the SSW and after passing some High Rocky Land which rises abruptly from the Sea[6] we steer'd more to the Westward. Just to the NE of this Land appear'd a small Opening that appear'd favorable for ⚓ but open to the Trade Wind, the entrance.[7] At Noon we observ'd in the Lat*itude* of 21°56′18″N in Wymea Bay, the Extremes from S75°E to N65°W Off shore 2 miles. We had 13 f*atho*ms after rounding the Eastern Point which shoals gradually to within ¾ of a mile of the Shore. Soon after Noon we came to anchor in 29 fath*oms* Sandy & muddy Bottom the Extremes of the Bay N75°W to S64°E. The mouth of the River[8] N52°E and our distance from the shore about 1½ miles. The Isl*and* of Oneehow[9] bore N79°W[10] and Tahoora[11] due West.

[1] Waimea.

[2] In annotated *Voyage*, p. 43, corrected in pencil to '33″'.

[3] Ibid, corrected in pencil to '39″'.

[4] Crossed out in pencil and replaced by 'Ta maah-maah'.

[5] Vancouver had in fact been told of its existence. 'The other opening to the eastward [of Pearl Harbor], called by the natives *Honoonoono, Tomohomoho* represented as being much more shallow and a smaller place; this induced me to pass it without examination ...', Lamb, *Vancouver's Voyage*, III, pp. 884–5.

[6] Haupu, a dome-shaped mountain 695 metres in height, the best landmark in the SE part of Kauai.

[7] Nawiliwili Bay.

[8] Waimea River.

[9] Niihau.

[10] 'S79°W' in the MS, which must be a transcription error as Kaula lies SW of Niihau.

[11] Kaula.

We had the misfortune to find the Island at War. A Chief from Wohahoo whose name was Taava[1] had taken up Arms against the Young King Timooree[2] (Son of Tayo)[3] and at present possessed the district about Wymea. Two Europeans of his party came off to us & with their assistance employing the Natives we compleated our watering which we had not been able to do at Wohahoo & receiv'd the next day from Taava a great abundance of Vegetables & 40 Hogs of various Sizes. He came off himself in the afternoon and I made him proper return for his civility.

In the afternoon we perceiv'd several Canoes under Sail coming round the East point of the Bay. These we understood were the opposite party coming to pay us a Visit. Taava immediately left us & the other Canoes return'd on Shore. I sent the 1st Lieut*enant* to meet the Young King[4] & soon after he return'd on board in the Pinnace his Squadron following. In one of them a Swivel was mounted. They had also their Arms with them. He present'd me with a Small feather'd Cloak & remain'd on board all the Night which prevented our having any Visitors from Wymea. Early in the morning of the 18th we got under way with a moderate Land wind which soon died away & was succeeded by a Light Air from the Sea with which we kept plying to Windward. In the Evening the Wind came from the North & we steer'd towards the Island of Oneehow. The Young King[5] & his party now left us seemingly satisfied with their treatment & Presents. He was very anxious to have some Powder But every application during my Residence among the Islands for that Article has been constantly denied. I used my endeavours to make Peace between the contending Parties but without Effect.

The Lat*itude* at Anchor mean of 2 Altitudes on the Meridian 21°56′18″N

Long*itud*e No 1 200°12′48″E.

At Anchor off Oneehow

Friday 19 February

Light Breezes from the NE quarter with very pleasant weather. At Noon we pass'd the SE point of Oneehow[6] in 35 f*atho*ms and observ'd in the Lat*itude* of 21°45′50″N when the SE point bore N77°30′E 2 or 3 miles the extremes to the North N10°W. We steer'd along shore for Yam Bay[7] and at 2 PM we came too in the Bay in 29 f*atho*ms coarse Sandy bottom. The extremes bearing from S25°E to N30°E. The Island of Tahoora S43°W. Orehoua[8] was just shut in with the North Extreme. Our dist*anc*e from the Shore 2 miles. The following day we had Fresh Breezes from the SE quarter And some Canoes came off bartering Yams, Potatoes, Water Melons and Pumpkins. Our Boat in the Evening also came off laden with Roots. The next day we had promises of our wants being fully supplied.

Unfortunately the Weather prevented our receiving any further Supplies. Throughout the Night we had Squally & rainy weather with the Wind at South which occasion'd a great Swell

[1] Not identified.
[2] Amended in pencil to 'Tamoerrie'; Kaumualii.
[3] Amended in pencil to 'Taio'; Kaeo.
[4] Crossed out in pencil and replaced by 'Chief'.
[5] Ibid.
[6] Kawaihoa Point.
[7] Probably Nonopapa Landing situated 5½ miles NNW of Kawaihoa Point.
[8] Lehua.

in the Bay. The Boat went on shore in the morning very early with one of the Europeans who had come from Atooi with us for the purpose of collecting Yams. But nothing being ready, I thought it necessary to get under way & stand out to Sea. The Wind Kept increasing & we had strong Squalls with rain which Split the Main topsail. After Hoisting in the Boats and securing the Sheet Anchor we stood in Shore & bent another Main topsail. At Noon the Extremes of the Island bore from S5°E to N42°E Off shore 2 or 3 miles. The European[1] came off saying everything was ready but the surf was so great that Canoes cou'd not come off without risque. I therefore gave up the idea of again Anchoring as there was no probability of doing it with safety while this Wind continued; and I cou'd not see any chance of its changing. He soon left us in his Canoe with some Recompense for his attention and I fear found great Difficulty in getting on Shore from the violence of the Squalls. This man had been transported to Botany Bay & came from thence in an American Vessel called the Mercury.[2] He deserted from her at this Island & is now of some consequence to Taava whose cause he thought proper to adopt after leaving the Young King Timoree.[3]

The Latitude at Anchor in Yam Bay was 21°51′28″N, mean of 2 sextants.

Longitude d*itt*o p*e*r No 1 199°37′24″ East.

Variation of the Compass 10°54′29″, Mean of Adams & No 3.

NB Walkers Compass wou'd not stand Steady on account of the Ships motion. The others were not so affected.

[1] Not identifed; but perhaps Hughes the European, encountered in Niihau the following July (see p. 59).

[2] The American brig *Mercury*, Captain William Barnet, sailed from Port Jackson on 7 December 1794, taking with permission five persons belonging to the colony; Collins, *Account of the English Colony*, p. 334

[3] Amended in pencil to 'Chief Tamoerrie' (Kaumualii).

CHAPTER 3

From the Hawaiian Islands to Nootka Sound and thence to Monterey

Sail'd towards Nootka Sound.

Monday 22 February
Strong Breezes & hazey weather at ½ past 3 PM we pass'd the NW Point of the Island Orehoua which bore SE 3 or 4 miles. I take my departure from it allowing its Latitude 22°02′N & Long*itude* 199°48′E. We steer'd NE carrying 3 reefs in the Topsails. At 6 PM the Extremes of Atooi bore from N82°E to S45°E. Under the Land we had the Wind moderate & we let the reefs out and steer'd NEbN. In the Forenoon we had Strong Breezes with rain from the SE quarter which prevented any Observations. Serv'd Fresh Pork for the last time to the Ships Company which has been their constant diet since the 1st of December 1795.[1] Our Ships Crew continued very Healthy Having only one Sick Person if we exclude the Venereals which were a numerous Body, tho' slightly infected. Unbent the Sheet & Small B*owe*r Cables. Course N42°E, 110′, Lat*itud*e account 23°24′N, Long*itud*e acc*oun*t 201°8′E, Thermometer [*blank*].

Tuesday 23 February
Strong Breezes & dark cloudy weather with frequent Squalls of rain. We steerd NNE with the Wind at SE and a large Swell from the same quarter. In the afternoon we got the Top gall*ant* Yards up but before midnight the weather oblig'd us to take in the Second Reefs. Course since leaving Orehoua N36°E, 310 [miles], Lat*itude* ⊙ Indiff*erent* 26°15′N, By account 25°56′N, Long*itude* account 202°47′E. Pump'd Ship every 4 Hours.

Wednesday 24 February
Strong Breezes & very Squally weather with heavy showers of rain & very dark cloudy weather with a large Swell from the Eastern quarter. We steer'd NbE. At 7 PM the Wind suddenly shifted to the North & we stood to the East. At Midnight it again shifted to the SE & we steer'd as before with a Strong Wind & Squalls of Rain. [In] the Forenoon the Sun appear'd & we cou'd see distinctly 8 or 10 Leagues around the Horizon. The Observation made the Latitude 28°23′N,. Course N23°E, 140′, Long*itud*e acc*oun*t 203°49′E, Lat*itude* acc*oun*t 28°15′N.

[1] On their arrival in Matavai Bay.

Thursday 25 February

Fresh Breezes & cloudy weather. At Noon we altered our Course to West to look for an Island call'd Don*na* Maria Lajara said to be discover'd by a Spanish Ship call'd the Hercules in 1781 & laid down in Arrowsmiths Chart from the authority of Mr Dalrymple.[1] The center of it is situated in 28°30′N and in the Longitude of 202°30′E. By the Chart it is of considerable Extent in a N*orth* & South direction. The afternoon Sights for the Watch made our Long*itud*e at Noon 204°1′30″E & we cou'd see half a degree to the East. We had a large Swell from that direction & no indication whatever to make us suppose there was Land in that quarter. Capt*ain* Cooke's Track in his return to the Sandwich Islands was in 206°E. In the Evening it became moderate & the weather clear'd away & the Moon rose at 7 PM & shone distinctly thro' the Night So as to enable us seeing Land 4 Leag*ue*s dist*ant*. At 8 PM being in 28°32′N we Steer'd WbS to pursue that Parallel. At 4 AM we haul'd up S*outh* under our Topsails with 3 reefs in them till daylight being within a few miles of its suppos'd situation. No Land being seen at 6 AM we steer'd W½S & made Sail with a Fresh Breeze from the SE & fine Clear weather. At Noon a good Observation made the Lat*itude* 28°34′N. Course since Yesterday N85°W, 146′, Lat*itude* acc*oun*t 28°34′N, Long*itude* d*itt*o 201°4′E. The mean of AM & PM Sights for the Watch deduc'd to Noon made the Longitude 201°12′ East. We still continued looking out for the Land.

Friday 26 February

Fresh Breezes and clear weather we pursued the same Course. At ½ past 3 No 1 made the Long*itud*e 200°44′E. We run till 4 when our Lat*itud*e was 28°37′N & Long*itud*e 200°40′E and we coul'd see distinctly any Land of moderate Elevation 12 or 15 Leagues. We had now Run & seen 5° of Long*itud*e nearly from 200°E to near 205°E in the parallel of 28°30′N. The situation of the Island is of course to the East or West of the above Long*itud*e most probably to the East of 206°E as Capt*ain* Cook pass'd this Parallel in the Long*itud*e of 200°15′E & many other Navigators to the Westward of that Longitude. I therefore did not think it necessary to cross their Tracks & alter'd our course to the North with a fine Breeze at ESE. The Ship made 4 Inches water p*er* Hour. Dark Cloudy threatening weat*he*r in the Forenoon. Course N2°E, 143′, Lat*itude* A*ccount* 30°58′N, Long*itude* 201°10′E. [*in margin*: Variation PM, No 3 10°39′E; Adams 13°1′E.]

Saturday 27 February

Strong Breezes with very squally weather and continual rain we steer'd NEbN the Wind at South. Got up our Winter Topgallant Masts, Jib boom &c. A large Swell from the SE quarter oblig'd us to pump the Ship every 2 Hours. At midnight the Wind shifted to the NW quarter & moderated gradually which brought us fine weather. AM We out all reefs got up the Top gall*ant* Yards & made Sail. Wash'd below & air'd with Stoves. Lat*itude* ☉ 32°17′17″N,

[1] Arrowsmith's 1794, 'Map of the World on a Globular Projection', which was dedicated to Alexander Dalrymple. See also Stommel, *Lost Islands,* p. 67, where it is stated that this island was taken from the Spanish chart captured by Anson. This non-existent island was said to have been named after a ' Spanish woman of that name who had cast herself into the Sea at this place'; E. W. Dahlgren, *Were the Hawaiian Islands Visited by the Spaniards before their Discovery by Captain Cook?,* Stockholm, 1916, p. 101. Also shown as St Maria la Gorda' on Henry Roberts's 'A General Chart exhibiting the Discoveries made by Captn Cook' and as Nuestra Señora de la Gorta, Doña Lajara or Laxara. Sought unsuccessfully by La Pérouse in October and November 1786; Dunmore, *La Pérouse*, I, pp. 202–4.

Account 32°04′N, Long*itude* 202°20′E, Course since last Observation N16°E, 232′, Longitude p*er* No 1 202°0′ East. [*in margin*: Variations AM [per] Az*imuth,* No 3 12°56′E; Adams 14°48′E; Inspection 13°30′.]

Sunday 28 February
Light airs & cloudy weather throughout the 24 Hours with the Wind variable from NNW to NNE. In the forenoon we stood to the NW. Mustere'd the Ships Company's cloathing. Indifferent Observation made the Lat*itude* 32°18′30″N, Account 32°20′N, Long*itude* d*itt*o 202°46′E, No 1 202°33′50″E.

Monday 29 February
Ditto weather with Calms at times. In the morning we saw several Birds chiefly Albatrosses & Sheer Waters. At Noon we had gloomy threatening weather. Course N24°W, 11′, Lat*itude* account 32°27′N, Long*itude* d*itt*o 202°42′E.

Tuesday 1 March
The wind came from the North & we stood to the East till 7 AM. We then tack'd & kept our Wind to the NW. The Breeze in the forenoon increas'd & gradually veer'd round to the NE a Large swell continued from the Eastward. Course N29°W, 84′, Lat*itude* ☉ 33°41′21″N, by account 33°45′N, Long*itude* account 201°55′E, No 1 201°43′22″E. Serv'd Slops to the Ships Comp*any*.

Wednesday 2 March
Fresh Breezes & fine weather. The Wind Kept veering gradually round to the SE'ward. At Noon we Observ'd in 35°16′41″N, By acc*oun*t 35°13′N, Course N24°E, 104′, Long*itud*e acc*oun*t 202°46′E. Wash'd below. No 1 202°51′46″E.

Thursday 3 March
Fresh Breezes & very pleasant weather Wind at SEbE. A heavy Dew fell in the Night. At Noon Lat*itude* ☉ 36°47′46″N, Course N52°E, 148′, Lat*itude* account 36°51′N, Long*itude* d*itt*o 205°10′E, No 1 205°03′E. AM we Observ'd Distances ☉ East of ☽ made the Long*itude* 205°38′E at Noon.

Friday 4 March
Fresh Breezes & hazey weather attended with some rain during the Night & morning. We steer'd NE with the wind at SEbE. Lat*itude* ☉ 38°12′N, Account 38°11′N, Course N53°E, 150′, Long*itud*e acc*oun*t 207°40′E.

Saturday 5 March
Strong Breeze & hazey weather, with Squalls of rain at times. AM we took the 2 reefs in the Topsails & got down the Topgall*an*t Yards. Throughout the Forenoon the Sea appear'd of a Whitish Colour & we saw many Sheer Waters & Grey Peterels. Wind & Course as yesterday, Lat*itude* account 39°47′N, Long*itud*e account 210°28′E, Course N53°E, 163 miles.

Sunday 6 March

Fresh gales & hazey weather with frequent rain. PM we took the 3d reefs in the Topsails & continued our Course to the NE. Winds at SE. Course N56°E, 168′, Lat*itud*e account 41°19′N, Long*itude* d*itt*o 213°28′E, Ther*mometer* 61°.

Monday 7 March

Fresh gales with thick Hazey weather & incessant rain. The wind veer'd more Southerly in the Night and in the Forenoon it became moderate with Foggy weather. At Noon it was nearly Calm. Course N61°E, 152′, Lat*itude* account 42°33′N, Long*itude* D*itt*o 216°25′E.

Tuesday 8 March

Foggy weather with rain continued. In the Evening the Wind came from the NW & gradually veer'd round to the North & we stood to the Eastward. The Forenoon was dry cold weather the Sun not visible for the Haze. Ther*mometer* upon Deck stood at 45°. Lat*itude* acc*oun*t 42°47′N, Long*itude* d*itt*o 218°30′E.

Wednesday 9 March

D*itt*o weather. Let the 3d reefs out of the Topsails. The wind veer'd to the NE and at 9 PM we tack'd & Stood to the NW. The Forenoon was moderate with the same Fog & we let the 2d Reefs out. No Observation. Lat*itude* account 43°29′N, Long*itude* d*itt*o 219°2′E, Course N29°E, 48 miles.

Thursday 10 March

Moderate Breezes with a constant Fog & Smooth Water. The Wind kept veering round to the SE. At noon it clear'd up a little But not sufficiently for any Observation. Lat*itude* account 44°48′N, Long*itud*e d*itt*o 220°50′E, Course NE, 111 miles. Aired with Fires & serv'd Sour Krout, Portable Soup and Vinegar to the People.

Friday 11 March

Moderate Breezes with Foggy weather & the Wind at South we steer'd NE & carried all our Small Sails. At Noon an indifferent Observation made the Lat*itude* 45°51′N, Account 45°33′N, Long*itud*e d*itt*o 222°50′E, Course N63°E, 100′ without correcting it for the difference in the Observation which in 6 days is 18′ to the North of account, No 1 from afternoon Sights made the Long*itud*e 222°19′E. Constant Fires between Decks on account of the Dampness of the Weather. Thermometer upon Deck 52°.

Saturday 12 March

Ditto wind & weather with a long Swell from the Southward. Course N63°E, 107 miles, Lat*itude* account 46°39′N, Long*itude* d*itt*o 225°9′ East.

Sunday 13 March

Moderate Breezes with foggy & hazey weather the Wind shifted to the SW & afterwards to NW. In the Forenoon it veer'd round to the South & continued a fine breeze with a clouded atmosphere. An indifferent Observation made 47°17′N, Account 47°20′N, Long*itud*e d*itt*o 227°07′E, Course N63°E, 90′. Muster'd the Ships Company as customary on Sundays.

Monday 14 March

Fresh Breezes with dark cloudy weather wind from the South. In the Evening it increas'd & we double reef'd the Topsails. The Night was dark and gloomy with rain. AM the Wind suddenly shifted round to the West and brought with it clear & pleasant weather. At Noon we observ'd in the Lat*itude* of 48°12′30″N, By account 48°25′N, Long*itude* by [account] 230°27′E, No 1 made the Long*itud*e 230°2′45″. Nootka at Noon N58°E, 51 Leag*ue*s.

Tuesday 15 March

Fresh Breezes and Cloudy weather. Our accounted Latitude not agreeing with the Observation we steer'd NE½N with a fine Breeze from the Westward. PM we had Squally weather & took 2 reefs in the Topsails & during the Night tried for Soundings with 100 f*ath*-*o*ms of Line. At Daylight we made Sail and at 8 AM we saw the Land about Nootka bearing NNE. The Wind gradually moderated towards Noon When we Observ'd in 49°9′42″N, Lat*itude* acc*oun*t 49°22′N, Long*itude* d*itt*o 233°17′E. Point Breakers[1] N8°W 4 Leag*ues* and the Extremes of Land from N28°W to N58°E. Port San Raphael[2] N35°E.

In the Entrance of Nootka Sound

Light Breezes and Cloudy weather with slight Showers of Hail and a very large Swell from the N'eastward. At ½ past 3 the South point of Point Breakers bore ENE and we had made a North course good having run about 12 miles. At 4 it became Calm & we had Sound*ing*s from 20 to 17 fathoms. Off shore about 3 or 4 miles. In the Evening we had Light airs from every direction of which we took the advantage of standing in for the Sound. At ½ past 9 the wind fix'd at NNE & we immediately anchor'd in 34 f*athom*s soft bottom Point Breakers bore S37°E and Breakers point[3] N78°W, Nootka N7°W 3 or 4 miles. In the Forenoon we had Fresh Breezes from the North with Showers of Hail, Sleet & Snow. The whole Country was cover'd with Snow presenting a most dreary Aspect. I sent an Officer[4] into the Cove for intelligence who return'd at Noon reporting that there were no Vessels there & that the spot where formerly the Spanish Settlement was fix'd was now Occupied by an extensive Indian Village.

[1] Estevan Point. Cook's name failed to supersede the earlier Spanish name given by Juan Pérez in 1774; John T. Walbran, *British Columbia Coast Names*, Vancouver, 1991, pp. 172–4.

[2] Sidney Inlet, named Port San Raphael on nineteenth-century maps; its SW entrance point is named Punta San Raphael on Vancouver's chart.

[3] Bajo Point. Broughton's confusion is understandable. On the engraved chart in Cook's *Voyage* Cook's name Breakers Point has been transferred across the entrance to Nootka Sound to the SW corner of Nootka Island to Bajo Point, a name given by Malaspina in 1791; Walbran, *British Columbia Coast Names*, p. 27.

[4] Zachary Mudge; see Mudge's letter to John Stuart, first Marquess of Bute, pp. 240–50.

Ther*mometer* at 42° upon deck. In the afternoon of the 17th the Wind suddenly Shifted to the Westward and we got under way. Maquinna the Chief of Nootka paid us a Visit. As we drew near the Sound the Wind died away. At 10 PM we Anchor'd in 60 fathoms soft bottom, the Cove of Nootka bear*ing* S20°W 2 miles. The Forenoon was very pleasant weather. At Noon the ☉ Alt*itude* was 39°3′30″ which made our Latitude 49°36′10″N. The Wind came in from the Sea & we again weigh'd & made Sail up the Sound for the Harbor of Maweena.[1] Maquinna & Clupananutch (another Chief) bro't me several Letters dated in March 1795 which inform'd me of Capt*ain* Vancouver having sail'd from Monterry the 1st Dec*embe*r 1794 for England & of the Spaniards having deliver'd the Port of Nootka &c to L*ieutenan*t Pearce of the Marines Agreeable to the mode pointed out for its Restitution by the two Courts. A Letter from the Spanish Officer (Brigadier Alava) inform'd me of their sailing from here March 1795.[2]

In the Harbor of Maweena Nootka Sound

In the Evening we had Calm weather which afforded us good Shelter between a Small Islot & the Nootka Shore, where we steadied the Ship with Hawsers from each Shore, the Ship laying in 7 *fathoms*. The next day having examin'd the Harbor which lay within us & finding the Beach convenient for laying the Ship on Shore, we Shifted our situation between two Small Islots & the Beach, which afford excellent Anchorage, & we had 3½ *fathom*s at low water, the ship being perfectly protected from all weather.

The 19th & 20th we had fair weather & we unbent Sails & Struck the Yards and Topmasts. The next day we had continual rain, but no wind. On the 22d we had again favorable weather and we fix'd our Tents on the Shore abreast of the Ship & commenc'd clearing the Holds, Sending the Provisions &c on shore to the Tents with an Establish'd Guard to protect them. In the Course of the Week we sent all our Provisions, Anchors, Cables & 14 Guns on Shore. The Carpenters having compleated a Stage of a convenient Size. On Sunday we read the Articles of War and punished W. Julford Marine for Drunkenness. After Dinner the People had leave to go on Shore.

The 29th Commenced with rainy weather & blowing Fresh from the SE quarter. On the 31st it moderated & having bro't the Ship to an even Keel drawing 13 feet we haul'd on shore with the morning's Tide. The Ship settl'd so much in the Bed of Mud we coul'd not reach lower than the Eleven Foot Mark. We repair'd the defective Copper and in the Evening we

[1] Marvinas Bay, situated on the west side of Cook Channel about 4 or 5 miles north of Friendly Cove.

[2] Lieutenant Thomas Pearce of the Marines had been appointed British Commissioner to implement the Third Nootka Convention in place of Vancouver. He left England in the late summer of 1794 and travelled to Nootka by way of Cadiz, Havana, and Vera Cruz on board Spanish vessels, from where he travelled overland to Mexico City, which he reached on 30 November. He sailed from San Blas on 13 January 1795 on board the Spanish frigate *Activa* to Monterey, where he was joined by Brigadier General José Manuel de Alava, the Spanish Commissioner, reaching Nootka on 16 March. On 25 April 1794, Pearce wrote to Philip Stephens, Secretary to the Admiralty, informing him of his proceedings at Nootka; copy in HO 28/20, ff. 192–3. Pearce also wrote to Secretary of State, the Duke of Portland, on the same date, which was published in the *London Gazette*, no. 13813 (12–15 September 1795). Since Pearce had left England before the *Providence* sailed, he was able to inform Alava that she was bound for Nootka, enabling them both to leave letters with Maquinna for Broughton; Cook, *Flood Tide*, pp. 421–3, 426, and Lamb, *Vancouver's Voyage*, I, pp. 186–7. Pearce had previously served with William Bligh in the *Providence*, during the latter's second breadfruit voyage, 1791–3. He was promoted to captain 9 May 1795 and placed on half-pay in 1801.

Haul'd off to Lighten the Ship more by sending some quantity of Iron & Shingle Ballast out of her. During the time we lay on Shore the Ship only made 2 Inches *per* Hour & when we pump'd her out in the Stream some Grass found its way thro' the Leak to the Limbers.[1] From feeling the Water running in under the Floor Timbers in the after Hold we had reason to suppose the Leak proceeded from the Garboard streak[2] or Seam, and during the time we lay on shore the Mud or some weedy particles might have been introduced into it which would account for the Ship not making so much water.

In the morning of the 1st of April we again haul'd on Shore but to no effect as the Ship settled so much in the Mud we coul'd not get within 9 feet of her bottom. The afternoon was rainy weather & in the Evening we haul'd off to our Moorings. The next Day was continual heavy rain which prevented our doing anything. The Ship in the Forenoon for 12 Hours only made 16 Inches water.

Sunday the 3rd of April was a fine Sunshining Day & after Mustering Everybody had permission to recreate themselves on Shore. Accompanied by some of the Officers I paid Maquinna a visit at the Cove, who return'd with us to Dinner. In the Night we had again rainy weather which continued incessantly for two days with very little wind. On the 7th we had again fair weather & we employ'd the People in light[e]ning the Ship & bringing her by the Head. Having reduc'd her Draught of Water to Ten feet & a half we haul'd on the Beach & heel'd to Starboard. The Tide left her within two feet of her Keel but not far enough forward to admit our finding the Leak. In the Night we haul'd off and the next day we again haul'd on Shore & haul'd her the contrary way. The Tide fell near Eleven Feet but we cou'd find nothing defective except some few Sheets of Copper which we Shifted. We hove off at Midnight.

Sunday the 10th was very pleasant weather & after mustering every body had leave to go on Shore. We made a Party & visited the Sound to the Eastward[3] returning on board to Dinner. In the afternoon some of the Gentlemen Visited Nootka & saw a Vessel in the Offing. In the morning I sent an Officer to gain some intelligence respecting her. The Boat return'd in the Forenoon with the Lady Washington Brig. She had been out 31 days from the Sandwich Islands & the cause of her arrival here was to repair her Leaks. Her coming in gave us great satisfaction as we might render ourselves mutually assistant to each other. Fine weather continued till Thursday and we were variously employ'd preparing the Ship for Heaving down as we cou'd not get at the Leak by any other means. The 14th & 15th we had a great deal of rain with a Strong Gale from the SE. Saturday again bro't us fine Weather and the Lady Washington came alongside. We put into [her] all our Spirits, Slops & perishable Stores & fitted our Capstan in board of her. Sunday Morning the Carpenters compleated a Wharf Ninety Feet Long and Twelve Feet in height the outer part From which we suspended a Stage Forty feel long for the conveniency of relading the Ship. At low water Spring Tides we lay in 10 Feet. Sunday afternoon the Ships Company as usual had leave to go onshore. The Evening set in with continual rain and on Monday it still continued blowing a hard Gale of Wind with the same weather. The Wind moderated the next day but the rain still continued. The 20th we had fair weather which enabled us to entirely clear the Ship of the Coals & Shingle Ballast.

The 21st was continual rain & we were variously employ'd preparing the Ship for Heaving down. The same weather continued to the Afternoon of the 23rd when it was fair & we sent

[1] Channels running fore and aft either side of the keelson, through which water drains off to the pump well.

[2] The garboard-strake or sand-streak, the first range of planks laid upon a ship's bottom.

[3] Tlupana Inlet.

our Powder on Shore to the Island where the Observatory was fix'd. We experienced in the morning a Heavy fall of Snow, Hail & Sleet. On Sunday the Ship was ready for heaving down the Fore & Mainmast properly secur'd with Shores & Out riggers 24 feet &c. In the morning the Lady Washington having been prepar'd for this purpose haul'd alongside & we rove the Purchase Falls which were 6 Inch Hawsers, our Blocks 24 Inch Treble & Double. At 10 AM we hove out the Larboard side as low as the Garboard Streak & found 14 feet of the False Keel carried away; and plugged up a Bolt hole in the Streak from whence we suppos'd our Leak to have proceeded. As the Carpenters cou'd not repair the Keel in the course of the Day we soon after Noon righted the Ship. The next day we had very unfavorable weather with Heavy & constant rain. The Main Shore appearing to [be] weak we got another up and in the interval the Carpenters were preparing for remedying our defects. On Tuesday we had fine weather and by 8 AM we had hove the Ships Keel out & repair'd the False Keel and also drove a New Bolt into the Hole which we suppos'd to have occasion'd our Leak. From the opinion of the Carpenter it seem'd that no Bolt had ever been drove thro' the planking to the Floor Timber as the boring auger hole remain'd perfect & no appearance of any decay'd Iron; neither was it large enough to mistake it for a Trunnel.[1] The Copper which cover'd it being very thin had crack'd round the Hole & by that means let in the Water which part (as before mention'd) we had frequently found while in Karakakoa Bay had been choak'd up, & after laying on Shore here we found Grass to find its way thro' it; and to be otherways so fill'd up as to admit very little Water for several Days. It was also exactly in the same place the Carpenter had suppos'd on examining the Limbers & from whence we cou'd feel the Water coming in on our first clearing the after Hold. Indeed there was no other part of her bottom we had reason to suspect being bad. The Copper looking very well though in some parts much Worn. Another Bolt below the Fore Chains was decay'd thro' the plankings; but the remaining part indicated its being decay'd, or corroded by the Copper. Before Noon we righted the Ship & Kept our 4 Pumps continually working for 3 Hours besides bailing with Buckets to free her from the Water she made thro' her Upper Works, while laying down on a Careen. Her Draught of Water was 9^F 6^I Forward and 10^F 3^I Abaft. Every thing in the Ship was 15 Chaldron[2] of Coals & about One Ton of Bread in the Bread Room when the Ship was Hove out. In the afternoon we took in the Stage from the Wharf & clean'd the Ship throughout & dry'd the Store Rooms & Magazine with Fires.

The next day we had moderate and cloudy weather & we were employ'd getting on board the Iron Ballast & Shingle (which the Beach afforded) for compleating the after Hold to receive our Spirits &c from the Lady Washington. On Thursday it blew a Fresh Gale from the SE with rain which much impeded our Operations. We notwithstanding continued receiving the Iron Ballast & by Friday Noon compleated our Coals & in the Afternoon receiv'd part of our Spirits. Slight rain still continued with moderate Breezes from the SE. The next day we clear'd the Lady Washington & she haul'd from us. Sunday the 1st of May commenced with fine weather. After mustering in Divisions the People had their usual leave to go on Shore. We made an excursion to Ship Cove[3] the place where Capt*ai*n Cook remain'd on his first entering the Sound during the Month of April. (We seem to have experienced similar Weather or rather worse.) No vestiges remain'd to discover that any Ship had ever been there.

[1] An archaic form of tree-nail, a long cylindrical oak or other hard wood pin, driven through the planks or timbers of a vessel to connect her various parts.

[2] A coal measure equal to 36 bushels.

[3] On the SE side of Bligh Island.

Throughout the Week we had a continuation of the finest weather with constant Land & Sea Breezes. Maquinna supplied us liberally with Venison & the Warm weather enabled us to make very good Spruce Beer for the Ship's Company, They had also Fish & Onions in great plenty. On Saturday we had compleated the Stowage of our Iron & Shingle Ballast & all our Water exclusive of a small quantity which we kept out to benefit the Cooks. Most of our Provisions we also received & got every thing in and the Lower Yards across. The Carpenters were employ'd Sheathing the Lady Washington. Our other Artificers variously employ'd. It was high Water at 00^h on the change of the ☽ & the Tide was 10 feet 3 In*ches*. The three following days the tide ebbed lower than we had ever seen it leaving the N*orth* & west passages between the Small Isles dry. It was the third after the Change 12 feet 6 In*ches* perpendicularly the day tide & the Night Tides were still higher.[1] Sunday we made a party up the Sound towards Tahsis returning late at Night. The Ships Company after mustering had leave as usual to amuse themselves on Shore. The Ship was wash'd and air'd with Fires. And for the first time since heaving down we pump'd the Ship after starting the Water which completed the Ground Tier.[2]

Fair and pleasant Weather enabled us to get every thing on board in the Course of the Week. On Thursday we bent Sails and got the Guns off and on Saturday we struck the Tents. On Sunday we hung the Rudders & hove up our Anchors & prepar'd for Sailing the next morning. The Lady Washington having been new Sheathed by our Carpenters & repair'd in other respects Sail'd in the morning. I also spared them a Kedge Anchor & Hawser for consideration of their Distress & the advantage we experienc'd in Heaving down by her.

Monday the 16th of May we warp'd out of the Harbor but the Land Wind failing us we were oblig'd to come to an Anchor in the fair way & steady the Ship with the Stream Anchor. Fair weather still continued with Strong Sea Breezes & we were detaind by them till Friday Morning when a Light air from the Land carried us down to Friendly Cove Off which we came too in 60 f*atho*ms the Point of Nootka bearing S36°W 1 mile. The Lady Washington having sail'd in the Morning was still in sight in the Offing.

On Saturday with a Light air from the Land we took our farewell of the Sound. As we open'd Breakers point[3] the Wind came in from the Sea & we steer'd to the SE'ward. At Noon we Obs*erve*d in the Latitude of 49°19′N and Point Breakers[4] bore N33°E 3 or 4 miles. The Point of Nootka Cove on with a Woody point to the NW of P*oin*t Breakers N28°W 5 Leagues. Fine pleasant weather. Mer*idian* Alt*itude* ☉ 60°45′ & this estimation from the point I conceive to be Point Breakers will place it in the Latitude of 49°21′35″N, Mean of 2 Sextants. Mr Crosley the Astronomer made the Lat*itud*e of the Observatory 49°39′39″·7N, Longitude from 90 Distances on each side of the Moon 233°25′11″ East.[5]

[1] The range of tide in Nootka Sound is indeed in excess of 10 feet, with a large diurnal variation, with the night tides much higher than the day tides.

[2] The lowest water-casks in the hold before the introduction of iron tanks.

[3] Bajo Point, the SW corner of Nootka Island.

[4] Estevan Point.

[5] In annotated *Voyage*, p. 55, corrected in pencil to:

Lat 49°39′36·1″ N per mean of 7 mer*idian* alt*itude*s of the ☉

Long	232 46 5	E by Time keeper No 1
	233 1 10	per d*itt*o No 2
	233 7 20	per d*itt*o No 56
	236 6 23	per d*itt*o No 248
	233 24 33	per mean of 36 sets of Lunars

From Nootka towards Monterrey Bay

Sunday 22 May

Fair & pleasant weather we Steer'd EbS & East till 8 PM. At 4 the entrance of Port Clioquat[1] bore of us N65°E 4 or 5 miles & 2 [canoes] Came off from Wicananish[2] pressing us strongly to pay him a Visit. At 8 the Extremes of the Land bore from N73°W to S30°E Off shore 4 miles. We steer'd SEbE during the Night and at Daylight we haul'd in for the Land which bore at 6 AM from N60°W to N75°E Land about Cape Claaset[3] S60°E. We Steer'd East & SE till Noon when Tatouche Island[4] bore N86°E, Island up the Strait call'd Nunez de Gaona[5] N81°E, Extremes of the Land from N55°W to the Frions[6] bearing SE, Cape Claasset was on with Isl*and* & distant about 6 miles. A good Observation made the Latitude mean of 2 Sextants 48°26′N, ☉ Alt*itude* 61°50′30″, Lat*itude* of the Cape will be 40°24′N and the Longitude [*blank*].

Monday 23 May[7]

Fair weather & the Ship surrounded with 100 Canoes laden with Halibut & Cod; they had been fishing upon a Bank that bears about WbN 3 or 4 Leag*ues* from Tatouche Island.[8] We sounded at Noon but cou'd get no bottom with 100 f*atho*ms. By having less Swell at that time I rather suppose we had pass'd over the Southern part of it as several Canoes still remain'd fishing to the North of us. I run in 3 miles & hoisted the Boat out & haul'd our Wind to the Southward but the Flood Tide setting in very Strong I found it necessary to bear up between Tatouche Isle and Duncan's Rock[9] on which the Sea broke with great Violence it was just visible. But the Tide running very Strong caus'd overfalls throughout the passage. The Boat got 13 f*atho*ms within 50 feet of the Rock but further from it no bottom with 30 fathoms. This rock is in one bearing with the Centre of Tatouche Island & Cape Claasset N36°W & S36°E about 1½ miles and bears from the Island of Nunez Gaona S80°E about 6 miles or very little more.

[1] Clayoquot Sound.

[2] Wickananish, chief at Clayoquot.

[3] Cape Flattery, the SW entrance point of Juan de Fuca Strait.

[4] Tatoosh Island, off Cape Flattery.

[5] Waadah Island, in the east part of the entrance to Neah Bay.

[6] Los Frailes or Frayles, wrongly transcribed as Frions by Broughton from a Spanish MS chart drawn up jointly by Hezeta, Bodega and Mourelle after the 1775 voyages of the *Santiago* and *Sonora*, and probably presented to Broughton in 1792 by Bodega. A copy of this chart is reproduced as Plate XXXIV in Henry R. Wagner, *Cartography of the Northwest Coast of America to 1800*, 2 vols, Berkeley, 1937, I, between pp. 176 & 177, where Los Frailes are shewn as two distinct islands in about 48°N, so named apparently in honour of the two Franciscan friars who accompanied the 1775 expedition. They are identified by Wagner in *Cartography*, II, p. 456 as James Island and Quilluyute Needle in 47°54′N, but from Broughton's account they can now be identified with some confidence as the two 58- and 60-metre high Bodelteh Islets, the largest of the Flattery Rocks in 48°12′N. However, as Broughton passed these islands on 28 May his bearings suggest that he considered the name applied to all the islands and rocks from Cape Alava in 48°10′N to Point of the Arches 5 miles farther north. We are grateful to Herbert K. Beals in helping to identify Los Frailes. For the voyages of the *Santiago* and *Sonora* see Herbert K. Beals, *For Honor & Country: The Diary of Bruno de Hezeta*, Portland, Oreg., 1985.

[7] Omitted 'At Anchor in the Bay of Nunez Gaona Juan de Fuca' as the statement is premature!

[8] Swiftsure Bank, on which there are several depth of less than 55 metres (30 fathoms), is situated about 15 miles NW of Cape Flattery.

[9] Broughton was clearly consulting 'Sketch of the Entrance of the Strait of Juan de Fuca ... by Charles Duncan, Master in the Royal-Navy, 15th August 1888', published by A. Dalrymple, 14 January 1790. The rock is Fuca Pillar, named Pinnacle Rock on Duncan's sketch.

In the Bay of Nunez Gaona[1] Straits of Juan de Fuca

We pass'd within a mile of the Claasset shore and at 4 PM came too an Anchor in the entrance of the Bay in 12 fathoms, the Western Ex*treme* towards the Cape bore S80°W and the North point of the Island which forms the Bay N76°E, Off shore nearly a mile. (It is in this Bay a Spanish Ship remain'd 4 months in the Year 1792).[2] It affords good Anchorage & plenty of Wood & Water with an extensive Beach for hauling the Seine. Off the Island are several reefs of rocks that shew themselves by the Weeds & there is a good passage between it and the Main carrying 5½ fathoms. The NW part of the Island is in the Latitude of 48°22′43″N & Longitude 235°15′45″E, Variation p*er* amplitude 22°34′ East and bears from Tatouche Island S75°W 6 or 7 miles. Duncan's Rock being kept well on with the Claasset Shore and the Island open with it will carry you up clear to the Northward of it when you may enter the Bay with Safety and Anchor where you please. It flows out at full & Change at 1^h 30′ Rise & Fall of Tide about 10 feet. There is little or no Tide in the Bay, but without it runs strong & causes a ripling which breaks when it blows Fresh.

Tuesday 24 May

Fresh Breezes & very pleasant weather. The Forenoon was Foggy & little wind. In the Morning we weigh'd with the Ebb and at Noon Tatouche Island bore S30°E 2 or 3 Leagues. No Observ*atio*n.

Wednesday 25 May

Thick hazey weather with Calms & light Airs we Sounded 70 f*atho*ms Rocky bottom, our Head to the NW'ward. At 3 we had 47 f*atho*ms & at 4 33 f*athom*s when Tatouche Island bore S5°E 3 or 4 Leag*ue*s & the Ex*treme* towards Berkeleys Sound N61°W. At ½ past 5 PM we stood towards the Island which at 6 PM bore S5°W. At 8 it bore S7°W, Off the N*orth* Shore 3 or 4 miles. In the Night approaching the Island we had no ground with 100 fathoms. It continued with Light Airs & Calms throughout. At 4 AM Tatouche bore S25°E 3 Leag*ue*s & we had 65 f*atho*ms, 56, 53 & 37 to 8 [fathoms] when Cape Claasset bore S43°E, N*orth* Extreme N58°W. At Noon Calm & thick wea*the*r 3 or 4 miles from the N*orth* Shore.

Thursday 26 May

Calm and thick Hazey weather Soundings from 37 to 27 f*atho*ms rocky & Stoney bottom till 5 PM when a light breeze from the SW D*irectio*n carried us to the SE'wards 9 miles when the Island of Nunez Gaona bore S85°E, Cape Claasset S65°W and Tatouche Isl*an*d S77°E 3 miles and we had no Soundings. In standing over when the Cape bore SSE½E the Pinnacle Rock without it appear'd very remarkable. At 8 we tacked & having run 4½ miles NWbN we had 43 f*athom*s coarse gravel. We run over 3 miles upon a NbW½[W] Course when it was Calm our Soundings gradually decreasing to 37 & 27 f*atho*ms fine Sand & coarse Sand but as we shoald

[1] Neah Bay, just inside the SW entrance to Juan de Fuca Strait.

[2] While he gives no reason for visiting the bay, Broughton probably wanted to confirm that the Spanish settlement, which had been established there in 1792, had indeed been abandoned.

our water we lost the coarse ground. It continued calm till 10 AM with very thick foggy weather. We cou'd hear the Surf break upon the North Shore but cou'd not see the Land. It was I suppos'd 2 or 3 Miles & we had 26 *f*a*th*oms. We kept our Wind SbW with the Westerly wind and at Noon having run 6 miles we had 30 fathoms rocky bottom with a very large Swell upon the Bank.[1] At Noon the same Thick foggy weather.

Friday 27 May

Fresh Breezes with very foggy weather. We continued our Course NW½W 5 miles further where we wore & stood to Northward having 65 fathom. We run across the Bank upon a NW course 10 miles having 55, 55, 30, 40, 30 & 28 *fathoms* rocky & gravelly bottom. At 5 PM the Fog clearing we found ourselves close over upon the North Shore at least about 3 miles from it when the N*orth* Ex*treme* bore N72°W & Cape Claasset S50°E 11 or 12 Miles. We tack'd at the same time but the Breeze left us with the Fog & we only made 2 miles to the SSE when it became Calm & our Soundings gradually increas'd to 34 *fathoms*. It was calm throughout the Night and our distance from the N*orth* Shore never exceeded 5 miles; we deepened our water to 40 *fathoms* which continued for several Hours when drifting a little further out we shoal'd our water very gradually to 30 fathoms. At 8 PM the Cape bore S46°E & the N*orth* Shore towards Berkeley Sound N64°W. At 8 AM the Cape boreS56°E and the Northern Extreme N65°W. At ½ past 9 we were 5 or 6 miles from the Shore with the same Bearings we steer'd SSE 5 miles till Noon increasing our Soundings from 30, 30, to 35, 38 & 42 *fathoms* Small Stones, the former & the latter Shells intermix'd with Stones. At Noon the Observation made the Lat*itude* 48°32′10″N, the Meridional Alt*itude* being 62°38′, Cape Classet bore S70°E about 4 Lea*gues*, Island of Nunez Gaona S82½°E, Extreme of the Land towards Berkeley Sound N62°W. Wash'd between Decks.

Saturday 28 May

Light Breezes & hazey weather. We steer'd SSE with the Wind from the Westward having run 5 miles [in] 42 & 53 *f*a*th*oms Stoney bottom. Having run 1½ further, we had no bottom with 80 & 90 *f*a*th*oms of Line when the Island of Nunez Gaona bore N82°E, part of it shut in with the Claasset Shore. When we had 53 *f*a*th*oms it bore due East which appears to be the extreme point of the Bank. By keeping the Island open with the Claasset shore is a good mark of being upon the Bank & you may carry Soun*dings* close home to the North Shore. We had no Soundings till 9 PM, the Frions[2] being East of us 3 or 4 miles [and] we had 33 *fathoms* Sandy bottom. At 8 PM having run 21 miles from Noon on a SSE course Cape Claasset bore of us N13°E, Tatouche Island N13°E, the Isthmus in Flattery Bay[3] which is not 2 miles through to the Bay of Nunez Gaona N33°E, NW Extreme towards Berkeleys Sound N49°W, Wooded point on the same Shore N35°W, Outer Frion Rock[4] S75°E & a rock just without the *South* Ex*treme*[5] S46°E, Off Shore 4 or 5 miles. It continued calm all Night & we had regular

[1] Swiftsure Bank; see p. 43, n. 8 above.

[2] See n. 6 on p. 43 above.

[3] No doubt Broughton was referring to the fact that from the south Cape Flattery looks like an island.

[4] Possibly Spike Rock, situated 1 mile west of Point of the Arches.

[5] Possibly Ozette Island, 52 metres high, situated 1 mile SW of Cape Alava or Bodelteh Islets, 60 and 58 metres high respectively, the largest of the Flattery Rocks, which extend about 1¾ miles WNW from Cape Alava.

Sound*ing*s from 33 to 26 *fatho*ms fine Sand. At 4AM the Frions bore N52°E [to] S43°E off shore 3 miles. It became afterwards very foggy weather. At 9 it clear'd up so as to give us sight of the Frions bearing N20°E & *South* Ex*treme* S65°E by which it appear'd a Current set us to the SE. At 10 a Breeze sprung up from the NW & we steer'd South to increase our distance from the Shore as the Fog still remain'd to prevent our seeing any thing. Lat*itude* acc*oun*t 48°7′N.

Sunday 29 May

Fresh Breezes & very thick foggy weather. We Steer'd South 10 miles, Wind NW, to increase our distance from the Shore. At 2 PM we Steer'd SSE. We tried for Soundings but found no bottom. At 8 PM we reef'd the Topsails & Steer'd SbE till midnight when the Fog dispersing we resumed our Former Course. At ½ past 3 AM we made Sail steering SE and at 7 not seeing the Land we Steered East. ½ an Hour after we saw the Land bearing East. At 11 AM the South point of Deception Bay[1] bore N35°E 2 or 3 Leag*ue*s when we alter'd our Course to SE. At Noon fine pleasant weather the Lat*itude* by Observation was 45°42′N the ⊙ Alt*itude* being 65°46′40″ + 30″ when Cape Look Out[2] bore S45°E 12 or 13 miles, Off Shore 4 or 5 Miles. Hazey weather with a Fresh Breeze. Muster'd the Ships Company's Cloathing &c. Off Shore as before. Lat*itude* acc*oun*t 45°48′, Long*itude* made 40′ E from Cape Claasset, Long*itude* per 1, 2 & 248 236°3′E means.

Monday 30 May

Fresh Breezes & Hazey weather. We steer'd SSE. Having run 11 miles upon that Course Cape Look Out bore N75°E which will place it in the Lat*itude* of 45°31′N, its Longitude 236°12′E. At 3 we steer'd SbE. At 8 PM Cape Foulweather bore of us N73°E about 2 Leag*ue*s. Our estimated Lat*itude* was 44°48′N which is nearly its situation. Throughout this days run we saw no opening whatever in the Land & we had 40 fathoms water sandy bottom. We pursued the same Course till 10 PM when the Wind suddenly shifting to the South with hazey weather we stood to the Westward close haul'd. At 9 AM we Tack'd to the East, but the Wind being SE by Noon we again Tack'd, at which time we had no Soundings with 80 fathoms. Lat*itude* per acc*oun*t 44°36′N, Long 235°16′E, Course since Noon of yesterday S27°W, 73 miles. Thick Foggy wea*th*er.

Tuesday 31 May

Fresh Breezes & thick Foggy weather. Wind from the SE. At 7 PM we tack'd to the Eastward. At 10 we had Light Winds from the South with Lightning & we were taken aback. At 4 AM we tacked to the Westward. In the Forenoon we had Light Winds from WSW & we steer'd SE. Course S68°W, 48′, Lat*itude* acc*ount* 44°21′N, [Longitude ditto] 234°24 E′.

[1] So named by Meares on 6 July 1788; Meares, *Voyages*, p. 167. Meares's name has not survived, but as he places the bay south of Cape Disappointment, it is almost certainly the entrance to the Columbia River. But if this is correct it seems strange that Broughton did not identify it as such, having entered and charted the Columbia River in 1792.

[2] So named by Meares in July 1788, who placed it in 45°30′N; Meares, *Voyages*, p. 169. Renamed Cape Meares by the US Coast and Geodetic Survey in 1857. The *Mesa* of Hezeta of 1775.

Wednesday 1 June

Light airs & Calms with dark cloudy weather. At 6 PM a Breeze sprung up from the Westward & we Steer'd SE with thick foggy weather. We had no Observation; but a large swell from the NW quarter gave us some hopes of an increase of Wind & dispersion of the Fog. Lat*itude* by account 43°45′N, Long*itude* acc*ount* 234°49′E, Course S28°E, 40 miles.

Thursday 2 June

Moderate Breezes and very hazey weather with rain. Wind from the NW. At 8 PM we steer'd SbE½E. At Midnight the Wind gradually increas'd & by Noon it blew a Fresh Gale with Hazey weather. At 7 AM we resumed our Course SE to approach the Land and at Noon we had fair weather with a following Sea when the Observation made the Lat*itude* 41°22′7″N. The ☉ alt*itud*e 70°39′40″ which is 28 miles to the South of our supposed situation, Lat*itude* p*er* acc*ount* 42°03′N, Long*itud*e 235°29′E, No 1 235°30′35″E.

Friday 3 June

Fresh Breezes & very hazey weather with the Wind from the NNW & large following Sea. At 7 PM having made [a] SSE Course since Noon we saw the Land bearing S50°E. At 8 it bore from S65°E to N45°E about 4 Leagues. We cou'd not distinguish any part particularly but knew it to be about the situation of Cape Mendocino.[1] At Midnight the Gale increas'd & at 4 AM we steer'd SE and at 8 SEbE. At 10 we saw the Land very indistinctly through the Haze. Our Latitude at Noon was 38°34′18″N, ☉ Alt*itude* 73°36′, Course S18°E, 167 miles, Lat*itude* acco*un*t 38°44′N, Long*itude* d*itt*o 236°37′E, No 1 236°17′15″E. Land from East to NE 4 or 5 Leagues. Fair weather.

Saturday 4 June

Fresh Gales & very hazey weather with a large Sea. We steer'd SEbE & ESE and at 5 haul'd up East [and at] ½ past we saw Punto [Punta] de los Reyes[2] in that direction: we passed within a mile of it carrying in regular Sound*ings*. At 7¾[h] we came to an Anchor in 15 fathoms when the Punto de los Reyes was just shut in with the SE point of Sir F*rancis* Drakes Bay[3] bearing S81°W. River entrance N6°W & the Extreme towards Port Francisco[4] bore S81°E our distance from the Point of [the] Bay about 1½ miles. At Noon the ☉ Mer*idian* Alt*itude* was 74°16′55″ which made the Latitude 37°58′46″N. The Breeze moderated in the Night and the next morning we had calm & very hazey weather. With two Boats we explor'd the Shore from what I conceive to be the place where Sir F*rancis* Drake Anchor'd in 1579 to a Remarkable Sandy Patch which bore N60°E from the Ship 4 or 5 miles & found no opening except the River.[5] A Bar broke entirely across & we had no more than 9 feet in passing it. It did not appear safe even to venture a Boat in. On the Hills and in the Vallies we saw Flocks of Deer feeding and before

[1] Cape Mendocino is in 40°26′N, 124°25′W, so Broughton probably made the land a little to the south of the cape.

[2] Point Reyes.

[3] Drakes Bay as named on US charts and maps. Its identity, however, is still a matter of controversy; see Lamb, *Vancouver's Voyage*, II, p. 701, n. 1.

[4] San Francisco Bay.

[5] An entrance to several lagoons including Drakes Estero.

our return we had some communication with the Natives who we perceiv'd walking along the Shore. The Surf prevented our landing, but one of the Seaman swimming to them convey'd some Beads, Knives, Scissors & other Trinkets with which they were much pleas'd. They were Stout & well made Men very dark color'd & perfectly Naked. The Women we saw at some distance; they were Cloath'd or at least cover'd.

This Bay affords very good Shelter except from the Easterly Winds. But the Sea cannot be great even with them as they blow over the Land about San Francisco. You may Anchor in [*blank*] water when the South point of the Bay will bear S50°E. Offshore [*blank*].

Sunday 4 [and Monday] 5 June

Fresh Breezes and hazey weather. At 1 PM we got under way & steer'd SE. At 6 we were within 2 Lea*gue*s of the Land to the South of Port San Francisco but we cou'd not see the entrance on account of the Haze nor the Rocks which lay to Seaward call'd the Farillones.[1] We now steer'd SSE & we had 30 fathoms water with a large Swell. At ½ past 7 PM the Extremes of the Land bore from N21°E to S76°E. Off shore about 8 or 9 miles.

About 9 PM we had the misfortune to lose Patrick Sherry Seaman by some unaccountable reason he fell from the Main Top down upon the Quarter Deck & was found dead upon the Spot, This is the Second unfortunate Accident we have experienced since leaving England. He was the only Man on board who did not voluntarily enter into the Service but was sent on board the Flag Ship at Plymouth by the Civil Power from whence he came to us.

At ½ past 1 AM it blew Strong with a great deal of Sea & we haul'd our wind under an easy Sail to the Westward. At ½ past 4 we wore & steer'd ENE for the Land. At 6 we saw it bearing N48°E 5 or 5 Leagues. Squally & very unsettled weather still continued & we took 3 Reefs in the Topsails. At 9 being well in with the Land about Punto [Punta] Nuevo[2] we wore & stood to the SW in hopes of an Observation as I did not think proper to run into the Bay with the Wind at NW blowing hard. I cou'd not distinguish Point Pinos.[3]

After mustering the Ships Company we buried the Deceas'd. At Noon the weather moderated & we observed in the Lat*itude* of 36°45′32″, ⊙ Alt*itude* 75°38′40″, & we saw Pinos Point bearing S73°E 4 or 5 Leagues. We bore away & Steerd East for the Bay of Monterey. The Wind moderated & veer'd to WSW. At 3 PM we came to an Anchor in 11 fath*oms* & moor'd Ship after saluting the Fort with 11 Guns; they returned an equal Number. Punto de Anno Nuevo bore [*blank*] to the Pinos shore, Presidio S13°E and the Flagstaff on the Fort S30°W, ½ a mile from Shore. We remained here a Fortnight during which time we were amply supplied with most excellent Fresh Beef, Mutton, Vegetables & Milk besides carrying to Sea a considerable quantity of the above Articles. The People were also serv'd Spruce Beer & Every Man in the Ship [had] permission to recreate themselves on Shore.

In Monterrey Bay

The Governor Senor Don Diego Borica, a Colonel of cavalry in the Spanish Service, was absent on our first arrival but he return'd two days after when I requested of him to erect a Tent

[1] The Farallon Islands in the approaches to San Francisco Bay.

[2] Point Año Nuevo, a low sandy point projecting 1 mile from the general run of the coast, about 16 miles NW of Point Santa Cruz, the northern entrance point of Monterey Bay.

[3] The southern entrance point of Monterey Bay.

for the Astronomer to settle the Rates of the Watches which he refus'd to grant, Saying their Orders were to comply with our absolute Wants with respect to Supplies agreeable to the laws of hospitality but in no other instance coul'd they condescend to the Contrary. A Jealousy pervaded their conduct which excluded us from riding on Horseback or walking any distance into the Country. We had not any intercourse beyond what became necessary nor did we receive the smallest attention or Civility from any Officer in the Settlement. They were not permitted to visit the Ship & we did not obtrude upon the Settlement. To the disgrace of Sociability was their conduct & we were happy to depart. On Sailing I did not think proper to Salute them tho' it appear'd by their preparations at the Fort they expected it.[1] Except this Fort which at present is very incomplete the Guns being mounted in barbet to the number of Eleven 12 pounders, I see not the smallest difference since 1793. They complain'd much of the dryness of the Season, indeed it evidently appear'd by the Country being burnt up how they had suffer'd for want of rain. The Water we procur'd was oblig'd to be roll'd down near ½ a mile which gave us a good deal of Trouble. I procur'd some Bushels of Indian Corn at [*blank*] per B*ushel* & a kind of callivances[2] which they call Frigoles.[3] Flour they had none & exclusive of their Cattle & Sheep the other articles we procur'd were very expensive. Vegetables alone during our Stay cost 40 dollars.

We had continual Foggy & Hazey weather with the Wind in general from the Westward & calm during the Night. By the 18th we had receiv'd on board the Cattle &c & having compleated our operations of Fishing the Main Mast (which was taken out for the purpose) & Caulking the Ship w[h]ere necessary, we prepar'd for Sea; and on Sunday morning the 19th of June with the Sea Wind we got under Way. Ships draught of Water 15^{F}. 9^{I} [forward] – 15. 4 aft. We made two Tacks and at Noon Point Pinos bore S6°E, 3 miles, from where I take my departure; its Latitude 36°38′30″N, Long*itud*e 238°22′E. Although it was the middle of Summer we in general found the air very cool so much so as to render Fires of comfort both in the morning & Evening. The Sun seldom appear'd to enliven the Scenes of Beauty that the surrounding Hills presented. I receiv'd a Present of a Bullock with a large quantity of Vegetables but the rigid conduct of the Gov*erno*r prevented me making any Return. The Padres who shew'd me their attention were the same I found at the Mission of St Carmelo in 1792. The Watches shew'd the Long*itude* of Monterrey as follows:

No 1	238°	49′	6″	NB The True Longitude is 238°25′E by Capt*ain* Vancouver
2	238	30	36	
56	237	26	16	
240	238	25	27 [4]	

No 1 Slow for Mean time at Monterrey June 17th at Noon 16^{h} 30′33″·78 and gaining from five days rate 6″·582 p*e*r day on mean time. No 248 Slow for Mean time at Monterrey June

[1] Borica's conduct towards Broughton was in contrast to the friendly reception he gave to Vancouver. As Kaye Lamb pointed out, this 'was undoubtedly due to a marked deterioration in the relations between Spain and Great Britain of which Broughton knew nothing'; Lamb, *Vancouver's Voyage*, I, p. 190.

[2] An obsolete form of calavances, a name for certain varieties of pulse.

[3] Possibly *frijoles*, a kind of kidney bean.

[4] In annotated *Voyage*, p. 63, corrected in pencil to:

No 1	238°	48	14
2	238	29	59
56	237	25	58
248	238	24	35

17th at Noon 1h39′4″·73 and gaining from five days rate 14″·625 per day on Mean time. The other Watches No 2 & 56 I deliver'd to the Governor with some Nautical Instruments &c who had an Order to receive them from the Executor of my much esteem'd deceas'd Friend Don Juan de la Bodega y Quadra who had commissioned me to procure him the above Instruments in England. I most sincerely lamented the occasion.[1]

It was now necessary I should come to some determination respecting my future proceedings. My orders from the Admiralty were, that I should survey the southern coast of the south-west part of South America, upon the idea that Captain Vancouver, who had similar orders, would not be able to fulfil them.[2] But as I now had certain intelligence that he had left this port eighteen months before, and that both the ships, Discovery and Chatham, under his command, were in a good condition, I had not the smallest doubt of his ability to comply with his instructions; particularly as I had information of his sailing from Val. Paraiso [*sic*], in lat*itude* 33°S for that purpose. As this was the case, my proceedings in future depended upon my own discretion; and I wished to employ his Majesty's sloop, under my command, in such a manner as might be deemed eligible for the improvement of geography and navigation. I therefore demanded of the officers their sentiments in writing, respecting the manner in which these discretionary powers allowed to me might most effectually be employed. The result of their opinions, I was happy to find, coincided with my own, which was to survey the coast of Asia, commencing at the island of Sakhalin, situated in 52°N lat*itude*, in the southern part of the sea of Ochotz,[3] and ending at the Nanking river[4] in 30°N lat*itude*. My intention was also to complete the survey of the adjacent islands, viz. the Kuriles, and those of Jeso[5] and Japan,[6] left unfinished in Captain Cook's last voyage. I thought such a survey would be very acceptable to geographers; for the limits of both the continents of Asia and America would then be known as far as navigation was practicable, and a knowledge of the Northern Pacific Ocean would be completed. Another reason for my undertaking this voyage was, that as yet the astronomer had met with no opportunity of complying with his instructions from the board of longitude,[7] in making observations and ascertaining unknown places, our line of navigation having hitherto led us to follow the track of Captain Vancouver. In any other part of the Northern Pacific Ocean there appeared little to be done, except that of settling the position of a few islands, in themselves of little consequence; but a survey of the coast of Asia and the adjacent islands, promised to be of more service to the science of geography than that of any islands left unexplored in these seas. As this survey could not probably be completed before the middle of the year 1798, I proposed spending my time in that pursuit till Christmas, then to go to Canton for stores and provisions, and to continue the survey early in the year.[8]

[1] 'I most sincerely regretted his loss, and was much hurt to deliver to his executor what I hoped to have done to him in person'; Broughton, *Voyage*, p. 64.

[2] In this respect Vancouver had been ordered by the Admiralty that 'It seems doubtful, at present, how far the time may admit of your making any particular examination of the western coast of South America; but, if it should be practicable, you are to begin such examination from the south point of the island of Chiloe, which is in about 44° south latitude; and you are, in that case, to direct your attention to ascertaining what is the most southern Spanish settlement on that coast, and what harbours there are south of that settlement'; Lamb, *Vancouver's Voyage*, I, pp. 285–6.

[3] Sea of Okhotsk.

[4] The Yangtze or Chang Jiang.

[5] Hokkaido.

[6] Honshu.

[7] See Appendix I.

[8] The preceding paragraph taken from Broughton, *Voyage*, pp. 64–6.

CHAPTER 4

From Monterey to the Hawaiian Islands and thence to Muroran Ko in Hokkaido

Monterrey towards the Sandwich Islands

Monday 20 June
Light Airs & hazey weather continued until ½ past 5 PM when being within 2 Leag*ues* of Punto de Nuevo Shore we stood to the SSW with the Wind from the West gradually increasing with thick foggy w*eather*. At 8 it clear'd up but we cou'd see no Land but about Punto de Nuevo which bore NW. The Wind drew to the NW and at 4 AM we steer'd SW with a Fine Breeze & a large Swell. Hazey weather, no Obs*ervation*. Course S60°W, 74′, Lat*itude* account 36°01′N, Long*itude* account 237°03′E.

Tuesday 21 June
Fresh Breezes & hazey weather Wind from the NW & we steer'd SW. At Noon we observ'd in the Lat*itude* of 34°0′N. Course from P*oint* Pinos S62°W, 280 miles. Lat*itude* account 34°28′N, Long*itude* acc*oun*t 234°15′ East, No 1 234°10′40″E.

Wednesday 22 June
Strong Breezes very Hazey weather, at times Foggy, Wind & Course the same as Yesterday. In the Night we took 2 Reefs in the Topsails but moderating in the Forenoon we again made Sail. Course S53°W, 189 miles, Lat*itude* ⊙ 32°6′40″, Account 32°19′N, Long*itude* account 231°15′ East. [*in margin*: Variation p*e*r Azimuth, No 3 PM 13°01′E.]

Thursday 23 June
Fresh Breezes & very pleasant weather with the Wind at NNW. We steer'd the same course as before, the Water was very Smooth and the Horizon Clear. At Noon we Observ'd in 30°5′45″N, Long*itude* acc*oun*t 228°19′E, Course S51°W, 191 miles, No 1 228°43′40″E. Wash'd between decks.

Friday 24 June
Fresh Winds & fair weather from the North & NNE. We steer'd the same Course with all Sail set. Lat*itude* Observ'd 28°22′N, Account 28°30′, Long*itude* acc*oun*t 225°33′E, No 1

226°6′55″E, Course S53°W, 182 miles. [*in margin*: Variations PM per mean of 10 Azimuths each compass; No 1 9° 47′E; No 2 11° 06′E; No 3 10° 11′E, Adams 10°02′E; very smooth water.]

Saturday 25 June

Ditto weather and a fine clear moonlight Night. At 11 PM we were in the Lat*itude* of 27°40′N & Longitude by the afternoon Sights 225°0′E when we alter'd our Course to South to see if we cou'd make an Island call'd Paxaros[1] in the Charts in about 27°N & 225°E. It was daylight at 4AM but we run on till Six AM when our Lat*itude* was 26°50′& Long*itude* 224°53′E & we cou'd see 6 or 8 Leagues round the Horizon but no Island appear'd nor anything to indicate the probability of Land being near, nor indeed we had scarcely seen a Bird since leaving the Coast of California. At 6 we alter'd our course to WbS with a fine Breeze from the NE. In the Forenoon we had slight showers & the Horizon clouded at times which prevented our having a good Observation. The Alt*itud*e just before the Sun was up[2] made [the latitude] 26°51′N & the next time we saw it it was dip't considerably. As the Sun might have rose a little more the Lat*itude* will be less. It was within 3 or 4 minutes [of noon] also by the Timepiece, ☉ A*ltitude* 86°21′, Course S52°W, 142 miles, Lat*itude* account 26°49′N, Long*itud*e 223°26′E by account. [*in margin*: Variation per Azimuth PM No 3 8°57′E, per Amplitude d*itto* d*itto* 8°4′E.]

Sunday 26 June

Fresh Breezes & cloudy weather these 24 Hours, Wind at NE. We steer'd WbS. AM muster'd the Ships Company & perform'd divine Service. Course S88°W, 168 miles, Lat*itude* acc*oun*t 26°46′N, Long*itude* d*itt*o 220°17′ East.

Monday 27 June

Fresh Breezes & cloudy weather without any variation. At 6 AM we steer'd South on the supposition of Land being seen in that bearing. At 8 having run 14 miles the mistake was evident & we resumed our former Course WbS. The Forenoon was pleasant Sun shining Weather when the Observ'd Latitude was 26°3′30″N, account 26°24′N, Longitude Account 217°32′E, Course S83°W,[3] 148 miles, Longitude by No 1 218°30′ East.

Tuesday 28 June

Fine and pleasant weather with very Smooth Water. The Wind continued at ENE and we steerd West. Lat*itude* Observ'd 26°15′57″N, Account 26°24′N, Long*itud*e account 214°38′E, No 1 215°37′E, Course N85°W, 158 miles.

[1] Paxaros is the archaic Spanish spelling for Pajaros = Birds, hence the alternative name 'Bird Island' inserted below between the lines and in RUSI/NM/132 (7); Pasaros on Henry Roberts's 'A General Chart'; searched for in vain in 1873 by USS *Portsmouth*; Stommel, *Lost Islands*, p. 113.

[2] I.e. on the meridian.

[3] 'S83°E' in the MS, which is surely a transcription error.

Wednesday 29 June

Moderate Breezes with the same fine weather, we steer'd the same Course. AM wash'd below & exercis'd Great Guns & Small Arms with Powder. Course N86°W, 111 miles, Lat*itude* ⊙ 26°24′6″N, Account 212°34′E, No 1 213°14′E, 248 212°57′E. [*in margin*: Variations the 29th AM p*er* Azimuth No 3 7°14′E; Adams 9°49′E; Walker 10°40′E; Longitude at Noon by Mr Crosley 213°4′30″E; d*itt*o by myself 213°3′15″E.]

Thursday 30 June

Ditto weather. In the Forenoon we saw some Tropic Birds & Flying Fish. Both Yesterday & today we Observ'd Distances ⊙ East of the Moon. Course S81°W,[1] 116 miles, Lat*itude* ⊙ 26°39′N, Account 26°42′N, No 1 210°51′45″E.

Friday 1 July

Fresh Breezes & fair weather with slight squalls of rain at times the Wind at EbN & we pursued our west course. At Noon the Lat*itu*de made 26°54′N, Account 27°01′N, Long*itude* d*itt*o 207°47′ East, No 1 made the Longitude 208°19′45″E.

Saturday 2 July

Fresh Breezes & Squally weather with the Wind at East. We steer'd West till dark when our Lat*itude* was 27°2′N & the mean Longitude by the Watches & distances 207°E. Capt*ai*n Cook in the same Lat*itude* was 35′ to the West of us at Noon *&* we cou'd[2] see any distance from the clearness of the Horizon. We had now preserv'd this parallel for near 20° of Longitude without any appearance of Land whatever. St Maria la Gorta[3] has been so frequently pass'd over I thought this Track the best to pursue. If the Islands exist I think their situation so erroneously plac'd as to render looking for them any more a loss of time. I therefore gave up the pursuit having no time to spare & steer'd South for the Sandwich Isles. Lat*itude* ⊙ 25°20′N, account 25°18′N, Long*itude* d*itt*o 206°33′E, Course S35°W, 115′, No 1 206°59′.

Sunday 3 July

Fresh Breezes & fine weather with the Wind at East we steer'd South. In the Forenoon after mustering performed Divine Service. Vertical Sun made the Observation indifferent.[4] It gave 22°50′N, Account 22°57′N, Course S10°W,[5] 150 miles, Longitude account 206°4′E, No1 206°18′20″E, No 248 Long*itude* 205°59′2″ East.

[1] 'S81°E' in the MS, which is surely a transcription error.

[2] 'not' inserted here above the line, which is clearly an error, perhaps 'not fail to' was intended.

[3] Also searched for by Broughton on his voyage north from Niihau to Nootka Sound; see p. 35 and n. 1. This time Broughton appears to have been consulting Roberts's chart and not Arrowsmith's 1794 'Map of the World'.

[4] The sun's declination would have been about 22°N and thus almost overhead at noon, making it very difficult to observe an accurate altitude.

[5] But possibly S10°E, the final letter being lost in the margin.

Monday 4 July
Fresh Breezes & Squally weather with the Wind from the same quarter. We steer'd South till 7 PM and afterwards SbW. At ½ 7 AM we saw the Island of Owhyhee bearing SSW and at 8 the Mountain of Monah Kaa[1] SbW. We pursued the same Course with a Strong Breeze and at Noon the Extremes of Owhyhee bore from S82°W to S39°E, Offshore 10 miles. Latitude by Observation 20°17′N, Account 20°14′N, Course S17°W, 160 miles, No 1 at Noon 205°22′13″E, No 248 made 205°1′34″E. Latitude of the N*orth* point of Owhyhee[2] 20°17′N by our Noon Bearings. Our passage must be consider'd a very good one having run 757 Leagues in 15 days without our finding it necessary to carry less than our Plane Sails with One reef in the Topsails &c. Smooth Sea. [*in margin*: Variations AM July the 4th, No 3 8°12′E; Adams 10°53′E.]

Off the Island of Owhyhee

Tuesday 5 July
Strong Breezes & Squally weather. At 1 PM we were within 3 miles of the Shore and we alter'd our course to the westward Keeping that distance from the Land. The Country presented a most Fertile and cultivated appearance; the Houses beautifully interspers'd among their Plantations on the Rising Grounds had a pleasing effect. The violence of the Wind prevented any Canoes coming off till we were near the NW point of the Island[3] which at 4 PM bore S76°W. At ½ past 7 PM the Extremes of Owhyhee bore from N4°E to S8°W and the North point of Toe Yan[-yan] Bay[4] S47°E 3 miles. We had Squally weather with hard Showers of Rain till 9 PM when the Wind became variable with Calms throughout the Night. At Daylight we were in the Center of Toe Yan[-yan] Bay and at Noon we observ'd in 19°48′03″N, Mer*idian* Altitude 86°49′15″. The NW point of the Island bore N17°E, South Extreme of d*itt*o S1°E, Longitude p*er* No 1 204°34′E, 248 204°11′15″ East.

Wednesday 6 July
Fresh breezes & fair weather with the Wind from the SW. We stood off till ½ past 2 when we stood in Shore after observing an Eclipse of the Sun. We had slight winds till 6 PM when we made Sail to the South*war*d with a strong Breeze from the ENE. By 7 PM it was nearly calm which continued with intervals of Land Winds till Daylight when Karakakoa Bay bore of us EbS 2 or 3 Leagues. At 8 a Light air from the NW sprung up & with the Boats ahead towing we made Sail in for the Anchorage. At Noon we came too & moor'd Ship in the Bay with the Bowers – the points bearing from S7°W to N77°W, Offshore ¼ of a mile [in] 14 fathoms. My inducement for stopping here was to ascertain the Rate of the Watches as they appear'd to have alter'd very considerably since leaving Nootka & Monterrey & to compleat our Water. Unfortunately the Taboo which had already existed 5 days continued to the 10th July which prevented our having any communication with the Shore. In the interim we unbent Sails & black'd the Yards & Tarr'd the Rigging & prepared for painting the Ship. An Operation which had become absolutely necessary from the Heat of the Climate.

[1] The 4,205-metre Mauna Kea, the highest peak in the island of Hawaii.
[2] Upolu Point.
[3] Upolu Point, which is also the north point of the island of Hawaii.
[4] Kawaihae Bay.

In Karakakoa Bay July 1796

On Sunday the 10th the Tents were pitch'd in their old situation & the Ship surrounded with numerous Visitors who seem'd much rejoic'd at seeing us. After performing divine Service the People had leave to go on shore, & in the evening we were crowded with Females, their Old acquaintance. Many unable to find Gallants were mortified at returning on Shore. The next morning we sent our Water Casks on Shore. They were fill'd by the Natives who brought it to us in Callibashes from the Hills 4 or 5 miles off. It cost 100 nails to fill a Hogshead & became so expensive we were unable to compleat our watering. The astronomer had equal altitudes the first day but a clouded atmosphere prevented any more for 7 days. He also got them on the 21st but the 22nd was again cloudy & we gave over our pursuit & struck the Tents & got every thing from the Shore. The Ship had been painted, Sails bent & ready for Sea some days before. The same evening with the Land Wind we sail'd from the Bay & took our final leave of our Friends who follow'd us till near Dark.

We experienced during our Stay the same kind attentions from the Natives & were amply supply'd with Hogs & Vegetables for which we barter'd Sheets of Copper, Iron Hoops & Nails. Copper was much esteem'd in preference to any thing else as Tai maah maah wanted some to Copper a Small Vessel at Wo[h]ahoo which he was building. We had the satisfaction to find the Ducks had bred & the Cattle left by Capt*ain* Vancouver much increas'd but all the Garden Seeds fail'd thro' inattention. Some Roots of Horse Raddish were in high Vegetation, Cabbages flourish'd in the more interior parts as well Potatoes & many other Fruits which we did not extend our Walks far enough to see. Goats are in great plenty. They have also several Sheep that breed. The British Seaman (Jo*h*n Young) who went down with me last time to Wohahoo had return'd but Tai maah maah & all the Chiefs still remain'd to Leeward. Their absence had much increas'd the power of a Chief named Naa-mete haw[1] (a Brother of Tianna's[2]) who was in a state of revolt & acquiring by degree the whole Island. By the last account out of the Six he now possess'd 4 districts & was approaching fast towards Karakakoa. At which place he was sure of meeting no Resistance as the People were averse to fighting without a Chief in whom they had confidence & the only Man of the Name[3] was Mahoa who had lost his Eye sight. He wanted to go with me to Wohahoo to explain matters to the King but the people wou'd not admit his absence having no better Commander, they even thought him of some consequence. In the last Skirmish with Naa-mete Haw they kill'd an European & there now only remained 4 to take part with the blind Chief Against whom Naa-mete haw vow'd vengeance in case he cou'd catch them. It appear'd extraordinary that Tai-maah maah shou'd neglect his own Island & by Pursuing his Conquests leave it entirely expos'd to any Usurper, but still more singular that while a Chief of Atooi was taking possession of Owhyhee he was preparing for an Expedition to that Island from whence Namete-haw [*sic*] & his Brother Tianna first came. Such was the absurd conduct of this Man Ruining himself & Friends by his Ambition and fearing to send any Chiefs from him to Owhyhee under the idea of them also taking part against him. A constant jealousy of each other power seems to be universal amongst the Islands in the South Sea. It did not appear since our absence that any vessel

[1] Laamaikahiki.

[2] Kaiana.

[3] 'that Rank' in Broughton, *Voyage*, p. 69.

had been in the Bay & we found every thing in great plenty excepting Cocoanuts. They were tabooed for the purpose of making Cordage.

We had the misfortune to find many of them infected with the Itch[1] which they had catched from the different People who came up from Wohahoo. Their other Complaints were trifling notwithstanding the promiscuous intercourse between them & our Ships Company they seem'd but very slightly affected with the Venereal Disease.

[Saturday][2] 23 July

Light Winds & variable continued throughout the day. At 8 PM we were well clear of the Bay the Extremes of Owhyhee bearing from N½W to SSE. At Noon the South point of Toe-yan [-yan] Bay bore N64°E 4 Leagues, Latitude Observed19°43′N, Karakakoa Bay S64°E, 10 Leagues. Calm & fair weather. ⊙ Altitude 89°17′20″.

[Sunday] 24 July

It continued calm with light airs alternately till midnight when a Fresh Wind from the NE enabled us to make Sail upon a NWbW course. At 8 AM the Extremes of Ranai[3] bore from N15°W to N5°E 5 or 6 Leagues. At Noon the South point of Morotoi bore N18°W & the Extremes of Tahoorowa[4] from N85°E to S86°E & Ranai from N2°E to N48°E. Lat*itud*e 20°34′N, ⊙ A*ltitude* [*blank*]. Very pleasant weather wash'd & clean'd below.

[Monday] 25 July

The Wind continued variable with light airs & showers of rain till 8 PM. At 7 the South point of Morotoi bore N12°W & Ranai from N23°E to N73°E. The Breeze from the NE gradually increas'd & we steer'd NW & NWbW for Wohahoo. At ½ past 5 AM we saw the Island bearing from N51°W to N80°W 3 or 4 Leag*ues* dist*an*t. At 8 AM we Hove too in Whyteetee Bay & continued standing off till Noon when Tai-maah-maah & the rest of the Chiefs paid us a Visit from the Harbor of Fair Haven[5] in a large sailing Boat built by the Europeans. I was very happy to find they had given over all thought for the present of going to Atooi & were all prepar'd to return [at] the first opportunity to Owhyhee, but strong winds had hitherto prevented them. It appear'd they had made [an] attempt to proceed to Atooi but finding the Wind & Sea too much for their Canoes they were glad to put back. The Island with respect to Provisions was worse than before for on their leaving it to go to Leeward they had destroy'd all the Hogs; & Vegetables were equally scarce from the neglect of cultivation. This unhappy want had been the death of many poor Creatures who being oblig'd to steal for the preservation of their Lives were at times caught in the act and murder'd by the Orders of the Chiefs in the most barbarous manner, even to burning them alive. It was computed he had lost by his conquest [of this island] & the subsequent causes near 6,000 People. The remainder however were perfectly

[1] Scabies.

[2] Broughton has lost account of the day, giving the symbol for Friday in his MS, which was not corrected until Monday 25 July off Atooi.

[3] Lanai.

[4] Kahoolawe.

[5] Honolulu Harbor.

recover'd of the Itch which seems to have left them & to have been carried to the Windward Islands more particularly about Karakakoa [where] we observ'd it. The news of Nametehaw's success at Owhyhee was the reason of their return & Tai maah-maah was very desirous I shou'd take them over, which from existing circumstances I cou'd not comply with though otherwise it wou'd have given me pleasure. The Vessel we left building in the Harbor remain'd in the same State & wou'd not be probably finish'd this Year. His wants were in her account as before, Cordage, Sails, Guns & Powder, Nails &c. I presented him with several Articles more useful & we parted very friendly. The Queen gave me her own beautiful Canoe for which I made a proper return & we all parted most amicably with their good wishes for our return & Tai-maah-maah's permission to make a Settlement on any of the Islands. Wohahoo seems more favorable for a first establishment than any other on account of the Harbor in Whyteetee Bay. There has since been found another Forming within a large Bason & here it is they collect the Pearls in great Numbers.[1] Owhyhee has two or three Harbors for Small Vessels. The best is in the district of Ahudo on the East side of the Island,[2] the entrance being form'd by the Reef off which Capt*ain* Cook speaks of his Ship being in danger from.[3] Since our last sailing from hence Tiytaboree[4] had been found Dead near a Cave, suppos'd to have been starv'd.

Off Atooi

Monday 25 July [*continued*]
Fresh Breezes & fair weather. In the Evening we took leave of our Friends and at 6 PM made Sail for the Island of Atooi steering WbS till 10 PM when having pass'd the West point of the Bay[5] we steer'd NWbW½W. This point was entirely over-run with Fire clearing the Ground, which illuminated the Coast to some distance. At ½ past 7 AM we saw Atooi bearing WbN to NW½N. At Noon we Observ'd in 21°51′N when the East point of Wymea Bay[6] bore N50°W 5 or 6 miles. Meridional Altitude 87°33′.[7]

[Tuesday 26 July]
In standing into the Bay after Noon we Shoal'd our water suddenly to 4½ [fathoms] the River[8] mouth bearing to the West of N*orth*. This is a very dangerous Reef with 2½ f*atho*ms only which breaks at times.[9] To avoid it You shou'd open the River with the Starboard Cliff point

[1] Hence its name Pearl Harbor.

[2] Hilo Bay, which is partly obstructed by Blonde Reef.

[3] At 1 AM on 19 December 1778, when off the east coast of the island of Hawaii, the *Resolution* was 'left to the mercy of a North Easterly swell, which impelled us fast toward the land ... At day-break, the coast was seen extending from North by West, to South West by West; a dreadful surf breaking upon the shore, which was not more than half a league distant. It was evident, that we had been in the most imminent danger'; Cook and King, *Voyage to the Pacific Ocean*, II, p. 537.

[4] Kalanikupule.

[5] Barbers Point.

[6] Puolo Point.

[7] In annotated *Voyage*, p. 72, position of anchorage inserted in pencil:
Lat 21°56′28″N, Long 200°14′16″ per mean of 3 Time k*eeper*s, 200°23′17″E per mean of 16 sets of Lunars.

[8] Waimea River.

[9] Makaweli Reef, with a depth of 5.5 metres [3 fathoms] or less over it.

about NNE & Anchor with it bearing NE any distance from the Shore you please as the Soundings are regular going in: we came too in 23 F*atho*ms the Extremes of the bay from S60°E to N72°W 1½ [miles] from the Shore. As my only motive for stopping here was to procure Water, early the next morning the Boats were sent with an Officer for that purpose. At first there was some Objection by a Person who claim'd the property of it, Saying that it was taboo and that we cou'd not have it without powder. They however permitted this first Time & the next I sent a greater Force with another Boat to insure our success & to explain my determination of taking it shou'd they refuse it. On the contrary I had the satisfaction to find they prevented the People coming to the River while we were filling except a few who assisted us who were paid for their troubles. By this means in 24 Hours we compleated our Water. I rather imagine their intentions were to see how we shou'd act & finding our determination they acted accordingly.

Since our leaving this Island Teavee[1] Grandson to Perrorannee[2] had been successful in conquering the Island & now possessed the whole of it having put every Chief to Death of the Young King's. Timoree, the depo'd King liv'd with him divested of all power. Korokranee Brother to Trytaboree (who we now found was secreted in the Young King's Canoe when he paid us a former visit) was also among the Slain. He was fearful of our putting him to death on account of the part he took in the Murder of Brown & Gordon. Teavee was under some Restriction of the Taboo at Tipoony[3] on the other side of the Island & had sent his People down hearing there was a Vessel arriv'd to taboo Hogs being sold unless for Guns & Powder. We of course did not get any as these Articles of Barter were strictly prohibited on board. A Ship from Bristol call'd the Ruby[4] had purchas'd Hogs & it appear'd with nothing else, & was unfortunately for us the only Vessel that had touch'd here since our departure, except the Lady Washington Brig.

At daylight on the 28th we got under weigh & hove too in the Bay expecting some Provisions from the Shore & the European who we saw there before. He came off with an excuse that they were not yet collected bringing 2 Pigs & begging some Files for Teavee which I sent him; & he went again on shore to expedite the Hogs &c promised (to Oneehow) where he expect'd they would be shortly & a Man who was come down expressly from Teavee to supply us with every thing (being unacquainted at first with the Ship). As I did not give implicit faith to this Fellows account after waiting a due time we made Sail for Oneehow which bore from S65°W to N85°W 6 Leag*ues*, Lat*itude* ⊙ 21°53′N, Mer*idian* Alt*itude* 86°50′.

At Anchor in Yam Bay Oneehow

We had bafling Winds till 2 PM when the Breeze came strong from the Eastward. At 6 PM we came too in 15 fath*oms* Sandy bottom [in Yam Bay],[5] the Two points forming the Bay

[1] Not identified, but perhaps Tennavee encountered by Vancouver in Oahu, Lamb, *Vancouver's Voyage*, III, p. 875.

[2] Peleioholani.

[3] Not identified.

[4] The *Ruby*, Captain Charles Bishop, had called at Waimea seven days after the *Providence*'s departure in 18 February 1796. Broughton met Bishop in Macau in June 1797, when he agreed to take Bishop's journal to England for Sydenham Teast, Bishop's employer in Bristol; Roe, *Journal of Charles Bishop*, pp. 135–7, 262.

[5] Probably Nonopapa Landing situated 5½ miles NNW of Kawaihoa Point.

from N29°E to S10°E Off shore 1½ miles. The next day we stay'd the Masts & set the Rigging up Fore & aft. Ships draught of water 15^{F} 8^{I} [Aft] 16^{F} 3^{I} Forward. As I intended to remain here only 48 hours to procure Yams I sent the Boat on Shore after breakfast with a Small Tent & 3 Marines Arm'd to protect the Trade they might procure, as we made few purchases on board. I was in hopes they wou'd be more successful. In the Evening I landed & found their collection very small & wishing to satisfy myself I walk'd to some of their plantations to see if they had any to dispose of. What I saw were very Small & they explain'd them to be the same all over the Island, owing to the want of rain. Returning to the Tent I met a party of Men who had just come from Atooi & with them Tupararo[1] the very Person who was directed to come over for the purpose of supplying us. I enquired after Hughes the European[2] who they said was not yet come. On our arrival at the Tent he assur'd me we shou'd have plenty of Potatoes tomorrow & as many large Yams as they cou'd procure & begg'd of me to come on Shore in the morning & to bring him some Red Cloth. I told him the Tent wou'd be on shore in the Morning with the Boat to bring off the things, when he cou'd come on board to receive his presents. I desir'd the Midshipman to go on board at Sunset with the Tent & Boat and I walk'd along shore to the South where the Pinnace waited for me about One mile distant. Only one Native accompanied me & with him I walk'd unmolested meeting several people by the way till we join'd our Boat (which had been more detach'd from me than I had any idea of by a misunderstanding of the Plan). Having visited this Island twice before, When our Gentlemen had made many Shooting parties unmolested I had not the smallest fear respecting my own safety but still was not pleas'd with the situation I was thrown into by the absence of the Boat; the unhappy event that took place the next day will point out my fortunate escape as will the Officers who were prevented recreating themselves on shore by indisposition & those remaining to attend their Duty were equally excluded from going. I found the other Boat safe return'd before me having experienc'd every civility from the Natives who crowded round them on shore seemingly from motives of curiosity both Men, Women & Children. In the forenoon of the 30th I receiv'd some Yams from an elderly Man who they said was Father to Teavee & more from another Man whose name was Row Row. Having made them proper returns they went on shore for more. I also rec*eive*d a Canoe full from Tupararo; he also went on shore with the same view. Just before the Canoes came I sent the Cutter on shore & to bring off whatever Tupararo might have to send, with Orders to the Mate to land as before with two Marines Arm'd,[3] himself, & another Man to trade; & the Boat's Crew with a Midshipman to remain off at a Grapnel to assist them in case of an accident & to be upon their Guard; they were direct'd if any thing occurr'd to make a Signal. They had not been on shore an Hour when I was acquainted by the Officer upon Deck that most of the Canoes were gone ashore. I therefore soon afterwards gave directions for the Boats signal to be made; this was at Eleven O'Clock. We saw them strike the Tent & immediately afterwards heard some firing from the Boat. As no Signal was Made I consider'd this to be a Signal to those who might be absent but soon after they hoisted the Signal when the Pinnace with an Officer & some Marines were sent to their Assistance. On the return of one of the Boats I had the misfortune to learn the two Marines were killed & the Mate with the other Man had much

[1] Not identified.
[2] Not identified.
[3] Mate J. John Cawley, and marines Daniel Dell and Jonathan Bird.

difficulty in escaping to the Boat. The Pinnace remain'd on Shore to protect the Dead Bodies from the Natives who seem'd anxious to get them, though they were sunk beyond the Surf. As this unhappy transaction had taken place without the smallest reason on our side I consulted with the Officers on the measures necessary to pursue. Their Opinion was at all events to prevent the Natives getting the Bodies & to proceed immediately to Atooi where by getting some of the Chiefs in our power we shou'd oblige them to deliver up Tupararo & the other principals of the Horrid Deed as an atonement for their Guilt as we conceiv'd the Atooi Men who came over with him were the cause of it. I agreed with them in that respect but still thought it necessary to make some example on the Spot & immediately gave Orders to the 1st Lieut*enant* to proceed on Shore with the Boats completely Mann'd & Arm'd with Seamen & all the Marines, the indisposition of the Officer prevented his Commanding his party. The Master therefore went in addition. They had Directions to burn & destroy every House & Canoe, Plantations &c within a certain distance but not to go more than One Mile from the Beach where the Boats attended & to be on board in time before Sunset for sailing & to send me off a Canoe or Boat with information of their proceedings & to give me the opportunity of sending them any additional Instructions. Before their Landing we heard some firing from the Pinnace & soon after the Boats landed without any Resistance the Natives keeping without Shot or whenever they approach'd within it they escap'd by dropping on seeing the Flash & then running away. The whole Village was soon in Flames & every Canoe on the Beach in number 16 burn't or destroy'd. They had mov'd nothing from their Houses not expecting us on Shore & every Article belong*in*g to them suffer'd in this conflagration. During this time the Natives had assembled in great Numbers, Arm'd with Spears &c and two of them with the Marine accoutrements & Musquets. As our People advanced they retreated so as to prevent any personal satisfaction for their Villany. In the Mean while the Boats had found the Dead Bodies, seeing them in about 9 feet water they got them easily into the Boats and about 3 PM Mr Mudge return'd on board with the Party. The Natives rush'd into the water immediately searching for the Deceas'd & we saw them find a Grapnel our People had search'd for in vain which had been lost from the Cutter in the attack. Of all the Murders which have been committed at these Islands this seems to be the most unprovoked as it was to me unexpected. The occasion of it I cannot imagine, but the manner of executing it was related to me in the following manner by Mr Cawley, the Mate, & Alex*ande*r Bishop,[1] the Men who escap'd. The Mate on seeing the Signal order'd the Boat in & struck the Tent. The Marines unfix'd their Bayonets & one of them gave his Piece to the Trading Man[2] while he put the Tent into the Bag. At this Period without any suspicion whatever & looking carelessly about them the Trading Man was Knock'd down from behind & Tupararo ran away with the Musquet. The Marines were I suppos'd serv'd in the same manner & without firing the remaining Musquet they instantly after recovering themselves from the Blows retreated to the Surf the Natives pursuing them with great eagerness. The Gentleman reach'd the boat safely & the Seaman[3] likewise escap'd by stabbing a Man in the Water who

[1] Described in RUSI/NM/132 (7) as 'the man who went for the purpose of bartering', but in several places in Broughton, *Voyage*, p. 77 as 'the botanist'. In the 'List of Names' in Broughton, *Voyage*, p. xv, Bishop is given as the 'Cook', but in the log as 'the Gardiner' (TNA, ADM 55/147, f. 3), which may be the cause of the confusion. No record has been found to suggest that Bishop was appointed to the *Providence* as a botanist or botanical collector in addition to being appointed the ship's cook.

[2] I.e. Bishop.

[3] Ibid.

had hold of him, but the unfortunate Marines with their accoutrements on were Murder'd by the Savages in the Water. On inspecting them one appear'd to have rec*eive*d several stabs with his own Bayonet, the other who cou'd not Swim had rec*eive*d a violent contusion on his Head & seem'd to have been drowned. They were taken up perfectly naked excepting part of a pair of Trowzers. All this time the Boat was within 50 yards of the Shore by the account of the Midshipman in her, Keeping a constant Fire upon them with two Musquets while the remaining Crew where [*sic*] pulling the boat closer in. Notwithstanding most unaccountably to me they did not save these Men from being sacrificed by the Fury of the Natives & only wounded 1 Man out of Twenty that followed them into the Water. I fear the Men were to[o] much agitated to exert themselves without confusion & the Boat too far from the Shore to give that Assistance these poor unfortunate Men requir'd. I thought I had taken every precaution against accident knowing the Boat to have been a certain resource, as their absolute distance from the Tent which was upon the Rising Bank cou'd not exceed 100 Yards if she had been properly plac'd & the Natives expos'd to her Fire if they follow'd down the Bank. It therefore appears most probable that the Boat must have been at a greater distance & which was confirm'd by the Mate who call'd to them (tho' he cou'd swim very well) that if they did not pull in towards him he shou'd be drowned & when he reach'd the Boat they then cut their Grapnel rope having 24 fathom out & the Wind from the Land. He also I am sorry to add found them in great confusion & he desir'd the Signal to be made to the Ship which the Midshipman of the Boat (W. Haywood)[1] had Neglected.

Before our Boats reach'd the Shore they had fir'd some Musquets from the Bank at the Pinnace that went thro' her without hurting any body. After firing her Swivel they were soon quiet & retreated. It was extraordinary that the two Women Rahina & Tymarow that Capt*ain* Vancouver brought from the NW coast[2] sho'd have come with us from Atooi & that the very Villain who executed the principal part of the Murderer shou'd be the Husband (Tupararow) whose Child had come off on purpose for us to see him. They appear'd to be much alarm'd & having made them some presents they went on shore in the Evening. I gave each of them a letter requesting them to give it to the first Vessel & by no means to let the Europeans at Atooi see the Contents.

As I was anxious to reach Atooi by the morning which was possible with the Winds we had found hitherto prevailed as they had been Westerly lately. After hoisting the Boats we got under way & made sail to the North of Orehoua which at 7 bore of us N45°E 5 or 6 miles. We kept close to the Wind all Night which continued at NEbE directly opposite to our Course. Towards Noon it Veer'd to East & appear'd invariably fix'd in that quarter. This change was more unfortunate still for we cou'd not even fetch Oneehow. It was therefore impossible to reach Atooi before the melancholy news, which render'd our Plan useless as we cou'd not expect those concerned in the Deed wou'd trust themselves on board & I strongly suspected the business must have been settled at Atooi & the Villain Hughes privy to it by his not coming over with Tupararo. I therefore was oblig'd to drop the idea for the present taking the Chance of a Wind shou'd it come within a few days & in the mean while proceed to settle the

[1] John J. Haywood in the crew list in Broughton, *Voyage*, p. xii.

[2] Raheina and Tymarow were taken from Niihau to Nootka Sound in the *Jenny*, Captain James Baker, having been detained on board without Baker's knowledge. In October 1792 Vancouver took them on board the *Discovery* in Nootka Sound at Baker's request so that they could be returned to their native country. In March 1793 they were settled in Kauai as Niihau was drought-stricken; see Lamb, *Vancouver's Voyage*, II, pp. 688–9, III, pp. 892–6.

situation of two Islands call'd Bird & Montague Island.[1] Accordingly after Noon we made Sail to the Westward. Course from Orehoua N21°W, 78′, Lat*itude* ⊙ 23°15′N, Lat*itude* account 23°09′N, Long*itude* acc*ount* 199°18′E, No 1 199°08′E. Fair & pleasant weather. In the Forenoon we performed divine Service. We understood from the People who came off there were 4 Men Kill'd on Shore.

Monday 1 August

Moderate Breezes & fine weather. At 6 PM we saw an Island bearing S65°W 10 or 12 Leag*ues*. We run on till Ten PM when we plyed under Topsails. At Daylight the Island bore of us S62°W 8 or 10 Leagues. At Noon we pass'd it within two miles when it bore from S23°E to S41°E, Latitude Observ'd mean of 3 Sextants, 23°5′50″N. This Island is not more than a mile in extent in any direction rising at the extremes & low in the middle. It appeared inaccessible to boats & entirely cover'd with various oceanic Birds who built their nests in the cavities of the Rocks that were nearly perpendicular. It was seen in 1789 by Mr Douglas & is well describ'd by him.[2] The Charts place two Islands nearly in this situation we saw only One & cou'd see distinctly many Leagues around. We made its Latitude to be 23°3′50″N, Longitude 197°53′53″ East by No1. Course these 24 hours W8°S, 73 miles, Lat*itude* Obs*erved* 23°6′N, Latitude acc*ount* 23°4′N, Long*itude* d*itt*o 199°00′E, By No 1 mean of AM & PM Sights 197°52′E.[3]

Tuesday 2 August

Pleasant Breezes & fine weather, the wind still continued in the same quarter & we kept our Wind close to the Northward. At 6 PM Bird Island bore S32°E 9 or 10 Leagues. AM we wash'd below & servd Salt Provisions with ⅔ of a pound of Flour in lieu of Biscuits (which was entirely expended). Lat*itude* ⊙ 25°3′41″N, Account 24°54′N, Long*itude* d*itt*o 197°28′ E, Course N14°W, 122 miles, No 1 197°29′E. Many Birds.

Wednesday 3 August

Ditto Wind and weather with very smooth water. Several Tropic Birds about the Ship.[4] AM Serv'd Slops to the Ships Company. Course N8°W, 103′, Lat*itude* ⊙ 26°47′N, acc*ount*

[1] Nihoa, discovered by James Colnett on 21 March 1788 in the *Prince of Wales* with the *Princess Royal*, Captain Charles Duncan, in company. Colnett named his discovery Bird Island; Galois, *Journals of James Colnett*, p. 204. There are the remains of some ancient Polynesian stone buildings on the island; *Pacific Islands Pilot*, vol. III, 10th edition, Taunton, 1982, p. 280. The twin islands of Bird I. and Montagues I. are shown on 'A Chart of the Northern Pacific Ocean ...' in Meares, *Voyages*, f.p. i, with a view of the island, f.p. 80, where it is named 'The new discover'd Sandwich Island'. As Montagu is the family name of the Earl of Sandwich this may account for Bird Island's twin island with that name.

[2] Nihoa was visited by William Douglas in the *Iphigenia* in March 1789, who described it as 'a barren rock, perpendicularly steep, and did not appear to be accessible but to the feathered race, with which it abounds'; Meares, *Voyages*, pp. 359–60.

[3] In annotated *Voyage*, p. 81, corrected in pencil to:

Lat 23°4′45″N, Long 198°4′7″E per mean of 2 Time k*eepe*rs, 198°10′58″ per mean of 16 sets of Lunars.

and in ink:

Lat 23°4′6″N, Long 198°4′8″ per mean of 2 Time keepers and 198°10′20″E per 16 sets of Lunars.

[4] Both Red-tailed Tropicbirds (*Phaethon rubricauda*) and the smaller White-tailed Tropicbird (*Phaethon lepturus*) breed on the Hawaiian Islands.

26°35′N, Long*itu*de d*itt*o 197°12′E, No 1 197°11′ East. [*in margin*: Variations per Amplitude PM on the 3rd, 8°36′E; Azimuth by No 3 10°44′E; D*itt*o Reversed 9°20′E.]

Thursday 4 August

Moderate & variable weather with Squalls of Rain & Wind. After midnight we kept our Wind all Sail set till Noon when we Observ'd in 28°14′N, Lat*itude* acc*oun*t 28°06′, Long*itude* d*itt*o 196°56′E. No Birds seen. This situation is near the parallel we last search'd for the Island of Donna Maria Lagara[1] & 4° East of it. As several Vessels bound to the NW Coast have pass'd in that space of Longitude shou'd it exist we may probably find it as we pursue our Course to the Westward.

Friday 5 August

Moderate and variable weather we steer'd NW till 6 PM when being in the Latitude of the Island we alter'd our course to WbS.[2] Throughout we had Showers of Rain with Calms & Breezes alternately from the SE quarter. Lat*itude* Obs*erve*d 28°23′N, acc*oun*t 28°30′N, Long*itu*de 195°41′E, No 1 195°33′40″ East. No Birds or any indication of Land.

Saturday 6 August

Moderate wind from the SE quarter with very clear warm weather. We steerd as before to pursue the same parallel of Latitude. At Noon we observ'd in 28°25′22″N, Account d*itt*o 28°23′N, Long*itud*e acc*oun*t 194°08′E, No 1 193°52′E. [*in margin*: Variation AM by the Compasses this morning made, No 3 10° 17′E; Walker 12°41′E; Adams 12°22′E; No 1 11°42′E; No 2 12°48′E.]

Sunday 7 August[3]

Light winds & fair weather without any variation. AM Muster'd the People & perform'd divine Service. At Noon the Lat*itude* 28°36′N, By account 28°26′N, Long*itud*e account 192°57′E. Our course as before with the Wind varying from SE to East, No 1 at noon 192°22′ East.

Monday 8 August

Moderate Breezes & very pleasant weather. In the afternoon we saw several Tropic Birds & in the morning a Turtle with several Sea Parrots[4] & Men of War Birds.[5] Lat*itude* at noon 28°41′N, Account 28°36′N, Long*itud*e d*itt*o 191°21′ East, No 1 190°36′ East.[*in margin*:

[1] In annotated *Voyage*, p. 81, 'Lagara' corrected to 'Lajara'. See also p. 35 and n. 1 above.

[2] To run down the parallel of Donna Maria Lagara.

[3] On this and the next two days' entries the month has been transcribed incorrectly as October.

[4] Probably Tufted Puffins (*Lunda cirrhata*) which range a little farther south than the Horned Puffin (*Fratercula corniculata*).

[5] Could be either Lesser Frigatebirds (*Fregata ariel*) or Great Frigatebirds (*Fregata minor*) whose ranges overlap in this vicinity.

Variation *per* Azimuth on the 8th, mean of 14 by No 3 12°31′E ought to be 9th;[1] On the 8th at Noon Longitude observ'd ☉ to ☽ 190°34′40″E; D*itto* by Mr Chapman 190° 21′34″E, mean of 3 Setts.][2]

Tuesday 9 August

Fresh Breezes & variable wea*the*r with Squalls of Rain attended with Thunder & Lightning. We steer'd with the Wind at East WbS¼S. Lat*itude* ☉ 28°39′N, Account 28°36′N, Long*itude* d*itt*o 189°26′E, Course W1°N, 106 miles.

Wednesday 10 August

Fresh Breezes and fine weather with the wind at East we pursued the same Course. Lat*itude* ☉ 28°36′N, Acc*oun*t 28°34′N, *Longitude* 187°26′, Course W1°S, 106 miles. Longitude at Noon shewn by No 1 186°20′ East.

Thursday 11 August

Ditto winds and weather without any Alteration whatever. At Noon Lat*itude* ☉ made 28°41′N, Account 28°30′N, Long*itude* 185°22′E, No 1 184°9′05″ East. Mean of 30 dist*ance*s ☉ to ☽ & ☽ from ∗ made the Long*itud*e 184°43′27″ by Mr Chapman. [*in margin*: Variation p*er* Amplitude 11th PM 14°04′ East; Longitude p*er* Mr Chapman ☉ to ☽ 184°43′27″E.]

Friday 12 August

Pleasant Breezes & very fine weather Wind & course the same as yesterday. In the Forenoon we pass'd two Turtles with some Sea weed. Many Birds about us & Bonitos, of the latter we took four. Latitude Obs*erve*d 28°36′40″N, Acc*oun*t 28°43′N, Long*itude* 183°27′E, Course S88°W, 101 miles, No 1 made 182°13′21″E, No 248 182°52′55″E, Long*itud*e mean of 25 distances of Stars East & West of the Moon 182°51′21″E. NB the Watches have differ'd since leaving Oneehow. [*in margin*: Longitude observ'd ☉ to ☽ the 12th 182°52′55″E.]

Saturday 13 August

Light winds & very fine clear weather. The Birds still remained with us. Course West, 67 miles. [Latitude] 28°34′38″N, Acc*oun*t 28°36′N, Long*itude* acc*oun*t 182°20′E, No 1 180°43′30″E. The Sea continued perfectly Smooth with a declining Breeze. [*in margin*: AM On the 13th [Variation] by Azimuth No 3 10° 23′E; Adams 13°28′E (mean of 7 observations); Walker 13°

[1] But entered under the 8th in Broughton, *Voyage*, p. 382.

[2] Further observations are given in Broughton, *Voyage*, p. 82, partly annotated in pencil and partly in ink:

At 20h. 45m. I made the long. Mean of 3 sets ☉ west of ☽ 190°50′30″E

Mr Chapman mean of 5 sets	191° 3′36″E
Mr Crosley mean of 4 d*itto*	190 22 47
No 1 same time	190°52′30″E
No 248	191 14 5
Mean	191 3 17

20′E (mean of 7 observations); No 2 13° 33′E (mean of 7 observations); [No] 1 12° 20′E; Inspection 12° 30′.]

Sunday 14 August
Light Breezes & fine weather. After mustering we perform'd divine Service. Our Fresh Provisions being expended we serv'd Sour Krout. On opening the Mustard Seed to issue at the same time we found it entirely Spoil'd. Lat*itude* Obs*erve*d 28°34′30″N, Account 28°34′N, Long*itude* 181°31′E, No 1 179°48′ East.

Monday 15 August
Light airs and calms with sultry weather and wind Variable from SE to SW. Course N64°W, 26′, Lat*itude* ☉ 28°45′32″N, Acc*ount* 28°42′N, Long*itude* d*itt*o 181°03′E, No 1 made the Longitude at Noon 179°08′E. No Birds seen to day.

Tuesday 16 August
Calm & clear weather without any Variation whatever. The Observations for the Latitude & Longitude made the same as Yesterday. A proof of no Current affecting our Course.

Wednesday 17 August
We had the same weather till 6 AM when a Light air sprung up from the Eastward & we steerd to the Westward. Lat*itude* 28°44′N, Long*itude* acc*oun*t 180°53′E, Latitude account 28°47′N.

Thursday 18 August
Light Breezes & fine Weather with some Squalls of rain we steer'd West and at Noon observ'd in 29°02′N, Account 28°58′N, [Longitude account] 179°44′ East, 177°29′ No 1. [*in margin*: Variations on the 18th AM by Azimuths, No 3 13°31′E; Walker 13°12′E; Inspection 13°00′; 18th Observ'd the Longitude by Stars East & west of Moon 178°24′47″E); Mr Chapman 3 sets on one side 178°0′0″E.]

Friday 19[1] August
The same continuance of uncommon fine weather with faint Winds from the Eastern quarter. At Noon the Lat*itude* made 29°20′N, Account 29°14′N, Long*itude* d*itt*o 178°45′E. Several Fish & birds attending us.

Saturday 20 August
After midnight the Wind increas'd from the SE & we steer'd West. Several Fish & some Tropic Birds still remain'd & Many Flying Fish. Lat*itude* at Noon 29°47′N, Acc*oun*t 29°38′N,

[1] '18' in MS corrected in pencil to '19'. This error continued until 27 August. Having now recrossed the 180° meridian Broughton's dates were once again correct.

Long*itude* 177°15′E, No 1 made 174°16′E and 248 175°14′E. Wash'd & air'd below. Mr Chapman made the Longitude by Stars East & West of the Moon 174°31′20″E.

Sunday 21 August

Fresh Breezes and fine weather with the wind gradually veering to the Southw*ar*d. We steer'd West and at Noon prevented taking any Observation by a heavy squall of rain. We had a swell from the SW which we had not experienced before. AM Muster'd the Ships Company and perform'd Divine Service. Lat*itude* acc*oun*t 30°16′N, Long*itude* d*itt*o 174°55′E, Course N78°W, 121 miles. At Noon Squally weather.

Monday 22 August

Fresh Breezes & Squally weather at times. At 2 PM we alter'd our course to WNW to pass to the North of Cap*tain* Clerkes Track. In the forenoon we saw a Bird similar to a Plover. Flying Fish and Tropic Birds were also seen. At Noon the Lat*itude* was 31°25′43″N, Acc*oun*t 31°24′N, Long*itude* d*itt*o 171°59′E, Course N56°W, 130′, No 1 made 170°0′ East, No 248 170°43′0″ East. Fair & pleasant weather.

Tuesday 23 August

Fresh Breezes & fair weather the Wind continued steady at S*outh* & we steeer'd WNW½W. This day we saw many Flying Fish. The Thermometer continued at 80 & 81½ without any variation but being so much accustom'd to it we did not feel it oppressive. Lat*itude* by Observation 32°26′18″N, Account 32°33′N, Long*itude* 169°34′E, No 1 167°36′ East. [*in margin*: Variations p*er* Azimuth, No 3 5°55′E; Adams 9°18′E; Walker 11°52′E; Inspec*tio*n 10°30′E.]

Wednesday 24 August

Fair and pleasant weather with smooth water, we steer'd as before with a fine Breeze from SSE. Latitude by Observation made 33°30′N, by Account 33°22′N, Long*itude* d*itt*o 167°23′ East, No 1 at Noon 165°49′4″E, same time ☽ to the West of the Sun mean of 3 Sets of Distances166°20′45″E, Mr Chapman 166°20′. This being the Parallel in which Rica de Plata is laid down in[1] and the Longitude of 164°E by Cookes Chart[2] we alter'd our course to WbS in hopes of seeing it shou'd it be so situated. We saw nothing to indicate Land being near us this day.[3]

Thursday 25 August

Moderate Breezes & fair weather we pursued the same Course Wind from the Eastern quarter. At Noon the Lat*itude* was 33°11′N, Acc*oun*t 33°27′N which is a difference of 16′ &

[1] '33½°N' in Henry Roberts, 'A General Chart'.

[2] Henry Roberts, 'A General Chart'.

[3] For the origin of this non-existent island see Stommel, *Lost Islands*, pp. 1–5. See also the extensive note in Beaglehole, *Journals of Captain Cook*, III, p. 638, n. 2.

shews how strong we must have felt the force of a Southern Current having only run 90 miles nearly due West.[1] Course made good S78°W, 92 miles, Long*itude* by account 165°36′ East, No 1 made 164°25′25″ deduc'd from afternoon Sights.

Friday 26 August

Fresh Breezes & fair weather, the Wind veer'd to ENE & we steer'd to the North till ½ past 3 PM when our Lat*itude* being 33°26′N & Long*itude* 164°25′25″E we steer'd West to avoid the effect of the Current which had before set us to the South. During the Night we run under an easy Sail & the ☽ rose at ½ past 10 so that we cou'd see distinctly some distance. At daylight nothing to be seen we again made Sail and observ'd at Noon in 33°44′N, the Observation was indifferent but indicated by our situation we had lost the Current. Lat*itude* by account 33°41′N, Long*itude*163°28′E, No 1 162°18′ East, Course N73°W, 111′. [*in margin*: Variations p*e*r Azimuth PM, No 3 8°58′E; Adams 9°16′E; Walker 10°28′E; Inspection 10°00′E; AM 26th, No 3 7°18′E, Revers'd 7°46′E; Adams 8°58′E, Revers'd 8°36′E; Walker 9°50′E, Revers'd 9°46′E; Inspection 8°30′.]

Saturday 27 August

Fresh Breezes and cloudy weather, we steer'd WSW till 8 PM to regain the parallel of the Island and afterwards WbS. During the night we proportion'd our Sail so as not to exceed running more than 5 knots p*e*r Hour. At daylight we made Sail and at Noon we Observ'd in 33°30′N. The Horizon throughout the Forenoon was very Hazey but the atmosphere above 20° very Clear. Wash'd and air'd below. Lat*itude* account 33°30′N, Long*itude* d*itt*o 160°42′ East. [*in margin*: Variations p*e*r Azimuth, No 3 mean of 10 sets 7°8′ East 27th.]

Sunday 28 August

Pleasant Breezes & clear weather in the Forenoon. We steer'd W½S with the wind easterly. AM after mustering the Ships company perform'd Divine Service. Lat*itude* Obs*erve*d 33°28′43″N, Account 33°32′N, Long*itude* 158°19′E, No 1 157°29′E. [*in margin*: On the 28th the Longitude observ'd ☉ to ☽ made 157°45′E; Mr Chapman d*itt*o 157°57′15″E.]

Monday 29 August

Moderate Breezes & clear weather, we steer'd West after 7 PM and at Noon observ'd in 33°57′30″N which is 22′ North of the estimation, Lat*itude* acc*oun*t being 33°35′N, Long*itude* d*itt*o 156°16′, which is 24′ more west than No 1 which shew'd 155°48′E. [*in margin*: On the 29th Mr Chapman made the Longitude 156°08′35″.]

Tuesday 30 August

Light winds, Smooth water & pleasant warm weather. Lat*itude* Obs*erve*d 33°56′N, Account 34°04′N, Long*itude* 155°44′E, No 1 154°3′25″,[2] Course due West, 77 miles. [*in margin*:

[1] Little is known of the currents in this region.

[2] '¼' in the MS; '25″' taken from Broughton, *Voyage*, p. 383.

Variations August 30th AM, No 3 5°18′E, Revers'd 4°6′E; Adams 5°36′E, d*itt*o 5°48′E; Inspection 5°30′.]

Wednesday 31 August

Moderate Breezes from the Eastward with fine clear weather. At Noon we steer'd WbN. Lat*itude* by Observation 34°01′N, Account 34°22′N, Long*itude* acc*oun*t 153°53′ East, No 1 152°36′E. A Current seems to have drifted us to the SE'ward by the variations in the Observations & Estimation very considerably.

Thursday 1 September

Moderate Wind from the SE which decreas'd towards Midnight. We steer'd NWbW at 7 PM & in the Forenoon the Wind veer'd to the West with a Light air which continued. We Kept to the North and at Noon the Lat*itude* made 35°19′18″, Account 34°46′N which is 34′ more than the [Traverse] Board gives, Long*itude* account 152°45′E and No 1 151°31′4″ East.

Friday 2 September

Moderate Breezes and cloudy weather the wind gradually veer'd from the West to North & NE with squally & unsettled weather which induced us to take 3 Reefs in the Topsails for the Night. At 7 PM we tack'd and steer'd NWbW. At Daylight we had fine clear weather & we made Sail. The Forenoon we clear'd the Slop Room & Sail Room. The Heat had been so great as to have damag'd many of them & indeed we had universally found a degree of dampness in the Air that affected every thing with Mould. After puryfying the Fore Cock Pit with Fires & extracting the Foul Air by the Ventilator we stow'd them away again. At Noon we observ'd in 36°39′N, Account 36°18′N, Long*itude* d*itt*o 151°41′ East, No 1 made 151°11′40″E. The Current this day has set us to the NW. [*in margin*: Variations p*e*r Azimuths the 2nd AM, No 3 1°58′E, Rev*ers'd* 2°49′E; Ada*ms* 4°18′E, [ditto] 5°6′E; Walker 4°22′E; Inspection 4°.]

Saturday 3 September

Clear and pleasant weather with the Wind at East which in the Night veer'd round to South & SW and continued variable in that course during the Forenoon. At Noon the Latitude made 37°35′36″N, Account 37°33′ North, Long*itude* d*itt*o 150°11′E, by No 1 149°02′E which is 19′ less westing than the estimation. [*in margin*: Variation p*e*r Amplitude PM 1°25′ East.]

Sunday 4 September

Fresh Breezes & fine weather. In the Night we had a heavy dew and the Forenoon was very hazey. We steer'd NWbW with the wind from the SW quarter. Several Whales, Porpoises, & Sheer Waters were seen. Muster'd the Ships Company's Cloathing &c. At Noon observ'd in 39°06′N, Account 38°46′ which is 22′ N*orth* of account, Long*itude* 148°01′E, No 1 146°33′E, 20′ to the West by the Watch. Smooth water and fine weather.

Monday 5 September

Fresh Breezes and very hazey throughout. During the night we felt a very heavy dew that penetrated into every part of the Ship. The Wind continued steadily fix'd in the SW quarter & we kept our wind to the Westward to ensure fetching the Northern part of Japan. At Noon with a bad Horizon we made the Latitude 39°49′50″N, Account 40°00′N, Long*itude* d*itt*o 146°00′E, No 1 144°30′46″E and 248 145°44′15″E. In this days run we appear to have experienc'd very little current & that southerly.

Tuesday 6 September

Ditto weather, if anything more hazey. We sounded hourly during the Night & at Noon, but did not reach the bottom. In the forenoon we pass'd several Trees floating in the water at intervals & many Spermaceti whales of a very large Size. An indifferent observation made the Lat*itud*e 40°28′N, Account 40°34′N, Long*itud*e d*itt*o 144°E. Course as before N67°W, 100′. No 1 made 143°0′E by PM sights which is 30′ less longitude than Estimation.

Wednesday 7 September 1796

Fresh Breezes and hazy weather with a Swell from the SW D*irectio*n. We kept our wind to the westward in hopes of seeing the Land shou'd the Haze clear away & the Barometer falling very suddenly seem'd to indicate a change of weather, which did not deceive us. At 5 the breeze increas'd & we took 3 reefs in the Topsails & got the top gallant yards down. At 6 the wind veer'd to South & increas'd to a hard Gale; we Close reef'd the Topsails & haul'd the Mainsail up to Sound. Before we cou'd try for Soundings the Foresail & Foretopmast Stay sail were split to Pieces – and the Squalls coming more violent. Before the Ship was Wore we Furl'd the Mainsail at which period the Fore topsail blew fairly away from the roping. Having wore under the Maintopsail we bro't to the Wind upon the Starboard Tack. The Wind suddenly veer'd from the S*outh* to NW & came with such violence as to blow the Maintopsail entirely away from the Yard. This was our second best sail and was only bent the day before. The Ship's Careen was so great & the Wind blowing so tremendously hard I was much afraid of our Masts giving way & having no Sail to set, we were left entirely to its mercy in the Hollow of the Sea. The water blew over us with such force we cou'd not stand against it. Luckily we experienc'd no other misfortune than the total loss of our Sails & which in our situation was to be much regretted. In contradiction to former Observations (more particularly Captain Cooke's) it prov'd in this instance the Rope was better than the Canvas, as no part of it gave way tho' they had never been strengthen'd to prevent it (which is not unusual in these Voyages). In no instance did I even ever experience so uncommon a Gale for the Hour it lasted. It was unattended with Lightning, Thunder or Rain & towards 8 PM it gradually abated; the wind remaining steady in the NW quarter, a moderate Breeze & fine Night. We employ'd ourselves bending other Sails & standing to the SW as the wind permitted, and at daylight we descry'd the Land of Japan bearing from WbN to SW 7 or 8 Leagues. The Wind was so moderate as not to permit our approaching the Land more than 4 miles till Noon when we observed in the Lat*itude* of 39°55′N[1] with the following bearings: Extremes from N69°W

[1] In annotated *Voyage*, p. 85, corrected in ink to 39°55′37″N.

to S46°W, an Inlet with an Island in its Entrance suppos'd to be Cape Nambu N52°W,[1] Off Shore 5 or 6 Leagues.[2] Lat*itude* in acc*ount* 40°12′N, Long*itude* d*itt*o 143°0′E. No 1[3] made 141°54′E & [No] 248[4] 143°12′E. Our situation with the bearings of Noon will place us in 142°50′E (by the Chart in Captain King's 3[d] Voyage of Cooke).[5] Allowing it to be right the Mean of the two Watches will be nearest the truth and 248 more so than No 1.[6] Our distance from the shore prevented any further Remarks & it evidently appear'd both by the observation at Noon & the alteration of our situation that a current had set us strongly to the South.[7] At Noon light winds and & fine weather.

Thursday 8 September

Light Winds and fair weather from the Southward. We steer'd to the NNW & in the afternoon we set the Rigging up Fore and aft & were otherways emp*loye*d repairing our defects & drying sails &c after the late Hurricane. At 6 pm the extremes bore from N40°W to S40°W off shore 5 or 6 Leag*ues*. In the night we frequently tried for Soundings. At 4 AM we lost the Southerly wind & we had Foggy & Calm weather with rain till Noon when the extremes bore from N42°W to S16°W & the South point of entrance into Nambu inlet[8] S35°W 3 or 4 Leagues. Instead of being to the NNW 30 Miles by our Account we found our selves more to the SW than we were Yesterday the Current setting along shore in that direction.

Friday 9 September

Calm and rainy weather continued till 4 PM when a Breeze sprung up from the SE quarter with which we made sail to the North at this time we had drifted near the entrance of Port Nambu the South point bearing from us S35°W 3 Leagues. It appear'd extensive & capacious but the Wind wou'd not permit us of approaching any nearer & the gloominess of the Sky prevented our seeing any thing distinctly; the Barometer likewise falling seem'd to imply bad Weather coming on. At ½ past 6 the NE[9] extreme of Japan bore N50°W 6 or 8 Leagues. At 10 PM the Wind shifted round by the South to NW and soon increas'd to a Fresh gale with cool weather; at ½ past 5 AM we cou'd just discern the Land bearing S30°W 10 or 12 Leagues.

[1] In annotated *Voyage*, p. 85, 'dist 25½ miles' added in ink.

[2] Broughton was probably consulting the 'Chart of the Coast of Japan', plate 79, in Cook and King, *Voyage to the Pacific Ocean*, III, f.p. 397, where the inlet is charted in 39°49′N, with its northern entrance point named C. Nambu. The inlet is clearly Miyako Wan and King's C. Nambu is possibly Ma Saki, a protruding 79-metre high wooded headland which appears as an island from seaward, or Anega Saki, about 3 miles south, the northern entrance point of Miyako Wan, ¾ of a mile south of which lies the island Hide Shima. Cape Nambu is not shown on Roberts's 'A General Chart', but Shiriya Saki, the NE extremity of Honshu is named C. Nabo. To add to the confusion, on Broughton's 'A Chart of the N.E. Coast of Asia, and Japanese Isles', Broughton *Voyage,* f.p. i, the inlet is named Port Nambu and the NE extremity of Honshu C. Nambu.

[3] Earnshaw No. 1.

[4] Earnshaw No. 248.

[5] I.e. 'Chart of the Coast of Japan'.

[6] In annotated *Voyage*, p. 85, inserted in ink 'Lat & Long of Nambu Entrance 142°19′12″E p*er* mean of Time keepers & p*er* mean of 89 sets of Lunars 142°27′39″'.

[7] The *Providence* was probably under the influence of the south-going Oya Shio current which runs along the east coast of Honshū until 36°N when it meets the northern edge of the north-going Kuro Shio.

[8] Hei Saki, a rounded thickly wooded cape.

[9] 'NW' in the MS, which is clearly a transcription error; Shiriya Saki.

Lat*itude* at Noon 40°44′36″N acc*ount* 40°55′N, Long*itude* in from Cape Nambu 143°14′E, No 1 142°06′E. NB the watch is 58′ to the West of Nambu. By the observations we have experienc'd the Current setting us to the SW about 15 Miles.

Saturday 10 September

Fresh gales and Hazey weather continued till near Midnight when it became moderate with the Wind from the same quarter NW. We kept our Wind to the NNE under Close Reef'd Topsails and Courses. At 3 AM we saw Land bearing N*orth*[1] and at daylight it bore from N5°W to [N]40°W about 5 Lea*gues*. To the North it form'd a low point[2] & the SW part was so distant we cou'd not distinguish whether it join'd the NW part or not. The double Land[3] within the first range of Hills was of considerable Elevation & the Eastern Shore appear'd bold and Cliffy. It was Calm throughout the Forenoon. An indifferent Observation made the Lat*itude* 41°32′22″N when the Island bore from N12°W to N38°W about 6 Leagues. Course N44°E, 61′, Accounted Lat*itude* 41°31′N, Long*itude* 144°08′E, 142°44′E No 1. The current set us to the SW by the shore from which we gradually increas'd our distance. In the Afternoon we tried & found the Current set South 3 fath*oms* p*e*r Hour.

Sunday 11 September

In the Evening a Light air from the SE carried us towards the Land. At Midnight it veer'd to NNE and at daylight we were near the SE part of the Island which we pass'd at 8 within 1½ miles having 22 fathoms Sand & Small Stones.[4] We steer'd N50°W which was nearly the direction of this side of the Island. This point is low & flat & several Rocks above water project from it to Seaward, two of them resembling Asses Ears they bear N15°E & S15°W from the NE side of the Island, within them to the North the shore forms a Bay. The low land continues for 2 or 3 Lea*gues* when you come to a Sandy Beach the Country rising gradually from the Sea to a great height in various ridges Cloath'd with Wood & presenting a beautiful appearance. Several Houses were scatter'd along the Shore with Boats haul'd up on the beach. Several People were observ'd Fishing & walking about them. Their Habitations appear'd to be made of reeds & cover'd with the same resembling in Shape small Barns with doors & Windows to them. At Noon we observ'd in 41°56′35″N and No 1 made 142°14′45″E, No 248 143°48′2″E. The SE point bore S73°E 3 Leagues & the Extremes N 53°W Offshore 3 or 4 miles, 30 fath*oms*. In the morning we pass'd a very large Turtle. Latitude of SE P*oint*[5] 41°[6]54′9″N, 12′ East of the Watches at Noon. [*in margin*: Variations PM p*e*r Azimuths, No 2 1°55′E, R*eversed* 1°34′E, Ad*ams* 1°27′E, R*eversed* 0°50′E, Inspect*io*n 1°45′E]

[1] Hokkaido.

[2] Possibly Erimo Misaki, the SE point of Hokkaido.

[3] 'The appearance of the coast when the shore line is bounded by parallel ranges of hills, rising inland one above the other': W. H. Smyth, *The Sailor's Word-Book*, London, 1867, p. 259.

[4] In annotated *Voyage*, p. 87, added in ink, 'At 20h. Rocky Point or SE point of Insu [Erimo Misaki] bore N80°E dist*ant* 3 miles. Its Lat*itude* 41°55′11″N and Long*itude* per mean of 2 Time keepers 143°20′11″E Long*itude* per 89 sets of Lunars 143°28′38″E.'

[5] Erimo Misaki.

[6] '42°' in the MS, which is clearly a transcription error.

Monday 12 September

Light Breezes & very fine Weather with the Wind from the SW quarter. We stood along shore to the WNW till ½ past 6 PM when we Tack'd in 42 f*atho*ms 4 or 5 Miles off shore, the Extremes bearing from N55°W to S64°E. In the afternoon three Fishing Boats came off to us & we had the good fortune to induce them to come on board after they had previously saluted us in a Solemn manner agreeable to the Oriental Custom of Salams. They were of a light copper Colour with dark Hair very thick & cut short behind & all of them with Long beards, fine Black Eyebrows, tolerable Teeth with expressive good natur'd countenances. There were two in each Boat. They were of a midling Hei[g]ht & Shape Cloath'd in a Gown wove from the bark of a Tree edg'd at the Cuffs and Collar with blue Linnen & wore a piece of white Linnen similar to the Maros[1] at the Sandwich Islands. We presented them with Beads & Trinkets with which they seem'd pleas'd but we cou'd not understand what they said. I should imagine by their behaviour they had never seen a Ship before as they appear'd alarm'd & sat down cross Legg'd close to the gangway where they enter'd without ever speaking a word. We show'd them our animals but they shook their Heads implying as we suppos'd they had not any on shore. We enquired frequently if the Island was called Matsmai[2] & they invariably pointed to the West, & as we understood this Land, they pronounced Insu.[3] Having remain'd with us an Hour they took their leave Saluting as before till they were at some distance. Their Boats were Hollowed out of a Tree like Canoes with upper Streaks sew'd on round them which increas'd their width aloft & spread Fore and aft being the same Shape both ends. They were pull'd in the same manner [as] the passage Boats at Plymouth One person making use of a pair of Skullers with the difference instead of pulling together they pull'd each oar alternately. During the Night we plied to Windward to get round the west end of the Island and in the Forenoon it was Calm. An indifferent Lat*itude* 41°46′25″N, the SE point bearing N65°E 6 L*eague*s & the NW point[4] N28°W 8 or 10 Leag*ue*s. Light Airs & Hazey at Noon. NB These People wore Ear Rings & each of them had a Knife in a sheath tied round their Waists. In

[1] A type of loincloth.

[2] Hokkaido.

[3] In the margin of RUSI/NM/132 (8) Broughton writes, 'It should be remember'd that we had no account whatever relating to these parts excepting Cooke's 3d *voyage* & general chart, in which Matsmai is laid down as an island N*or*th of Japan, which we suppose this to be'. Also in a lengthy footnote to the published account of his voyage, Broughton writes, 'Insu is the famous land of Jeso, only known to Europeans before this and La Perouse's Voyage, by the journals of the *Castricom* and *Breskes*'; Broughton, *Voyage*, pp. 89–92. He then continues with a lengthy and slightly misquoted untranslated extract from a book by Johann Reinhold Forster: 'In 1643, the Dutch East-India Company gave orders for two ships to be sent from India to the North, in order to explore the route from Japan northward, and even to go as far as North-America, and to seek for a passage there. In consequence of these orders, the two ships set sail together on the 3d of February, 1643, from the harbour of the island of Ternate: these were the *Castricom*, commanded by Captain Martin Herizoom van Vries, and the *Bresces*, under the command of Captain Hendrick Cornelis Schaep. On the 14th of May, the two ships were separated by a storm, at the distance of 56 leagues from Jeddo, the capital of Japan; and both of them saw the land of Jeso. The *Bresces* sailed through the strait between Jeso and Japan'; John Reinhold Forster, *History of the Voyages and Discoveries made in the North,* London, 1786, pp. 244–6. In his misquote Broughton changed 'the strait between Jeso and Japan' to 'Straits of Sangaar' and commented 'It does not seem to be the fact that the *Bresces* ever passed the Straits of Sangaar, as she was cast away on the north part of Japan attempting it. Broughton then continues with another lengthy and slightly misquoted extract on the same theme from Gerhard Friedrich Müller, *Voyages et Découvertes faites par les Russes le long des côtes de la Mer Glaciale & sur l'Océan Oriental, tant vers le Japon, que vers l'Amérique*, 2 vols, Amsterdam, 1768. For another account of the voyages of the *Castricom* and *Breskes* see James Burney, *A Chronological History of Voyages and Discoveries in the South Seas*, 5 vols, London, 1803–17, III, pp. 154–66. See also p. 85, n. 3 below.

[4] Chikiu Misaki.

pulling they made use of their Skulls not together but first one hand & then the other which prevented their moving in a strait Line, but still they went very fast.

Tuesday 13 September

Moderate breezes & dark cloudy Weather with the Wind from the NW. We stood within 2 miles of the Shore when we tack'd in 42 *f*a*th*o*m*s the Ex*tremes*[1] bearing from N61°W to S62°E. As we stood in we perceiv'd two large Boats at Anchor. They had each one Mast rigg'd a midships on which they made use of a Cloth or Canvas Sail very square. One of them got under way & ran along shore; they were similar to a small Chinese Junk in Shape, but we were not near enough to be very particular. In the Night the Wind gradually veer'd round by the North to the SE with gloomy & rainy weather. At Daylight we saw more Land bearing from N15°W to N50°W & a High Bluff S69°W whether they were distinct Islands we cou'd not distinguish.[2] The Western part of Insu bore N18°E for which we steer'd but the Hazey weather at Noon obscur'd it from our sight. Course since Noon of Yesterday N45°W, 48 Miles, Lat*itude* acc*oun*t 42°20′N, Longitude [*blank*], 46′ West.

Wednesday 14 September

Moderate Breezes from the Eastward with thick rainy weather. At 2 PM we tack'd in 25 *f*a*th*-o*m*s[3] & stood off till 4 when we again stood in shore. At ½ past 6 we tack'd in 42 *f*a*th*o*m*s within 4 or 5 miles of the Land which bore from N22°W to N70°E & from N35°W to N75°W, the Land to the N*orth* apparently joining by the Low Shores. We kept under an easy sail till daylight when the Land bore from S78°E to N7°W and from N22°W to N87°W, Bluff point set Yesterday S38°W. We were nearly midway in the Bay both shores being distant, the East 7 and the West 5 Leag*ues*. From the Deck we cou'd see to the North the Bay was clos'd by low Land. We therefore steer'd to the West with a light Air from the North. At Noon the Lat*itude* made 42°19′35″N, the Bluff point bearing S18°W 6 or 7 Leag*ues*, apparent Island[4] S80°W 5 Leag*ues*, Extreme of Insu East and a Volcanic Mountain[5] S63°W.

Thursday 15 September

Moderate Breezes and Hazey wea*the*r with a large Swell from the Southward. The Wind from the same quarter enabled us to weather by 6 PM an Island set at Noon bearing west; & at 9 we had fetch'd well up with the Volcanic Mountain carrying in from 36 to 45 *fathoms,* when it became Calm which continued during the Night. In the morning we found we were in a deep Bay[6] the Volcano bearing S10°E 2 or 3 miles, the Island N64°E about 5 Leag*ues* & the Extremes towards the bluff point S45°E. To the North & west we saw no Opening. A great

[1] In annotated *Voyage*, p. 92, 'in one at 4h' added in ink.

[2] 'High Bluff S69°W whether they were distinct Islands we cou'd not distinguish' crossed out in RUSI/NM/132 (8) and 'very conspicuous head-land apparently unconnected with each other' substituted, which has been copied to Broughton, *Voyage*, p. 93, with the addition of 'bore S70°W' added after 'head-land'. Probably Esan Misaki.

[3] In annotated *Voyage*, p. 93, 'tacked in 42 fathoms' crossed out and replaced in ink by 'wore', followed by 'Query! The ship tack'd at 5½ whether the bearings were taken then or at 6½ hours.'

[4] An unidentified feature on the west side of Uchiura Wan.

[5] The 1,133-metre high Komaga Take.

[6] Uchiura Wan.

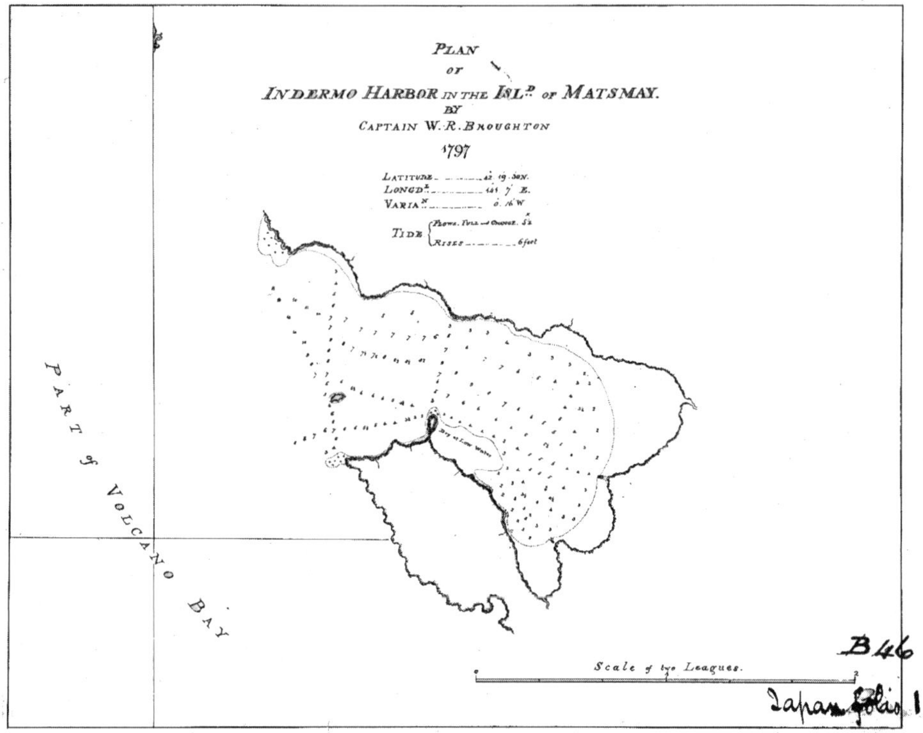

Figure 4. Plan of Indermo Harbor in the Isl[d] of Matsmay, by Captain W. R. Broughton 1797.
Courtesy of the UK Hydrographic Office.

quantity of Smoke issued from the North side of the Mountain but we saw no Flames. To the SW of it were several Houses scattered along shore & the Inhabitants before noon paid us a Visit. We found them to be the same we had before seen exclusive of several, who resembled Chinese their Hair excepted, being shav'd from the Crown and Forepart of their Heads remaining long from the sides & lower part & tied up behind, Smear'd with Oil or Grease. Every Man was furnishd with his Pipe and Tobacco Box & seem'd much entertained in visiting the internal parts of the Ship. A Breeze springing up from the Eastward they went away & we stood over for the North Shore. An indifferent Observation made the Latitude 42°16′N, the Island bearing N87°E and the shore towards Bluff point S47°E. We had 50 f*athoms* water about 4 Leagues from the Volcanic Mountain. No 1 made the Long*itud*e 139°32′55″ East and 248 141°22′30″E, their difference daily increasing so much prevented our having any faith in them, or which to give the preference to.[1] At 3 PM of the 16th having run on an NE course 14 Miles we were abreast of a Village off which lay a Junk at an Anchor in the opening seemingly of a small Harbour.[2] A Reef of rocks extends from the Entrance two or three miles. We tack'd in 10 f*athoms* to avoid them & sending a Boat ahead we bore up to Leeward of them for a Bay that promis'd good Anchorage.

[1] In annotated *Voyage*, p. 94, added in ink, 'The altitudes of the Sun observ'd by Mr Hayward [Haywood] the time per watch are very erroneous', followed in pencil by, 'which is the cause of the watches giving the Long*itud*e so different this day from the preceding ones'.

[2] Muroran Ko, Broughton's Indermo Harbor.

In Volcano Bay[1] Island of Matsmay

We came too an Anchor in 7 *fathoms*, soft muddy bottom, about ¾ of a mile from the Shore along which extended a Scatter'd Village.[2] The Outer Rocks bore South & we remain'd perfectly Land lock'd from the Sea in the North part of a Spacious Bay, the Volcanic Mountain bear*ing* from us S5°W 10 Leagues. We had several visitors but a Japanese coming off soon sent them away. He remain'd with us some time conversing (we suppos'd) as to our arrival, but we cou'd not well understand each other or indeed at all, excepting with regard to water. At Sun Rise the next morning we soon found several runs of Water thro' the Marshy grounds emptying themselves over the Beach & which appear'd to be very good (The Japanese assur'd us the same), so we immediately prepared to compleat our watering having between 30 & 40 Tons of empty Casks. Several of the Natives attended but were not permitted by the Japanese to approach us. He spread some Matts for us and while he smok'd we enter'd into Conversation, the result seem'd to tend to our departure, when we shou'd sail &c. On our going towards the Village he pointed to the Sun & was very expressive in pointing out to us we were not to go that way. We row'd along the Beach to the westward about 2 miles where the Country gradually rises with sloping Hills covered with Verdure & variegated with wood, landing at a small Village near a fine run of water, which induced us to give it a preference to those we had before seen. The Natives attended us with great humility bringing us Mats to sit upon & we had no Japanese to interrupt their civility. After Observing the Sun's Meridional Altitude we return'd on board. During our absence it appear'd several Japanese arriv'd at the Village on Foot & follow'd by 16 Horses that were laden seemingly with merchandize belonging to them. The Junk also stood out to Sea at the same time. In the afternoon these Strangers accompanied by the former Japanese came on board. They were cloath'd in the Chinese manner with Dark colour'd Cottons with Silk Sashes round their Waists & each of them arm'd with two Sabers ornamented with gold & Silver & the Scabbards Japanned. They wore half boots of Cotton & to their feet Sandals of wood & matting. They shew'd us some writing in Chinese characters on which I wrote the Ships Name &c. Their questions related to who we where, whence we came &ca and as we made them understand, one of them who had a Pencil with Ink immediately noted it. After smoaking their pipes & tasting our Liquors they took their leave. Another Junk anchor'd near us & sail'd the same Night. She was laden with sea weed (Fucus sacharinus).[3] These same people came off with two more Strangers before Night but I did not see them being absent from the Ship. [*in margin* in pencil: Lat*itude* at Observatory Point 42°33′11″N, Longitude 140°50′32″E, Variation [*blank*]]

The next day being moderate weather we haul'd the Seine with little success & employ'd the Long boat watering. After Breakfast we visited a Small Harbor[4] the entrance being between the Rocks & we had 3 *fathoms* water about half Tide. It's a very convenient Port for small

[1] Uchiura Wan.

[2] The modern town of Muroran, which extends over most of the bay.

[3] Now *Laminaria saccharina*, which does not occur in Japanese waters, though *Laminaria japonica* and several related species are found there. From these species a seafood product known to the Japanese as Kombu or Koba is produced, enormous quantities of which are exported to China and other parts of the East Indies.

[4] This and other features in the bay are no longer identifiable, the whole of Muroran Ko being extensively developed.

Vessels. Several Houses are scatter'd about it & One of a larger kind inhabited by Japanese with a small garden near it in which were growing Turnips, Raddishes & French beans. Before we return'd on board we landed at the large Village & met our Japanese acquaintance. They appear'd to be uneasy at our coming & beg'd us to return on board. Finding we wou'd not comply with their wishes & looking about the place they accompanied us, keeping as far from the Houses as they cou'd till we arriv'd at the watering place, where they appeared very happy seeing us embark for the Ship. In the morning we saw some Women rowing with the Men on their fishing parties, their Hair was cut short round their Heads like a Mop & their Lips punctured with some Bluish colour. Their Cloathing was to appearance the same with the Men and their behaviour becoming their Sex.[1] On Sunday we had the same fine weather & we mov'd our Ship to the best watering place which lay 2 Miles further to the North. Here we came too in Eleven fathoms. We tried the Seine once more without any success but our Boat fishing with Hook & Line was more fortunate. The same party of Japanese came down to us seemingly to know what we were about, but did not come on board. The next day were employ'd wooding and watering.

On Tuesday [20 September] we had Cloudy and variable Weather & we made a party to explore the NW part of the Bay; about 3 Miles from the Ship. We came to a Village situated on a fine Beach with a small River to the NW which empty'd itself into the Sea. It took the direction of the Bay & must probably have its source some distance to the N*orth* from the force of the Stream & its depth of Water. The Country thro' which it Run presented a beautiful Autumnal prospect & the Hills were cloath'd with Verdure & interspers'd with various clumps of Trees, so well disposed in the Vallies & sides of the Hills as to give them the appearance of Art. In no part did we see any Cultivation. The people of the Village both Men & Women received us very civilly. In the afternoon we had a Fresh Breeze from the SE quarter with dark threatening weather & a large Swell which increas'd during the Night altho' the Wind moderated. Wednesday was dark gloomy weather & a Surf on the Beach so great as to prevent our getting off any wood or Water after Noon. Thursday brought with it fine weather but the violence of the surf prevented our Working parties going on Shore. Friday was equally unfortunate as we cou'd not trust our Boats to bring off our Wood & Water. One Boat being already stove we were employ'd repairing her and putting the Rigging to rights. Fine Weather throughout the day with the Swell decreasing. It appear'd to have blown hard at Sea though we did not feel the effects of it beyond the swell it occasioned. On Saturday we had fair weather with a moderate land & sea wind which enabled us to compleat our watering and to get off some wood. The Natives on shore attended us in small parties bartering Grapes for buttons & now and then a Canoe wou'd sell us some fish on board the Ship. [*in margin*: Mr Crosley made the Longitude on Shore opposite the Anchorage 140°50′32″ East, being the mean of observed distances east and west of the Moon. Mean of the Observed Latitude on Shore 42°33′11″ North. Mean Variation by all the Compasses on board 1°27′20″ East. The Variation observ'd on shore by 3 Compasses 0°16′30″ West. It appear'd to be high water at the full and Change at 4^{h} 30′ apparent time, rise and fall between Four and Five Feet. We experienced no Tide at Anchor.]

[1] In annotated *Voyage*, facing p. 97, is an engraving titled 'A Man & Woman of Volcano Bay' (see Fig. 5, p. 80), above which is inserted in pencil: 'The original drawing from which this engraving was made were designed by the Natives themselves, one containing the figure of the Man, the other of the Woman. Had the bookseller made a facsimile of them they would have been curious specimens of the people in the arts of these rude Islanders, but this engraving is a very imperfect representation of the manner of the originals.'

Sunday 25 September

Calm and Cloudy weather. After mustering the Ships Company they had leave to go a fishing and a Boat was sent under the direction of the Master to examine the SE shore towards West Island. Before Noon we were visited by a new party of Japanese superior in Dress to any we had seen & in behaviour equally so. We derived not only pleasure but information from them. They brought a Chart of the World which by the writing appear'd to be Russian & perfectly understood from whence we came. Having a Book with them on which was pencil'd the Arms of the different Countries in Europe & immediately [they] pointed to us the British Arms. They had also a Book with the Russian Alphabet & Figures to 100,000 & speaking the Language imperfectly we had the good fortune to comprehend them by means of a Russian Seaman. They permitted me to copy a large Chart of the islands North of Japan & promis'd to bring me one the next day of their own doing. In the course of the night the Boat return'd reporting they had found an excellent Harbor near the apparent Island & which they found connected with it by low land. Monday was fine weather with Smooth water which enabled us to compleat our wooding & watering. Our Friends came & dined with us bringing the promis'd Chart for which I presented them with Capt*ai*n Cookes General Chart of the World which appear'd to gratify them exceedingly. They were curious in taking an account of every thing & what they cou'd not understand they represented in India Ink drawing. The same fine weather on Tuesday permitted Mr Crosley to compleat his observations for the Time Piece & we prepare'd for sailing. The Cutter was sent to examine the Head of the Bay where they landed near a extensive Village; contiguous to which they saw many Horses grazing & the appearance of ploughed Ground. Early in the morning of the 28th we got under way & the wind coming Easterly we stood over the Bay towards the Volcanic Mountain. At 11 AM we tack'd and at Noon observ'd in the Lat*itude* of 42°18′20″, apparent Island bearing N89°E 4 Lea*gue*s, Harbor entrance N76°E Volcano mount S2°W Old Volcano N50°E. We kept plying to the SE till near Dark when we came too an Anchor in 10 f*athoms* soft Bottom in the entrance of the Harbor (seen by the Cutter) Outer part of apparent Island bearing S22°E a small island on with a large Village S37°E Head of the Harbor ESE 5 miles, Larb*oar*d p*oin*t of Entrance N10°W off the North shore 1 mile. Throughout the Night it blew strong from the SE quarter. Thursday continued with an increase of Wind and dark threatening weather with Thunder & Lightning & some Rain. In the afternoon it was more moderate which enabled me to examine the Harbor which affords good Shelter from all Winds. Bringing the Bluff on the Isthmus to bear NW you will have 4 & 5 f*athoms* and the Larboard p*oin*t of Entrance will be just on with the Bluff. In running in you must keep the Island open with the Bluff till you are within ½ a Mile of an Island join'd at low water to the Land & steering to the SW you will gradually shoal your water to the Berth you prefer. You carry in from the Entrance Regular Sound*ing*s from 10 f*athoms* to 5, 4 & 3 soft Bottom. A few Houses are scatter'd on the Starb*oar*d Shore within the Harbor & towards the Head of the Harbor the Shores are low and flat. Wood and water is to be had in great abundance. [*in margin*: This Harbor which the Natives call Endomo[1] is situated in the NE corner of the Bay Form'd by Apparent Island which is an extensive peninsula of a circular figure. It lies in the Latitude of 42°19′29″N, Longitude of 141°7′36″ East, Variation the same as before. Flow full & change at 5^h 30′ apparent time, rise & fall about 6 Feet.]

[1] Endermo in Broughton, *Voyage*, p. 104 and on plan facing p. 96, etc.; Muroran Ko.

Friday 30 September

This morning died Hans Oldson Seaman he was a Native of the Kingdom of Denmark, his death was occasioned by the fall of a Tree. After lingering some days a mortification[1] took place & he became delirious some Hours before he departed this life. He was buried at the SE point of the Small Island opposite a Village. To the Island in consequence of this unhappy event I gave his name.

[1] Gangrene or necrosis.

CHAPTER 5

Remarks on the inhabitants of Muroran Ko and vicinity and thence to Ostrov Shpanberga, one of the Kurile Islands

Remarks on Endermo harbour, Volcano bay and the natives of Insu

The entrance into Volcano bay is form'd by the land making the harbor call'd by the natives, Endermo, and the south point call'd Esarme:[1] they bear from each other N17°W and S17°E eleven leagues. There are no less than three volcanoes in the bay which induc'd me to give it that appellation. There are 50 fathoms of water in the centre which gradually decrease on nearing either shore. During our stay at the period of the Equinox the weather was generally fine with gentle land and sea-winds from the NW & SE. Nor was there any swell to prevent a ship riding in safety even in the bay, and the harbour of Endermo is perfectly shelter'd from all bad weather. I have seen few lands that bear a finer aspect than the North side of this bay, exhibiting an agreeable diversity of rising grounds and a most pleasing variety of deciduous trees shedding at this season their foliage.

The Natives answer'd to the description given by the Russian navigator Spanberg, who visited an Island in Lat*itude* 43°50′N.[2] The men were of short stature, their legs inclining to bend outwards, and their arms rather short in proportion to their bodies. Their beards were thick and large, covering the greater part of their face, and inclining to curl. The hair on their head was bushy, which was cut short on the forehead and below the Ears, behind it was strait [*sic*]. Their bodies were almost universally cover'd with long black hair and even in some young children we observ'd the same appearance. The women had their hair cut short round their heads tho' it was longer than the mens. The back of their hands, & their foreheads were tatoo'd as well as round their mouths: strings of glass beads and other ornaments were tied round their necks. The dress of the Men consisted of a loose gown made of the inner bark of the lime tree, it reach'd down to their knees and was fasten'd round the middle by a belt in which they carried a tobacco box, pipe and knife. Some had silver rings in their ears with beads hanging from them. Their dresses are only worn in cold weather, which they slip off and on as may be convenient, at other times they only wear a piece of linen round their waists. The dress of the women differ'd but little from the men excepting that their gowns reach'd to the middle of their legs. Their garments in some cases were of seal and Deer skins ornamented with pieces of blue Cloth. Many of the Women had pleasing features tho' much disfigur'd by their mode of cutting the hair; their behaviour was modest, reserv'd & becoming their sex. The children went entirely naked. The men saluted us in the most humble manner, sitting down cross-legg'd,

[1] Esan Misaki.

[2] Ostrov Shpanberga.

Figure 5. A man and woman of Volcano Bay by an unknown artist.
From William Robert Broughton, *A Voyage of Discovery to the North Pacific Ocean*, London, 1804, f.p. 97.

stroaking their beards after stretching forth their hands and bowing their heads nearly to the ground. The houses were built of wood, and the walls form'd of reeds, which were thatch'd with the same materials. In the centre was the Fire-place, and small openings at each end of the roof to carry off the smoke. Over the Fire-place were suspended copper saucepans for cooking and slight scaffolding for drying fish and grain. A platform was rais'd cover'd with skins and matts on which they slept, and the form of the house was generally oblong. Their food consists chiefly of dried Fish boil'd with sea-weed mixing with it a little oil procur'd from the liver of the Sun-Fish. They also eat fruit and vegetables, such as Grapes, winter berries with millet seed &c. They had a custom at each village of keeping & feeding young bears and Eagles in cages, probably for the purpose of food as we could not persuade them to dispose of any. Indeed their scarcity of provisions allowed them but little opportunity of bartering, for even fish we could seldom procure tho' it was their constant employment to catch them. Wild grapes were plentiful and a species of Garlic or Chives. Their boats were made of Fir and other wood with upper streaks increasing their width and continuing fore and aft making them sharp at each end. They were secur'd together by twisted willows and the seams were fill'd up with moss. In rowing they made use of skullers but instead of pulling them together they mov'd one after the other, and consequently did not go in a strait line. Their nets were made of the twisted bark of the lime tree dy'd with oak bark; the Japanese supplied them with Hooks and Harpoons. Their chief employment seems to be in repairing their boats, and collecting sea weed which is exported in great quantities to Japan where it serves as an Article of food. The bark of the birch tree is also exported.

The women are employ'd in weaving cloth for garments and other domestic concerns. Smoking tobacco is a favorite amusement with both sexes. The natives of Insu are tributary to the Japanese and in great subjection to them. They were prevented as much as was possible by the Japanese officers from having any intercourse with us, and were instantly driven away. The natives appear'd a most harmless, inoffensive race of men. We therefore much regretted the jealousy of the Japanese, which entirely precluded us from acquiring the information we were desirous of gaining respecting their customs and manners. They speak in a slow timid manner and several Japanese words are mix'd with their language. We saw some trifling plantations of Indian corn & millet but no other signs of cultivation. This created our wonder as their diet seem'd very precarious and scanty, nor is there the least deficiency in natural means for producing vegetables which the Japanese who reside here are well aware of by taking advantage of them. We saw no animals but horses, dogs, Deer, Bears, Foxes and Rabbits, but as many of the natives had skins of other animals we may conjecture that the island is not deficient in peltry.

The soil is very good, and vegetables grow very luxuriantly. The woods produce Oak, Ash, Elm, Birch, Lime, Yew, Silver Fir, Fir, Pines, Poplars, Hornbeam & the Sassafras tree[1] with a great variety of shrubs. Indeed there are few plants which grow spontaneously in England that are not to be found here. The birds observ'd were, one yellow of the Linnet kind, house sparrows, Doves &c, Wildfowl, Shags, Herons, Sandpipers, Seagulls, two sorts of wagtails, Curlews &c. The Fish were more numerous than the Birds, such as Whales, Porpoises, Turtles, Cod, Flounders, Rock Cod, Sword Fish, Stringara,[2] Sun Fish, Star Fish, Bonetas &c, Seal. Of the Shell Fish there were the Crab, Oysters, Musels, Cockles, Limpets &c.

[1] A small North American tree yielding a bark used medicinally.

[2] Not identified and not included in Broughton, *Voyage*, p. 109.

The Astronomer made the following Observations on shore opposite the Anchorage for the Longitude &c &c

Mean of Observ'd Latitude	42 33 11 N*or*th
Long*itude* ded*uce*d from many observ*ation*s	140 50 32 East[1]
Variation on Shore by 3 Compasses	0 16 30 West
Ditto on board by all the Compasses	1 27 20 East

High water at the full & change of the moon at 4h 30′ app*aren*t time
Rise and fall between 4 &5 feet. We felt no tide when at anchor.[2]

Saturday 1 October
We had the Wind variable from the SW to the NW throughout the day blowing strong at times & in the Forenoon it rained incessantly. The afternoon brought with it clear weather and a strong breeze from the westward. As I was anxious to get to Sea we got under way carrying 2 Reefs in the Topsails & Steer'd SSE for the Bluff point[3] which forms the South entrance into the Bay. At ½ past 6 PM we were within 2 or 3 Leagues of it but it being near dark we cou'd not distinguish the Shore to particularize anything. At 7 we steer'd EbS for the ESE point of Insu[4] which at Noon bore N50°E 3 or 4 Leagues having run 73 miles on that Course. Indifferent Lat*itud*e made 41°48′N.[5] At Noon Fresh Breezes & cloudy weather wind at West.

Sunday 2 October
Strong breeze & Squally weather the wind from the western quarter. At ½ past 2 PM we rounded the Rocks off the SE point steering to the NNE in the direction of the Coast. At 6 PM the point bore S25°W 5 or 6 Leagues a Rocky head[6] west 3 Miles & the Ex*treme* to the North N30°E & we had 20 fathoms water. The Night was moderate & we kept plying under an easy Sail having regular Sound*ing*s to 46 f*athom*s as we increas'd our distance from the Shore. At daylight the SE point preserv'd the same bearing about 6 or 7 Leag*ue*s & the Shore to North of the West became low & Cliffy. To the North of the SE Point it was high & rocky & indifferently wooded which had the same appearance for 4 or 5 Leagues. At a considerable distance to the NW over the low Land the Mountains were entirely cover'd with Snow. Perform'd divine Service and muster'd the Ships company's Cloathing. At Noon Lat*itude* Obs*erve*d 42°29′N, Acc*ount* 42°24′N, No 1 at Noon per Mean of AM & PM sights 143°48′E, Ex*tremes* from S37°W to N30°E 3 Leag*ue*s off.

[1] In annotated *Voyage*, p. 109, corrected in ink to:
Lat 42°33′19″N
Long p*er* mean of 2 Time keepers 140°42′46″E
Long p*er* mean of 89 sets of Lunars 140°51′13″E

[2] The preceding five paragraphs from RUSI/NM/132 (8).

[3] Esan Misaki.

[4] Erimo Misaki.

[5] Actually 41°38′N with +10′ added below.

[6] Rubeshitbetsu Misaki, situated 17 miles north of Erimo Misaki.

Monday 3 October

The wind came in from the SE and we steer'd NE. The shore continued low & level & in the NW quarter appear'd to be the entrance of a River or a deep Bay clos'd by low land.[1] I rather think it took a winding course we cou'd not perceive. At Sunset it was nearly Calm & the land still open'd more to the Eastward, the Ex*tremes* being N62°E to S36°W, an Opening N30°E, off shore 3 or 4 Leag*ues*. At 7 PM we were taken aback with the wind from the NE and we stood to the ESE till 4 AM when we stood in Shore. At Noon an indifferent Observation made 42°48′N when the East extreme bore N70°E, Starb*oar*d point of apparent Opening[2] N21°W & the SW Ex*treme* S76°W off Shore 4 or 5 Leag*ues*.

Tuesday 4 October

Light winds from the Southward & we steer'd East. At 2 PM the opening bore of us NNW & appear'd clos'd by low Land. It much resembled the appearance of a River similar to what we had seen in the Forenoon of yesterday & the Coast from it trended to the Eastward, Flat Land with steep clayish Cliffs. At 7 PM we were abreast of an opening which promised good Shelter;[3] an Island lay in the Center with a Hummock on its East extreme; a smaller Isle was to the west of it near which the Sea broke. At Sun set the N*orth* Extreme bore N50°E the Island N35°E & the South Ex*treme* N80°W. At 8 PM we haul'd our wind for the Night and in the morning steer'd NE. At Daylight the Island bore N84°W 4 or 5 Leag*ues* & the Ex*treme* to the N*orth* bore N30°E. Some low flat Islands[4] appear'd between us & the Shore, but we cou'd not positively ascertain them for the Haze. We steer'd NE till Noon. The shore presented the same bare prospect with a few scatterd Trees but we saw no high land. The lat*itu*de with a bad Horizon made 43°0′N and No 1 145°5′[5] East when the Extremes bore from N40°E to S74°W off Shore 3 or 4 Leagues.

Wednesday 5 October

Moderate winds and fair weather. We steer'd along shore NEbE and at Sunset we passs'd two Islands near the Main[6] bearing from N41°W to West 5 or 6 miles, Extremes from N55°E to S64°W. At 6 we haul'd off the Land under an easy sail and at 2 AM we steer'd NNE to draw in with it. At 6 AM the extremes appearing like Islands bore from N11°E to N31°W. The wind freshening; by 8 AM we were nearly up with them. To the west we saw the Flat Land which we suppos'd the Main[7] & the Islands presented a more barren Surface with projecting rocks from their extremes to a considerable distance. They were cover'd with [a] Variety of Oceanic Birds & the Sea seem'd equally well stor'd with Whales. Land now appear'd to the NE'ward[8] & we alter'd our course to NEbE. By Noon we were nearly abreast of several Rocks detach'd from

[1] The mouth of the Tokachi Kawa, the largest river on this stretch of coast.
[2] Akkeshi Wan.
[3] Hamanaka Wan.
[4] Yururi Shima.
[5] Actually 144°55′ with +10′ added below.
[6] Ostrov Zelenyy, known to the Japanese as Shibotso To, and Ostrov Yuri.
[7] Ostrov Kunashir, known to the Japanese as Kunashiri To.
[8] Ostrov Shpanberga, known to the Japanese as Shikotan To.

the low flat land[1] & to which they were connected by a reef. At Noon they bore from N*orth* to N34°W 3 or 4 miles & we had 45 f*athoms*. To the North we saw high land[2] though we cou'd not ascertain the continuation of the coast and some nearer to us bore from N15°E to N40°E. Latitude by Observation 43°30′N, Long*itude* by No 1 146°20′[3] East. Wash'd below &c.

Thursday 6 October

Squally weather with heavy showers of rain throughout the afternoon. We steer'd NEbE till 4 PM when we had variable winds from the NE quarter. At 3 PM we were abreast of the extreme set at Noon;[4] to the NW of it we saw very high Land at a great distance & not being apparently connected with it we rather suppos'd this to be a new Island.[5] At Sunset it bore from N25°E to N60°W & the Rocks we were abreast of at Noon S70°W[6] off Shore 4 or 5 miles. We stood to the SE till 4 AM when we haul'd in North to make the Land but thick & foggy weather prevented our making it distinctly till near Noon. By which it appear'd we had been considerably affected by a Current setting to the WSW. The Lat*itude* made 43°35′N and the same Land set last Evening bore from N12°W to N42°E, the Rocks also bore from S67°W to S80°W and we were distant 3 or 4 miles from the Land having 41 fathoms, Sandy bottom, Light Airs & Clear from the Eastward.

[1] Ostrov Poloskogo, known to the Japanese as Taraku To.

[2] Probably the NE end of Ostrov Kunashir.

[3] Actually 144°10′ with 10′ added below.

[4] The southern extremity of Ostrov Shpanberga. In annotated *Voyage*, p. 114, inserted in ink, 'Spanberg's Island south point Lat. 43°39′32″N, Long. 146°31′12″E *per* No. 1'.

[5] The NE end of Ostrov Kunashir.

[6] Ostrov Poloskogo.

CHAPTER 6

From Ostrov Shpanberga to Ostrov Simushir and thence to the islands south of Tokyo Wan

Friday 7 October

At Noon as the wind wou'd not permit our steering to the NE'ward we steer'd to the NW'ward to confirm our opinions respecting this New Land which presented a most barren Aspect. It was of moderate elevation very rocky towards the Sea & in the Hollow of the ridges we saw some Trees. Passing the *South* point we also perceived a Small Village; we had a very light air from the SE but the Current assisted our Motion. The Land we soon found trended to the NE'ward and at 4 PM it was Calm. At Sunset the Island bore from S5°E to N74°E 4 or 5 Miles & we had 37 *fathoms.* With a light air we steerd NE till midnight & afterwards preserved our situation by making easy boards till daylight sounding every hour. At Sunrise the Island bore from S22°E to S42°W off shore 4 or 5 miles. The wind continued South-west & we made sail to the NE'ward. High Land was seen to the NNE & NW indistinctly. At 8 AM we pass'd to the North of the Island when we tack'd & stood to the Northward. When we came within two miles of its NE extreme we bro't too & sent the Boat on Shore. They found the Country more productive than we expected abounding (in this part) with springs of good water & very good pasturage. The Trees were the Silver Pine & the Savina species of Juniper. The Soil was rich mould compos'd of decayed Vegetables. Crow Berries & a species of Cranberry they found in abundance. The Shores abounded with wild Fowl & there appear'd in Land many paths of Animals; they saw no Inhabitants. A good observation being the mean of 3 Sextants made the Lat*itude* 43°50′30″N when the Northwest point bore N80°W 3 or 4 miles and the other Extreme S23°W Long*itude* per No 1 146°50′E. High Land bore of us NbE[1] & we had 45 *fathoms* of water. [*in margin*: NB Spanberg a Russian Navigator is said to have water'd upon an Island in this Latitude & had some intercourse with the Inhabitants. It lies to the South of Ooroop and is call'd Nadeegsda. We found both water & Inhabitants. It is most probably the same Island.[2] Vide Cookes 3d Voy*age.* [*vol. 3, page 377, note*][3]

Saturday 8 October

Moderate Breeze & very pleasant weather we steer'd NNE for the Land ahead. At Sunset we had little wind the High Land bore N10°E and a remarkable High Peak'd Hill cover'd with

[1] The southern part of Ostrov Iturup, known to the Japanese as Etorofu To.

[2] For possible identification of Nadeegsda, see p. 91, n. 6 below.

[3] The note reads 'Spanberg places the island here spoken of in 43°50′ North latitude, and mentions his having watered on it; and that this watering party brought off eight of the natives ...' Broughton's latitude agrees with that given by Spanberg. Hence its present name Ostrov Shpanberga.

Snow N65°W.[1] The Island we had left extended from S30°W to S50°W about 5 Leag*ues*. We kept plying under easy Sail till Sunrise when the Peaked Hill bore N70°W & the High Land from N8°E to N30°E which appear'd like an Island. We steer'd NNW for the Passage which was open for 45° to the west of North.[2] At 9 AM the Island we left yesterday bore S12°E & the Peak west. We tried for Soundings in the passage which is 4 or 5 Leag*ues* wide but did not get any & passed several strong Riplings. At Noon we were well through the passage when we hove too and Sounded in 45 f*athoms*. The Lat*itude* made 44°32′N when the Island (on which we landed) bore S17°E 41 miles. A Rocky point which forms the west point of the passage[3] & from which extends a Reef some distance S20°W 5 or 6 m*iles*. The High Land from S61°E to N48°E, Peaked Hill S52°W & the west point of it which falls from a Saddle Hill S82° West. We were now satisfied of the High Land being an Island but cou'd not equally judge of the Peaked Hill Land tho' I rather think it is separate.[4] [Longitude p*er*] No 1 146°24′25″E.[5]

Sunday 9 October

Fresh Breezes and hazey wea*ther*. We steer'd NE for a Hill which rose with a steep ascent to a very great Height.[6] After running 20 miles we were abreast of it and found the Island trend to the NE'*ward* from it. This Hill which was like a Haycock & presented in every point of view the same appearance was evidently a Volcanic Hill cover'd entirely down to its Base with Stones of the color of Cinders.[7] Ragged points appear'd round its Crater & some small Shrubs grew in the lower part of its SW side. It was connected with the Land by a low Isthmus which form'd a Circular Bay[8] on each side of it the Land continuing low to some distance. We steer'd NEbN and at Sunset the Extreme being two High Hills[9] bore N55°E off shore 2 or 3 Leag*ues* and the Volcan*ic* Hill on with SW part of Island S24°W. We had strong breezes from the Southward throughout the Night & stood to the West till 4 AM when we made Sail for the Land which bore of us N65°E. At Noon the Extremes bore from S5°E to N80°E off the distant points set last Night 3 or 4 Leag*ues*. Dark cloudy & threatening weather with heavy Squalls

[1] Vulkan Tyatya, with a truncated cone surmounted by a secondary cone, the highest mountain in the Kuriles, situated at the NE end of Ostrov Kunashir, known to the Japanese as Chacha Take.

[2] Proliv Yekateriny, known to the Japanese as Kunashiri Suido. Proliv Yekateriny was named by Vasily M. Golovnin in 1811 after Vasilii Fedorovich Lovtsov's ship *Ekaterina* in which he took part in Adam Laxman's embassy to Japan in 1792–3; Lydia T. Black, *The Lovtsov Atlas of the North Pacific Ocean*, Kingston, Ont. & Fairbanks, Alaska, 1991, pp. iv–vi.

[3] Mys Lovtsova, known to the Japanese as Atoiya Misaki; clearly named after Vasily Fedorovich Lovtsov; see n. 2 above. In annotated *Voyage*, p. 116, inserted in ink, 'Pocky point Lat. 44°27′43″N, Long. 146°24′42″E p*er* No 1. Peaked Hill NE of Insu Lat. 44°20′17″N, Long 146°3′22″E p*er* No 1', and in pencil 'Peaked Hill Lat 44°20′18″N, Long. 146°3′25″E p*er* Time k*eeper* No 1'.

[4] Ostrov Kunashir is indeed a separate island.

[5] In the margin of RUSI/NM/132 (8) and reproduced almost verbatim in Broughton, *Voyage*, p. 116: 'As I conceivd the Peaked Hill Land to be the NE part of Insu it plainly taking a direction to the westward and the wind being fix'd at SW we embrac'd the opportunity of proceeding to the NE so as to explore these islands more further. In the bearing of the Peaked Hill the Coast form'd a bay with a fine sandy beach, and, the mountain, in this point of view, forming a Saddle Hill, presented a most magnificent spectacle from the great height and extensive base.'

[6] Possibly Gora Golubka. In annotated *Voyage*, p. 116, inserted in ink, 'At 3h 16′ a volcanic hill bore N80°E distant 3 miles'.

[7] In annotated *Voyage*, p. 116, inserted in ink, 'Volcanic hill Lat. 44°44′4″N, Long. 146°51′7″E p*er* No 1'.

[8] Zaliv L'vinaya Past'.

[9] Broughton thought that Ostrov Iturup consisted of two islands and that the southern part terminated at Mys Bol'shoy Nos.

from SSE. No Observation. [*in margin in faded pencil*: I presume this Island to be Ooroop of the Russians and the Straits of De Vries are N*orth* of it in 45°N pass'd in the Castricom and Breskes.[1]]

Monday 10 October
Fresh Gales with thick hazey weather Winds at SE, and we kept our Head to the NE'ward. At 4 PM the Gale set in with constant rain & strong Squalls & we furl'd the Topsails; the Land then bore of us SE, the North Ex*treme* 2 or 3 Leagues & we had 95 fathoms. At Midnight it was more moderate and at daylight we made sail: the Wind shifted by the South to NW & we had Hazey & foggy weat*he*r and the Land bore of us from S61°E to S27°E, a low point S8°W, off Shore 4 or 5 Leag*ues*. We soon lost sight of the Land in the thick wea*the*r and the weather not appearing likely to clear away we again Close reef'd the Topsails & stood to the NE'ward. Dark gloomy weather at Noon prevented any Observation.

Tuesday 11 October
Strong breezes and very dark cloudy weather. In the Evening the Gale increasing we struck Topgallant Masts and before midnight we took in the Fore topsail and wore the Ship to the SW. Throughout the Forenoon we had Fresh Gales with strong Squalls of wind, Hail and rain from the NW quarter. At Daylight the Land bore of us from SSE to S5°W 8 or 10 Leagues, another Island.[2] At 8 AM we wore Ship to the Northward with a large Sea from the NW quarter. Lat*itude* at Noon 46°06′N. The same weather.

Wednesday 12 October
Fresh gales & Squally weather with a large Sea. Towards midnight it came moderate & we wore Ship to the SW. At Daylight we saw Land extending from N70°E to a low point S30°E.[3] The Island seen Yesterday bore from S20°W to S37°W and soon after a Third appear'd bearing S52°W this we conjectured to be the same we pass'd on Sunday.[4] The Wind continued from the Westward & we stood to the South till 8 AM to satisfy ourselves that the Islands were not connected,[5] when we steer'd NE. The Wind gradually decreas'd & we had a very heavy Swell. At Noon the Lat*itude* made 46°1′30″N, No 1 149°5′50″ East by what it appeared we were 12′ South of our estimation. The Extremes of the 1st Island[6] bore from N70°E to South it appear'd very high & broken Land in Parts projecting Rocks & bare with Snow in the Cavities of the Hills & we were 3 or 4 Leag*ue*s off it. The 2nd Island[7] appear'd equally High & barren, it bore from S32½W to S46°W but was of no great extent. The other island we had lost sight of.

[1] Broughton was, of course, mistaken as the island he was off was not Ostrov Urup but Ostrov Iturup, which lies south of Proliv Friza, the strait which De Vries passed through; see also p. 92, n. 5 below.

[2] The northern part of Ostrov Iturup.

[3] Ostrov Urup.

[4] Yet another part of Ostrov Iturup.

[5] I.e. towards Proliv Friza, the strait separating Ostrov Iturup and Ostrov Urup in 45½°N, through which De Vries passed in 1643.

[6] Ostrov Urup.

[7] Northern extremity of Ostrov Iturup.

Thursday 13 October

Light Airs & Calms throughout from the South and East we steer'd to the NE but the swell being very heavy against us we made but little way. At Sunset the Extremes bore from N79°E to South. At Daylight we saw more Land to NE but cou'd not distinguish their connection by a thick Fog bank.[1] At Noon it was Calm with dark gloomy weather an Indifferent Observation made the Lat*itude* 46°32′N 15′ to the North of our account and the Extremes of the Island set last Night bore from S40°E to S20°W. A High round Hill making like an Island[2] N63°E, Off shore 6 or 8 Leagues. Longitude East by No 1 149°36′ East. No Soundings with 115 fathoms.

Friday 14 October

It continued Calm and threatening weather till ½ past 4 PM when a strong wind sprung up from the SE quarter & we made Sail to the NE'ward. At 5 the round Hill bore N67°E; at ½ 9 we shorten'd Sail & continued under easy Sail upon different tacks. At Midnight it blew very hard in Squalls & we close reef'd the Topsails. At Daylight the Round Hill which we now ascertain'd to be a small Island bore East. The wind continued blowing strong at SE with thick Hazey w*eathe*r & we made sail to the NE'ward. At 10^h 30′ AM we had nearly lost sight of the Island bearing SbW. At Noon we Tack'd & stood to the SW. Lat*itude* by estimation 47°05′N No land in Sight & the same unfavourable weather. Cold and damp air.

Saturday 15 October

Fresh Breezes & Squally weather with a dark & Hazey atmosphere. At 3 PM we fetch'd well up with the Island when we again stood to the Eastward under easy Sail. In the evening the Wind moderated & we had heavy showers of rain. At Sun Rise Round Island bore S10°W 4 or 5 Leag*ues* & we saw more Land to the Eastward[3] with the Wind at SSW we kept our Head to the SEastward and at Noon observ'd in 46°55′N Round Isl*an*d bearing S60°W 5 or 6 Leag*ues*. An Island S24°W[4] and the New Land seen this morning S84°E to N56°E. No 1 at Noon 150°30′26″E. Fair weather & hazey, no Soundings.

Sunday 16 October

Light winds & fair weather from the Southern quarter. At 2 PM finding we cou'd not pass to the South of the New Island we steer'd a course along Shore to the NE. This point was very high sloping gradually from its elevation & its Base terminating in broken ridges a Neck of low Land connected it with the other part. At Sunset it bore S36°E to N61°E. Another Island indistinctly seen N50°E[5] and Round Island S63°W. We ran on till 10 PM when we haul'd off for the Night and had no soundings. At Sunrise we steer'd ENE our situation being S77°E from a Circular Hill near the Center of the Island. At Noon Squally weather prevented any

[1] Ostrova Chernyye Brat'ya, comprising Ostrov Chirpoy and Ostrov Brat Chirpoyev. These islands together with Ostrov Iturup, Ostrov Urup and Ostrov Simushir were sighted by La Pérouse in August 1787.

[2] The 801-metre-high Ostrov Broutona.

[3] Ostrov Simushir.

[4] Ostrov Chirpoy, the NE island of Ostrova Chernyye Brat'ya.

[5] Ostrov Ketoy.

Observation and the NE part which seem'd to form a small Harbor bore of us S48°E about 3 miles & which I concluded to be the settlement of the Russians which is said to be situated on the NE part of the Is*lan*d of Marukan[1] in 47½°N by Captain Cooke.[2] An Island three or four Leagues dist*ant* bore from N47°E to N67°E & between that and Marukan we found a clear passage.[3] Latitude by estimation was 47°20′N.

Monday 17 October

Moderate Breezes & very pleasant weather. The Harbor entrance at 1 PM bore SSW near 2 miles & we bro't too having 55 f*athoms* under easy Sail.[4] An Officer in the mean while went on shore in the Cutter & we continued plying off the entrance till Sunset when it bore S5°W 3 m*ile*s and we open'd the ex*treme* SW point which was in the direction of S40°W on with the NW point which fell abruptly to the Sea. Some flat Rocks are scatter'd off it. The NE point bore S30°E about 4 or 5 Miles. The Island which makes the passage[5] S36° East to S66°E the Eastern part runs out low land but the Western part forms a High rugged Bluff Head it was 3 or 4 Leag*ue*s off. At 7 PM the Boat return'd & with a light breeze from the SW we made Sail to the NW. It was calm after 9 PM which continued to 4 AM. The Officer reported the Harbor to be only calculated for small Vessels having a Bar in the entrance on which there was only 2 fathom water, but within it form'd a spacious Bason with 5 & 7 f*athoms*. The Settlement of the Russians they found abandoned; but they cou'd not learn from the Natives at what period. There still remain'd the Russian Arms carved in Wood and painted, and Crosses erected in different places. [*in margin of* RUSI/NM/132 (8): I conjectur'd this to be the harbour which is mentiond by Capt*ai*n Cook as situated in the NE side of Marukan, and where the Russians are said to have a settlement.[6]]

The Natives were similar to those at Insoo, but spoke apparently a different language. They were cloath'd in Bear Skins & wore Boots of Russian Manufacture with cotton handkerchiefs round their Heads. They were equally gentle in their manner & the same appearance of Poverty in their Habitations and manner of requiring subsistance. They were not numerous.[7]

[1] Ostrov Simushir.

[2] In October 1778 Cook was told by Grigory Izmaylov in Unalaska that 'he sailed from Bolscheretzk [Bol'sheretsk], in a Russian vessel, to one of the Kuril Islands, named Mareekan, in the latitude of 47°, where there is a harbour, and a Russian settlement'; Cook and King, *A Voyage to the Pacific Ocean*, II, p. 500.

[3] Proliv Diany.

[4] The harbour is now called Bukhta Broughton or Broutona, having been named by Adam J. Von Krusenstern in *Atlas k puteshsetviiu vokrug svieta Kapitana Kruzenshterna* ..., St Petersburg, 1813, plate 104, 'Plan eines Hafens an der Nordspitze der Insel Simusir gefunden in Kamtschatka: Dieser Hafen ist im Iahre 1793 von Capt. Broughton und im Iahre 1811 von Capt. Golownin untersucht worden' (see Fig. 6). Vasily M. Golovnin (1776–1831) visited this harbour in the *Diana* but shortly afterwards he was captured by the Japanese and kept a prisoner in Hokkaido for two years. He subsequently had a distinguished career; Richard A. Pierce, *Russian America: A Biographical Dictionary,* Kingston, Ont. and Fairbanks, Alaska, 1990, pp. 175–7.

[5] Ostrov Ketoy.

[6] This was from information supplied to Cook in October 1778 in Unalaska by Grigory Izmaylov, who told Cook 'that on the 12th of May 1771, he sailed from Bolscheretzk, in a Russian vessel, to one of the Kuril Islands, named Mareekan, in the latitude of 47°, where there is a harbour, and a Russian settlement'; Cook and King, *Voyage to the Pacific Ocean*, Vol. 2, p. 500.

[7] 'By the Officers account these people were Cloathed and in other respects similar to those at Matsmay [i.e. the inhabitants of Uchiura Wan]. It therefore plainly appears that all the Islands to North of Nipon or Japan are Inhabited by the same Race of Hairy Men. The Northern Isles being subject to the Russians, & the Southern Ones to the Japanese'; UKHO, OD177.

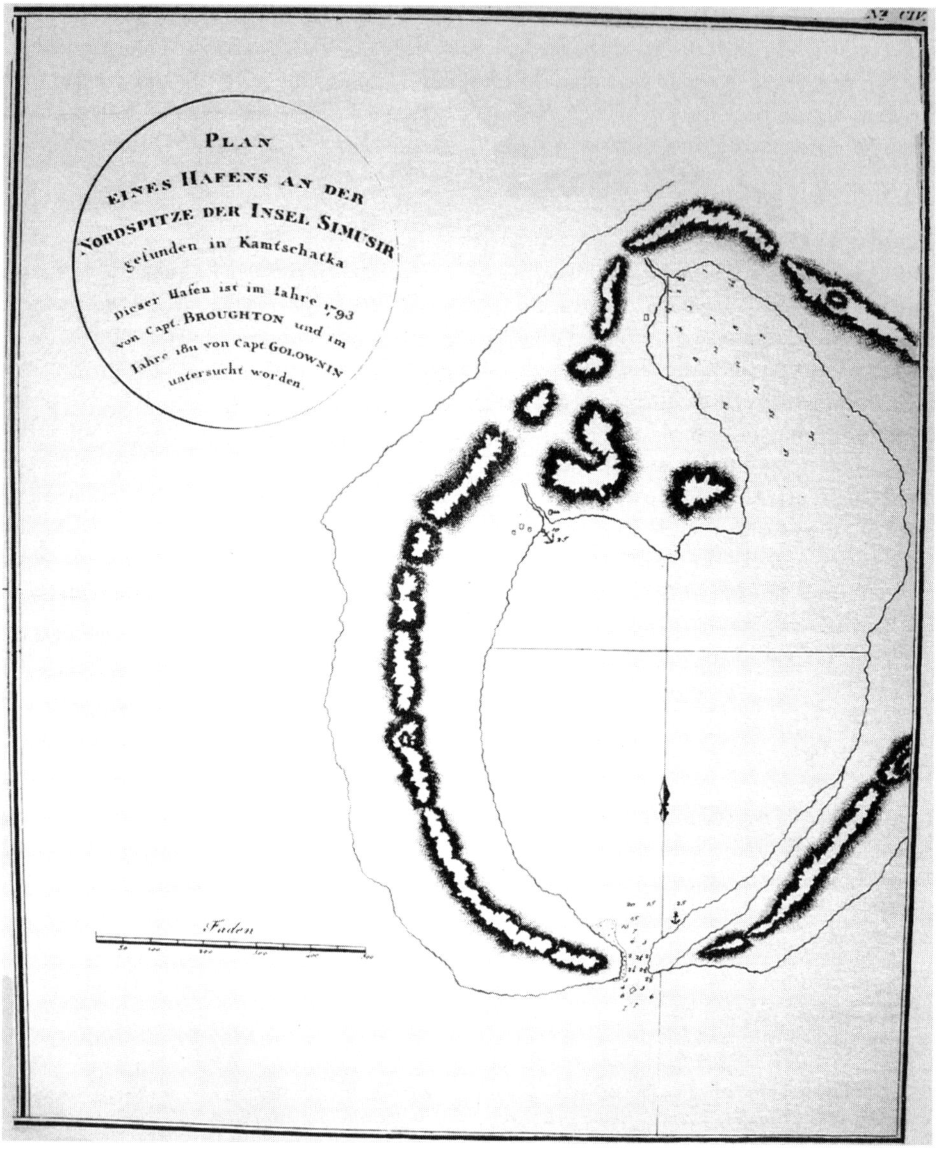

Figure 6. 'Plan Eines Hafens an der Nordspitze der Insel Simusir gefunder in Kamtschatka'; I. F. Kruzenshtern [A. J. von Krusenstern], *Puteshestvie vokrug svieta v 1803, 4, 5, i 1806 godakh*, St Petersburg, 1809–13, Atlas, plate 104.

The Land was cover'd with long grass and moss. Some Trees of the Dwarf Pine, Alder, Bill & Crow berries. In the Harbor they saw many Seals & abundance of Sea Eggs. These were the only remarks we had an opportunity of making as the weather next day and afterwards prevented any further intercourse with the Kurile Islands.[1] On the 18th off Round

[1] See also Broughton's letter to Evan Nepean dated 20 December 1796 in Appendix I.

Island[1] we experienced a very heavy gale of wind from the SE quarter and at 9 PM when lying too under a Mizen staysail I had the misfortune to fall upon deck and Fractur'd my right Arm which confin'd me to my Bed & Cabbin till the 10th of November & totally depriv'd me of the use of my arm for these Months which prevented the continuance of this Journal by me. In three months tho' the Journal was carried on by an Amanuensis, daily under my inspection.

Macao Jan*uary* 1797 W. R. Broughton

It should remember'd that we had trac'd the Land of Insu or Jesso from Volcano bay to the NE part of what we conceiv'd to be one Island, extending from 41°49′N to 44°30′, and from 140°30′E to 146°22′E,[2] being an extent of land of its SE Coast for 100 leagues. This statement agrees well with the account of De Vries's voyage[3] supposing he made the land in the same situation we did about the SE point. Capt*ain* King, in the notes to Cookes 3d voyage,[4] remarks that in this part, Spanberg's discoveries place the Islands of Matsmai, Kunashir & Zellany[5] and that probably the mistake of De Vries arose from the foggy weather in supposing it one land. I am of a different opinion for we saw no Island of any size till we came to one in 43°50′N & 146°50′E, and which I suppose to be the same Spanberg water'd at, & which is call'd in Cooke's Voyages Nadeegsda.[6]

On Oct*obe*r 12th we were in the Lat*itude* of 46°1′N and 148°45′E, which is nearly the situation of the straits of De Vries[7] and we conjectur'd the Northern Land to be what he calls the Company's land[8] and the Southern one to be Staten Island,[9] but which is named Nadeegsda in the Cooke's voyage and charts. Capt*ain* King supposes the Company's Land to be the Ooroop and Nadeegsda of the Russians. It is most probable that this land is Ooroop in which there is said to be a good harbour,[10] and Staten Island is the Nadeegsda of the same nation. The violent gales we met with after leaving Marukan prevented our examination of the East side of these Islands, and also our passing thro' the Straits of Sangaar that divide Insu from Japan.[11] Capt*ain* King also imagines Staten Island seen by the Castricom[12] to be the three Sisters;[13] and for which reason he has plac'd it as such in the Chart. See Cooke's 3ᵈ Vol., Page 391, &c.[14]

[1] Ostrov Broutona.

[2] Hokkaido.

[3] For De Vries see p. xlii, n. 2 above. After the *Castricum* and *Breskens* lost contact while searching for islands reputed to be rich in precious metals, De Vries in June 1643 became the European discoverer of Ostrov Iturup, which he named Staten Eylandt, and Ostrov Urup, which he named Compagnies Landt, two of the Kurile Islands; Burney, *Chronological History of Voyages*, III, p. 156.

[4] 'page 377' added in Broughton, *Voyage*, p. 126, but Vol. III omitted.

[5] Hokkaido, Ostrov Kunashir and Ostrov Zelenyy.

[6] The island that Spanberg watered at is, according to his latitude, Ostrov Shpanberga, while King's Nadeegsda is Ostrov Nadezhdy, an earlier Russian name for Ostrov Urup.

[7] The strait between Ostrov Urup and Ostrov Iturup.

[8] Ostrov Urup.

[9] Ostrov Iturup.

[10] There is no good harbour in Ostrov Urup.

[11] Tsugaru Kaikyo.

[12] Maarten Vries's ship.

[13] 'This land [Staten Island] was also discovered by the *Castricom*, and from its situation, as described in the journal of that vessel, it appears to be the islands of the Three Sisters'; Cook and King, *Voyage to the Pacific Ocean*, 3, p. 388.

[14] From an examination of Roberts's 'A General Chart ...', it is clear that Roberts and not King has placed the Three Sisters in the location of Ostrov Shpanberga.

It was my intention to have pass'd thro' the strait which divides Marukan from the Island North of it, but the wind being unfavourable, after the return of the Boat I proceeded to the Southward intending to pass through de Vries Strait to the Southward of Marukan & survey the East side of it and I did not propose this Year to go further to the North owing to the advanced State of the Season, but to explore the Eastern Sides of the Kurile Islands towards Japan & if possible to find shelter in those of Ooroop or Nadeegsda where the Russian are said to have found good Harbours ... [*the remainder crossed out*][1]

Continuation of Journal from breaking my arm on the 17th [18th] October 1796 off Marukan

[Tuesday] 18 October

Moderate & fine weather. The Cutter return'd from the Harbor at 8 Oclock & finding we cou'd not get thro' the Straits north of the Island[2] we stood to the SW towards Round Island.[3] In the afternoon the wind increased to a Gale from SE with thick rainy & foggy weather & we ran between Round Island & Saddle Isl*and*[4] at ½ past 5 PM, about 3 Leag*ues* dist*ant* from each. At ½ past 7 it blew a strong gale of wind and we brought too under the Fore & Mizen Stay Sails. In the morning the gale subsided & towards Noon it was moderate with a heavy swell from the E*ast*ward. Round Isl*an*d bore from S40° to S64°W 3½ miles, Lat*itude* 46°47′N, Long*itude* 150°3′E. Calm and light airs prevented our progress thro' the Straits [*in margin* De Vries] which divide Marukan & which we conceiv'd the Island Ooroop.[5]

[Friday] 21 October

In the afternoon we took our departure from the Kuriles when Marukan South extreme bore N27°E 10 or 12 Leag*ues*, extremes of Ooroop from N84°W to S66°W 8 or 10 Leag*ues*. The Wind came to the Westward. It is to be observed that in consequence of my fractur'd Arm I had no opportunity of making any remarks myself. Therefore little is said of these Straits which the Chart will best explain. Wind from the westward continued & we kept our course to the SSW as it permitted. We had clear weather but no Land in sight. Our Lat*itude* 45°22′40″, Long*itud*e 150°26′E. Since the last observation we have felt the influence of a strong Southerly current.[6]

[Saturday] 22 October

The wind veer'd to the Southward after midnight and we wore & stood in for the Land on a west course. Lat*itud*e 44°42′N, Long*itude* 149°15′E.

[1] The preceding three paragraphs from RUSI/NM/132 (8).

[2] Proliv Diany.

[3] Ostrov Broutona. In annotated *Voyage*, p. 127, inserted in ink, 'Lat. 46°42′49″N, Long. 150°9′46″E per No 1'.

[4] Ostrov Chirpoy, the NE island of Ostrova Chernyye Brat'ya.

[5] Proliv Bussol' which separates Ostrov Simushir from Ostrova Chernyye Brat'ya, which in turn is separated from Ostrov Urup by Proliv Urup. De Vries in fact passed through Proliv Friza; see p. 87, n. 1 above. La Pérouse passed through Proliv Bussol' on 30 August 1787, naming it Passage de la Boussole after his ship.

[6] The Oya Shio, which runs south past the Kurile Islands and Honshu until it meets the northern edge of the Kuro Shio in about 36°N.

[Sunday] 23 October

The breeze increas'd at SSE with continual rain and after dark we kept plying to the Eastward under easy & reduced Sails. The following day drew the wind to the Westward & it soon after blew a hard Gale with heavy Squalls & frequent showers of Hail. We lay too under Storm Staysails. Indiff*eren*t observation made 44°15′N.

[Tuesday] 25 October

It moderated towards Noon and we had continual rain with the wind at SW.

[Wednesday] 26 October

The wind suddenly flew round to NW & blew strong with clear weather & frequent hail showers. Lat*itude* 42°23′47″, 148°23′35″E Longitude.

[Saturday] 29 October

The wind came to N*orth* and NE and we steer'd West to make the Land with a strong breeze in Lat*itude* suppos'd 41°43′N. After running 130 miles the wind came to East & increas'd to a fresh gale with continual rain. We frequently tried for Soundings & during the Night we stood to the SSE under reduced sails & at midnight the wind veer'd to South and became light and variable. The Barometer was lower to day than we have yet experienced it. By an indifferent observation we found ourselves in 40°46′, considerably S*outh* of acc*oun*t.

[Monday] 31 October

Soon after noon the wind veer'd to West and blew a hard gale. At Sunset we were nearly in the Lat*itude* of the Northern point of Nipon & being considerably to the Eastward we steer'd South to range the Southern parts of the Coast which Capt*ai*n King had not an opportunity of observing owing to the tempestuous weather they experienced nearly at this Season of the Year. The continued western gales had in a similar degree prevented our making the Land to the North of Nipon [*in margin*: or Japan] & take shelter as I intended in the Harbors of Nadeegsda or Ooroop as described by the Russians. I was necessitated therefore on account of my own misfortune to give up my plans very early in hopes next Season to compleatly accomplish the same. As Gales blew directly from the Land we were constantly attended by various kinds of small birds many of which took refuge in the Ship & were frequently taken.

[Tuesday] 1 November

At noon it was calm, temperate, and very pleasant weather. Good observations made the Lat*itude* 37°6′33″N and Long*itud*e 144°9′40″ E, still a strong southerly current. The Calm continued till the following morning when the Breeze springing up from the Eastward brought with it continual & heavy rain. We directed our course to the SW attended by some few Sheerwaters and Peterels.

[Friday] 4 November

The timepiece was neglected to be wound up[1] & at Noon the wind came to the North with dark cloudy wea*the*r. We shap'd a course to fall in with the Land of Nipon about White point.[2]

[Monday] 7 November

Strong gale from the SW reduced us to our Storm Staysails. Towards Noon it moderated and veered to the NW. [A] great variety of small land birds, Butterflies &c about the Ship.

[Wednesday] 9 November

At 2 PM we discernd the Land of Nipon to the westward & a Sail in the same direction. We continued under easy sail during the Night & early in the morning made Sail. The Land extending from N42°W to N80°W. At Noon rainy w*eathe*r.

[Thursday] 10 November

By our making the Land it appears we must have felt the effects of a considerable current setting to the NE[3] as the SE point of Nipon which I take to be White Point of Capt*ain* King, bore of us at 9^h 20′ AM N5°E on with a distant hummock.[4] The Shore from thence to the westward was of moderate elevation with chalky cliffs to a small opening from whence came out several fishing Boats.[5] They proceeded to their employ without taking any notice of us. The Land from the opening took a SW direction. It appear'd bare of wood and destitute of Verdure.[6]

At Noon we were abreast of a very extensive town. The Land round it was well cultivated tho' with much apparent labour for the Hills appear'd very steep and craggy towards the top but the sides were cut into regulars Plats[7] one above the other. In some of the hollows between the Hills was an inconsiderable quantity of wood.

The afternoon was calm & the fishing Boats came to us & threw on board their Fish & were much amused seeing out people scramble for them. They gave all gratis & did not seem to expect any return. We considered our Long*itude* from White point as ascertain'd

[1] In annotated *Voyage*, p. 138, inserted in pencil, 'The Time keeper was set going on the 4th at Noon', and on p. 134, also in pencil, 'The weather was cloudy till the 7th therefore the Time k*eepe*r was of no use before that time. On that day altitudes of the Sun & several sets of Lunars were observed from which the error of the Time k*eepe*r from a mean time at Greenwich was found with which and the rate found at Volcano Bay the Time keeper was as useful as before it was let down.'

[2] Taito Saki, the SE extremity of Honshu; a prominent wooded point whose southern side consists of a vertical white cliff.

[3] The *Providence* was now under the influence of the NE-going Kuro Shio, which in extreme conditions in November can reach a rate of 5¾ knots.

[4] Inubo Saki, Sandown Point on Broughton's 'A Chart of the N.E. Coast of Asia, and Japanese Isles', Broughton, *Voyage*, facing title page.

[5] Katsuura Wan, situated 11 miles SSW of Taito Saki.

[6] In annotated *Voyage*, p. 138, inserted in ink, 'At 21h 9′ the SE Point of Japan [Taito Saki] bore N8°E dist. [*cropped*] miles. Its Latitude 35°13′17″ North, Long*itude* 140°33′47″E p*er* mean of 31 sets of Lunars.'

[7] Archaic for plot of land.

Figure 7. 'Japanese Boats'; Broughton, *Voyage*, f.p. 139.

by Capt*ain* King & our timepiece again became useful. Lat*itude* 34°54′52″, Long*itude* 140°25′53″E, South extreme[1] S79°W, Round Hill of Capt*ain* King[2] N58°W, Land about White point forming the N*orth* extreme N34°E, an Island[3] S64°W to S71°W, 4 or 5 miles off the Land.

[Friday] 11 November

Light airs from the NE carried us along the shore & we saw Islands to the SSW. We tried but had no Soundings with 100 f*atho*ms. Towards Sunset our friendly fishermen left us in number about 200. Their boats were about 36 feet in length & 8½ feet wide, 2½ in depth, neatly built of Oak, Elm & Fir, very sharp forward with a rising Stem, the sides are continued much beyond the Stern which gives them the appearance of being open abaft. [See Fig. 7.] The ends of such parts as were exposed to the water were neatly cover'd with Copper & in every respect they were ingeniously constructed. About halfway down was a false bottom to contain the Fish. Each Boat had one Mast with an Oblong square Sail of cotton some of which were

[1] Nojima Saki, the SE extremity of Boso Hanto.

[2] Myoken San, a 418-metre high wooded peak, situated 16 miles SW of Taito Saki, the highest hill on Boso Hanto and the first object sighted by a vessel approaching from seaward. Mount King on Broughton's 'A Chart of the N.E. Coast of Asia' and not King's round hill, which was Fuji-san.

[3] O Shima.

painted red about 2 or 3 feet from the Yard downwards. Others had 2 long Stripes of dark blue cotton stitch'd into them. They had 12 men to each boat & Seines of 80 f*atho*ms long. They mov'd with great velocity by 6 or 8 Skulls acting on out-riggers about 2 feet from the side and steered by a rudder laying in an Oblique direction from the Stern something in the Chinese Manner.

At 5 PM the City Point[1] bore N30°W 6 miles and a high Mountain[2] N60°W. Several Islands lay scatterd from S20°W to S86°W[3] 10 or 12 Lea*gue*s. Between the Mountain & the City point we saw no Land. It appear'd to be an extensive Bay or the entrance of a River or harbor, most probably the entrance into the famous Port of Jeddo.[4] We had a fresh breeze in the forenoon when we steer'd for the Outer Islands. At Noon they bore from S8°W to N32°W, the nearest 5 miles. Lat*itude* 34°1′40″N, Long*itud*e 139°58′47″E.

[Saturday] 12 November

Fresh breezes & fine weather from the NE quarter. After passing thro' the Islands[5] we steer'd a West course for the Main Land and at daylight NW, leaving the Islands to the NE.[6] At Noon we had the appearance of blowing weather & we hauled off to clear the Land which extended from N53°E to WbN 5 or 6 Lea*gue*s, Humm*ock* Island[7] only visible bearing N71°E. The Main Land was of moderate elevation & much broken. Near the Sea it seem'd low & level but the interior forming double Land rose to considerable Mountains extending from N17°E to WbN. The weather prevented any Meridian Observations & tho' we frequently tried we had no Soundings.

[Sunday] 13 November

The Night was moderate but after midnight it freshend up from the SW quarter. At 5 PM we sounded in 45 fath*oms* brown Sand, Gravel & Coral Rock. We kept our wind to the SE under reduced canvas.[8] At 7 AM we discover'd a bare Rock[9] bear*ing* SE & we had a strong gale of wind, continual rain and a confused Sea. We soon after discover'd Outer Island[10] by which it appear'd to our great Surprize to be the Rock laying South of that Isl*and*. We wore Ship to avoid it & the wind flew round to the WSW in a heavy Squall which bro't with it clear weather & we steer'd thro' another clear passage among the Islands.[11] At Noon the Lat*itude* was

[1] Suno Saki, the SW extremity of Boso Hanto and the eastern entrance point of Uraga Suido, leading to Tokyo Wan.

[2] Fuji-san.

[3] The northernmost islands of Izu Shoto from the 853-metre high Mikura Shima, which would just be visible above the horizon, to O Shima.

[4] Uraga Suido.

[5] Between Miyake Shima and Mikura Shima.

[6] In annotated *Voyage*, p. 141, inserted in ink, '8 in number'. During the night the *Providence* passed a short distance north of Zeni Su, a group of islets and reefs, the highest 12 metres high, situated 34 miles WSW of Miyake Shima, without sighting them.

[7] To Shima; it has a slightly conical summit.

[8] During the night the *Providence* passed a short distance south of Zeni Su, again without sighting them.

[9] Inanba Shima.

[10] Mikura Shima.

[11] Between Kozu Shima and Miyake Shima.

33°59′N. The Islands extending from N15°W to S76°E, High Mountain in Nipon[1] N20°W. As the Japanese names were unknown to me I cou'd only distinguish them by their appearance. The Passage we first passed thro' was between Outer Island[2] & Stone Henge[3] and to day between Stone henge & Broken Isl*and*[4] which appeared to be about 6 Leag*ue*s wide. The N*orth* part of Broken Island is rather high with perpendicular white Cliffs. Off the SW point is a large rock[5] with several smaller ones round it. The greatest extent of the Island seemed to be in a NE & SW direction; between it & which from its form I term'd Wedge Island[6] but to the westward is a long low flat Island.[7] Wedge Island appear'd of considerable extent of Moderate height connected at intervals by low land which at some distance makes it appear like three separate Islands with a remarkable white patch on the South side and a Rock laying off it to the N'Eastward. Off it are two Hummock Islands, one large.[8] Between them and the Main land seem'd a good passage.

[Monday] 14 November

It continued to blow a strong gale from the NW which prevented our approaching the Mainland. At ½ past 5 Mount Fusi in Jeddo Bay bore N40°W and we wore & steer'd to the SbE in search of an Island seen from the Masthead to the South of Outer Island the first day. In the forenoon we were up with the Land we were in pursuit of composed of two Islands. These are called in the Charts Fatsisio[9] & we understood them to be inhabited by exiled Noblemen who were banished from Nipon. The Westernmost is a small high peaked Island[10] not more than a league in circuit and to appearance inaccessible. Notwithstanding which on the Northern point we observed several habitations surrounded with cultivated patches of Land as before observed at Nipon. It lies about 2 Leag*ue*s to the SW of the large Island; between them to appearance the passage was perfectly clear. After passing Peaked Island which bears NbW to SbE from Outer Island 13 or 14 Leag*ue*s we kept our wind to examine the other Island[11] which appear'd of considerable extent & presented a fertile appearance as far as the haziness of the weather wou'd permit us to judge but the wind veering to the Eastward prevented our approaching it within 3 or 4 Leag*ue*s. Unfortunately we cou'd not determine its situation exactly not being able to see the Sun. By our run & its bearing from Outer Island we estimate it to be situated in 33°6′N lat*itude* and Long*itude* 140°E. The unpromising appearance of the weather gave us no hopes of a change, we however steer'd to the westward on a parallel with it in hopes of a Meridian Observation the next day but the Weather growing worse and worse in the forenoon having increased to a strong gale with constant rain we haul'd up WSW to make the Coast of Nipon.

[1] Fuji-san.
[2] Mikura Shima.
[3] Miyake Shima.
[4] Kozu Shima.
[5] Onbase Shima, a group of rocks, the largest of which is 64 metres high, situated 2 miles west of the SW extremity of Kozu Shima.
[6] Nii Shima.
[7] Shikine Shima, a small flat-topped island, 101 metres high, situated 7 miles NNE of Kozu Shima.
[8] To Shima, the northernmost and largest and Udone Shima, the smallest.
[9] So named on Henry Roberts, 'A General Chart ...'; present day Hachijo Shima and Ko Shima.
[10] Ko Shima.
[11] Hachijo Shima.

CHAPTER 7

From the islands south of Tokyo Wan to Sakishima Gunto and thence to Macau

[Wednesday 16 November]
Strong gales and thick hazey weather with continual rain. At Sunset 5 PM we bro't too under a Foresail & Main topsail with our head to the SE. In the forenoon the Gale subsided & at Noon we had Light airs from the SW quarter.

[Thursday] 17 November
We steer'd to the WSW, Breeze increasing, & which soon reduced us to a low Sail for we cou'd not carry our topsails. At ½ past 3 we saw the Land extending from W*est* to WNW 3 or 4 Leag*ue*s. We wore Ship as the Weather appear'd still more threatening & at 5 PM tried for Soundings. [At] ½ past the Wind veering to the North of west gave us a more distinct view of the Land as the Sun set extending to NNE & ending in a low point at west. We kept our wind to the SW but tho' we had uncommon clear weather we saw no land in the Morning. I therefore suppose it was the S*outh*ern part of the Coast we had fallen in with.[1] At Midnight the Moon being very bright we saw the Land from NbE to WNW and 1^h 30′ afterwards the Mer*idian* Alt*itu*de of the Moon made our Lat*itud*e 32°51′34″N. At Noon the Gale had moderated & our Latitude was 32°14′19″N & Long*itud*e 135°48′15″E.

[Friday 18 November]
Moderate and very pleasant weather from the West & NW. In the Night it was Calm. Our course was to the SW and by the difference in our observations & estimated account we must have experienced a current in the direction S18°E, 31 miles.

[Saturday] 19 November
Same weather & nearly the same effect of current. Wind came NE'ly & we steer'd WSW to make the land. Lat*itude* at Noon was 31°3′53″N, Long*itude* 134°45′E. The current this day has set us to the westward 25 miles.

[Tuesday] 22 November
Squally weather with heavy rain at times. At ½ past 4 we saw the Land to the WNW and at dark we bro't too for the Night. No Soundings. At daylight the same land bore of us WSW 10

[1] Broughton was correct, it was Shiono Misaki, the southernmost point of Honshu.

or 12 *Leagues* by which it appears we must have been set to the Eastward by a strong current. At Noon the Land extended from N40°E to WbN, 8 or 10 Leag*ues*. Good observations made our Lat*itude* 32°7′7″N,[1] by Timepiece Long*itude* 132°42′E, By distances ☉ to ☽ 132°17′E. Near this situation in Lat*itude* 32°47′ Count Benyowsky water'd on an Island Fruitful but not inhabited[2] in 1771. Vide his Voyage.[3] I imagine the Land to day seen to be Xicoco or Sikoff the Southernmost of the three Islands forming the Kingdom of Japan, Nipon, Ximo & Xicoco or Sikoff, in which is Nangasaki.[4]

[Wednesday] 23 November

Fresh Breezes and cloudy weather with small rain with the wind at SW. We stood in for the Land. At 5 PM having run in 30 miles we tacked but though we were within 1 or 2 Leag*ues* we cou'd not discern the Shore. At 4^{h} 15′ a bluff point[5] bore of us N12°W 2 or 3 Leag*ues*. To the NE of this point the land receded considerably having the appearance of an Opening but the indistinct view we had of it caused by the haze & rain prevented our making any satisfactory observations,[6] nor were we able to get Soundings. During the Night we had strong breezes & a heavy head Sea & we kept our wind to the S*outh*ward. At daylight the Land extended from N40°W to N65°W[7] and at Noon N60°W 10 or 12 *Leagues*. Our Lat*itude* 31°34′24″N & [Longitude] 133°1′20″E. Current these 24 hours N76°E, 30 miles.

[Thursday 24 November]

In the afternoon the wind veering to the N*orth* of west kept us in sight of Land tho' at a great distance nor cou'd we approach it any nearer from the violence of the wind. We however conjectur'd the whole to be connected. It was of considerable height in detach'd peaks & in the morning we were well up with that part which made the S*outhern* extreme last Evening bearing from S80°W to N12°W, offshore 6 Leag*ues*, and the Land abreast form'd a deep bay.[8] Throughout the forenoon we tried for Sound*ings* without reaching the bottom. At Noon our Lat*itude* was 31°13′55″ N, Long*itude* 131°55′40″E. The Extremes bear*ing* from N*orth* to

[1] '32°17′17″' in the MS, which is clearly a transcription error, since both the log, UKHO, OD 177, and Broughton, *Voyage*, p. 384, have 32°7′17″.

[2] Probably Okino Shima, a detached island off the SW coast of Shikoku.

[3] Mauritius Augustus Benyowsky, *Memoirs and Travels of Mauritius Augustus Count de Benyowsky*, 2 vols, London, 1789. On 9 August 1771 Benyowsky (Móric Ágost Benyovszky) was in the vicinity of Okino Shima, where he anchored, but a proposed landing was opposed and to avoid bloodshed was abandoned; *Memoirs*, I, p. 413. Benyowsky was born in Verbó, Hungary (now Slovakia), in 1746. His involvement in politics resulted in his exile to Kamchatka, from where, assisted by others, he made his escape by capturing the *Sv Petr* and in her visited a number of places in eastern waters before reaching Macau. Here he hired two vessels and returned to Europe. He was eventually killed in Madagascar on 23 May 1786. His *Memoirs* were translated into several languages and widely read.

[4] Broughton is confused. Nipon is Honshu; Ximo is Kyushu, the southernmost of the three islands in which is Nagasaki; and Xicoco or Sikoff is Shikoku, the land they had just sighted. Broughton shows the islands in their correct position in his 'A Chart of the N.E. Coast of Asia, and Japanese Isles', Broughton, *Voyage*, facing title page.

[5] Possibly Sen Saki on the east coast of Kyushu almost opposite Okino Shima.

[6] The opening was almost certainly Bungo Suido, the strait separating Kyushu and Shikoku.

[7] N60°W in the journal, but N65°W in the log, which makes more sense in view of the bearing of N60°W at noon.

[8] Shibushi Wan.

[N]75°W. Starb*oar*d point of bay[1] N53°W 2 or 3 Le*agues*, a Distant Island[2] S40°W 10 Lea*gues*. This Land was high and rugged & much broken into ridges, the Hills generally bare of Wood & Rocky. Other parts again cloath'd with wood up to their very summits. We saw some Boats fishing in Shore and a Junk standing into the Bay & we observed a Volcanic Hill[3] situated in the interior of the Bay which emitted vast columns of Smoke. Curr*ent* N14°E, 22 miles.

[Friday] 25 November

Light airs from the ENE carried us to the westward and at 3 PM we had 25 f*atho*ms, Sandy bottom. At 5 PM we were within 6 miles of the Shore. The South point of the bay[4] N50°W & we cou'd just discern some low land from the deck in the direction of S68°W,[5] Distant Island S42°W, connected with which appear'd more low land S25°W.[6] The Bay seem'd capacious & likely to afford good anchorage.[7] We kept our wind off the Land under easy sail but had no Soundings after 6 PM. In the morning we saw the low land forming the S*outh* extreme of last Evening which now bore S67°W & the distant Island appeared over it S50°W. We stood for it with the wind from SSE, a moderate breeze & cloudy weather. At Noon the extremes of Land to the North of Bay N15°W, South extreme to the Southward of the Bay N72°W and the low land from S68°W to S82°W, Offshore 5 or 6 Leag*ues*. The Supposed Isl*an*d still appear'd bear ing, its S*outhern* extreme S58°W. Our Latitude was 30°50′15″N, Long*itud*e 131°58′45″E and the Current appears to have set North 25 miles.

[Saturday] 26 November

We kept our wind in for the low land, more appearing as we approached. Having run in 7 Leag*ue*s we tacked in 45 fathoms, fine sand & we were within 4 or 5 miles of the Land which extended from S20°W to N75°W of very moderate elevation, cover'd with scattered trees & towards the sea were white Sandy beaches above which we saw Smoke in several places. Distant Island still appeared over the Low Land bearing S50°W making very high & at a considerable distance. South extreme at noon now bore N55°W. To the west of it as far as N68°W it made in Hummocks, South point of bay N2°W, N*ort*h Extreme N17°E. In our present situation we cou'd not determine whether the low Land was an Island or connectd with the distant high land (supposed before to be an Island) which I suppose to be the case.[8]

[1] Toi Misaki. In annotated *Voyage*, p. 151, inserted in pencil: 'Starboard Point of the Bay Lat 31°19′4″N, Long. 131°42′53″E p*er* Time keeper No 1, 131°36′40″E per mean of 31 sets of Lunars.'

[2] The 1,935-metre Yaku Shima.

[3] The 1,118-metre active volcano in the centre of Sakura Shima.

[4] Hi Saki, the southern entrance point of Shibushi Wan.

[5] The northern extremity of the 301-metre Tanega Shima.

[6] The southern extremity of Tanega Shima.

[7] 'There are depths suitable for anchorage [in the bay] ... but the anchorage should be avoided during winds from between E and S. A swell frequently runs into the bay and vessels at anchor usually ride uneasily'; *Japan Pilot*, Vol. 2, 7th edition, Taunton, 1979, p. 163.

[8] Tanega Shima and Yaku Shima are not connected. At the head of this page of the MS is an editorial comment in pencil, 'NB Had Payne's Epitome of History, which explains the names of the Japanese Isl*ands*. You have an extract from Payne's among my Papers with you' (John Payne, *An Epitome of History*, London, 1794). The extract is held in NMM, RUSI/NM/132 (3).

The opening between the lands was so much contracted & making in low points from each shore I conceived the whole to be a continuation of Ximo or Sikoff. That is the same Land or Isl*and* we made on the 22d Instant and that the Opening South of the Capacious & Volcanic Bay probably terminated in a Harbor imperceptible to our Observations, which point we were prevented clearing up by the wind in the Night shifting to the SW directly in our course.[1] We therefore stood to the SE'ward & at noon found ourselves 21 miles N*orth* of our Reckoning. Lat*itude* was 30°36′N & we had no Land in Sight. The Land we left last evening must be the Southern part of the Japanese Islands.[2]

[Sunday] 27 November

The wind continued invariably fix'd in the SW quarter & we saw no more of the Japanese Islands, nor had we any Observations.

[Monday] 28 November

Same weather & wind brought us into the Lat*itude* of 28°30′N & Long*itude* 135°4′E & we found the Current for the last three days had set N63°E 38 m*iles*. In our situation of today are several Islands laid down in Arrowsmiths Charts[3] supposed to be discover'd by the Spaniards. We saw nothing to indicate our being in the vicinity of any.

[Tuesday] 29 November

In the evening at 6 PM as appearances foretold & in Lat*itude* 28°10′N & Long*itude* 135°15′E the wind shifted to the NW with rain, & we kept upon a Wind to the westward having dark gloomy weather.

[Wednesday] 30 November

The wind veering more Northerly at 9 PM we steer'd West to make the Island of G*rea*t Liqueo[4] being nearly in the Lat*itude* of its center as laid down in the Charts. In the Night we tried for Soundings unsuccessfully.

[Thursday] 1 December[5]

Before Noon we discern'd Land to the westward which we supposed to be Liqueo. At Noon the wind veer'd to the SE with Squally weather which oblig'd us to keep our wind under reef'd

[1] Osumi Kaikyo, separating Tanega Shima from the southern extremity of Kyushu, leads towards the major port of Kagoshima.

[2] Broughton was of course correct; it was the islands of Yaku Shima and Tanega Shima off the southern estremity of Kyushu.

[3] The non-existent Todos los Santos in 30°45′N, 139°22′E; see Stommel, *Lost Islands*, pp. 108, 132, 133. Possibly a second issue of Aaron Arrowsmith's, *Chart of the World on Mercator's Projection*, on 8 sheets, 1790. Also shown on Arrowsmith's 1794, 'Map of the World', which has Zuyden Island in about 32°N and 138°E.

[4] Okinawa Shima.

[5] From here in a different hand.

topsails to the SW, our distance from the shore being 5 or 6 Lea*gue*s. It appear'd rather low & even, the western point falling abruptly. As we stood to the Southward (after noon) it was more broken, the higher parts being connected by low Land. Towards the Southern extreme it presented a pleasing prospect of cultivated lands diversified with wood. On the Coast the Land rose in clayey cliffs of a bright Yellow with sandy beaches. Our Lat*itude* was indiff*eren*t. It made 26°47′N but we had no bearings at Noon with any distinctness. The Extremes were S76°W to West.

[Friday] 2 December

We stood to the SW as the wind permitted, the Land being only visible at times owing to Squally weather. At 2 PM Land appearing like a Small Island from the Masthead[1] bore NWbN about 8 Lea*gue*s & the extremes of Liqueo N65°W to S55°W. At 5h 15′ they extended from S80°W to N24°W, the latter being the same point set at Noon West,[2] and which seem'd to be the NW part of the Island as the Land fell back considerably within in it. The Coast still presented the same level appearance. We stood to the SW till we supposed ourselves off the extreme South part of the Isl*an*d in 26°N & kept plying under easy sail till daylight at which time the Wind shiftd to the NW & continued in that quarter very squally. The nearest land bore N30°W about 3 Leagues & the Extremes of Liqueo from N10°E to N67°W. We saw no Land in any other direction tho' many Islands are laid down in the Charts near our present situation. We stood to the Westward and at 8 AM we discern'd distant Islands to the WNW of Liqueo. To the Northward on the Western side of Liqueo the rising grounds were cover'd with trees. Towards & at Noon it blew a Fresh Gale & direct from the Islands off Liqueo which at Noon bore from N46°E to 68°E 6 or 7 Lea*gue*s, a Low Island North 8 Lea*gue*s and the Islands from thence extended to N34°W 5 Lea*gues*. We cou'd not ascertain the Number of these Islands but we counted 5 of them.[3] They were of considerable elevation & appear'd to form a Group closely connected with each other. They are named in some Charts Matchi Islands. Lat*itude* Obs*erve*d 25°53′N.

[Saturday] 3 December

We steer'd SWbW with as much wind as we cou'd carry 3 reefs in the topsails in search of some Islands supposed to be situated between Liqueo & Formosa. The Gale increased with heavy Squalls and Rain. Yet the Barometer did not show any alteration, therm*omete*r at 72°. Not wishing to pass any Land unseen we brought too as usual for the Night under a low Sail and at daylight resumed our former course. At 9 AM we discover'd low land to the westward[4] which we hauled up for. Before Noon a distant Isl*an*d[5] bore North on with a reef of Rocks which extended from the SE point of the low Land first seen,[6] and another like a

[1] Possibly low-lying Yoron Shima.

[2] Hedo Misaki, the northern extremity of Okinawa Shima.

[3] Kerama Retto; consisting of ten moderately sized islands and a number of smaller ones.

[4] Mikayo Shima, Typinsan on Broughton's 'A Chart from the South Point of Formosa to Great Lieuchieux', Broughton, *Voyage*, f.p. 147. The easternmost island of Sakishima Gunto.

[5] Possibly Ikema Shima; a conspicuous rounded hill, 27 metres high, stands at the SE end of the island.

[6] Henna Saki.

Humm*ock* bore N38°W on with the Same point[1] and a third was seen from the Masthead to the NW.[2] We had Soundings in a Corally bottom of 60 & 65 fathoms. At Noon we were well up with the Land seen 1st in the Morning. Our Lat*itude* 24°39′20″, Long*itude* 126°E when the South[-east] extreme bore N9°W 2 or 3 miles, the other extreme forming a low point N85°W, extent of reef from SE point to N22°E 3 miles, composed of rocks above and under water.

[Sunday] 4 December

Just within the South[-east] point to the westward we observ'd a small Sandy bay above which was a Village interspersed among the trees with some cultivated ground about it. The Island appeared universally barren & rocky & of little elevation. On the summit of the rising grounds were some groves of trees but chiefly Heath & coarse grass seemed to abound. The Gale blew strong, we cou'd not risk any communication with the Shore but we ranged the South side at the distance of 3 or 4 miles. As we ran to the westward the Shore became more rocky & much lower & we observed many scatter'd villages on the more elevated Parts. Our Soundings were from 18 to 38 f*atho*ms corally bott*o*m. Our course was WbS and at 1 PM the west extreme formed a small Island.[3] At 2 having run 13 miles the west extreme forming a low Sandy point bore N11°E on with some distant land[4] and low Land bore from N3°E to 18°W.[5] Between this New Land & the west extreme point of the other was an extensive reef and we had 18 f*atho*ms about 3 miles without it. At Sunset having run 35 miles on a west course a very low Island[6] bore from N65°W to 87°W[7] 5 or 6 miles, High Land from S70°W to 85°W,[8] distant Land N28°W[9] 5 or 7 Leag*ues*. We plied under easy sail during the Night with Soundings of 70 f*athoms* rocky bottom. At daylight the low Island bore of us N66°W 4 or 5 Leag*ues*. Land we had passed in the afternoon of yesterday NNE to ENE. At 8 in the morning we were within 2 Leag*ue*s of the low Island. In the middle of it are some trees which give it a little elevation; every other part except the White Beach is cover'd with brush wood. Its direction is ESE & WNW 5 or 6 miles. We now steer'd to the WSW towards the high land.[10] At Noon the Lat*itude* was 24°23′38″N, Long*itud*e 124°53′55″E. When the South extreme making a low point bore N54°W,[11] the other extreme North, Center of a Bay N30°W,[12] High Peaked Hill N50°W. Our distance off was between 3 & 4 miles. We had no Soundings. This Island is of considerable elevation much broken and the Hills connected by low land.[13] In general it was

[1] Part of the main island.
[2] Possibly Irabu Shima seen over the main island.
[3] Kurema Shima.
[4] Ikema Shima.
[5] Irabu Shima.
[6] Tarama Shima.
[7] But N85°W in Broughton, *Voyage*, p. 157 and the log.
[8] Ishigaki Shima, Patchusan on Broughton's 'A Chart from the South Point of Formosa to Great Lieuchieux', Broughton, *Voyage*, f.p. 147.
[9] Minna Shima.
[10] Ishigaki Shima.
[11] 'S54°W' in the MS, which is clearly a transcription error.
[12] Miyana Wan.
[13] Changed to 'low isthmuses' in Broughton, *Voyage*, p. 158, after which is inserted in ink in annotated *Voyage*:

South Extreme	S54°W
North Extreme	North
North Isthmus	N30°W
Peak	N50°W
Center of the Bay	N30°W

well cloathed with wood but in some parts the Hills were bare particularly to the NE. The SE point is very low & off it was a small reef. Above the beach was an extensive Village. This side of the Island took a SSW & NNE direction. We saw Smoke ascending in every direction.

[Monday] 5 December

Fine pleasant weather from the NE quarter enabled us to continue our course along the shore. After passing the Village or SE point we open'd some distant high land bear*ing* S75°W[1] leaving a passage between it & the Island we were now passing open 15° & between these disjointed lands & in the direction of our course was a continuation of Breakers & rocks with small coral reefs or Flat Islands just even with the surface which seem'd to preclude any passage between the two Islands. We steer'd more to the Southward but had no bottom with100 f*atho*ms of line. Indeed from the Village or SE point where these insulated spots were not joined they were connected by Reefs which broke entirely across the space between the two Lands. At ½ past 5 or Sunset we approached another low [Island] bearing from S50° to S56°W[2] 2 or 3 leagues, fall of high Land being the Extreme of second high Island N65°W. This was the South p*oin*t which rose abruptly from the Sea. Between the two was seen another Island making in two hummocks N85°W[3] & we were now abreast of the westernmost of the low Islands[4] extending from N40°W to N40°E 2 or 3 Lea*gue*s off the nearest.[5] We plied under easy Sail during the Night without having any soundings. At daylight the low Island set last night to the SW bore NNW 2 or 3 Lea*gue*s and an hour afterwards the two Hummock Isle was seen to the NW'ward. At which time the Abrupt point of Second Large & high Island was in the direction of North, an Opening N15°E and the other extreme N34°E. The low Island trended nearly NW & SE, 3 or 4 miles in extent leaving a clear passage between it and the high land. Since daylight the Wind had rapidly increas'd from the NNE with thick hazey rainy weather and no appearance of its altering for the better to permit our communicating with the Shore. Nor cou'd we at times see more than a mile ahead which induced us to lay too. At ½ past 8 the Hummocks bore of us North 3 or 4 Lea*gue*s and the weather prevented our seeing any more of the land.

[Tuesday] 6 December

It clear'd up a little before Noon but the gale kept increasing. We however bore up steering to the westward but we cou'd not see more than 3 or 4 Lea*gue*s in any direction and at 5 we furled the topsails & brought too for the Night under low canvas. The rain & wind continued incessant throughout. At daylight it was more moderate and the wind veering to North we cou'd see more distinctly except in the frequent Squalls of hail and Sleet. We therefore set the close reef'd Topsails and bore away to West. At 9 AM the Gale increased with heavy Squalls, hail & rain (We had furled the fore topsails) which entirely obscured the Horizon. [At] ½ past the violence of the Squall being over we discover'd the Land of Formosa[6] about 3 Lea*gue*s. It

[1] Iriomote Shima; Rocho-o-ko-ko on Broughton's 'A Chart from the South Point of Formosa to Great Lieuchieux', Broughton, *Voyage*, f.p. 147.

[2] Hateruma Shima.

[3] Nakanogan Shima.

[4] Aragusku Shima.

[5] Kuro Shima.

[6] Taiwan.

appeared very Mountainous rising abruptly from the Sea to a prodigious height. An uncommon fall of Water descending from the Hills nearly in a perpendicular direction pointed out the Coast before we cou'd discern it thro' the gloomy sky. Broach'd our ground tier of water this day being 68 days from compleating it in Volcano B*ay*.[1] At intervals we perceived the land extended from NNW to SW. We steer'd SSW & South till Noon having a large following Sea and a favorable prospect of the gale subsiding. Lat*itude* 23°45′N, Long*itude* 122°22′14″E. We made the land 24°2′N.

[Wednesday 7 December]

The Gale still continued without any abatement and we directed our course to the South with the Wind at NE. The high Land appear'd at times as the Squalls clear'd away. At 4 PM it bore S65°W. We ran on till ½ past 8 PM when being by estimation in 22°45′N we bro't too for the Night under a Foresail and Mizen Staysail. The Gale broke up after midnight & at daylight we set the topsails steering to the SW. Formosa bore from West to NNW, a Low Island[2] S80°W 4 Leag*ue*s and Botol tobago Xima[3] SWbS 10 or 12 leag*ue*s. Towards noon it became moderate & pleasant weather but very hazey over the Land of Formosa. We steer'd more to the Westward. Our Lat*itude* 22°15′48″, [Longitude] 121°57′45″E, when the extremes of Botol tobago South to S23°E, the S*outh* point 8 or 10′, Low Island North on with Formosa, the extreme point of Formosa to the Southward forming a low point[4] S70°W. Soon after noon little tobago open'd with S*outh* ex*treme* of the G*rea*t Island S20°E. This is nearly the direction which the North side of the Island takes.

[Thursday] 8 December

Fresh Breezes continued at NNE. We steered WSW towards Formosa South point. At ½ past 5 the center of Botol tobago bore N75°E, Formosa [South] extreme S78°W 2 or 3 miles. At 6 PM the South point bore WNW½W 3 or 4′ when we steer'd WSW to pass between it and the Vele Rete Rock.[5] We cou'd see nothing of it from the Masthead before dark. At 7 the extremes of the Southern point of Formosa making low Land bore from NWbN to NE when we alter'd our course to West; ½ an hour after we had Soundings of 55[6] f*athom*s gravelly bottom. Extremes ENE to NbW 4 miles. At 8 the Land from N½E to EbN when we hauled up NW. NB the bearings were of course taken by Moonlight. We sounded continually but only [occasionally] reached the bottom. During the Night we plied under easy sail and in the Morning Formosa extended from N10°E to 60°E. A Round & high topp'd Mountain[7] made the w*ester*n extreme. We made Sail to the Eastward in hopes of seeing the Vele Rete Rocks but seeing nothing of them at 8 AM we again stood to the NW, the Round Mountain being North of us. The Breeze again increased at NE & blew strong reducing us to our Courses &

[1] Since watering was last completed on 24 September in Volcano Bay it was 73 and not 68 days

[2] Lü Tao.

[3] Lan Yü

[4] O-luan Pi.

[5] Chi-hsing Chiao.

[6] '35 fathoms' in Broughton, *Voyage*, p. 161, which appears to be a transcription error as the log also gives 55 fathoms.

[7] Ta-chien-shih Shan, a 318-metre-high peak, situated 5 miles NW of O-luan Pi.

Main topsail. At Noon we had an observation and the extremes of Formosa extended from N20°E to 64°E. Our Lat*itude* was 21°45′N.

I made the Lat*itude* of Botol tobago 26°6′N, 121°22′E
Formosa S*outh* point 27°54′, 121°47′
Botol tobago bears from S*outh* point N74°E 43 miles.

[Friday] 9 December
Strong breezes and cloudy weather. At 4ʰ 40′ [PM] we lost sight of Formosa bearing N50° to 75°E. We steer'd WNW & towards noon it was more moderate. Lat*itude* 22°12′N, South Current 12 miles.

[Saturday10 December]
Strong wind & we steer'd WbN. At 6 PM we had struck sound*ings* in 45 f*athoms* fine sand & after midnight we passed several Chinese fishing boats. We carried a Light at the bolt Sprit end that they might avoid us. Our sound*ings* decreas'd to 28 f*atho*ms by daylight, having fine grey Sand. [At] ½ past [10][1] AM we saw Pedro Blanco[2] N63°W. At Noon N23°E 3 or 4′. Lat*itude* Obs*erved* 22°13′14″N, Long*itud*e 115°46′E. I make its true Lat*itud*e 22°17′N & Long*itude* 115°19′E, 29 miles. Southern Current this day.

[Sunday] 11 December
Fresh Breezes and hazey weather. We continued a Course for the Lima Islands[3] under double reefed topsails and at 7 PM we came too in 20 fathoms water, Clayey bottom, between the Great Lima[4] & Pulo Toy[5] the first bearing S45°E and the other N45°E at 1½ miles distant. The Wind remaining at NNE we rode very well. At Midnight it increased to a Strong Gale & we veer'd to a whole Cable.

In the morning we made the sig*na*l for a Pilot and at a General Muster informed the Ships company & Officers it was expected they would deliver up their Journals or any Remarks they may have made in the course of the Voyage, at the same time enjoining them secrecy as to our proceedings.

[Monday] 12 December
Soon after Noon the Gale moderated & having agreed with a Chinese Pilot to carry us to Macao for 30 Dollars we got under way and at Sunset or 6 PM came too again in 13 fathoms

[1] 22ʰ30′ in the log and in Broughton, *Voyage*, p. 162.

[2] Zhenyantou, a rocky islet with a white summit (hence its name), which when seen from the south appears as two rocks, making it a useful landmark. Known variously in the past as Pedra Branca, Pedro Branco, Pedro Blanco and Tai-Sing-Chau. It is situated 40 miles east of Hong Kong Island.

[3] Dangan Liedao, a group of five islands, 11 miles south of Hong Kong Island, usually known in the past as Lema Islands.

[4] Dangan Dao, the largest and NE island of Dangan Liedao, usually known in the past as Great Lema.

[5] Po Toi Island, the largest of the Po Toi Group, 6 miles NNW of Dangan Dao from which it is separated by Dangan Shuidao (Lema Channel).

amongst the Islands. We got under way at daylight & work'd through the Islands with some difficulty owing to an under tow or Current affecting the Ship so much we cou'd not Steer & at times wou'd not stay. At 10 in the Morning we anchor'd in Macao Roads in 4½ *fatho*ms muddy bottom, the Town bearing N60°W, Peak of Lantow[1] N75°E, Cabarita point[2] S20°W. We found riding here an East India Packet just arrived from England,[3] which gave us a late account of Public events, but private ones we were debarred the pleasure of Knowing as it was impossible our Friends cou'd know where to address us. We had now been two Years from England without any communication. The Ships Crew were in perfect health & in the course of our Voyage we had only buried one Man.[4] Nor had we any appearance of the Scurvy. We had not carried away or Sprung any Mast or Yard in the Ship which I attribute to her good qualities & her not being too heavily rigged the cause of it.

December 1796

In the Typa Harbor[5] w[h]ere my Lord Anson refitted after his Voyage in the South Seas – with frequent accidents to Ships in European Seas.

I sent an Officer to acquaint the Portuguese Governor Don [*blank*] Pinto[6] & request a Pilot to carry the Ship into the Harbor called the Typa. He very obligingly sent the Master of the Port the following Morning to conduct us & as we got under sail we Saluted the Port which was returned by an equal number of Guns.

The Longitude of Macao Anchorage is	113°	35′E
The timepiece shew'd the Longitude	114°	01′
Error in Long*itude* since leaving White point 10th November		26′

NB The Constant polite attention of the Governor I ought particularly to notice for his Gentlemanly conduct and pleasantness of manner.[7]

At Sunset we came too in the harbor and the next morning moor'd the Ship with her Stream anchor to the Westward in 4¾ fathoms. The Town of Macao just open to us, and the Harbor N25°W 2 or 3 miles.

It was the end of the Month before I had sufficiently recover'd the use of my Arm to leave the Ship when hearing there was a Small Vessel in Larks bay for sale I though it would be highly advantageous to the Voyage to purchase her and on an examination finding she was very well calculated for the purpose, I made the agreement on the Spot, and gave the Master 1500 £ Sterling for her. She was Schooner rigg'd & registered at 87 Tons.[8] On the 30th she

[1] Tai Yue Shan, formerly Lantau.

[2] Formerly the eastern extremity of Ilha de Taipa, which has disappeared under the construction of Macau airport.

[3] She is identified as the *Crescent* in Broughton, *Voyage*, p. 164. She is not, however, listed in Farrington, *Catalogue*.

[4] This is misleading and has been amended to 'the loss of one man by sickness since we sailed from Plymouth Sound'; Broughton, *Voyage*, p. 165. Five others had died as a result of accidents or had been killed.

[5] At the time of Broughton's visit the Typa was an important anchorage for Macau, between the islands of Taipa, 1½ miles south of Macau and the Island of Coloana, 1 mile farther SSW. By 1834, however, the Typa had shoaled to such an extent that it was no longer usable as an anchorage.

[6] Jose Manuel Pinto.

[7] From here taken from TNA, ADM 55/147, ff. 40–41.

[8] For details of this purchase see p. 215 below.

joined us in the Typa. On which day arrived Rear Admiral Rainier in H.M. Ship Suffolk with the Swift Sloop of War from Amboyna,[1] Having come from thence through Gilolo Passage,[2] and bound to Madras on the coast of Coromandel. Four homeward bound East Indiamen joined them in Macao Roads from Canton[3] and sailed under their Convoy on the 7th of January 1797.

January 1797

The weather throughout the last month was cold and pleasant, varying in the night the Wind from North to NE in the day, with fine weather but very Hazy and generally blowing Strong from Sunrise till Sunset when it usually moderated. The Carpenters were busily employ'd caulking the Ship & repairing our other defects where most necessary.

Throughout the Month we were employed repairing our rigging, and in other respects refitting the Ship, and repairing the defects of the schooner, altering the Step of her Mainmast, and making new Sails both for her & the Ship.

On the 18th fired 21 Guns in compliment to her Majesty's Birthday. The 20th & 21st were very squally, with rain & at times Foggy weather. An English Brig call'd the Experiment,[4] was nearly lost in the late blowing weather and only saved by the assistance of our Boats. The winds were very regular from the NE quarter, and we had with very little variation constant dry and cool weather.

February

Having compleated the caulking of the Ship the Carpenters were employ'd in repairing the Copper under water and other necessary Avocations. Chinese Painters were engaged painting the Ship and Boats; and the Ships Company compleating the rigging and receiving some provisions from Macao, Viz. Biscuit, Callivances, Rice and Arrack. Sailmakers at work on Shore opposite the Ship, where we pitched some Tents for their conveniency and others for the Astronomer to regulate the going of a Time piece we had upon a Trial.[5] Winds and weather continued much the same and at intervals when the wind was variable we had some Showers of rain.

The Governor of Macao with several Gentlemen din'd on board. His Excellency was saluted with eleven Guns.

March

The Month set in with variable and Squally weather attended with frequent rain. The Schooner on the 8th went up to the Boca Tigris[6] to bring us down some Stores & Salt Provisions which she was to receive there by a Chop Boat[7] from the East India Ships lying in

[1] Ambon.

[2] Selat Jailolo, the passage between the islands of Halmahera and Gébé, two of the Maluku islands.

[3] Guangzhou.

[4] Not identified.

[5] This was Arnold 45 which Broughton purchased from the *Henry Addington*; see pp. xlv and lxv above.

[6] Hu Men, a narrow passage in the entrance to Zhu Jiang, once guarded by forts on either side.

[7] Sampans that lightered goods between Guangzhou and Huang-pu, formerly Whampoa, the main anchorage below Guangzhou.

Canton river, with orders by no means to enter the Bogue,[1] for fear of giving offence to the Chinese Government. Several of the Officers took this opportunity of visiting Canton for a few days.

[Monday] 13 March

A spanish Brig arrived from Manilla and went into Macao Harbor. As the weather permitted we compleated the watering and in other respects prepared for sea. In the morning of the 18th the Schooner returned from the Boca Tigris; and we receiv'd by her Cordage, Canvas, Pitch, Tar, &c. with Salt Provisions, and 20 chests of tea. The middle and later parts of the Month the wind was chiefly from the SE quarter with rain. On the 26th sail'd a small Cutter called the Dragon for the NW coast of America.[2] The following days we had a great deal of bad weather with thunder and Lightning; and we parted our Stream Cable. Wind at South & SSW.

[1] Mouth of a river; Bogue forts, China: Smyth, *Sailors Word-Book*, p. 115. Presumably an instruction not to proceed above Hu Men.

[2] The sole record of the *Dragon*, Captain Lay, on the NW coast of America, was that she was found at anchor in Norfolk (Sitka) Sound in March 1797; F. W. Howay, ed. Richard A. Pierce, *A List of Trading Vessels in the Maritime Trade, 1785–1825*, Kingston, Ont., 1973, p. 33.

Book Two

Second Expedition to the North and Return to England

CHAPTER 1

From Macau to Sakishima Gunto, and the loss of the *Providence* off Miyako Shima

Observations for the Longitude and the Rate of No 45 in the Typa

By Single Altitudes of the Sun taken by Mr Crosley the Astronomer in a Small Bay at the back of the Village near Rocky Head.[1] No 45 Box Time Keeper of Arnolds was Slow for Mean Time 7^{h} 32′ 42″·11 [on] April the 8th at Noon, and gaining on Mean time 0°″419 p*er* day in Deadman's Cove.[2]

Latitude by Mean of Meridian Observations 22°9′40″[3]N

Longitude by mean of distances on each side of ☽ 113°32′8″[4] East

The Tides were so variable between 9^{h} and 11^{h} Full & Change [that] there was no absolute fixing the time of High water.

We had provisions for 15 months on board both vessels, and consider'd ourselves perfectly well equipp'd for the prosecution of the voyage. The crews of both ships were in perfect health. During our long residence we had been constantly supplied with fresh beef, Pork, and vegetables. The ships company had also bread bak'd for them every day by a copper oven brought with us and now made use of for that purpose. The Chinese gave us no trouble but assisted us in making Cordage for a reasonable Consideration. In the last three months I resided chiefly on Shore, compleating a general survey of the Kurile Islands, and transmitting them with a copy of my proceedings and further intentions to the Lords of the Admiralty.

I should be extremely deficient in gratitude were I to neglect expressing my sense of the mark'd & constant attention I receiv'd from Messrs Drummond[5] & Arthur,[6] two gentlemen of the English factory, and I ought to speak in the highest terms of the respectful politeness of the Portuguese Governor Signor Don [Jose Manuel] de Pinto.[7]

In the Typa and Macao Roads 1797

[Saturday] 1 April 1797

Moderate weather with the Wind chiefly from the NE Quarter. Having completely caulkl'd & refitted both the Ship and Schooner we prepar'd for Sea with 15 Months

[1] The SW extremity of Ilha de Taipa.

[2] Not identified, but clearly one of the coves on Ilha de Taipa.

[3] In annotated *Voyage*, p. 170, amended in ink to '32″'.

[4] In annotated *Voyage*, p. 170, amended in ink to '38″'.

[5] James W. Drummond, who in 1801 was President of the Select Committee of Supercargoes at Canton. He later became Lord Strathallan.

[6] His name does not appear in *A List of the Company's Civil Servants at their Settlements in the East Indies*.

[7] The above paragraph taken from RUSI/NM132 (8).

Provisions on board the Vessels. On the 6th the schooner proceeded to the Boca Tigris to procure some Stores &c from the East Indiamen lately arriv'd. On the 10th in the Morning we warp'd out of the Typa and came too in 5 F*athoms* water in the Entrance of the Harbor, Macao Fort bearing N45°W and Cabarita point S10°W 2 or 3 miles, the Bar Fort being shut in with the Island that forms the Starboard Entrance into the Harbor. This Anchorage is most Eligible for the conveniency of communicating with Macao & running into the Port for Shelter when Necessary & you have no more water if you lay 2 Miles further to the Eastward. In this situation also you may water your Ship with great ease attending to the Tides on the Larboard Island[1] going into the Port.

We lay at Anchor in the Typa from the 13th of December to the 10th of April moor'd in 5 *fatho*ms with the Town just Open. The Ship was steadied with a Stream Anchor which Cable we parted in a Strong Gale from the Southward on the 28th of March. This was the only bad weather we experienced during our Stay. We were generally very well supplied with Refreshments & on the Island where there is a Chinese Village[2] we pitch'd our Tents for the Astronomer & Sailmakers who met with every Civility from the Inhabitants. The least water we carried in warping out was ¼ less 3 fathoms. It was high water at the Time; but in no parts can you carry more than 3 fath*oms*. [*in margin*: At 5^{h} on the 11th Astronomical Full Moon]

Tuesday 11 April

Pleasant Breezes & fair weather from the North. In the Forenoon we got under weigh and Saluted the Fort with 11 Guns which they return'd with an equal Number. We stood towards Lantoe[3] but the Ebb Tide making strong against us we cou'd not weather the Small Isle of Chang Chiou,[4] but were oblig'd to Anchor in 4 fathoms within ¼ of a Mile of its Northern part, Macao bearing West and Linting Island[5] N15°E.

Wednesday 12 April

In the Evening with a Light air from the Southward we again Weigh'd and the same Night brought up in 5 f*atho*m, Linting bearing N27°E, Lantoe South point S60°E to N62°E the extremes. Here we remaind expecting the return of our Schooner. Calm and moderate wind South Easterly. Seeing nothing of the Tender throughout the day we made a Party in the Pinnace to examine a Harbor said to be situated on Macao Island[6] & which by appearance bore from us N41° West. As the Wind permitted we steer'd to the Westward and at 8 PM we reach'd some small Islands.[7] Here we pitch'd & took our residence in a Tent for the Night. In the morning for several Hours we had excessive heavy rain which subsiding after breakfast we proceeded along shore to the North, the Eastermost of the Nine Islands bear S11°E from those where we Slept 2 or 3 L*eague*s. Having a strong Flood in our favor we soon pull'd up with

[1] Xiaohengquin Dao.

[2] Ilha de Taipa, where considerable reconstruction has taken place.

[3] Tai Yue Shan, formerly Lantau.

[4] Ch'ing Chou, 105 metres high, and the northernmost of a group of small islands in the SE approach to Macau.

[5] Nei-ling-ting Tao.

[6] Possibly the shallow bay between the north end of Macau Harbour and a point 3½ miles NNE.

[7] Ta-chiu Chou, known at the time as Cow-chow or Nine Islands. Anchorage for small craft can be obtained between Ta-chiu Chou and the coast affording some shelter from the NE monsoon.

Calm & cloudy weather & reach'd a Small Islot laying in the Center of the Entrance about 2 miles or less within the Points; and which bears from the South point of Lantoe N50°W. The Channel on either side is perfectly Clear. We pass'd between it & the North Shore having 7 & 5 f*athoms* & nearly 1 mile across. On the Island[1] is a large Town call'd Keeough which we had no opportunity of visiting. We row'd to the Westward of this Island till we came to a Smaller One inhabited by Fishermen which bears S55°W from the N*orth* point of Keeough Island 5 or 6 Miles. To the Westward of Keeough Island Shutting the Small Islot in with it you are perfectly Shelter'd from Southerly Winds & you may ride in perfect Safety. The Channel towards the Nine[2] Islands seem'd passable for any Ships. They bear about North from the Fishing Island 2 or 3 Lea*gue*s. Having din'd at Keeough Point which is 4 or 5 Miles to the North of the Harbor Entrance and bears from the *South* point of Linting S77°E we pursued our Course to the Ship, the Schooner appearing in sight on our departure. In the Night she anchor'd close to us. At 9 AM we got under weigh and steerd for the passage between Lantoe and Las-sam-me[3] but the Wind baffling we were soon oblig'd to Anchor.

Saturday 15 April

Light breezes and Fair, the Wind South Easterly. With the Ebb Tide we weigh'd in the afternoon working thro' between the point of Lantoe & the Islands South of it,[4] our Soundings from 15 to 7 Fathoms. We came too at Night and heaving up [the anchor] the next Morning we broke the Iron Stock of our Best Bower Anchor. At 8 AM we bro't up in 10 fath*om*s South Ex*treme* of Lantoe N46°W 3 or 4 leagues.

Sunday 16 April

Fresh Breezes and hazey weather. In the afternoon we took the advantage of the Tide and turn'd to Windward. At Dark we came too in 10 fathoms. Throughout the Night we had a Fresh easterly Wind with a large Swell. Ling-ting Island[5] bore of us S15°E and the Lamma[6] on N85°E, off some small Islands 3 or 4 Miles. In the morning the Tender having parted her Cable we were employ'd sweeping for it the whole day, and at last we had the good fortune to find it. Muster'd the Ships Company.

Monday 17 April

Calm weather with no Tide prevented our getting under weigh.

Tuesday 18 April

The next day we still remain'd Stationary with Foggy and Calm weather a great dew falling & very Sultry weather.

[1] Ch'i-ao Tao.

[2] Nine crossed out and replaced by White, which appears to be an error.

[3] Guishan Dao, the largest and SE-most of 3 islands forming the SW boundary of Lantau Channel.

[4] Soko Islands.

[5] Wailingding Dao.

[6] Pok Liu Chau.

Wednesday 19 April

Light Airs from the Westward induc'd us to profit by the occasion & we got our Anchor up and made Sail to the East. At Dark the Wind having fail'd us we bro't up in 14 fathoms, East Extreme of Lamma[1] N12°E 5 or 6 Miles, Poo tooy[2] N85°E and the Grand Lema[3] S60°E.At 7 AM we again got under [weigh] but it soon after falling [calm] we brought up nearly in the same place.

Thursday 20 April

Moderate Breezes and hazey weather which soon increas'd to a thick Fog with an Easterly Wind. At 1 PM we took up our Anchor and stood towards the Grand Lema. We tack'd close in with the passage[4] between it and the Second Lema[5] and at 7 PM we came too in 13 fathoms near the Island Poo Tooy.

Friday 21 April

Throughout we had a moderate Breeze from the ENE with the Weather so Foggy we cou'd not see the points forming the passage between Poo Tooy and the Grand Lema.

Saturday 22 April

Light Winds and a thick Fog still prevented our getting to Sea, nor had we from appearances any reason to expect an Alteration while the Wind remain'd from the Eastern Quarter to our great disappointment.

Sunday 23 April

Moderate Breezes from the Eastward and the Fog having dispers'd induced us to get under way. We stretch'd over to the Grand Lema and tack'd close in with a small Bay within the west point where [are] some Houses when we tack'd towards Poo Tooy and as we drew in with it we had Squally and thick weather from the NE quarter; seeing no prospect of getting to Sea before Night we came to an Anchor in 15 f*athoms*, Poo Tooy extreme bearing N80°E 1 Mile and the East extreme of the Grand Lema S45°E. Throughout the Night and Forenoon it blew strong from the same quarter which occasion'd a large Swell. AM Muster'd the Ships Company and read the Articles of War.

Monday 24 April

Thick gloomy and unsettled weather with a heavy Swell still rolling in; the Wind continued moderate from the Eastern quarter.

[1] 'Lammon' in the MS, which is clearly a transcription error.

[2] Po Toi Island.

[3] Dangan Dao.

[4] Dangan Shuidao.

[5] Dangan Liedao.

Tuesday 25 April

Fresh breezes and pleasant Weather with a very heavy Swell from the Eastward in which quarter the wind seem'd invariably fix'd. We employ'd the Boats watering at the head of a small Bay within a Rocky Islot on the West point of Poo Toy. Several Fishing Boats took shelter in the Bay for the same purpose where they had good Anchorage & protected from all Easterly Winds.

Wednesday 26 April

Fresh breezes and clear weather. In the Forenoon at 7 AM the Wind being at NNE we got under weigh notwithstanding the heavy Swell against us which still continued. At Noon we were in the Channel turning to wind*war*d between the Grand Lema and Poo toy (a strong breeze having set in from Sea with an increase of Swell).

Thursday 27 April

We continued plying throughout the afternoon but finding we cou'd not weather the Grand Lema before Dark and the Anchorage not being eligible for the Schooner we bore up and came too an Anchor in 12 f*athom*s, the point of Poo toy bearing ESE 3 or 4 Miles. During the Night we had constant Lightning in the Western quarter. The Morning was Moderate and pleasant weather.

Friday 28 April

Towards Noon it became Calm and soon after a breeze sprung up from the SW. We immediately got under way & made Sail to the Eastward having the Ebb Tide in our favor. The Schooner not being able to weather Poy Toy pass'd between it and another Island to the North and East of it, Wag-loang,[1] carrying 12 and 14 Fathoms without reaching the bottom. At ½ past 4 the East point of the Grand Lema bore South 3 or 4 Miles and we steer'd East to pass to the Southward of Pedro Blanco.[2] Towards daylight the Wind gradually decreas'd and at 6 AM we saw Pedro Blanco bearing N18°E 3 or 4 Miles having run 39 miles from ½ past 4 PM to the time of our seeing it. Dark & Cloudy with Squally weather. The Wind suddenly Shifted to the Eastward and we stood to the SSE till 10 AM when we tack'd to the Northward. At Noon Lat*itu*de acc*oun*t 22°07′N, the Rock bearing by estimation N18°W 3 or 4 Leagues.

Saturday 29 April

Fresh Breezes and dark cloudy weather the Wind fix'd in the NE quarter. At 4 PM we pass'd within ½ a mile of Pedro Blanco and we had 22 fathoms when it bore West. We stood in Shore 3 Leagues to the North of it but the gloomy weather prevented our seeing the Land. At 6 PM we tack'd in 14 fathoms and at 7 the Rock bore South 4 or 5 Miles. During the Night we had constant Rain & thick hazey weather. At 8 AM we tack'd & stood to the Northward. Pedro Blanco at Noon bore N52°W 3 or 4 leagues. Cloudy and moderate. No Observation.

[1] Written in a later hand, but transcribed as Way-loang in Broughton, *Voyage*, p. 176. The schooner would first have passed south of Beaufort Island, then south of Sung Kong, and then south of Waglan Island.

[2] Zhenyantou. For a full description of this feature see p. 106, n. 2.

Sunday 30 April
Moderate and hazey w*eath*er. At 2 PM the rock West 3 or 4 Leagues. At 6 the Extremes of the Land from N33°W to ENE 4 or 5 Leagues & we had 13 f*atho*ms. At 7 with a Light Air we stood to the SE. In the Morning the Wind increas'd from the NE quarter with gloomy weather. [At] ½ past 10 AM we tack'd, Course S75°E, 39′ from Pedro [Blanco]. Lat*itude* acc*oun*t 22°6′N, Long*itude* 115°52′E. When we tack'd in the morning we had 45 fathoms, the Sounding gradually decreasing as we approach'd the Land. The weather still continued gloomy and very hazey without the Sun even appearing either to admit of our having a distinct view of the Land or having any Observations for the Latitude. Although it is universally acknowledged the change of the Monsoon takes place in April or at least the Winds become Variable this Journal will prove the mistake as in no one Instance have we experienc'd any other than a regular Easterly Wind excepting only for 12 Hours which continued Westerly during the Run we made from the Lema Islands to Pedro Branco. Our anxiety for a change daily increas'd from our People being in general afflicted with a Dysentery, some of them in a violent degree. We cou'd only attribute it to the Hazey and Foggy weather but whether we were right in our conclusions I am rather at a loss to determine.

In respect to our Sea Stock we were more at a loss having since our departure lost the whole of our Pigs with three breeding Sows having each eight Young ones in perfect health when we sail'd. They were the produce of Otaheite and the Sandwich Islands & of course accustom'd to their diet. Notwithstanding the Chinese Pigs remained in health and we lost every one of the Others in all 33. Their complaint was sickness and no desire to eat, upon opening some of them were found large Worms. They were (as we term'd it) equally with ourselves troubled with the Dysentery and pains in their Bowels.

Monday 1 May
Fresh breezes and very hazey weather with the Wind at ENE. At 6 PM the Land bore from West to N40°E 5 or 6 Leagues the west extreme made a Small Island.[1] [At] ½ past we tack'd and stood to Sea for the Night [*in margin*: Fathom 14]. At 5 AM we again stood to the Northward. At 8 we saw the Land indistinctly and by Noon we tack'd in 10 fathoms and the Center of a large Sandy Bay[2] the Points forming it bore S72°W and N72°E from each other. A Small entrance to a river or Harbor for Boats bore N35°E[3] and we were 2 Miles from the Shore. A Reef of rocks above water lay off the West point and others were scatter'd off the NE part. We counted no less than 300 Boats em*ploye*d Fishing as we stood in. Latitude Acc*oun*t 22°40′N, Long*itu*de account 116°02′E, Course N17°E, 31′.

Tuesday 2 May
Fresh breezes and thick hazey weather with rain. We stood to the SE till daylight when we tack'd in shore. The Forenoon we had strong breezes with heavy rain & we close reef'd the Topsails and got the Topgallant yards down. At Noon the Latitude by account 22°18′N, Long*itud*e d*itt*o 116°18′E, course S37°E, 23 Miles.

[1] Zhugau.
[2] Jieshi Wan, on the south coast of China, 35 miles NE of Zhenyantou.
[3] The entrance to Jieshi Wan on the eastern side of the bay.

Wednesday 3 May

Fresh Gales and thick hazey weather with incessant Rain. At 6 PM we had 15 fath*oms* and the haze clearing up for a short time we just got an indistinct view of the Land bearing North 3 or 4 Leagues. [At] ½ past 6 we wore & stood to the SE, the Wind remaining steady in the NE quarter. The Gale increas'd during the Night with a head Sea which reduc'd us to our Courses & Maintopsail and oblig'd us to strike the Topgallantmasts. At 8 AM we had 21 fathoms, fine Sand. The Gale now broke up and we had the wind variable from SE to North. It however soon fix'd in its Old quarter & blew Fresh. We got the Topgallant Masts up and made sail to the ESE. At Noon as usual no Observations but dark threatening weather. Course S38°E, 13 Miles, Lat*itude* acc*oun*t 22°08′N, Long*itud*e d*itt*o 116°27′E.

Thursday 4 May

Fresh breezes and dark hazey weather. At 2 PM we had the wind variable with the appearance of heavy Squalls round the Horizon which only produc'd incessant Rain. At 5 PM we had the Wind Easterly, we stood to the N*orth* till 7 when we tack'd to the Southward. At 9 the Wind veering to the SE we tack'd & stood to the ENE. During the Night we had Soundings from 29 to 26 fathoms, fine Sand with Shells. At daylight the Wind suddenly increas'd to a strong Gale from SSE with heavy Rain & we cou'd only carry our Courses & Main topsail Close Reef'd, the Schooner being press'd with Sail to keep up with us. It gradually clear'd up to the N*orth* and West and towards Noon moderated when for the first time we observ'd the Latitude which made 22°38′N, Latitude acc*oun*t 22°20′N, Long*itud*e d*itt*o 117°15′E, Course since yesterdays Estimation N53°E, 52 Miles and we had 25 fath*oms*, fine Sand.

Friday 5 May

Soon after Noon we lost the SE wind & we had light airs & Calms with clear pleasant weather till 9PM when a gentle breeze sprung up from the westward and we steer'd East having 26 f*atho*ms, broken Shells with fine Sand. Throughout the Night we had a great deal of Lightning with frequent Claps of Thunder & incessant heavy rain. At daylight we made Sail the Breeze increasing attended with Rain & we had 24 fathoms. At 10 AM we pass'd over strong riplings and the appearance of Shoal water & our Soundings varied from 10 fathoms to 15, Coarse gravel and broken Shells.[1] We therefore sent the Schooner ahead to Sound while we follow'd under an easy Sail. At Noon moderate breezes from the SW with dark gloomy weather. No observation, Course since yesterday East, 48 Miles, Latitude 22°38′, Long*itud*e account 118°7′E.

No 45 at 4 PM made the Longitude 117°35′E.

Variation p*e*r No 3 [compass] mean of 7 Azimuths 3°42′ East.

Saturday 6 May

Variable weather with Rain, the Wind from the SW quarter & we steer'd East till 7 PM (30 Miles). Our Soundings were equally irregular from Ten to 17 fathoms. At 7 we alter'd our course to NE, the Schooner keeping ahead of us with a Light during the Night. We had

[1] Broughton was over Taiwan Banks, with general depths of less than 18 metres (10 fathoms).

baffling winds with a great deal of Lightning and rain. At last we had a strong breeze from the NNE which continued till Noon, raining incessantly. During the night till 7 AM we had from 12 fathoms to Twenty varying every Cast, but the same hard gravelly bottom. At 9 we had no bottom with 30 fath*oms*. [At] ½ past we had 45 which increas'd to 56 at Noon, Black Sand, Course S84°E, 67 Miles, Latitude account 22°33′N, Long*itud*e 109°20′ East.

Sunday 7 May

Moderate Breezes & thick weather with continual Rain. The Wind remain'd in the same quarter and we kept our Head to the Eastward. Our Soundings were 55 fath*oms* till 7 PM, afterwards we had no bottom with 100 Fathoms. At daylight we saw the High Land of Formosa[1] extending from SE to NE, the breeze gradually died away and we had Calm and Sultry weather with showers of rain which prevented any Observation. As we approach'd the Land till 8 AM we were able to discover several other Islands Some of moderate elevation & others very low but all forming a Group which led us to imagine they were the Piscadores,[2] situated the center of them under the Tropic by the Charts. [*in margin*: NB The Isl*and*s call'd Piscadores are laid down from 23° to near 24°N. Of course we ought to have seen them had they been properly plac'd in the Charts.] At Noon the Southernmost bore S68°E and to the North they bore N40°E the High Island S85°E 4 or 5 Leagues. The High Land of Formosa was Obscur'd by the Hazey weather and we had in this situation 114 fathoms, gravelly bottom. Course S70°E, 32′, Lat*itude* 22°22′N, Long*itude* acc*oun*t 119°52′E. Soon after Noon our mistake was clear'd up as the whole of the Suppos'd Islands were connected by low Land.[3] [*in margin*: No Observation]

Monday 8 May

Light breezes from the North & we stood in shore till 5 PM when being within two Leagues of the Land we tack'd in 23 fathoms muddy bottom. The Suppos'd Islands of the Piscadores we now plainly discern'd were part of the Main connected by low Land well wooded with Sandy Beaches & many Habitations scatter'd along Shore. The weather clearing up we had a fine Moonlight Night and ½ past Nine PM we made the Latitude 22°50′N, the ☽['s] Alt*itude* being 68°45′ L[ower] L[imb]. It remain'd Calm from that time till the Morning when a breeze springing up from the same quarter we stood in for the Land & fetched in nearly to the same part we had quitted last evening. Finding there appear'd no probability of the Winds changing to enable us making our passage to the North thro' the Straits of Formosa,[4] at 7 AM we steer'd a long Shore Course SSE. At noon we had light breezes & very fine clear pleasant weather but by mistake lost our Observation. The Schooner being very near us made 22°35′N which agreed very near with mine, although I was to[o] late to see the Sun rise. The Extremes of Formosa N12°E to S46°E, High rugged Land from N35°E to N75°E 4 or 5 miles. Within this point I suppos'd to be the Harbor w[h]ere formerly the Dutch had a Settlement[5] as over the Beach we saw the Masts of several Junks & the Country

[1] Taiwan.

[2] P'eng-hu Ch'un-tao.

[3] Since P'eng-hu Ch'un-tao, as the Pescadores are now called, lie between 23°30′ and 23°40′N and between 119°30′ and 119°40′E, it seems unlikely that Broughton sighted these islands.

[4] Taiwan Strait.

[5] Fort Zealandia, near present day T'an-nan in 23°01′N.

appear'd to be well inhabited. An Island Lamay[1] of moderate elevation bore S35°E 4 or 5 Leagues and we had 75 fathoms, blue Mud. [*in margin*: In the Morning we saw several Boats Fishing & some of them came near enough to remark their curious appearance. They were simply Floats compos'd of Bamboo's lash'd together and carried three Men. Their Mast was fix'd in a wooden Step in the Center & they sail'd very Fast. They were about 20 feet by 6 and very light.]

Tuesday 9 May

Light Breezes and fine Clear weather. We steer'd a Course to pass without the Island which at ½ past 4 bore East, the *South* point. This Island is detache'd 3 or 4 Leagues from the Main & is 5 or 6 miles in circuit of moderate elevation & well cultivated. We saw several Houses and Cattle grazing on the Downs. It appear'd in many places to be easy of access. By our run since Noon I make its Latitude 22°22′N. It bears from the Bluff Land of Fort Zealand N18°W & S18°E. We had no Soundings at Sun set, when the Extremes of Formosa bore from N16°W to S50°E, the Island from N10°W to N7°E off Shore 4 or 5 Leagues. Very Light Airs continued from the NW throughout the Night. At Daylight the extremes bore from S72′E to N20°W 5 or 6 Leagues. We steer'd more to the East but cou'd not reach the bottom with 100 fathoms. At 21^{h}30′ we saw the Vele rete Rocks[2] from the Masthead and at Noon we pass'd between them & the South part of Formosa[3] with a gentle Breeze at West; Observ'd Latitude made 21°50′N when the SW point bore N35°W, SE point N63°E ,Vele rete Rocks S34°E 7 or 9 Miles, Five Miles from the Southern part of Formosa. We were very fortunate this day in having good Observations for the Latitude which enabled us to determine the situation of the Rocks & Formosa South Point very exactly. We experienc'd a very strong Current in our favor with great riplings but had no bottom with 75 fathoms at noon; half an hour before we had 53 fathoms, gravelly bottom. The SW & SE points of Formosa bear from each other S85°E and N85°W 12 Miles. Vele rete Rocks & SW point N34°W & S34°E 15 miles, SE point and Vele rete Rocks N12°E & S12°W 12 Miles. The Shore between the two points is low with Sandy Beaches & some high Black Rocks are scatter'd at unequal distances. The Land rises gradually from the Sea & presents a fertile country interspers'd with some Rocky Eminences. One of which is remarkable making like a Friars Hood,[4] it bore N14°E of us at Noon. Course S33°E, 54′, Lat*itude* acc*oun*t 22°00′N, Long*itud*e acc*oun*t 120°37′E.

Wednesday 10 May

Fresh breezes and fine clear weather. At ½ past 1 the SE point bore West 4 or 5 Miles & we steer'd NNE. At 4 the Wind suddenly died away & we had Light breezes from the Northward. At sunset the South point of Formosa S46°W and Botol tobago Xima[5] from S48°E to S53°E, a Low Island (observ'd last year)[6] N35°E off the High Land of Formosa 4 or 5 m*ile*s. Although it was calm mostly throughout the Night we found the effects of a current setting to the

[1] Liu-ch'iu Yü, situated about 40 miles NW of O-luan Pi, the southern point of Taiwan.

[2] Chi-hsing Chiao.

[3] O-luan Pi.

[4] Ta-chien-shih Shan, a 318-metre-high chimney-shaped peak, 5 miles NW of O-luan Pi.

[5] Lan Yü.

[6] Lü Tao.

North.[1] At daylight we were well up with the Island & we had a gentle breeze from the Southward. At 8ʰ it bore West and we steer'd N*orth*. At Noon it bore S21°W about 9 Leag*ues* & the Extreme land we cou'd see bore N*orth*; our distance off Shore was 3 or 4 Leagues & we had no Soundings. Course N11°E, 74′, Lat*itude* Observ'd 23°06′N, Accounted Lat*itude* 22°36′N, Long*itude* 121°E, allowing a new departure from Formosa SE point. The Land from 22°30′N continued of an Amazing height Rising very abruptly from the Sea & the very tops of the Mountains cloath'd with Wood. They form'd in the direction of the Shore nearly North & South having between the Ridges most fertile Valleys. We this Day set up the Rigging Fore & Aft. [*in margin of* OD 177: Variat*io*n 3°30′E']

Thursday 11 May

From Noon to 4ʰ it remain'd Calm when a breeze springing up from the NE we stood in Shore. At ½ past we tackd within a Mile of the Land and had no Soundings with 100 fath*oms*. The Extremes of Formosa seen indistinctly thro' the Haze from S18°W to N8°E. A small stream thro' the Valleys bore S80°W. Our acc*oun*t Lat*itude* 23°12′N* [*in margin*: Near this place Count Benyousky found a Har*bo*r, Vide Vol. III Page [*blank*]],[2] but still we experienced a Current setting us to the NE which will make our real situation more Northerly. The Forenoon we had a steady NE wind with gloomy weather & Squalls of Rain. Course S83°E, 35′, Lat*itude* acc*oun*t 23°4′N, Long*itude* d*itt*o 121°53′E. We had a sight of the Sun about Noon its Alt*itu*de being 83°43′, which makes 24°04′N, but we cou'd not depend upon it.

Friday 12 May

Fresh Breezes & gloomy weather with Small Rain, the Wind steadily fix'd in the NE quarter. This Day we had the Misfortune to find the Foremast very rotten, extending through the Center of the Mast from the Heel upwards above the Mainstays. The Mast to appearance outside was perfectly Sound & we only discover'd it after the Rigging had been set up which occasion'd it to be rent & splinter'd in the Fore Cockpit below the Wedges, Opening so much as to discover the internal part was entirely decay'd leaving Six or Seven Inches Diameter of Sound Wood Outside. The Mast had also settled a little. We prepar'd by Fishes[3] & every other means to secure the Mast for the present carrying very little Head Sail to affect it & wedging the Mast in the upper Deck.[4] At 9 PM we tack'd & we had moderate weather throughout. At 9 AM we saw Land bearing NbE. At Noon we were within 3 or 4 leag*ue*s of the Island[5] bearing from N*orth* to N23°E. No observation to depend upon. Course N45°E, 18′, Lat*itude* acc*oun*t 23°16′N, Long*itude* 122°04′E.

[1] The Japan Current sets north along the east coast of Taiwan attaining a rate of 1½ to 2 knots.

[2] 'Saturday, August the 27th [1771], moored in a bay at the mouth of a river, on the island of Formosa'; Benyowsky, *Memoirs and Travels*, II, p. 27. Possibly Ch'eng-kuang ao in 23°10′N, 'a small cove in which small craft can shelter, even in the NE monsoon'; *China Sea Pilot,* Vol. III, Taunton, 1982, p. 125.

[3] Long pieces of hard wood, convex on one side and concave on the other, bound opposite each other to strengthen the mast, secured by bolts and hoops or stout rope called woolding.

[4] Broughton adds in Broughton, *Voyage*, p. 188, 'and got the top-gallant mast down upon deck'.

[5] Yonakuni Shima, an isolated island, hilly at either extreme, situated in 24°27′N, 123°00′E, about midway between Taiwan and Iriomote Shima.

Saturday 13 May

Moderate Breezes & Cloudy weather. At 4 PM the Island bore from N55°E to N74°E when we tack'd in hopes of getting close to it before dark, in which we were disappointed. At 7 it bore from N18°W to N20°E about 3 Leagues. At the same time several Fires were lighted which gave us reason to suppose it was Inhabited; in this situation we had no Soundings with 100 fath*oms*. We stood to the SE till midnight when we tack'd. At 5 AM the Extremes bore from N38°W to 53°W 5 or 6 Leag*ues*. We also saw more Islands to Windward the Largest[1] bore N70°E and two Hummocks[2] S85°E 7 Leag*ues* distant. As the Wind veer'd to the Eastward we stood to the North with light airs & pleasant weather. At Noon we observed in 24°27′45″N when the Western Island bore from S86°W to S76°W about 10 miles, the Hummocks S46°E and the large Island from S65° to S85°E near 6 Leagues. Course N28°E, 12′, Lat*itude* acc*ount* 23°22′N, Long*itude* 122°30′E, No 45 made 124°2′E. By which it appear'd we had experienced a Current setting us N43°E, 88 miles, the first day we stood to the East from Formosa. As the Schooner being to Windward had made the Signal for seeing Land bearing NEbE in the Forenoon of the 11th which prov'd to be the Western Island. The Others appear'd to be the same we had notic'd last year. The West Island's greatest direction was SE & NW about 5 or 5 miles. It was of moderate elevation falling to abrupt points at each extreme with a peaked Hill in the Center of the Island. The Country presented a fertile prospect. It bore from the large Island nearly East and West about 10 Leagues. [*in margin of* OD 177: Variation: 1, 3°18′E; 2, 40′; 3, 4′ do; 4, 1°10′ by 4 Compasses.]

Sunday 14 May

Till Sunset we had Light airs from the NE and throughout the Night it was Calm & Sultry weather. At Sunrise the wind came very gently from the SW quarter and we steer'd SE to make the Land plainer which bore from S49°E to S83°E, Hummock Isl*and* S22°E. At 10 AM it bore South 8 or 10 miles & appear'd to be a broken Rocky Island of very little extent. We also just discern'd from the Main Top a Sandy Island notic'd last Year[3] near them which bore S30°E on with the West point of the large Island. At Noon the same point bore due South 4 or 5 miles, Hummock Isl*and* S22°W 10 miles. A small Bay S45°E 3 Miles & the Extremes of the Island N60°E. Lat*itude* Observ'd 24°25′N, Course S81°E, 15′, Lat*itude* acc*ount* 24°20′N, Long*itude* 122°45′E. No bottom with 100 fath*oms* of Line. The Shore was highly diversified by Woods & presented a very fertile prospect.

Monday 15 May

Light breezes & mild pleasant weather from the SW quarter, we steer'd NE along shore keeping about 3 Miles off. Within the small bay we saw some scatterd rocks on which the Sea broke. Some Boats with matted Sails were close in Shore. By 4 the Small Island (set at Noon)[4] bore of us SE 2 or 3 miles. It was low & Flat cover'd with brush wood & Coarse Grass, an extensive Reef partly dry projected from each Ex*treme* to the NE & SW. It is very little

[1] Iriomote Shima, Rocho-o-ko-ko on Broughton's 'A Chart from the South Point of Formosa to Great Lieuchieux', Broughton, *Voyage*, f.p. 147.

[2] Nakanogan Shima.

[3] Hateruma Shima.

[4] Hatoma Shima.

distance from the large Island which now took a SE'ly direction from its N*orth* Ex*treme*. We steer'd East for the other Island[1] which bore in that direction 4 or 5 Leag*ues*. We now discover'd the range of low Islands joining each other by Reefs (seen last year) that seem to exclude any passage between the two large Islands & yet we had no Soundings. At 7 the Small Flat Island[2] bore from S55°W to S65°W 4 or 5 miles on with N*orth* Extreme of Rocho.o.koko. East Ex*treme* Rocho.o.koko S18°E, a peaked Island among the reefs[3] S25°E. The other Island[4] extreme about Village point S55°E and the Extremes to the North of it N75°E. After running 15 Miles we plied off & on till Daylight when the Extremes of Pa-chu-san [bore] from N80°E to S10°W 3 or 4 Leagues off, the Isthmus Rocho-o-koko from S30°W to S60°W. We had light airs & sultry weather in the Forenoon & we sent our Boat on Shore to acquire some Knowledge of the Islanders who we plainly discern'd with our Glasses though we cou'd not see any Habitations. At Noon we observ'd in 24°33′15″N, Course N77°E, 40′, Lat*itude* acc*oun*t, 24°31′N, Long*itude* 123°29′E and we had 125 fathoms water within 4 Miles of the Land.

On the 5th of December last year we pass'd on the East side of this Island, but blowing Weather coming on the next Morning we were unable to acquire any informations relat*ing* [to] them, nor cou'd we at present find any safe place to Anchor in shelter'd from the reigning Winds.

In the afternoon the Boats[5] return'd; the Schooner having accompanied them close in Shore for their protection. They landed on the NW side of the Island within a Coral Reef which extended nearly ½ a mile off and afforded excellent Shelter for Boats within it. The Soundings were irregular without it from 5 to 40 fathoms.

They had the good fortune soon after disembarking to fall in with a Path-way which following soon led them to a Small Village consisting of about 30 Houses. On entering one of them they found an Elderly Man who treated them with the greatest civility and gave them some refreshments. The rest of the Inhabitants seem'd to have deserted their Houses on their Landing, but on our Gentlemen Reembarking several of them came down to the Beach. These Islanders were similar in appearance to the Chinese & Japanese, their writing in characters. But their Hair was roll'd up to the Crown of their Heads & secur'd by two Metal pins. They wore loose linnen Gowns & Trowzers & paid their respects by joining their Hands & lifting them up to their Heads.

These Islands produce Tarro, Sweet potatoes, Rice, Millet, Peaches, Limes &c. They have also black cattle & Horses. Their Houses were of one Story of a Square Form, the roof coming up to a point & thatch'd with Indian Corn Leaves.[6] The inside of them were Neat & Clean & cover'd with Matts. The Natives appear'd harmless and in-offensive. Their Women were not seen nor any Arms or Weapons of defence.

Tuesday 16 May

Light airs from the North which increas'd to a Fresh Breeze towards Night. At 2 PM we were within 2 or 3 miles of the shore which seem'd to promise anchorage but we soon after observ'd

[1] Ishigaki Shima; Patchusan on Broughton's 'A Chart from the South Point of Formosa to Great Lieuchieux', Broughton, *Voyage*, f.p. 147.

[2] Hatoma Shima.

[3] Kuro Shima, which is 12 metres in height.

[4] Ishigaki Shima.

[5] Broughton mentions only a single boat in Broughton, *Voyage*, p. 191.

[6] 'coarse reeds' in Broughton, *Voyage*, p. 192.

a Reef of Coral Rocks extending some distance from the shore across the Suppos'd Bay and we had no bottom with 100 fathoms. We therefore Tack'd and the Boat returning at 5 PM we stood to the WNW till Midnight when we tack'd to the East*ward*. At daylight Extremes of Pachusan S65°E, the North point. At Noon Lat*itud*e Observ'd 24°55′50″N when the N*orth* part bore S14°W to the other Extreme about Peaked Hill[1] S30°W about 6 Leagues, Course N39°E, 30 miles, Latitude account 24°55′N, Long*itud*e 123°49′E.

Wednesday 17 May

Fresh breezes and cloudy weather with a large swell from the Northward. We Steer'd East as I propos'd examining the North side of the Islands we had seen last Year & the Schooners signal was made to steer EbS. At 2 PM we saw a Sandy Island the same seen last year.[2] It was just visible from the Deck bearing S15°E about 5 Leag*ues*. At 3 PM we saw the Land ahead[3] and we haul'd up ENE and at 4 made the Schooners Sig*na*l to haul her Wind on the Larboard Tack. At ½ past 5 PM we haul'd our Wind to the NE, wind at NNW, & Shorten'd Sail for the Schooner to come up with us. As the Carpenters had not yet compleated Hooping the Fishes on the Foremast we carried only a Reef'd Foresail & close reef'd Foretopsail, the Topgallant mast being down upon Deck. At 7 PM An Hummock making like an high Island[4] beyond the other Land bore S60°E and the other extreme S10°W Off nearest shore 4 or 5 Leagues. At this period the Mate of the Watch (Mr Cawley) reported there was no Land to be seen beyond the Hummock Hill Nor any appearances of Land &c ahead or to Windward. We therefore stood on meaning to tack at 8 Oclock & try for Soundings and afterwards ply to Windward for the Night the Ship lay up NE½N going 4½ Knots p*e*r Hour.

Almost immediately after, I left the Deck to protract our situation, which I found to be on the North side of an Isl*an*d call'd Typinsan[5] & which we pass'd on the South side of in December last year; within 3 miles of the Shore.

At ½ past 7 PM White water was seen ahead and upon each Bow and reported to the Officer of the Watch (L*ieutenan*t Vashon) and almost immediately after the Ship Struck upon a Reef of Coral Rocks.[6] Having plainly felt the Shock I instantly went upon Deck (& met in the passage Mr Vashon coming to acquaint me with our disaster). The Officers & Seamen were upon Deck also. At this Period the Ships Head was ENE (Sails all Full) and as I understood the Helm was hard a Weather by the Man at the Wheel). [*In margin of* RUSI/NM/132 (8): Had the Helm been put a Lee on seeing the danger, I think we should have escap'd it.] The Sails were directly brac'd Aback, the Anchors order'd to be clear for Running. The Signal of Danger [was] made to the Schooner and the Master went to her in the Cutter to bring her to an Anchor near us if possible to heave off by. The Ship soon after paid off with her Head to the

[1] Minanosoko Mori, a grassy hill 86 metres high, with a remarkable rock precipice on its summit, stands near the SW extremity of Ishigaki Shima; *Japan Pilot*, Vol. II, 7th edition, Taunton, 1979, p. 71.

[2] Kurema Shima; Ashuma on Broughton's 'A Chart from the South Point of Formosa to Great Lieuchieux', Broughton, *Voyage*, f.p. 147.

[3] Irabu Shima; Eraboo on Broughton's 'A Chart from the South Point of Formosa to Great Lieuchieux', Broughton, *Voyage*, f.p. 147.

[4] Ikema Shima; Corumah on Broughton's 'A Chart from the South Point of Formosa to Great Lieuchieux', Broughton, *Voyage*, f.p. 147.

[5] Miyako Shima; Typinsan on Broughton's 'A Chart from the South Point of Formosa to Great Lieuchieux', Broughton, *Voyage*, f.p. 147.

[6] Yaebi Se, an extensive area of coral reefs extending 9 miles north of Ikema Shima.

East & we haul'd up the Mainsail and Shiver'd the Sails to let her go round without acquiring head way. Before her Head paid off to South She again Struck Fore & Aft & remained fast her head then due South. Breakers were then on the Starboard Bow & we had from 5 to 15 f*ath*-*o*ms in the Starboard Chains & 2½ at times both ahead and Astern. We chock'd the Rudder, Got the Maintopgallant Yard & Mast down & Struck the Topmasts & began hoisting the Boats out having kept the lower Yards up for that purpose. The Ship during this period did not strike violently & had only made 19 inches water. Unfortunately the Wind freshend from the NNW & the Surf broke with great Force which soon knock'd her Rudder off and we secur'd it by Hawsers. We now only waited for the Schooner & in the mean time began hoisting out the Longboat, this was after 9 Oclock. The Ship began to make water fast and the Violent Shocks she receiv'd render'd it doubtful whether the Masts wou'd stand while getting the Longboat into the water. The Water increas'd fast upon the Pumps & before the Longboat was out we had 7 feet water in the Hold. The Schooner had just anchor'd near us in 25 f*atho*ms & the Master return'd on board, when the Ship suddenly chang'd her Position swinging round from the South to North by the East & striking more violently than ever. Before we cou'd carry our Hawzers out to the Schooner the Carpenter reported the Water up to the Orlop Deck and the Ship having bilged forward. We therefore gave up the idea of attempting to heave the Ship off the Rocks as she wou'd inevitably have founder'd with us.[1] It as now past 10 Oclock & we had got the Spare Pumps down the Fore hatchway but the Water increasing fast upon the Gun Deck render'd all our exertions ineffectual. The Officers were unanimous with me in agreeing nothing cou'd be done to save the Ship & to cut away the Masts cou'd have no effect. As the Ship was settling fast forward from her being bilg'd as we imagin'd in her Larboard Bow.

I therefore directed the Officers to save their Commissions & to prepare the Boats for embarking the People and to collect Arms & Ammunition with all the Armourers & Carpenters Tools. But the Ship laying nearly on her Beam Ends & the Gun Deck being full of Water with the washing to & Fro of the Bulk heads, Chests &c prevented our saving many. We had at this time 6 feet water on one Side and 3½ f*atho*ms on the other, the Ship nearly immers'd forward in the Sea & the Surf breaking in upon the Upper Deck. As nothing further cou'd be sav'd at present we began to send the People in[to] the Boats which was happily affected without any accident & soon after Eleven PM they reach'd the Schooner in safety. The Pinnace having return'd for myself and the remainder of the Officers we quitted the Ship at ½ past Midnight & Mustering the Crew on board we found no one absent.

The Moon rose about 12 PM & at that period it blew a Fresh breeze & occasion'd the Sea to break very high upon the Reef. The Schooner having come to with both Anchors we hove up one and at 4 AM we had the misfortune to part our Cable. Fortunately we cast the Right way. If we had not most probably we shou'd have experienc'd a Second Shipwreck & to many of us more Fatal. As in our present instance we were thankful for our escape although we had lost our property, but in the second disaster I fear we shoul'd have lost our Lives. Thus were we thankful in our present situation miserable as our prospects were but by comparison of what it might have been we had every reason to be contented with our Lot, and to rejoice in our personal Safety.

At daylight we had Fresh breezes & hazey weather from the NNW & we dispatch'd our Boats under the Masters directions to save every thing that might be of future use from the Wreck. They did not return till after Noon having the Longboat full of Cordage they had

[1] 'supposing it possible to heave her off' added here in UKHO, OD 177.

unrove & Sails they had unbent. Also the Kedge Anchor & a 7 Inch Hawzer which was most acceptable to us having already lost an Anchor & Cable. The Ship lay nearly on her beam Ends her Larboard side up to the Hatchways entirely immersed in the water and the Sea breaking entirely over her. The Decks were partly blown up & the Bulk heads fore & aft entirely wash'd away. Every moveable Article was floated away Nor was there any signs of Books, Mathematical & Nautical Instruments of any kind remaining to my great disappointment and Mortification. The Officers & People were equally unfortunate with myself in finding their property either destroy'ed or below the Decks of the Ship where they cou'd not be got at. In the Cabin they found several Stand of Arms, Cutlasses and Pistols which gave us great satisfaction.

The Master finding no Provisions, Liquor or any other useful Stores cou'd be got at by our remaining near the Wreck in the Schooner & our Water on board being very trifling for our Numbers it became a principal consideration to procure some at the Islands, as our existence depended upon it and we knew the Island to be inhabited we accordingly cleard the Boats. At Noon we observ'd in the Latitude 25°2′N, the Island Hummock[1] bearing S35°E about 4 Leag*ue*s, South Extreme S10°W, Shipwreck East 2 Cables, Outer part of the Reef N40°E and we had 56 f*atho*ms rocky bottom. Course from Noon of the 16th to the Ships striking N73°E, 35 Miles.

Thursday 18 May

Moderate Breezes & fair weather. After having clear'd the Boats we steer'd to the Westward. At 6 PM we bore up to the Southward rounding an Island of moderate Elevation[2] at the distance of 5 or 6 Miles which bore WSW[3] about 12 miles from the Island we observ'd at Noon bearing from us S35°E. It also bore NW[4] from the Sandy Island[5] observ'd last Year. At dark we bro't too for the Night having 50 fathoms close into the shore. At daylight the Extremes of the two Islands[6] from NNE to East 3 or 4 Leag*ue*s. We steer'd NE to pass between them supposing we shou'd find anchorage. As we approach'd the Land we had great overfalls from 15 to 2 & 3 fathoms but by frequent tacking & wearing to escape the Reefs we at last brought up in 13 f*atho*ms near the point of the 1st Island[7] where there was a prospect of getting Wood & water. We soon perceiv'd the Island to be inhabited & a Canoe paying us a visit we express'd to them our wishes. They directly left us to comply with them & seeing a Village at some distance & another on the opposite shore to the East[8] we sent a Boat to each place. They were receiv'd in the most friendly manner and water immediately brought down to the Boats, and in the afternoon [19 May] they sent us in their own Canoes & Boats a larger quantity with Wood & some packages of Canary Seed without asking for any return or indeed in any manner seeming to expect it, but strongly express'd a desire for us to proceed to the Eastern Village where we shou'd have our wants supplied. In the morning at daylight the

[1] Ikema Shima; a conspicuous rounded hill, 27 metres high, stands at the SE end of the island.
[2] Irabu Shima.
[3] 'ENE' in the MS which is clearly a reciprocal bearing.
[4] 'SE' in the MS which is another reciprocal bearing.
[5] Kurema Shima.
[6] Irabu Shima and Miyako Shima.
[7] Irabu Shima.
[8] I.e. on Miyako Shima.

Longboat and Pinnace were dispatch'd to the Wreck to see if anything more cou'd be sav'd. At 9 AM the Flood Tide making we got under weigh and turn'd to Windward towards the Village situated 2 or 3 Leagues to the East of us.[1] In our traverses we met with a succession of Reefs with deep water between them and at last with great difficulty (and once touching upon a Coral Rock) came to an Anchor in 3½ fathoms near several Junks and abreast of the Village. We had no sooner arriv'd than our Friends brought us a Load of Water and Three Hogs.

[1] Present day Hirara.

CHAPTER 2

Events in Miyako Shima and return to Macau in the schooner

Fresh Southerly winds after a Rainy Night enabled the people to dry their Cloathing and in the morning we paid a Visit to our Friends who receiv'd us with great civility in a large & convenient House. The Floors were well matted for our reception, on which we sat & partook of the refreshments of Tea, Pipes & tobacco. We made them comprehend our wants by Signs and we had great pleasure finding they were desirous & had the power of satisfying them. This House which appear'd to belong to the Principal People was in an Elevated Site environ'd by a Square Wall of Stones 12 feet in height and a gateway through it with a Guard House over it. The Rooms were spacious open at the Sides with balcony's to keep them cool. Several Venerable Old Men encircled our party dress'd in large loose Gowns of fine Workmanship similar to Tiffany[1] and of various Colours and paterns: they were tied round the middle (their flowing garments) with a Sash. They also wore Trowzers & Sandals. The Crowns of their Heads were Shav'd & the Hair was brought up to a Knot on the Top & securely fasten'd by Metal Pins in the Malay stile. They universally made use of Fans & some wore Neat Straw Hats which tied under the Chin; the Aged men had most respectable Beards. Having taken our leave we were desirous of walking about the Town which they seem'd very unwilling to permit nor cou'd our persuasions induce them. Not wishing in our present circumstances to give them any Offence we did not persist but proceeded in our Boat some distance to the Watering Place where the Natives were readily employ'd assisting our people in their Work drawing Water from a Stone well which appear'd to have been sunk for the purpose of watering the adjacent plantations. This Water they assured us was only fit for washing and they promised to send us better.

We had a fine afternoon and they sent us off Water, Rice, & Wood. The night was rainy with unsettled weather. Towards the morning we had the Wind SW and a prospect of a fine day which induc'd us to send another Boat with a Party of Marines to assist our other Boats shou'd it be necessary. They were soon oblig'd by strong Squalls of wind to return & we were oblig'd to let go a Second Anchor to secure our not going on shore. In the Afternoon we had more moderate weather but to[o] boisterous to admit of any communication with the Shore. This variable weather made us uneasy for the absent Boats, which was soon done away by their joining us in the Evening to our great satisfaction. They had visited the Wreck but without any success, nor cou'd they procure any provisions whatever either from her or by tracing the Shore where they might have drifted to. The Wreck remain'd nearly the same & they cut her Masts away. Her Starboard Side was just out of the Water. The Master reported that off Hummock Island[2] the Natives bro't them Potatoes & water and in every other part of their

[1] A kind of gauze muslin, originally the dress for Twelfth Night.

[2] Ikema Shima.

excursion did they receive the same Kind attentions. These good People were fully acquainted with our misfortune & naturally conceiv'd our greatest wants were the Articles of Life which they parted with in a most humane Manner. Part of their Houses containd pieces of Ship Timber & in surrounding the Island they saw several human Skulls in the Caverns of the Rocks, probably the remains of some Ship more unfortunate than ourselves.

As it plainly appear'd from the Masters representation we had nothing to expect from the Wreck any unnecessary delay here became dangerous from the want of Provisions of which we had a very insufficient quantity (even at ⅔ allowance) for so numerous a Crew & considering we had to expect constant SW winds we cou'd not possibly make our passage with any degree of certainty. As we cou'd only carry twenty four days Water which wou'd not admit our risking anything by keeping to[o] long at Sea – added to which the smallness of the Vessel preventing more than ¼ of them being below we had to dread every complaint among the People who were without Cloathing (like ourselves) & destitute of every comfort Shou'd we have to encounter bad weather. If we had been fortunate enough to have procur'd Provisions &c from the Wreck I purpos'd if we cou'd prevail upon the Inhabitants to have left Seventy Men till my return in the NE Monsoon and in the interim employing the Vessel surveying &c. However I found from the dislike the Inhabitants even had to our seeing the Country militated against my wishes & I believe wou'd have render'd my Scheme totally impracticable without Force. A Power in any situation I shou'd have been sorry to have exerted, more particularly in our forlorn State.

In the morning of the 23rd we receiv'd the remainder of our Supplies which amounted to 50 bags of Wheat, 20 of Rice & 3 of Sweet potatoes each bag containing 100 lbs.[1] Also one Bullock of 300 lbs weight 6 pigs and one dozen of Fowls. Indeed every thing we ask'd for they immediately sent us and more we cou'd not take for want of Room. They also even gave us Many of the Jars they brought the Water off to us in, but still objected to us seeing the Country. Being ready for sailing we paid our last visit carrying with us some trifling presents (the most acceptable we had). We endeavour'd to make them comprehend how sensible we were of their Kindness & they receiv'd our Gifts with great satisfaction particularly a drawing of the Ship & a Spying Glass. After receiving refreshments as usual they accompanied us down to the Beach and hoisting a Union Jack in the Long Boat (which was compleat with her Sails &c) I made them a present of her to their great Joy. They immediately took possession of her and we then parted with these humane & civilized People not unaffected by the favors we had receiv'd from them in our distress'd situation.

The evening before we sail'd the Heel of the Providence's Foremast drifted pass'd us. It appear'd entirely rotten throughout and nothing held it together but the Fishes. We tow'd it on Shore & took off the Iron Hoops. Some of the Ships planking also drifted into the Bay.

As the Master reported the passage to Sea between Eraboo & Coore-mah[2] ineligible from the numerous reefs we piloted the Vessel out to Sea between the Shoals having two Boats ahead to warn us of danger, pursuing nearly the same Track we came in. At Noon we observed in 24°42′30″N when the Rocks off Eraboo bore North 2 or 3 Miles and we had 12 fathoms which continued varying to 3 f*atho*ms till we brought Sandy Island or Achuma[3] on with Eraboo West point SE & NW when being clear of the Shoal water we call'd the Boats on board & made sail to the Northward passing about 3 miles on the west of Eraboo. At noon Fresh breezes & fair w*eathe*r from the SE quarter.

[1] 1 cwt [hundredweight] in Broughton, *Voyage*, p. 207.

[2] Corumah in Broughton, *Voyage*, p. 208.

[3] Ashumah in Broughton, *Voyage*, p. 208; Kurema Shima.

Wednesday 24 May

Fresh Breezes & fair weather. At 4 PM we were in sight of the Wreck & passd several planks floating and a Cask of Hops which the Boats pick'd up. The Hull of the Ship was but just visible & neither Masts nor Bowsprit standing. The Sea made a continual breach of her. It appear'd to me she had alter'd her Position as the Hummock Island bore of her S31°E. We steer'd NE till 6 PM when the outer breakers (within which were Rocks above water) bore S15°E on with Hummock Island 2 or 3 miles and we had no Soundings with 100 fathoms of Line. In this situation we saw no Land to the East of the Hummock. Our Estimated Latitude was 25°13½′N & Longitude 125°E. The Reef broke very high & long before we discover'd the Ship we saw the Breakers yet there was no more Wind (only in a different quarter) when the Ship struck & yet inconceivable as it may & does appear. The Breakers were hardly seen before the Ship went on shore. Such was our unhappy fatality to be cast away even when we had not or cou'd not apprehend the least danger from appearances or from any human Foresight of mine depending on a proper lookout being kept after dark agre[e]able to my written Orders & constant directions. Finding there was no prospect of procuring any thing further from the Ship & make[1] every necessary Remark on the Reef &c We secur'd the Pinnace & Cutter astern meaning to carry them with me in case of any unfortunate Accident happening to the Schooner. [*in margin of* RUSI/NM/132 (8): In addition they carried part of the crew as the Schooner was not large enough for such a number (112 Souls). By night I did not chuse to risk them.] We steer'd WbS till 6 AM when the wind veerd more Southerly & continued Squally with Rain, Thunder & Lightning. [*in margin of* RUSI/NM/132 (8): The group of Islands, off one of which the ship was lost, is compos'd of seventeen, of different sizes, many of them small and uninhabited. They extend from 24°3′N to 24°52′30″N and from 123°2′E to 125°37′ East Long*itud*e. The Inhabitants distinguish'd them by the name Mad-ge-co-simah, and as we understood them they are tributary to the Liquieux Islands.]

Thursday 25 May

Squally and variable weather excessive Sultry and continual Rain. The wind at last fix'd in the North and we made sail to the SW. Before Midnight we saw the Island of Pachusan.[2] At daylight the Extremes bore from East to SSW 6 or 7 Leagues & we alterd our course to SWbS. At noon the South[3] point of Rocho-o-ko-ko[4] bore S15°E 3 or 4′. Wind Easterly and moderate throughout the Forenoon.

Friday 26 May

Cloudy weather. In the afternoon the Cutter return'd having examin'd a small Bay which afforded Anchorage.[5] A Canoe brought them off water & potatoes. Surely these Islanders were not acquainted with our misfortunes. Towards noon the wind veer'd to South & SSW.

[1] 'made' in the MS, which is clearly a transcription error.

[2] Ishigaki Shima.

[3] Possibly a transcription error for south-west.

[4] Iriomote Shima.

[5] Funauke Ko, on the NW side of the island, affords anchorage for vessels with local knowledge.

At 12 PM we tack'd to avoid West Island[1] which at daylight bore of us N57°W 6 or 7 Lea*gues*. At Noon the Rocky Island[2] off the S*outh* point of Rocho-o-koko bore N65°E 6 or 7 Lea*gues*, Lat*itude* acc*oun*t 24°7′N, [Longitude] 123°34′[E].

Saturday 27 May
Variable & Squally weather with a great deal of Rain which rendered our situation truly uncomfortable. The wind continued SSE till near Noon when it veer'd to the west & we stood to the South. Latitude acc*ount* 23°49′N,[3] Longitude 123°0′ East. Squally at Noon with frequent Showers of Rain.

Sunday 28 May
The Wind in the afternoon veer'd to the North and blew a Fresh Breeze and we steer'd SSW. At 10 AM we saw the Island Tabaco[4] (situated near Formosa) & soon after Botol Tobago Xima.[5] At Noon the former bore N65°W to N75°W and the latter S25°W to S30°W 10 or 12 Leagues. Lat*itude* Obs*erve*d 22°32′N, account 22°21′N, Long*itude* 122°11′E. Fair and pleasant weather.

Monday 29 May
Fresh breezes and fair. At Sunset Botol tobago Xima bore SE, South point of Formosa[6] SW½W about 10 Leagues. At 10 PM we were by estimation in the Lat*itude* of it when we steer'd South not intending to pass between Formosa & Vele rete.[7] At daylight we had run 25 miles when to our great surprize we saw the South point of Formosa bearing SWbW 4 or 5 Lea*gues*. We steer'd SW till Noon 21 miles, when it bore S40°W 3 Lea*gues*. The Wind having moderated we cou'd not make any way against this strong Northern Current[8] but remain'd nearly stationary. A Vertical Sun made our Latitude 22°1′30″, Lat*itude* acc*oun*t 21°22′N, Longitude D*itt*o 121°11′E when Botol tobago Xima bore from N84°E to S88°E, Formosa N*orth* extreme N10°E, 3 Miles offshore, and we had no Soundings.

Tuesday 30 May
We insensibly drew near the Land in the afternoon & had the good fortune to experience a counter Current in our favor, which set us along shore to the South. The Cutter having been sent in to examine the Shore within us found good Anchorage in 20 &15 fathoms about ½ a miles off. They saw Many Runs of water & several Herds of Cattle but no signs of Habitations

[1] Yonakuni Shima.
[2] Nakanogan Shima.
[3] '24°07′N' in the MS, which is clearly a transcription error as it seems unlikely that the noon latitude would be the same as the previous day; 23°49′N taken from Broughton, *Voyage*, Appendix I, p. 386.
[4] 'Tobago' in Broughton, *Voyage*, p. 212; but surely Lü Tao.
[5] Lan Yü.
[6] O-luan Pi.
[7] Chi-hsing Chiao.
[8] The north-going Japan Current attains a rate of 1½ to 2 knots, but may exceed 3 knots.

or any Cultivation. At Dark the South point which is low, Flat and Rocky [bore] S40°W 3 miles. At 8 it bore West. It continued calm till ½ past 9 with strong Lightning in the NE quarter from whence there suddenly sprung up a strong Breeze & we steer'd WSW. At 12 PM the Land about the South Point bore North and we steer'd WbS. The Wind now moderated & we thought ourselves most fortunate in having so speedily enter'd the China Seas & the prospect of the Easterly Winds continuance. At Noon the High Land on the Southern part of Formosa bore N65°E 10 or 12 Leag*ues*. Vertical Sun made the Latitude 21°44′N, Long*itude* account 120°21′E.

Wednesday 31 May
Moderate breezes from the Eastern quarter gradually decreasing. Throughout the Night we had incessant Rain with Thunder and Lightning, the Heat excessive. At daylight the breeze increased & we steer'd west from 8 AM.[1] Latitude observ'd 21°37′N by vertical ☉ , Longitude 119°09′E, we pass'd strong riplings with the appearance of a strong current, accounted Latitude 21°37′N.

Thursday 1 June
Light airs and calms from the Southward throughout these 24 hours. Lat*itude* observ'd 21°53′N, Lat*itude* acc*ount* 21°40′, Long*itud*e 118°26′E. No 45,[2] allowing its error 1°30′ at Formosa, shew'd 117°12′E.

Friday 2 June
Calms throughout till near Noon when we had a gentle breeze from the Eastward. When we sounded at 5 PM we had 53 fath*oms* and trying the current we found it set to the westward. Smooth water and excessive sultry wea*the*r. Ther*mometer* at 87° in the Cabin. Lat*itude* Observ'd 22°3′25″N, Account 21°58′N, Long*itude* 118°E, No 45 116°24′E & same Error allow'd.

Saturday 3 June
The breeze gradually increas'd to a gentle gale. We steer'd WbS & West, our Soundings continued the same & in the afternoon we saw some Fishing Boats. At 11 AM we saw the Lema Islands.[3] At Noon the Grand Lema[4] N47°W 4 or 5 Leag*ues*, Lat*itud*e obs*erv*'d 21°53′N, Acc*oun*t 21°55′N, Long*itude* d*itt*o 115°54′E. At ½ past 2 PM we pass'd the Great Lema, our Long*itude* per acc*oun*t 115°46′E, Lema Long*itude* 114°26′, by which it appears we have been set 1°20′ West of our account since leaving Formosa. Latitude of the Great Lema by our run since Noon 22°04′ North.

[1] 'PM' in the MS, which is surely a transcription error.
[2] Since Broughton implies that no instruments were saved from the wreck, Arnold No. 45 was presumably on board the schooner at the time.
[3] Dangan Liedao.
[4] Dangan Dao.

Sunday 4 June

Fresh breezes & very hazey weather. At ½ past 2 Great Lema bore West 1½ miles and we steer'd WbN 30 miles till 8 PM, when we haul'd up between Jagro[1] and Lantoe point[2] & having run 11 miles further, we came too in 12 f*atho*ms, Lantoe point SE 4 miles. In the morning the Wind continued a Fresh breeze from the Eastward & we got under way. At 8 AM we were abreast of Lintin,[3] from whence the Cutter was sent to Macao with an Officer & we steer'd NbE½E for Lankeet Island,[4] carrying 10, 8 and 7 fath*oms* till we came in sight of Sampanchow[5] when we had only 5 f*athoms* within 2 or 3 miles of Lankeet. We gave it a good berth to pass to the East of Sampanchow & soon increas'd our water after passing Lankeet to 10 fathoms. At Noon Sampanchow West 1 mile.

Monday 5 June

Moderate breezes and pleasant weather from the SE q*uarte*r. Having a strong tide in our favor we soon pass'd thro' the Boca Tigris[6] & discover'd 13 Ships at Anchor up the River. Not Keeping near enough to Tigre Island[7] we had very Shoal water on the extensive Bank which runs off on the opposite Side but hauling to the Westward we soon deepen'd our Water. At 5 PM we pass'd the Homeward bound Fleet of East Indiamen laying below the Second Bar. Many of their Boats visited us & offer'd any Assistance we might require. We pursued our course up the River and at 5 PM keeping too near the Starboard Shore we grounded on the Mud. At 10 we hove off into the Stream & remain'd at Anchor for the Night. At daylight we again weighed & at 8 AM we anchor'd at Whampoa[8] in 4 f*atho*ms water. Here we found two Botany Bay Ships[9] & pass'd coming up the River an East Indiaman & the Crescent Pacquet who were going down. After Anchoring I proceeded to Canton in the Pinnace to regulate the distribution of our Seamen and to procure Fresh Provisions, leaving directions for no Mandarine Boats to be admitted on board in my Absence. We reach'd Canton about Noon having a strong tide in our favor. The Supercargoes & the Captains of the Indiamen being on the Spot, I soon regulated my business & directed the Schooner to proceed from Whampoa to the Second Bar as I intended following in the Pinnace.

[1] Identified as Lassammu in Broughton, *Voyage*, p. 216; present day Guishan Dao.

[2] Present day Fan Lau Kok.

[3] Nei-ling-ting Tao.

[4] Longxue Dao, situated on the west side of Zhu Jiang estuary.

[5] Shanban Zhou, consisting of two islets ¼ mile apart and 1¾ miles NNE of Longxue Dao.

[6] Hu Men, formerly Boca Tigris, a narrow passage in the entrance to Zhu Jiang, once guarded by forts on either side.

[7] Dahu Dao, situated on the west side of the main channel, 1½ miles upstream from Boca Tigris.

[8] Huang-pu, the main anchorage below Guangzhou (Canton).

[9] Wrongly identified in Broughton, *Voyage*, p. 217, as 'two ships from Port Jackson, having laden with tea, on the Company's account, the *Alfred* and *Crescent*, East India ships, going down the river'. No ships of these names are listed in J. S. Cumpston, *Shipping Arrivals & Departures: Sydney, 1788–1825*, Canberra, 1977, as having arrived or sailed from Port Jackson in 1796 or 1797. In addition no East Indiamen of these names are listed in Anthony Farrington, *Catalogue*. Possibly the transports *Prince of Wales* and *Sylph* which had sailed from Port Jackson for China on 23 November and 6 December, respectively; Cumpston, *Shipping Arrivals*, p. 31. According to Farrington, *Catalogue*, p. 525, the *Prince of Wales*, Captain William Milner, left Port Jackson on 3 November, arrived at Macao on 7 February, at Whampoa on 10 April, and at Second Bar on 6 June, but Farrington has no record for *Sylph*. Farrington, Vol. I, gives no details for Milner except '1795/6 in command of the *P. of W.*'

Tuesday 6 June

The Night was rainy weather which continued at intervals during the day. Early in the Morning arriv'd the 1st Lieu*tenan*t of the Swift Sloop of war with dispatches from Admiral Rainier[1] which gave us the first account of a Spanish War. In the course of the day I explain'd thro' an Interpreter the cause of my proceeding up the river & my present Wants. This Man had been deputed by the Mandarins to inquire into the particulars of my arrival with the reasons thereof. Which being fully acquainted with by Mr Hall[2] (the principal Supercargo) & myself he went to report the same. This afternoon was cool and pleasant after the Rain.

Wednesday 7 June

Very sultry fair weather. Not finding it necessary to remain longer at Canton to expedite our receiving Refreshments which Mr Hall obligingly undertook, we proceeded down in the Pinnace to the Second Bar having been 4½ hours in our Voyage.

Thursday and Friday 8 & 9 June

Fresh Breezes from the Northward. These two days we were employ'd receiving Salt provisions from the India Ships and distributing the Officers & Men who were to take their passage with them to Europe. Early in the Morning we were visited by a Mandarin of the 1st Rank who assur'd us our wants shou'd be immediately supplied now he was convinc'd whom and what we were, which he was determin'd to satisfy himself of before the permission was granted. Their good intentions were of no use as we did not remain long enough to benefit by them. By the Lieutenant of the Swift I sent our Marines down to Macao to join the Sloop of War & a Letter to Capt*ai*n Hayward[3] acquainting him I shou'd follow immediately on the arrival of our Cutter whose safety from her long absence I almost doubted.

Saturday 10 June

Light breezes from the North. Some of the East india Ships Sail'd down the River and in the afternoon I sent the Schooner down to wait my coming in Anson's Bay.[4] Early in the morning I followed her in the Pinnace and very luckily met our Cutter in the Boca Tigris returning from Macao. The excuse for her long detention was strong Northerly Winds which I did not conceive a sufficient reason, but I had great pleasure in her safe arrival.

Sunday 11 June

Fresh breezes & hazey weather from the NE. At 2 PM we join'd the Schooner & got under way and kept turning towards Macao Roads as the Wind veer'd to the SE in the afternoon.

[1] Rear Admiral Peter Rainier, Commander-in-Chief East Indies Station.

[2] Richard Hall.

[3] Thomas Hayward, formerly a midshipman in the *Bounty*, who accompanied Bligh in the open boat voyage to Timor after the mutiny.

[4] On the east side of the estuary just above Boca Tigris, where Anson spent 2 weeks at anchor in 1743.

At ½ past 8 AM we came too an Anchor in 5 f*atho*ms the Town W¼N & Cabarita point[1] SbW½W. His Maj*esty*s Sloop Swift anchord near us & the Crescent Pacquet with two Indiamen were in the Roads. The Forenoon was attended with heavy Squalls of Wind and Rain. We remain'd here till the 14th and discharged the remainder of the Providence's Crew (except 35 in the Schooner) into the Swift for his Majestys Service. The Winds continued Southerly which prevented the arrival of the Indiamen before & in the Forenoon we got under way with them and at Noon Anchor'd, the tide having done, Cabarita point bearing NW 3 or 4′.

Thursday 15 June

Light breezes and pleasant weather, Wind SE'ly. In the afternoon H*is* Maj*esty*s Sloop Swift,[2] in Company with 14 Sail of East Indiamen, the Crescent Pacq*ue*t and two botany bay Ships (all laden with Tea) got under way & proceeded for England by the Eastern passage. We also got under way and stood towards the Typa. The next morning we anchor'd off the Watering place having the Nine Islands[3] just open with the Westermost of the Typa Isles[4] NE¾N, the Harbor of Macao NbW½W. Close in the entrance a Spanish Ship was moor'd. She was bound to Manilla but on seeing the Swift enter the Typa she ran for the Port. We remain'd here till the 26th having variable weather with frequent Rain & at times Strong Winds from the SE with severe thunder & Lightning. The Winds in general from the Eastward, Seldom or ever from the Westward. This I understood had been the case since our Sailing. The weather was very Sultry and if not refresh'd by the rain wou'd be almost insupportable. We here compleated our Provisions, water and Wood and on the 26th sail'd out of the Typa. [*in margin of* RUSI/NM/132 (8): The Schooner could not contain more than 5 months provisions. We however fill'd every part of her that we might keep the sea as long as possible to the N*or*th of Japan.]

On the 14th at $21^h20'$ A.S.[5] No 45 shew'd the Longitude of the Typa
115°19′ East
<u>113°32′</u> E true Longitude of d*itt*o
1°47′ difference in 67 Days

By Single Altitudes of the Sun observd between the 14th and 25th of June with a quick silver Horizon cover'd by a Glass Roof Her rate appear'd pretty Regular and on the 26th at Noon the same Watch was slow for mean time $7^h\,40'\,36''\cdot1$ and losing $6''\cdot651$ p*er* day on Mean time. Slow for Greenwich $6'30''\cdot4$

NB the Altitudes were taken from the same places as Mr Crosley formerly observ'd from.

We now sail'd a second time in the prosecution of the voyage, not indeed with the most flattering hopes of success, as the season was far advanc'd & the vessel in many respects ill-suited to

[1] Formerly the eastern extremity of Ilha de Taipa, which has disappeared under the construction of Macau Airport.

[2] In the margin of RUSI/NM/ 132 (8): The ship viz the *Swift* has never since been heard of and is suppos'd to have foundered at Sea in a Gale of Wind by which the Fleet severely suffer'd. [*and in brackets*: in all 43, 4 petty Officers, 24 able & ord*inary* Seamen, 15 Marines.]

[3] Ta-chiu Chou, known at the time as Cow-chow or Nine Islands.

[4] Xiaohengquin Dao.

[5] Not understood, but perhaps Apparent Solar Time or a corruption for A.T. = Apparent time.

the purpose. But still there was a prospect of acquiring a geographical knowledge of the Tartarean & Corean coasts; nor did I wish to fail in any endeavour that might promote the improvement of science by the discovery of parts of the world as yet but imperfectly known. The officers and men were equally well dispos'd with myself to do our respective duties and we departed in good health.[1]

[1] The above paragraph taken from RUSI/NM/132 (9)

CHAPTER 3

From Macau in the schooner to the east and north coasts of Taiwan and thence to Okinawa

Tuesday 27 June

Moderate breezes & pleasant weather from the SW quarter. At 2 PM we got under weigh with a strong Ebb Tide and made Sail out of the Typa. At 3 we sent a Boat on board a Portuguese Ship bound to Macao from Cochin China. At 5 we weather'd Potoe[1] & pass'd within ½ a mile carrying 6½ [fathoms]. As we pass'd to the West 2 or 3 miles of the Grand Landrone[2] we gradually increas'd our water to 12 *fathoms*, South Ex*treme* of Landrone G*ran*d bore S80°E on with the outermost rock SW of the Asses Ears.[3] Mr Dalrymple makes the Rock R̲ bear more Southerly.[4] It is probable therefore we did not see it.

At Dark we boarded an East India Packet call'd the Amazon four Months and Ten days from Falmouth.[5] At 8 the Grand Landrone bore N55°E, Potoe N5°W, East ex*treme* of Islands in sight E½S, wore & made sail to the SE'ward till 5 AM when we alter'd our course to the ENE, Grand Landrone N32°W, Ex*treme* of Islands N68°W. At Noon fine pleasant weather. Lat*itud*e Ob*served* 21°40′N, Course S70°E. 67′, Lat*itu*de acc*oun*t 21°49′N, Long*itude* account 115°4′E, No 45 made 114°51′25″ East. [*in margin*: G*rand* Ladrone Lat*itude* 22°02′N, *Longitude* 113°56′E]

Wednesday 28 June

Fresh Breezes & fine clear weather. At 6 PM we saw from the Fore Yard a Fleet bearing ENE consisting of 9 Large Ships. They were on a Wind with their Starboard Tacks on board and One Ship on the other Tack. Conceiving them to be the Spanish Squadron from Manilla cruizing for the East Indiamen [*in margin of* OD 177: NB On our return to Canton it appear'd these Ships were part of the Indiamen that sail'd in comp*any* with the Swift.[6]] we steer'd for[7] them till Dark, when having the Wind at WSW we steer'd NWbN in hopes of

[1] Hsiao-pu-tai.

[2] Dawanshan Dao.

[3] Two remarkable peaks near the SW end of Beijian Dao.

[4] Probably 'A Chart of Part of the Coast of China, and the adjacent Islands from Pedro Blanca to the Mizen. Laid down from observations made in 1759 and 1760, in the Schooner *Cuddalore* and the Ship *London*, 1764. By A. Dalrymple'; published 21 January 1771.

[5] The *Amazon*, Captain William Waring, left Falmouth on 16 February 1797, reached Whampoa on 11 July, returning to Long Reach in the Thames on 11 July 1798; Farrington, *Catalogue*, p. 18.

[6] In annotated *Voyage*, p. 224, added in pencil '4 East Indiamen The *Crescent* packet'.

[7] Possibly a transcription error for 'from'; it seems unlikely Broughton would have steered towards a possible enemy fleet; 'steered in for the land' in Broughton, *Voyage*, p. 224.

escaping by getting in Shore. At ½ past midnight we had run 30 miles & to our surprize we discover'd a Ship upon our Lee beam without her Topgallantsails set. We kept close to the Wind & Forereaching upon her we soon increas'd our distance. As she evidently took no notice of us or probably suppos'd us to be a Fishing Boat we bore away to the North*war*d & shooting across her Course we had the good fortune soon after to loose sight of her. At ½ past 2 we saw some Rockets from her which we imagin'd to be answering Signals. At Daylight we were within 3 or 4 Leagues of the Coast of China & abreast of Pissoang or the Great Bay.[1] The Ex*tremes* bore from West to N80°E and we alter'd our course to East, having no sight of the Fleet. At 8 we steer'd ENE 2 or 3 Leagues from the Shore which appear'd nearly its direction. The Coast seem'd to form several Bays: & scatter'd Rocks & Islots were detach'd at uneven distances along the Shore. At Noon the Extremes bore from N85°W to N42°E & we had 15 fathoms 2 or 3 Leagues off. Lat*itude* account 22°52′N, Long*itude* 116°7′E, Course N39°E, 93 m*ile*s.

Thursday 29 June

Fresh breezes and fine weather with smooth water. With the Wind from SW we pursued an ENE Course which increas'd our distance from the Shore. At 6 PM the Ex*tremes* to the East making high [land] bore N14°E 6 or 7 Leagues. At 10 the Moon being Set we haul'd our Wind to the South under an easy Sail having 24 f*atho*ms. At daylight we had no sight of Land when we made Sail to the Eastward. Course N81°E, 61 miles, Lat*itud*e acc*oun*t 23°17′N, Long*itud*e 117°54′E, 20 fathoms Water.

Friday 30 June

Moderate breezes from the SW with fair weather, Course East. In the afternoon we pass'd several strong riplings & continuing Sounding we had over falls from 13 to 20 fathoms.[2] At 12 PM we had 30 f*atho*ms. [At] ½ past 1 we saw Land bearing NE when we tack'd till daylight – 22 fathoms. At 5 the Piscadore Islands[3] bore from East to NbW; about 4 miles distant from the Eastermost, which is low and Flat. We steer'd NNW with a gentle breeze to pass on the West side of the Islands. At 10^{h}50′ AM we pass'd within two miles of a Rocky Island[4] & had 25 fathoms. This Rock is very conspicuous being much higher than any of the other Islands. Some Reefs connect its Eastern part with two low Islots which are surrounded by breakers. At Noon the Westermost of the Group bore from N76°E to N86°E 4 or 5 miles & we had 27 fathoms. This Island is of small Extent, moderate Elevation, Rocky & Barren & apparently uninhabited, though we saw several Fishing Boats among the Reefs. High Island in one bearing with the Eastermost seen in the morning now bore S48°E. Course N76°E, 45′, Lat*itude* Ob*serve*d 23°22′30″N, Long*itu*de 118°41′E, [No] 45 119°9′E. By these Observations it appears we have experienc'd very little current, the Accounted Lat*itude* being 23°28′N.

[1] Honghai Wan

[2] Once again Broughton was over Taiwan Banks.

[3] P'eng-hu Ch'un-tao, consisting of two main groups of islands.

[4] Mao Yü, a prominent dome-shaped islet 78 metres high, situated on the west side of the southern group.

Saturday 1 July

Moderate breezes & fair weather, the Wind from the SSW. We steer'd NE & NEbE till 5 PM when seeing Breakers ahead we alter'd our Course to N*orth* till 7 PM when we resumed a NE course till Midnight. At 3 PM we were abreast of Fishers Island.[1] On the SW extreme of it, which forms rather more elevated than West Isle, we observ'd an Obelisk of Stones rais'd I shou'd imagine for the advantage of the Junks who trade to these Islands. We saw several of them in the course of the day, making their coasting Voyages. This Island is of some Extent & is both inhabited & cultivated. We saw Cattle feeding as we pass'd & many Boats passing to & Fro'. At the NE Ex*treme* of this Isle the Land falls back & seems to promise Anchorage. But so many Rocks both above & under water must render the passage intricate. A Black Rock extends a League from the Entrance & seems connected by Reefs with more to the North of it. Indeed in this part the Rocks & Sand Banks cou'd not be enumerated. The large Island[2] which lies to the East I cou'd not observe so as to make any Remarks upon it. Our Soundings were from 15 to 25 f*atho*ms, Hard bottom. At ½ past 6 Bird Island[3] being the Ex*treme* to the North bore S58°E 3 or 4 miles and Obelisk point on Fishers Island S17°W. At Midnight we kept our Wind under easy Sail till daylight when we steer'd East to make Formosa. At ½ past 5 we saw the Island. At $10^h30'$ AM we were within 3 miles of the shore, the Extreme to the South making a low Sandy point S6°E. On the other side of the land we observ'd several Junks at Anchor. In this situation we had only 7½ fathoms & we steer'd NNE along shore. At Noon the Extremes of Formosa from N45°E to S7°E 3 or 4 miles offshore & we had 11 fathoms sandy bottom. Course N40°E, 71′, Lat*itud*e ⊙ 24°10′32″N, Lat*itud*e account 23°55′N, Long*itu*de acc*oun*t 119°39′E, No 45 120°16′E mean AM [&] PM [observations]. The Land towards the Sea presented a very barren appearance, Sand Hills being interspersed with coarse grass. Near the beach were some scatter'd Trees but more inland the Hills rose to a very great height and were richly cloath'd with woods. The Piscadores are a Group of low Islands & Rocks above & even with the Water. We only saw the western side of them and by our Observations they extend from the Latitude of 23°10′N to 23°40′N & from the Longitude of 119°20′E to 119°40′E p*e*r 45 watch & situated about 8 Leag*ue*s west of Formosa. The Dangers which surround them may be avoided by attention as they appear to Shew themselves & you have Anchorage in case of necessity. I lamented we cou'd not examine the Eastern side & have had some communication with them.

Sunday 2 July

Moderate breezes & fair weather. We steer'd NNE Keeping 3 or 4 miles from the Shore. At 6 PM it became calm but we still experienced a strong current setting us to the north. At 7 the extremes of Formosa bore N45°E to S13°W. The entrance of an Opening[4] bore S40°E off which were several Junks & Boats at Anchor. Our Soundings were 31 fathoms. It was calm throughout the Night and in the Forenoon we had the Wind from the NE q*uarte*r. At daylight the Ex*tremes* bore from N58°E to S18°W off shore 7 or 8 miles. The Land towards the North Ex*treme* gradually decreas'd in height but to the SE the Hills or rather Mountains

[1] Hsi Yü, the westernmost island of the northern group.

[2] P'eng-hu Tao.

[3] Probably Mu-tou Yü, a 16-metre high rock, rather than Chi-pei Tao, 18 metres high (on which there is a village), the northernmost island of the northern group.

[4] Possibly the mouth of the Hou-lung Hsi, which junks can enter at high water.

were very high. We stood to the NW and at Noon the Northermost Land we cou'd see bore East. Off the nearest 6 or 7 Leagues. Lat*itude* Observ'd ⊙ 25°9′35″N, Lat*itu*de acc*oun*t 24°35′N, [course] North, 59′, Long*itu*de account 119°40′E, No 45 120°36′ East.

Monday 3 July
Fresh breezes & hazey weather the wind gradually increasing from the NE quarter. [*in margin of* RUSI/NM/132 (9): This gale of which we bore a part was the same, which is suppos'd to have occassion'd the loss of our unfortunate companions in the Swift Sloop as she has never been heard of since that time.] At 4 PM we stood in shore. At ½ past 5 we had Squally wea*the*r with rain & so hazey we cou'd not discern the Land. At ½ past 6 we cou'd just discern it bearing S40°E 3 or 4 Leagues. At 8 we handed the Topsails & double reef'd the lower Sails, Wind & Sea increasing. In the Forenoon it blew a hard Gale with a great deal of Rain & we split most of our Sails. Thick hazey weather with a large Sea. Pump'd Ship every hour. At Noon Lat*itud*e by acc*oun*t 25°19′N, Long*itud*e d*itt*o 119°21′E, Course N59°W, 20 miles.

Tuesday 4 July
Strong gales of wind with heavy Rain & a great Sea running. At ½ past 2 saw Land (the Coast of China)[1], bearing WNW indistinctly thro' the Haze. We soon after wore & the Gale still blowing harder we reef'd the Boltsprit[2] & hove too under a close reef'd Mainsail. We had also the misfortune to spring a leak in the Starb*oar*d Quarter which Kept us almost continually pumping. The Wind kept gradually veering round to the west with increas'd Force which caus'd a confus'd & heavy Sea, but we did not Ship much water, the Vessel laying too very well. In the Forenoon, the wind having at last fixd in the SW, the Gale broke up and we had a Fresh breeze from that quarter. At 8 AM we were enabled to steer NEbE, the Sea having considerably fallen. Course N83°E, 30′, Lat*itude* 25°23′N, Long*itude* 119°53′E.

Wednesday 5 July
Fresh breezes and cloudy weather wind at SW. We steer'd East. At 2 we hove too while the Carpenters were endeavouring to stop the Leak in our starboard quarter which they had the good fortune to compleat in a short time. We again made sail and at ½ past 3 we saw the high Land of Formosa bearing S30°E. At ½ past 6 the extremes bore from S28°E to S23°W 5 or 6 Leagues. Toward Night the wind moderated and at 12 PM we saw a small rocky Island[3] bearing SEbS and we had no Soundings with 25 fathoms. We stood off and on till daylight when at 5 AM the Rock bore S50°E. The Ex*treme* of the North part of Formosa from S50°W to S10°E, Two Islands one N64°E the other S84°E.[4] We steer'd to the SE passing between Formosa & the Rock without having bottom with 25 f*atho*ms. At Noon the Observ'd Lat*itud*e made 25°18′44″N when the NW point of Formosa[5] bore S88°W, NE d*itto*[6] S48°W, Extreme

[1] More probably the off-lying island Haitan Dao (25°30′N, 119°45′E).
[2] An alternative name for bowsprit.
[3] Hua-ping Yü, a rugged rock 47 metres high.
[4] P'eng-chia Yü, 142 metres high; and Mien-hua Yü, a rocky islet 55 metres high.
[5] Pai-sha Chia.
[6] San-tiao Chioa.

beyond d*itto* which appear'd detached S2°W, the smaller Isle suppos'd Kelang[1] which lays off the Harbor[2] S72°W, Off shore about 4 or 5 Leagues. This part appear'd High Land and on the Larboard entrance as you sail towards the Harbor is a remarkable Hill shap'd like a Cone[3] from whence the low land commences South of it, till you near the NW part which is high tho' some low land projects from it. Our Latitude may err a little supposing we did not see the Extent of the low land. The Islands in the Offing bore N10°W, N13°W & N55°W from 4 to 6 L*eagues*. Course S86°E, 60′, Acc*oun*t 25°15′N, Long*itud*e 121°E, No 45 121°52′E. No current since last observations.

Thursday 6 July

Moderate breezes & fine weather from the NW quarter. At 6 PM it had decreas'd to nearly a Calm when Formosa NE part bore S67°W. [*in margin of RUSI/NM/132 (9)*: In the direction of the Island of Quelang the main land of Formosa appeard very high, and on the larboard entrance as you sail towards the harbour is a conspicuous hill of a conic form from whence the low land commences extending to the South. Some low land also projects from the NW part which is in general very elevated. At 6h the NE part of Formosa bore S67°W.] Course from Noon NE. It continued variable with light airs & calms throughout the Night & in the Forenoon freshen'd up from the South East & South. We steer'd EbN for Great Liquieux. At ½ past 10 AM we saw an Island bearing ENE.[4] At Noon it bore nearly the same. Lat*itud*e observ'd 25°35′25″N, Acc*oun*t 25°35′N, Long*itud*e 121°55′E, [No] 45 122°53′E, Course N72°E, 55 miles. Fair & pleasant weather.

Friday 7 July

Fresh breezes and fair weather. At 1 PM seeing some peaked high Rocks projecting to the East of the Island[5] we bore up and pass'd to Leeward within two miles of the Shore having no Soundings with 50 fath*om*s of line. This Island is very high forming two peaked Hills with a hollow between. Its direction is nearly East & West 3 or 4 Miles & entirely cloath'd with wood up to its Summit. To the N*orth* East of it for 3 miles is a continued chain of Rocks above & under water; they are connected seemingly with the Peaked Rocky Island which bears East from it about 1 mile. We make the Lat*itud*e of the Island 25°40′N and Longitude 123°14′E by the watch.[6] At 3 PM we saw another Island bearing N60°E.[7] At 4 we steerd East & at ½ past 5 it bore N*orth* 2 or 3 miles. It was of small extent & moderate elevation & like the other well cloath'd with wood & bounded by a Rocky Shore. It is situated in 25°48′N, Longitude 123°26′E. At 4 we steer'd East with a fine breeze from the South. At ½ past 12 we descry'd a

[1] Chi-lung Tao, a precipitous black rocky island 182 metres high, an excellent mark for making Chi-lung Kang. Quelang in Broughton, *Voyage*, p. 231 and below.

[2] Chi-lung Kang, now a major port.

[3] Probably Ch'iu-tzu Shan, a 133-metre-high hill on the west side of the harbour.

[4] Uotsuri Shima, the largest and SW island of Senkaku Shoto, a group of uninhabited islets and rocks, laying about 90 miles ENE of the NE extremity of Taiwan.

[5] Kitako Shima and Minamiko Shima.

[6] '123.39E returning' omitted as meaningless.

[7] Kobi Sho, the summit of an extinct volcano, 116 metres high, situated about 17 miles NE of Uotsuri Shima.

high Rock bearing E½S.[1] At 1 we steer'd ENE & pass'd about 3 Miles to the North of it having no bottom with 50 f*atho*ms. At Noon the Observed Latitude made 26°12′37″N, Account 25°43′N, Long*itude* d*itto* 123°27′E, No 45 124°59′E, Course N64°E, 72′, but by the Observation our Course has been N70°E, 115′, having experienc'd a NE Current of 42′.[2] This current is allow'd for in fixing the situation of the Rock, deduc*ing* our course back to 6 AM when it bore S37°West 2 or 3 Leag*ues*. Its Lat*itude* 25°58′N, Long*itude* 124°29′ East [*in margin of* OD 177: Variation 2° west]

Saturday 8 July

Light breezes & fine weather. We still continued an East course with the wind from the Southward. At 11 PM we saw land bearing E½S; at 5 AM it extended from S49°E to S77°E. It was low in the middle but rather high [in] the Southern part.[3] We sail'd within 3 miles of it but had no bottom with 50 f*atho*ms. From each extreme extended a long sandy Spit with a detach'd Rock South of each. It appear'd well cultivated & we saw several habitations, from whence the Natives soon paid us a Visit. They were the same race of Islanders we had seen at Typinsan[4] equally docile & communicative. At 8 we saw more Islands of which they gave us the several Names. On their first coming on board they enquir'd if we were from Hollandia. Their cultivation up the rising grounds were in Ridges, above which the Country appear'd well wooded. At noon we observ'd in 26°26′33″N, Account 26°12′N, Long*itude* 124°34′E, Course N77°E, 62′, No 45 at Noon, AM & PM 126°41′E. By this variation in Observed situation & estimation we find still the NE'ly current though not so much Northerly as yesterday.

Sunday 9 July

The Island bore of us at Noon from S64°W to S31°W about 4 Leag*ues*, They call'd it Koominsan.[5] Another Island form'd like a Quoin* [*in margin*: * NB this is one of the cluster seen last Year to the North of our situation at noon] S44°E, Tu-na-chi[6] – a third Karumah[7] S69°E, 4th d*itto* Agonhee[8] N64°E, a very small one Tu-si-mah[9] N31°W 3 or 4 Leag*ues*. Light

[1] Sekibi Sho, a steep-to islet, 83 metres high, resembling a junk under sail, situated 48 miles east of Kobi Sho. According to Broughton, *Voyage*, p. 233, 'it resembled the rock called Pula Sapula, in the Chinese seas'; possibly Poula Sepate in 9°59′N, 109°05′E, about 75 miles off the coast of Vietnam, which from some directions resembles a shoe.

[2] The Japan Current or Kuro Shio flows NE at a considerable rate in the vicinity of these islands.

[3] Kume Shima, the westernmost island of Okinawa Gunto; Koomi-sang in Broughton, *Voyage*, p. 234, and Komisang on Broughton's 'A Chart from the South Point of Formosa to Great Lieuchieux', Broughton, *Voyage*, f.p. 147.

[4] Miyako Shima.

[5] I.e. Kume Shima.

[6] Possibly Yakabi Shima at the western end of Kerama Retto; Tunatchi in Broughton, *Voyage*, p. 234, and on Broughton's 'A Chart from the South Point of Formosa to Great Lieuchieux', Broughton, *Voyage*, f.p. 147.

[7] Tonachi Shima; Karumah in Broughton, *Voyage*, p. 234, and on Broughton's 'A Chart from the South Point of Formosa to Great Lieuchieux'.

[8] Aguni Shima; Agenhu in Broughton, *Voyage*, p. 234, and on Broughton's 'A Chart from the South Point of Formosa to Great Lieuchieux'.

[9] Tori Shima, a 24-metre-high rocky islet, 12 miles north of Kume Shima; Tu-si-mah in Broughton, *Voyage*, p. 234 and Tusimah on Broughton's 'A Chart from the South Point of Formosa to Great Lieuchieux', Broughton, *Voyage*, f.p. 147.

breezes & very fine weather. With the current still Easterly we approach'd the Islands very fast. At 3 PM we kept our wind to the ESE and at 7 PM the Group of Islands (call'd Matchi in the Charts[1]) bore from S16°E to S24°W 3 or 4 Leagues. Tu-na-chi from S60°W, Agonhee N58°W and the Extremes of Land ahead (Great Liquieux[2]) from SE to NEbE. At 9 PM we Tack'd & plied to windward during the Night with a Fresh Southerly Wind. At 5 AM we were 4 or 5 leagues from Liquieux. At 9 AM we were in sight of the Harbor of Na-pi-chan[3] bearing from us S25°E 4 or 5 Miles in this direction (½ a mile from us) was a Chain of Breakers. We tack'd & plied to windward till Noon between a low Sandy Island[4] and a Hummock Point which makes the Westermost part of the island Liquieux[5] & bears S14°E and N14°W from the Southern part; an extensive Reef projects to the West. From this Point we had 28 & 30 fathoms corally bottom. At Noon we Observ'd in the Latitude of 26°12′N when the Harbor bore of us S64°E 3 or 4 miles.[6] Reef point which forms the Starb*oar*d Shore going in S15°E, outer part of the Reef S14°W, Matchi Islands from N81°W to S36°W, Low Sandy Island N45°W to N63°W 3 or 4′, Liquieux N*orth* Extreme N25°E. Course since yesterday S37°E, 20′, Lat*itud*e acc*oun*t 26°11′N, Longitude d*itt*o 124°47′E, No 45 127°22′E, which makes our true Course S69°E, 40′, the current being solely East. By which it appears the Current has set us 2°35′ E*ast* of our estimation since leaving Macao and 1°43′ since passing Formosa by the Watch No 45.

Monday 10 July

Fresh breezes from the SW quarter. After noon we bore up for the Harbor keeping the Starboard Shore on board but having very irregular Soundings from 15 to 2½ [fathoms] & seeing Rocks under us we haul'd out & brought too. Soon after we saw a large Boat coming out when we stood towards them. Having express'd a desire of going in to the Port to get Water &c, they most willingly conducted us, taking the Helm & Conning the Vessel from the Masthead. We pass'd close to a Reef on our Larboard Hand having no Sound*ing*s with 15 f*ath*-*o*ms. It bears N20°E from the Reef point about one or two Miles & is steep to. After passing it we steer'd ESE for what appears like a Block House on the Ex*treme* of some elevated Black Rocks, when opening the Harbor SEbS. We haul'd up for it and came too an Anchor in 12 f*atho*ms muddy Bottom ½ a mile from the entrance. Throughout the forenoon it blew strong from the SW with frequent Squalls. Our Friends after having enquir'd our wants sent us off Water, Rice & Fowls but did not seem to wish we shou'd go onshore.

Tuesday 11 July

It still blew very strong but we rode well shelter'd. Few Boats came off to us on account of the weather. Towards the Evening it moderated & the Master was employ'd in the Boat Sounding the Roads. At Daylight it was more moderate & the wind continued steady at South. Our Friends who seem'd very desirous we shou'd Sail brought us off a Bull, two Hogs, 10 bags of Potatoes, with plenty of wood and water. In the Forenoon a Junk arriv'd from Typinsan and a Japanese Junk

[1] I.e. Kerama Retto.

[2] Okinawa Shima.

[3] Present day Naha Ko.

[4] Kamiyama Shima, close NW of Naha Ko.

[5] Possibly Senaga Shima, a prominent island now connected to the mainland by reclaimed land, situated 1¼ miles south of Omine Saki, a low headland forming the western extremity of Okinawa Shima.

[6] In annotated *Voyage*, p. 235, inserted in pencil, 'Lat 26 12, Long 127 22'.

sail'd out of the Harbor & proceeded on her passage towards Nipon. It still blew Fresh & the Surf on the Shores so great as to prevent a Boat landing any where but in the Harbor, which to us was inadmissible, to our great regret. [*in margin of* OD 177: Variation 0°51′ w*es*t]

Wednesday 12 July

It was extremely mortifying (having Anchor'd here 48 hours) and not being able to acquire some information respecting the Government and trade of these People and yet we us'd every endeavour in our power, but cou'd not succeed. Having made use of entreaties to the principal people who came off, requesting two of our Officers might go on shore with them, which they constantly evaded I determin'd to send them in our own Boat. But the Person remaining with us wou'd not accompany them & seem'd much displeas'd at their going. The Boat however landed them without any Objection in the Harbor & came off again immediately. About an Hour after our two Gentlemen return'd escorted by their Friends in a Shore boat. It appear'd on their landing on the Pier Head they were met by several of the principal People, who led them about 100 yards along the Causeway which goes to the Town where, after some Consultation, they were not permitted to go any further but shewn into a large Dwelling House & offer'd refreshments of Tea & Tobacco. All the persuasions of the Officers cou'd not prevail upon their Friends to let them proceed and finding their efforts ineffectual, they thought it most prudent to return on board than persist in opposition which cou'd not have any good effect. The entrance of the Harbor was very narrow & in the fair way tho' just past low water they had 2½ f*atho*ms [and] within 7. The Harbor took a winding direction to the SW & enlarg'd considerably after getting in. There were about 20 Junks at Anchor chiefly Japanese. On each side going in was an apparent Block House (or place of defence) built of Stone in a Square Form with loop Holes. The causeway leading to the Town was 300 or 400 Yards long & at intervals were Stone Arches to admit the water. For as the tide Flow'd up to the Town over the Coral Reefs which projected from the shore I suppose it was necessary to give the water its course without any Obstruction. The Town extended on a line with the Shore to the Northward & seem'd from the Numbers of People we saw at different times to be very populous. The Houses were of one Story & all [of] them well cover'd with a Brick Tiling. Trees were interspersed among their Dwellings & the Country rose at the back of the Town interspersed with Cultivations & Clumps of Trees agreeably diversified the View from the Sea. At the Extreme of the Town to the North was a large burial Place. The Tomb Stones were Whiten'd & seem'd to lay in an East & West direction. They are conspicuous as you sail in. These People are in every respect of manners, Customs & appearances the same with those at Typinsan. This being the principal Island is the seat of Government & the center of Trade both from Japan & the Southern Islands & the advantage of possessing a good Harbor, besides. I imagine it to be more fruitful in Grain & Cattle. Their Boats were large & flat floor'd & they manag'd them with Skulls & Oars. In these they bro't off water in large Tubs & Casks. Their Canoes were very inferior, Hollowed out of Pine Trees. These seem'd to be chiefly employ'd Fishing. They made use of Light matted Sails & went very fast thro' the Water. They perfectly understood our having being at Typinsan but cou'd not comprehend well what we had done with our Men. This may be some reason for their not suffering us to go on shore. As they cou'd not well understand after leaving Typinsan why we shou'd come as they suppos'd to their Island in want of Provisions. All this we endeavour'd to explain to them. They were in every other respect extremely civil & attentive, bro't us off Vegetables & some of their Spirits (Sakki) & behav'd with a degree of Politeness which render'd their company pleasing. Every Boat brought

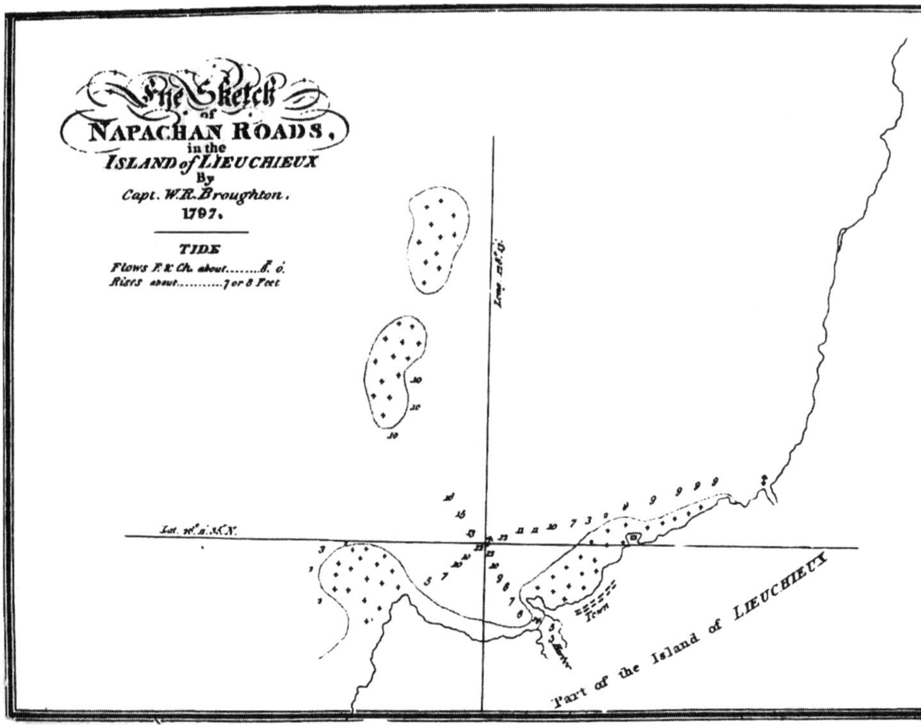

Figure 8. 'Sketch of Napachan Roads, in the Island of Lieuchieux by Captn W. R. Broughton, 1797'; Broughton, *Voyage*, f.p. 96.

off each time a different Crew to gratify their curiosity in seeing the Vessel & to observe our Manners of eating & drinking which much excited their attention. As they were very desirous we shou'd Sail, having everything we asked for now on board, & having no inducement to remain considering the advanc'd Season, we took our leave, presenting them with the only thing they seem'd desirous to possess, which was a Spy Glass and a Memorandum with a drawing of the Vessel expressing our Nation & arrival. By three Meridian Observations of the Sun we made the Lat*itude* 26°11′30″N, by three Days Observations for the Watch we made the Long*itude* 127°26′E.

This Island is call'd by the inhabitants Lieuchieux, or Loo choo. It extends from the Lat*itude* of 26°04′N to 26°50′N and from 128°12′E to 128°45′E. Its greatest direction is NE & SW nearly about 21 Leagues. The Harbour of Napachan is on the SW side in the Lat*itude* 26°11′N & Long*itude* 128°13′30″E. The inhabitants resemble more the Japanese than the Chinese, speaking with very little variation in the language of the former and writing after their manner. They were also well acquainted with China and show'd us several articles from that country, particularly English broad cloth which they seem'd to value.[1]

[1] In Broughton, *Voyage*, pp. 242–3, there is a footnote beginning 'Le Pere Gaubil, a missionary of the Jesuits at Pekin, wrote a memoir on these islands, called by the Chinese Lieou-kieou. See *Lettres Edifiantes*, tom. xxiii, p. 182.' This is followed by a lengthy extract from pp. 183–7, in translation from the French, of parts of the

At Lieuchieux they manufacture their own garments and their linen checks appear'd to equal any of ours. According to the weather they increase or lessen their apparel, and like the Japanese they carry a fan, Pipe and Tobacco box. As we were not permitted to visit the country we can give no account of many of their productions beyond Wheat, Millet, Indian Corn, Rice and sweet Potatoes in which they seem to abound. They had horses and Cattle with hogs of a large breed different from the Chinese and fowls of a large growth. We did not see any geese, but left them two couple.

The watch No 45 show'd the Longitude of the Harbor of Lieuchieux by two days observations to be 127°25′30″E. As its situation, or at least the South part of the Island was tolerably well fix'd last year, we were apprehensive our watch went irregularly and yet we could not well account for it in such a short time. Four days after sailing [from Napachan], when our situation was well known off the coast of Ximo[1] by many observations made this year as well as the last, No 45 made the Longitude too little by 47′45″[2] with every proper correction between the interval of time, which defect being applied to the Longitude shown in Napachan roads makes it to lie in 128°13′ East which I take as the true Longitude and as the watch did not alter for the three days we were of[f] & in the roads I think there cannot be any mistake, but rather suspect the violent motion of the Vessel in the gale of wind off Formosa must have affected it before our arrival.[3]

memoir describing the islands. Gaubil himself had not visited the islands. The footnote continues with a reference to remarks on the islands contained in Vol. II of Sir George Staunton, *An Authentic Account of an Embassy from the King of Great Britain to the Emperor of China*, 2 vols, London, 1797, again based on a secondary source.

[1] Kyushu.

[2] 'too little by 47′45″' taken from Broughton, *Voyage*, p. 244, as the original MS is corrupt.

[3] The above three paragraphs from RUSI/NM/132 (9). As it is probable that No. 45 was not set in gimbals, as were most box chronometers at the time, its performance would certainly have suffered in the severe gale they encountered.

CHAPTER 4

From Okinawa to the south and east coasts of Japan and thence to Muroran Ko in Hokkaido

Wednesday 12 July[1]

Fresh breezes & hazey weather. At ½ past 4 PM weigh'd & steer'd WbN to pass to the South of the breakers that bear N*orth* & S*outh* about 1 mile from the Bluff point in going in. At 5ʰ22′ PM the Harbor bore S58°E 2 or 3 miles when we steer'd NNE for the Extreme Land.[2] At ½ past 7 the extreme point off which were some breakers bore N60°E and to the East of it the Land trended forming a deep Bay.[3] We plied off and on till daylight when the S*outh* Extreme of Luchieux bore S26°W, [the] Bay East, and an Island N35°W,[4] in the center of which was a remarkable Sugar Loaf Hill[5] rising from the Flat Land. We stood into the Bay till ½ past 6 when the Rocks set last Evening now bore on the opposite bearing & plainly shew'd we had experienc'd a NE current. It blew strong in Squalls & the Shore on the South side being lin'd with Breakers we steer'd to the West to round Sugar Loaf Island. Two more Islands are between the Sugar Loaf & Luchieux but it blew too hard to risk running thro' between them as from the extreme points were extensive reefs.[6] At 8 AM having rounded the Sugar Loaf we steer'd ENE and at 9 we saw more Islands to the NE.[7] We pursued the same course to pass between them & Luchieux. At 11ʰ we were abreast of them. At ½ past we hove too for the Observation, a Fresh gale & a great deal of Sea. At Noon the Lat*itude* made 26°54′34″N when the Northern part of Luchieux[8] bore S80°E about 3 Leagues, Sugar Loaf just visible thro' the Haze S58°W, Extremes of Luchieux S45°E, an Island to appearance near the Shore[9] *South*, Extreme[s] of an High broken Island[10] N10°W to 46° West 5 or 6 L*eague*s, Low Distant Island[11] N56°W, Extremes of another Island[12] from N67°W to 81° west about 4 Leagues, of moderate Elevation. Course N7°E, 43 miles, Lat*itud*e account 26°42′N, Long*itud*e 127°33′E.

[1] When Broughton anchored in Naha Ko he changed from keeping nautical time, which begins and ends at noon, to civil time, which begins and ends at midnight, reverting to nautical time on sailing on 12 July.

[2] Sesoko Shima. [3] Nago Wan. [4] Ie Shima.

[5] Ikosuko Yama, a curiously shaped rocky hill, 163 metres high, in the eastern part of the island.

[6] There are no islands between Ie Shima and Bise Saki, the NW extremity of Okinawa Shima. Broughton must have confused them with Minna Shima, a flat islet situated 1¾ west of Sesoko Shima, which is fringed with reefs extending as much as ¾ mile offshore.

[7] The two large islands of Izena Shima, situated 13 miles NNE of Ie Shima and Iheya Shima, situated 4 miles farther north.

[8] Hedo Misaki.

[9] Kouri Shima, situated 7½ miles east of Bise Saki, the NW extremity of Okinawa Shima.

[10] Iheya Shima, which consists of hills separated by deep valleys.

[11] Noho Shima, a small island between Iheya Shima and Izena Shima.

[12] Izena Shima.

Thursday 13 July

Strong breezes & very hazey weather with the wind at South. We steerd East & EbS. At ¼ before 2 the North point of Lieuchiew bore South about 4 Miles. At ½ past 3 the Extremes bore from S25°W to S67° west, a Long Island about 5 or 6 Miles distant[1] from N20°W to N25°E. The Northern part of Lieuchiew from the Bay upwards is high Land & Much broken. The NE point is very Rugged & high; falling abruptly towards the Sea,[2] but the whole Island presents a fertile appearance & the South part is lower land & more richly cultivated. We make the Latitude of the North East point 26°50′N which is One degree less than the Charts place it in, Long*itude* 128°39′E, true Long*itude*. With a large following Sea at ½ past 3 we steer'd NE and at $5^h45'$[3] we saw an extensive Island bearing N*orth* to NNW.[4] At $4^h15'$ the Long Island South Extreme bore West about 3 Leagues. The Island was elevated near the middle but at each extreme very low with breakers off the Points. All these Islands we have hitherto seen are to appearance Fertile and in general inhabited. At 7^h the Island last seen bore from N*orth* to NW but very faintly for the Haze, which entirely Obscur'd the lower parts. We imagin'd it to be about 4 Leag*ue*s distance. At ½ past 7 we bro't too for the Night with our Head to the SE and at daylight we had no sight of Land. We steer'd N*orth* till 8 when we alter'd our Course to NE. Lat*itud*e Observ'd 27°44′N, Acco*un*t 27°48′N, Long*itud*e acco*un*t 128°33′E, No 45 then shewed 129°9′35″E. The Sights taken yesterday cou'd not be depended upon & are left out in the account.

Friday 14 July

Fresh breezes & fair weather Wind SSW. We steer'd NE. At 4 PM an Island was seen bearing NNE.[5] At $7^h30'$ it bore from N*orth* to N50°W 2 or 3 Leag*ue*s. At 8 PM we haul'd off for the Night. At daylight we had the same bearings & we steer'd NbE till ½ past 6 when the North part bore west. A distant Island bore from the North point N80°W 5 or 6 Leagues[6] & we altered our course to NNW. This 1st Island was between 3 & 4 Leagues in a NE & SW direction, low at each extreme but rising abruptly from the Sea on the East Side to a considerable elevation, very rocky. To the N*orth* of this fall of Land are several Villages scatter'd along shore & a Small Opening that afforded Shelter to some Junks riding within the Beach. The Houses were built among the Trees with thatch'd Roofs and the cultivated Land rose behind them. On the Higher part of the Island was no Wood. Nor did we see any Reefs breaking off the extreme points. At 8 AM the other Island bore from West to S66°W 3 or 4 Leagues. It was of moderate Height & more diversified in uneven ground but we were not near enough to particularize it minutely. The Hills however appear'd rather barren. At 20′ past 10^h we saw a Rock from the Main Rigging bearing S80°W 4 or 5 L*eagues*.[7] At 11AM we lost sight of the 2d Island bearing S22°W about 7 Leag*ue*s, Lat*itude* ☉ made 28°57′N, acco*un*t 29°3′N, Long*itud*e 129°5′E, [N° 45] 129°36′5″E, Course N20°E, 80′. [*in margin*: No current] We this morning buried Thomas Horn one of our best Seamen, a Fine active well behav'd Young

[1] Yoron Shima.

[2] Hedo Misaki lies at the end of a steep-to, flat promontory ending in a steep cliff 29 metres high.

[3] $5^h15'$ in Broughton, *Voyage*, p. 246.

[4] Okinoerabu Shima, situated 17 miles NNE of Yoron Shima. O Yama, its summit, is 245 metres high.

[5] Kikai Shima.

[6] Amami O Shima

[7] Sandon Iwa, a group of three rocks, situated 14½ miles NNE of the northern extremity of Amami O Shima.

Man. He had not been unwell more than five days. His death therefore most unexpected. The Doctor discrib'd his complaint to be a violent dysentery and Billious fever. He was nearly delirious the last 24 hours.

Saturday 15 July

Fresh breezes & hazey weather we steerd NbE till 7 PM when we saw two Islands uncommonly high bearing N47°W & N73°W 7 or 8 Leag*ues*.[1] The remainder of the Night & day we had moderate breezes from the SSW; our course was NE. At 8 AM we saw Land extending from N25°W to S78°W.[2] At 11^{h}20′ N*orth* Ex*treme* of Land N73°W 6 or 8 Leag*ues*. At Noon we observ'd 30°42′N, Acc*ount* 30°15′N, Long*itude* acc*ount* 130°E, Course N25°E, 116 miles, No 45 AM & PM 131°16′30″E, by which it appears a Current has set N59°E, 53 miles, since yesterday.[3] [*in margin of* OD 177: Variation 1°0′ west]

Sunday 16 July

Fresh breezes & hazey weather. By our situation at Noon we ought to have seen the same Land we visited last Year as we were at Noon about half a degree west of it. We therefore steer'd East till 4 PM when the watch making us to the Eastward 30′ since Noon & seeing no Land we were convinc'd the Land set in the morning must be what we were in search of. We accordingly tack'd and stood to the Westward. At 2AM we saw the Land to the NW and at Daylight it bore from S64°W to N10°W, a deep Bay or Opening N80°W.[4] Throughout the forenoon it was Calm & we had no bottom with 80 fathoms of line within 3 Leagues of the Land. By the Land appearing to the NE[5] we found ourselves setting fast to the North and at Noon a good Observation made our Lat*itud*e 32°9′N, Account 31°30′N, Longitude acc*ount* 129°54′E, Course N5°W, 86 miles, Current N34°E, 47′. The Extremes of the Land bore from N46°E to S56° West, Opening from S66°W to S77°W 5 or 7 Leagues.[6] Our present situation is nearly the same as on the 22d of November last Year when No 1 shew'd 132°42′E and Observ'd [lunar] Distances 132°17′E. No 45 at this time shew'd the Longitude 131°42′, Distances ☉ and ☽ 132°34′ East [*in margin of* RUSI/NM/132 (9): which will account of our not making the land we expected having no opportunity before of taking Lunar observations.]

Monday 17 July

Light airs from the Eastward till midnight & we steer'd to the SW. It continued calm till Noon when a breeze sprung up from the Southward. [*in margin of* RUSI/NM/132 (9): At noon the land from N22°W to S67°W four or five leagues.] The Latitude made 32°14′N, account 31°54′N, Long*itude* d*itto* 129°26′E, Course N80°W, 24′ and No 45 shew'd 131°39′, by

[1] The active volcanoes of Nakano Shima, 978 metres high, and Suwanose Shima, 972 metres high.

[2] Tanega Shima and Yaku Shima.

[3] The schooner was still under the influence of the Kuro Shio.

[4] Osumi Kaikyo, the strait between Tanega Shima and Kyushu.

[5] The SW part of Shikoku.

[6] There is no opening on this stretch of coast. Broughton may have sighted the mouth of the Omaru Kawa, which enters the sea in latitude 32°8′N.

which it appear'd we had scarcely alter'd our situation and the same bearings of the Land confirm'd it, the Current having set us N50°E, 31miles. Observ'd distances made the Long*itud*e mean of 3 Sets 132°30′E, Mr Chapman 132°22′E. [*in margin of* OD 177: Variation 1 2°42′, 2 1°06′ w*es*t]

Tuesday 18 July

Light Breezes & hazey from the Southward & we stood to the Westward towards the Bay or opening. We ran 27 miles to Sunset when we were within 2 Leagues of the Land and we had 38 f*atho*ms, Sandy bottom. We cou'd not see the extremes of the Shore for the Haze.[1] The Land near the Sea was tolerably level but the interior Parts rose to High Mountainous Land very easy to be discern'd 20 leagues off. The Bay or what I before suppos'd to be an Opening seem'd entirely clos'd by low Land well wooded,[2] with distant Hills at the back to the SW, which was nearly the direction to the center of the Bay. During the night and Forenoon it continued Calm with Light airs at intervals. Our Soundings according to our distance from the Shore varying from 21 to 60 f*atho*ms from 5 to 12 miles off. During the Night we cou'd distinctly hear the Surf beating on the Shore. In the Forenoon we had four large Junks in company, but they seem'd to take no notice of us. At Noon the Latitude made 32°11′N, Account 32°02′N, Long*itud*e d*itt*o 128°49′E, [No] 45 131°15′40″E and the Extremes of Land visible only thro' the Haze from NbW to SSW, off shore about 4 miles. Current these 24 hours N57°E, 16½ miles. The Observ'd Long*itud*e ☽ and ⊙ made 132°3′E mean of 3 sets, Mr Chapman 131°59′E 6 Sets.

Wednesday 19 July

Light breezes from the SE till 2 PM when it again became calm and we came too an Anchor in 28 fathoms, sandy bottom. The Ex*tremes* of Land as the Sun set bore from N18°E to S15°W, our distance 3′ off. A Fishing Boat came a board & gave us some Fish. She made use of Skulls & they Fish'd with Hooks & Lines for Bream &c. While at anchor the Current set to the NNE & East, 6 f*atho*ms p*e*r Hour. At 7 PM we weigh'd but had no Wind during the Night which was a continual Fog. In the Forenoon it clear'd up but it still remain'd Hazey. Lat*itude* ⊙ 32°15′N, Acc*oun*t 32°07′N, Long*itude* d*itto* 128°45′ E, Course S20°E, 4′, No 45 made 131°16′, Current North about 8 miles. We heard the Surf distinctly but cou'd scarcely see the Land for the Haze. ☽ and ⊙ 131°48′, W. C[hapman] [*blank*].

Thursday 20 July

Calm & hazey weather till midnight when a light air sprung up from the North. We steer'd SSW till Noon but our latitude & Bearings prov'd we had very little alter'd our situation except being further off shore. Lat*itude* ⊙ 32°14′N, Acc*oun*t 31°56′N, Long*itude* d*itto* 128°37′E, No 45 131°15′E and the Land from S30°W to N5°E,[3] Current North E*aster*ly 18 Miles.

[1] 'extremes extending from N.6°E to S.33°W' in Broughton, *Voyage*, p. 250.

[2] Broughton makes no mention of this opening or bay in *Voyage*, pp. 249–50.

[3] In margin of RUSI/NM/132 (9): 'N50°E, offshore three or four leagues'. However, the log has N5°E with 'E' crossed out and replaced by 'W' and Broughton, *Voyage*, p. 251, has 'N5°W, off shore three or four leagues'.

Friday 21 July

Light airs & calms with very hazey weather and no prospect of any wind, the Current still drifting us to the NE'ward. From midnight till Noon we had a Light breeze from the North & we steer'd SSW. We had now some hopes of clearing these calm Latitudes, but before Noon we were again disappointed as it remain'd only a very light air from the N*orth*. Lat*itud*e ⊙ made 32°06′N, Account 31°40′N, Long*itude* d*itt*o 128°21′E, [No 45] 131°20′E, Current NbE, 25 miles. [*in margin of RUSI/NM/132 (9)*: Ext*reme*s at noon from N5°W to S70°W offshore three or four leagues] The Center of the Opening bore of us WbS 4 or 5 L*eague*s, High Land obscur'd.

Saturday 22 July

Light Airs from the East continued till Sunset but by the Land it did not appear we deriv'd any advantage from it, being North of our situation at Noon. Thus were we continually disappointed, the Wind & Current opposing each other so equally we cou'd not gain anything either way. At Noon Lat*itude* 32°4′35″N, Acc*oun*t 31°45′N, Long*itu*de acc*oun*t 128°47′E, [No 45] 131°21′E, Course S61°W, 4′, Current North, 20 Miles. Our bearings were much the same as Yesterday. [*in margin of OD 177*: [Variation] 0°16′, 0°52′ w*es*t]

Sunday 23 July

Throughout the afternoon till 10 PM we had a Light breeze from the SE and we stood in Shore. At 7 PM we tack'd close in with the Low Land in 15 f*atho*ms, the high land to the Southward bearing S4°W[1] about 3 Leagues, an Inlet which bore WbS[2] and the Clump of Trees N4°E about 10 or 12 miles. This does not make a Bay when you are close in but each extreme being bounded by very high Land makes it appear so at a distance & between is a considerable extent of low Land which continues to the westward several Miles till you approach more high Land. Several Junks were at Anchor within the Inlet, which was very narrow in the entrance but seem'd to afford Shelter within the low points of [the] Beach. The Surf broke high along the Shore. We had no Wind the remainder of the day and a thick haze obscur'd the Land from our view. We had 80 fathoms water and the Lat*itud*e made 31°49′N, Acc*oun*t 31°49′N, Long*itu*de d*itt*o 128°12′E, Course S17°W, 16′, no current.

NB We had now been detained 8 days vainly trying to beat to the Southward of Ximo but all our efforts with the light winds & adverse Currents prov'd ineffectual. This I considered [unhappily] as my intention was to have pass'd thro' the Corean Straits[3] & keeping the Tartary Coast on board have proceeded in that manner to the North. The obstructions we had met with rendered any further attempts unnecessary. In the afternoon of the 24th we therefore directed our Course to the East, meaning to make the Southern part of Nipon.[4]

[1] 'S4°E' in Broughton, *Voyage*, p. 252.

[2] Possibly the mouth of the Oyodo Kawa, which enters the sea in latitude 31°53′N.

[3] The Korea Strait.

[4] The above paragraph from RUSI/NM/132 (9). Not included in Broughton, *Voyage*, p. 253. From this it is clear that it had been Broughton's intention to attempt to pass through Osumi Kaikyo, between Kyushu and Tanega Shima and then proceed to the north along the west coast of Japan.

Monday 24 July

Moderate breezes from the Southward after 2 PM. At 3 we tack'd. The Southernmost point set last evening now bore S44°W about 2 Lea*gue*s. We steer'd East. At 4 AM we had light airs & calms, the Wind Northerly. Towards noon it became a gentle breeze from the westward with very hazey weather. Lat*itude* ⊙ made 32°17′N, Acc*oun*t 31°48′N, Longitude d*itt*o 129°08′, Course N58°E, 35 miles, [No] 45 at 3 PM 133°20′E. The Current since we last saw the Land has set us N64°E, 67 miles, which accounts for us not seeing the Land having been set pass'd it in the Night at the rate of 3 Knots an hour. This Part which I expected to have made is in 32°30′N and ought to have bore to the NW of us about 5 Lea*gue*s. I imagine it be the East extreme of the Island Ximo.[1]

Tuesday 25 July

Moderate breezes & fine weather. We steer'd to the North to make the Coast and at 7 PM it was calm. Our Lat*itud*e per Estimation without Current 32°34′N. We cou'd see at Sunset 6 or 8 Lea*gue*s but we only discern'd three Junks steering to the East. The Night was Calm & we experienc'd frequently Riplings which indicate our being still at the Mercy of the Current. We had no bottom with 80 fathoms. In the forenoon we had a fresh wind from the North with pleasant weather tho' still hazey. Lat*itud*e ⊙ 32°54′20″N, Acc*oun*t 32°50′N, Long*itud*e d*itt*o 129°40′E, [No] 45 134°26′E, Course N39°E, 43′, Current N81°E, 28 Miles. [*in margin of* OD 177: Variation 43′ west, 43′ d*itt*o]

Wednesday 26 July

Fresh breezes & hazey weather wind at North. At 5 PM we descry'd the Land of Nipon[2] bearing N*orth*. At 7 PM it bore from N14°W to N25°E[3] 5 or 7 Leagues. The Wind in the Night veer'd to West and moderated. Having run in Six Leagues, at 2 AM we saw the Land and then bro't too till daylight when it bore from N35°W to N40°E 2 or 3 Leagues & we had no Soundings with 90 fathoms. As the Sun rose we observ'd many Boats under Sail in shore & several Junks who appear'd to come out of a Small Opening nearly abreast of us[4] form'd by a level projecting point off which were some Rocks. We steer'd NEbE with a Light air from the SE quarter. After passing what appear'd very much like an Island, but prov'd as far as we cou'd judge to be only a Peninsula,[5] the distant land opened with it N34°E. On the NE extreme of this Flat Peninsula are four remarkable Trees & some Rocks detachd from it to the East*war*d. N30°W from the Point of Trees was a small Harbor[6] in which several Junks were laying before a Town. On the Peninsula & in the Bays form'd by it were several Villages & Scatter'd Houses & great cultivation about the Country. Several boats were fishing & an infinity under Sail large & Small passing to the East & West. The interior Land was rather High and much

[1] 'read Xicoco or Sikoff' inserted here in the MS has been omitted, as it has in Broughton, *Voyage*, p. 253. It has also been crossed through in UKHO, OD 177. As Broughton had been set N64°E, 67 miles, he would have expected to have sighted the south coast of Shikoku, not the east coast of Kyushu.

[2] It was in fact the eastern end of Shikoku. 'Of Nipon' omitted in Broughton, *Voyage*, p. 254.

[3] 'N15°E' in Broughton, *Voyage*, p. 254.

[4] Kushimoto Ko, 2 miles north of Shiono Misaki, the southern extremity of Honshu.

[5] Shionomisaki Hanto, which is indeed a peninsula. The small island of O Shima, lies close NE.

[6] Kushimoto Ko.

broken, forming Hills with Sharp Ridges running parallel with each other NE and SW. Near the Sea was sandy beaches with intermingled Rocks and Clumps of Woods in the Hollows form'd by the rising grounds. Still we had no Soundings. At Noon we Observ'd in 33°28′17″N, Account 33°33′N, Long*itude* 130°47′E, [No] 45 135°23′E, Course N65°E, 70′. The Current to the WSW, 9 miles. It was nearly in this situation we made the Land last year in a hard Gale of wind. I make this part to be the South point of Nipon & lays in the Latitude of 33°25′N, Longitude 135°47′E.[1]

Thursday 27 July

Light airs from the SW quarter throughout these 24 hours. We steer'd to the NE. At 7 PM the Peninsula Ex*treme* S54°W 6 Leag*ues* & the North Ex*treme* N29°E, off shore 2 or 3 Leag*ues*. Several Fishing boats came off to us in the Afternoon. These Boats were beautifully shap'd with a projecting Stem; they were 39 feet in length & 6½ feet wide. They had 5 Skulls of a side & each boat had 13 or 14 Men. They were highly decorated with Paint, the ground work being black & mov'd through the water with great velocity. At 5 AM the Ex*tremes* from S69°W to N20° East. Lat*itu*de Observ'd 33°52′45″N, account 33°40′N, Long*itu*de d*itt*o 131°20′E, [No] 45 135°55′E. At last we have lost the Current having experience'd very little these two days. The Land continued High & much broken. An Opening with an Island in the Center bore N54°W at Noon,[2] about 4 Leag*ues* Ex*tremes* S65°W to N13°E.

Friday 28 July

Light breezes from the South with very fine weather. We pursued the NE Course. The Land continued high & the Shore indented by projecting Points. At 4 PM the extreme North which made like an Island bore N21°E.[3] To the NE of it the Land became very low. At 7 Apparent Isle N30°W, Outer low point N20°E about 3 Leag*ues*, South Ex*treme* S64°W, off shore 4 or 5 Leag*ues*. At 9 we were nearly abreast of the low point & we saw many lights upon the Shore, but we had no Soundings with 75 f*atho*ms. We steer'd East 8 miles after passing the point when we bro't too to explore it in the Morn*ing*. At daylight it was Hazey & Calm weather & we cou'd not discern the Land. At 9 we cou'd just distinguish it bearing from W½N to NW and a light air from the East enabled us to stand to the North. At Noon we observ'd in 34°28′N when the extremes bore from NE to WbN, Low Land to the North about 3 or 4 Leag*ues*. Accounted Lat*itu*de 34°20′N, Long*itu*de d*itt*o 131°56′E, [N°] 45 by AM & PM 136°59′E, Course N41°E, 46 miles, Current N75°E, 35 miles.

Saturday 29 July

Light breezes from the Southward till 7 PM when it became calm. We steer'd East from 2 PM along the Shore which was very low & Sandy towards the Sea. The interior Land we cou'd not distinguish for the Hazey weather. At 7 the extremes we cou'd see bore from N65°E to N70°W 3 or 4 Leagues. The Calm continued with very hazey weather and we cou'd not reach the

[1] Shiono Misaki is situated in 33°26′N, 135°45′E.

[2] The extensive inlet of Ise Wan, with the 170-metre-high Kami Shima situated 2½ miles SW of Irago Misaki, at the western extremity of Atsumi Hanto forming the eastern entrance point of the inlet.

[3] Probably Irago Misaki, which is a small rocky peninsula.

bottom with 80 fathoms of line. Several Junks were making their Voyage up and down the Coast. At Noon we had a light breeze from the SW and the Latitude made 34°30′N, Account 34°32′, Long*itude* d*itt*o 132°24′E, [No] 45 137°30′ East, Course East 4°N, 23 miles. The Land bore from N32°E to N40°W 5 miles. It was so hazey we cou'd hardly discern the Sand banks. The extreme Northern point was the extent seemingly of the Sandy Shore. From whence the Land trended more northerly falling back as if there was some opening round it,[1] in which direction we saw several Boats under Sail having been out to Sea a Fishing. No Current. [*in margin of* OD 177: Variation 0°49′ west, 0°20′ east, 2°3′ d*itt*o]

Sunday 30 July

Light breezes and very hazey weather. We steer'd East and at ½ past 1 the Point set at Noon N32°E now bore North 4 or 5 miles, Land open with it extending to the NE but most distant. At ½ past 4 we saw some high land with Clayey Cliffs which discover'd it thro' the Haze bearing N60°E.[2] At 6 PM it bore N50°E about 3 Leagues. The Night was calm and in the morning the Wind was from the NE we stood to the SE and at daylight we were well up with the Islands lying to the South of Jeddo Bay, bearing from NE to SEbS,[3] Ex*treme* of Nipon N60°E. During the Night we had thunder and a great deal of Lightning for the first time, the Heat most oppressive. At 5 AM we tack'd to the North and found the Current to set NE'ly. At Noon we had very fine weather but Calm and hazey which prevented our seeing many Objects around us particularly Mount Fusi which we were most anxious to distinguish. Our Latitude was 34°38′N, Account 34°32′N, Long*itu*de d*itt*o 133°9′E, [No] 45 138°55′E. The two Hummock Islands in on[e][4] bearing S30°E 7 or 8 miles, Island in the entrance of Jeddo bay[5] N32°E to N83°E and the extremes of Nipon from N10°W to S80°W. We had no Soundings with 90 f*atho*ms. Current N72°E, 15 miles. Muster'd the People and perform'd Divine Service as usual.

Monday 31 July

We had a calm afternoon and the Current drifted us up to the North between the Islands in Jeddo entrance and the Western Shore. At 4 PM the North part of the Isle[6] bore East 2 or 3 miles and on the West side saw a large Village. The Island was very high in the Center gradually rising from each extreme. In the afternoon at three different times we plainly discern'd large columns of black Smoke issuing from the East side of the highest part of the Mountain[7]. At 7 PM the Island bore from S5°E to S40°E 3 miles and a light breeze springing up from the South we steer'd East. At 5 AM the City point or Eastern Entrance into Jeddo bay[8] bore N5°W to 20°W. We had now variable winds & calms till 10 AM when the Wind set in from

[1] Suruga Wan.

[2] Probably Iro Saki, a high precipitous rock, the southern extremity of Izu Hanto, a mountainous peninsula forming the eastern side of Suruga Wan.

[3] Izu Shoto, a group of islands extending some 250 miles south from the entrance to Tokyo Wan.

[4] To Shima.

[5] O Shima.

[6] O Shima.

[7] O Shima is an active volcano rising to Mihara Yama, 754 metres high, from which smoke is always emitted.

[8] Suno Saki or Nojima Saki, the southern extremity of Boso Hanto, the peninsula which forms the eastern side of Uraga Suida, leading into Tokyo Wan (Jeddo Bay).

the Sea. At Noon Lat*itude* 34°56′N, acc*oun*t 34°45′N, Long*itu*de d*itt*o 133°38′E, [No] 45 139°25′E, Current [*blank*], the City point S80°W about 3 Leagues. Several Boats came off to us from the Town to barter Fish, Bonettas, Mackerel &c.

Tuesday 1 August

Light breezes & Hazey weather from the Southward. At 4 PM White point[1] bore N20° west. [*in margin*: When White P*oin*t bore N*orth* No 45 139°46′10″ [subtracted from 140°44′ = 56′ East of the true Longitude.] At 7 PM we were abreast of it bearing in one with Mount King[2] N87°W. We had 9 fathoms water and our distance from the Shore 3 or 4 miles. White point is remarkable & easily known in any direction being much higher than any Land near it & terminating to the Sea abruptly with white Cliffs cap'd with wood, the Land on each side receding from it. We steer'd NE and gradually deepen'd our water to 20 & 30 f*atho*ms. At 5 AM the Land bore from N*orth* to N30° west 4 or 6 L*eague*s and we steer'd NNE & NNW to Noon. This point which I suppose to be Sand-down point[3] makes conspicuous on this low Sandy shore (and bore from us at 7 PM NNE). The shore between it and white point is very low falling back to the west between the Points. We saw nothing of what he, Capt*ain* King calls Hummock Island & I rather suppose what he took for it was the elevated part of Sand-down point which has that appearance at a distance from its projecting beyond the Low land. At Noon it bore S15°W 5 or 6 miles. Our Latitude 35°51′N, Account 35°45′N, Long*itude* from White point allowing it to be in 140°44′E[4] was 141°9′E, [No] 45 140°28′E, very little Current NE'ly. Extremes to the North still continued a very low Sandy shore bearing N50°W off the nearest part 5 miles. During the Morning we had a great many fishing boats on board. They were much smaller than those from the Town near the entrance of Jeddo. They made use of fine Mats for Sails, the middle Cloth tho' was of Cotton. Each Boat admitted water as a reservoir to preserve the Fish. [*in margin of* OD 177: Variation 0°10′ east, 0° 46′ d*itt*o 1°1′ d*itt*o, 0°26′ d*itto*]

Wednesday 2 August

Fresh breezes and very hazey weather. We steer'd NNW till 3 PM, keeping 5 or 6 miles from the low land. We steerd more to the North after 3 on seeing high land in that direction. At 7 PM we were at the extremity of the low Land which bore N70° west 3 Leag*ue*s when the High Land commenced from a falling point with an easy descent. Ex*treme* North bore N18°W. At 8 commencement of high Land nearly west our estimated Latitude 36°31′N[5] which is nearly I image the Latitude of Low point or at least the beginning of the High Land. We steer'd NbE throughout the Night and at 12ʰ we had 50 fathoms which decreased to 25 at 7AM. The

[1] Taito Saki, the SE extremity of Honshu; a prominent wooded point whose southern side consists of a vertical white cliff.

[2] Myoken San, a 418-metre high wooded peak, situated 16 miles SW of Taito Saki, the highest hill on Boso Hanto and the first object sighted by a vessel approaching from seaward. Mount King on Broughton's 'A Chart of the N.E. Coast of Asia' and not King's round hill, which was Fuji-san.

[3] Inubo Saki.

[4] 'White Point, 140°40′0″E., Captain King; Latitude 35°13′N. I therefore take a new departure, allowing the point to be well fixed by the observations made in that voyage'; Broughton, *Voyage*, p. 360.

[5] '36°21′N' in Broughton, *Voyage*, p. 361.

morning was an excessive thick Fog which clear'd up at 7, the High Land bearing from SWbS to NWbN and our distance from the land abreast did not exceed 3 or 4 miles. Our Lat*itud*e 37°20′N which is just to the North of Cape de Kennis[1] but we cou'd not see in any direction more than 2 or 3 Leag*ue*s. At 9 the Fog entirely dispers'd and we steer'd North in the direction of the Coast which appear'd bold & Cliffy. At Noon the Observation made 37°44′21″N, Acc*oun*t 37°49′N, Long*itude* d*itt*o 141°18′E. The Extremes of Land bore from NW to SbW½W, 5 or 6 miles nearest shore, No 45 140°28′E. Therm*omete*r decreas'd from 85° to 75°, pleasant cool weather.

Thursday 3 August

Fresh breezes and very hazey. We steer'd N*orth* and NW, 4 or 5 miles from the Shore. At 35′ past 2 PM we saw the Islands of Matzima[2] bearing from N34°E to N55°E 7 or 8 Leagues. At 3 we saw the Main Land extending out to a low Hummocky point N15°E, High Land abreast of us then bore west, as the passage between the Islands & the Main form the approaching points coming so near must be very narrow & we had not daylight sufficient to get through before Night.[3] We haul'd up East. At ½ past 7 North part of Peaked Island[4] bore West 3 or 4 miles. Capt*ain* King places it in the Lat*itude* of 38°19′N.[5] Our estimated Lat*itude* at 7½ h was 38°15′N. This Island is of small extent & divided from the others by a narrow Channel. It presents Rocky shores but it is cover'd with wood up to the Peak. Some low Isles bear from it N4°E[6] but at no great distance. The direction of the East side of these Islands is nearly North & South and the West side NWbN & SEbS. At 12 PM we bro't too to examine these Isles more particularly as Capt*ain* King only saw them at a distance. At 5 the Peaked Island S35°W and Ex*treme* N30° west. At 8 the North[7] point off which are two or three high Rocks bore west 3 or 4 miles. Peaked Island S10°W, Low Islands without it. At 10h the passage made very clear S65°W. There are detach'd Rocks on each side [of] the entrance and the passage between the Main and the Island appear'd very narrow & must take a Winding course after running some distance.[8] To the North of these Islands is a Bay with Islots and scatter'd Rocks on which the Sea broke.[9] The North point is richly cultivated & on the Sea side is a very extensive Town. At Noon we Observ'd in 38°49′N, Peaked Island bearing S10°W, Rocks off the North point which forms the Passage S28°W, 9′, N*orth* extreme N30°E, South point of an Opening[10] N17°E 3 miles. Course N8°E, 64′, Accounted Lat*itude* 38°45′N, Long*itude* 141°29′E, [No] 45 140°51′E. The same cool pleasant weather, Southerly winds & Smooth water.

[1] Placed in 37°14′N on the chart titled 'Part of Japan or Nipon'; Cook and King, *A Voyage to the Pacific Ocean*, III, f.p. 395. Possibly Shioya Saki in 37°00′N. King did not sight this cape, taking its position from a chart in Jansson's atlas, based on the voyages of the *Castricom* and *Breskes*.

[2] Oshika Hanto and the off-lying islands of Tashiro Shima, Aji Shima and Kinkasan To. Matisma I[s] on Roberts's 'A General Chart' and on 'A Chart of the N.E. Coast of Asia, and Japanese Isles'; Broughton, *Voyage*, facing title page.

[3] Broughton is referring to the passage between Oshika Hanto and the islands of Tashiro Shima and Aji Shima.

[4] Possibly the 445-metre-high Kinkasan To.

[5] I.e. on Roberts, 'A General Chart'.

[6] Eno Shima Retto.

[7] 'South' in the MS, but crossed out and 'North' substituted; 'South' in Broughton, *Voyage*, p. 262.

[8] Possibly between Sameura Wan on the east side of Oshika Hanto and Oginohamo Wan on the west side of the peninsula, which are separated by low land.

[9] Onagawa Wan.

[10] Karakuwa Saki, the southern entrance point of Hirota Wan.

Friday 4 August

Fresh breezes and fair weather. We steer'd NE, Wind steady at South. At 1 PM we were abreast of a deep opening running in a NW direction. On the South Shore are two Hills bearing from each other [blank]; the Southermost is bare with a Clump of Trees upon it the other when to the North is not unlike a Sugar Loaf, richly cloath'd with Wood which extends to the point. This point is Rocky & of moderate elevation well cultivated. The North entrance is bare with a conspicuous Clump of Trees and off it are two Conical Rocks[1] that are not seen to the South of it. We saw many Habitations along the Shore & Boats fishing, some of them paid us a visit. I imagine this opening must afford good shelter. In the afternoon we had baffling Winds & Calms. The Land to the N*orth* & west rose to a prodigious height forming projecting Heads to Seaward & between them Bays in a NW direction running up a considerable way. All the Shores were bold and Rocky & we cou'd get no Soundings. At 8 PM we were abreast of the High Table Hill.[2] Our estimat'd Lat*itude* 39°16′N and it bore West several miles. Extremes N*orth* to Ex*treme* South (set at Noon) S53°W off shore 4 or 5 miles. We steer'd NNE till 5 AM when the extremes bore from N30°W to S42°W 2 or 3 Leagues off the Shore. At 8h AM we open'd the points forming a deep Inlet[3] bearing S75° west 4 miles. At the same [time] the East entrance of Port Nambu[4] N55°W 2 or 3 Leag*ues*. A Boat came off from the Inlet and by signs requested us to go into Nambu & not to go to the Northward. Finding they cou'd not prevail they soon departed. The Wind died away towards Noon & we remain'd about 5 or 6 Miles from the Shore. The Inlet South of Nambu bore S43°W 5 miles from the entrance. It ran up in a SW direction & promised to afford good shelter The SE point of it is very high & bold Land,[5] the Top cover'd with Trees, & is remarkable by being more elevated than the Land to the Southward. Some Rocks are off the NW point[6] which forms a projecting Hummock, at the extreme point of it. Low point of Port Nambu[7] N87°W 6 miles, Extremes from N28°W to S24°W, Lat*itude* ☉ 39°40′14″N, Account 39°56′N, Longi*tu*de 141°57′45″, [No] 45 141°23′E, Course N23°E, 55 Miles . We found ourselves 16 miles South of the Observation this day.[8]

Saturday 5 August

Light airs, calms and Cloudy weather throughout the 24 hours. The Southerly Current still prevail'd which prevented our getting to the North. To our surprize we saw no Fishing Boats nor any Junks trading up and down the Coast as we had so frequently Observ'd to the South. At Noon we Observ'd in 39°41′9″N, the Low point of Nambu S84°W 4 or 5′, Inlet South of Nambu S38° west 9′, the extremes from N32°W to S24°W. No 45 141°26′E. We frequently tried for Soundings without success.

[1] The conical-shaped 40-metre-high Tsubaki Shima off Hirota Saki, the northern entrance point of Hirota Wan.

[2] As depicted in 39°10′N on chart 'Part of Japan or Nipon' in Cook and King, *A Voyage to the Pacific Ocean*, f.p. 395. Possibly the 1,341-metre-high Goyo San, a prominent mountain which has been sighted from 55 miles offshore.

[3] Yamada Wan.

[4] Hei Saki, the eastern entrance point to Miyako Wan.

[5] Kone Saki, which rises to 504 metres a few miles inland.

[6] Ne Saki, which projects about ¾ of a mile SE.

[7] Hei Saki.

[8] The schooner was now under the influence of the south-going Oya Shio current which runs along the east coast of Honshu until 36° when it meets the northern edge of the north-going Kuro Shio.

Sunday 6 August

We had the same Calm weather attended with some rain in the morning. At 7 PM Port Nambu was well open bearing S58°W° 9 miles, the Inlet S28°W, the extremes N35°W to S22°W. At Noon it clear'd up and we observ'd in the Latitude of 39°43′N, Account 39°56′N, Nambu Port bearing S68°W 7 or 8′, South Inlet S31°W and the extremes as before, by which it appears we remain'd nearly stationary, the Current counteracting the Light Southerly Airs. [*in margin of* RUSI/NM/132 (9): At noon the inlet bore S31°W and Nambu entrance S60°W seven or eight miles.]

Monday 7 August

We pursued our Course to the North as the variable Winds admitted. They were fluctuating from all quarters and during the Night we had a thick Fog. At daylight it dispers'd & we had Cloudy weather with some rain which prevented our having a good Observation. Indifferent Lat*itude* 40°12′N, Account 40°26′N, Port Nambu on with low land S10°W, the High Land forming the South Inlet & extreme S3°W, North extreme of Nipon N51° west 5 or 6 Leag*ue*s, Off Shore 5 or 6 Miles. [*in margin of* OD 177: [Variation] 2°0′ 0°54′, 0°48′ west]

Tuesday 8 August

Light airs and calms alternately still continu'd, which made our progress very tedious. At 7 PM an Opening[1] bore of us S53°W 2 Leag*ue*s. An Island[2] is off the North point which makes it easily remark'd (the same bore west of us at Noon) & we had 55 fathoms, black Sand, the North extreme making a very low point N55°W and the High Land about Nambu S3°E. A Junk in the evening came out of the Harbor and Spoke us in the Night. We cou'd not understand them but our course laying the same way we accompanied each other throughout the day. Fine mild Weather during the morning gave us a distinct view of the Land which still extended very low to the WNW. The Ex*tremes* of High Land abreast of us bore from S50° to S80°W. It is the only rising Hills near the Coast between the North point of Capt*ain* King's Chart and Nambu Shore. The Country was universally laid out in Plantations & to appearance richly cultivated. At Noon it was very hazey over the Land but good Observations for the Latitude gave us 40°33′32″N, Accounted Lat*itud*e 40°39′N, Long*itud*e [d*itto*] 141°29′E, Course N29°W, 24 miles, [No] 45 140°51′45″E. The Extreme low land N75°W 4 or 6 Leag*ue*s, Near point which is the low point projecting from the High Land S81°W, a White Rock close to the Shore S68°W 2 or 3 Leag*ue*s. In the forenoon we had 45 f*atho*ms, sand & broken Shells. [*in margin of* RUSI/NM/132 (9): The N*orth* part of Japan was always suppos'd to be in 40°15′N by Jansen.[3] Capt*ain* King says nearly the same. See Cook's last voyage – 3d vol.[4]

Wednesday 9 August

Light breezes and Cloudy weather, Wind from the Eastward. We steer'd NW and as the Haze clear'd away we found the Land trend to the WNW and at 1 PM we distinguish'd High Land

[1] Kuji Wan. [2] Ushi Shima.

[3] The Dutch map-maker Jan Jansson (Janszoon/Janssonius) (1588–1664).

[4] In a footnote King writes, 'Jansen places the northern extremity of Japan in latitude 40°15′. The point seen by us was in latitude 40°27''; Cook and King, *Voyage to the Pacific Ocean*, III, p. 396.

bearing from N24°W to N34°W ,which seem'd to be a continuation of Nipon. Within the low point bearing west of us at Noon the low Land receded considerably & either form'd a large Bay or more probably from the appearance of the Shores the entrance of a River.[1] Our Junk enter'd it soon after Noon & we hauld up NbW for the Outer Land, the Wind a pleasant breeze from the Eastern board. In the Evening the wind died away & throughout the Night and Morning we had a thick Fog which clearing up at times gave us a view of the High Land to the NW of us. A Variety of Brown Peterels Tip'd with white under the outer part of their wings attending us from Sunset to Sun rise. These birds made a continual chattering Noise.[2] Before Noon the Fog dispers'd and we found ourselves well up with the Land, the Outer extreme[3] bearing N28°W about 9 or 10 miles. From the Mast head low land seem'd to extend from it in the same direction. This Head Land rose abruptly from the Sea to a considerable height, bold & Cliffy. Within the Head the Shore presented the same barren appearance but rather lower till it join'd the High land which was much broken, the extreme of it bore SW, from whence the low shore continued. An extensive Village bore WSW, the Cultivation mounting up the Hills in Ridges over one another. A Sandy Beach continued to the Head. Our distance from the Shore 4 miles. Lat*itude* ☉ 41°11′14″N, Account 41°25′N, Long*itu*de d*itt*o 141°18′E, Course N*orth*,[4] 39 miles. The same Southerly current makes us 14′ S*outh* of Obs*ervatio*n.

Thursday 10 August

Light Airs and Calm from the Eastward throughout the afternoon. At 7 PM we were only abreast the Extreme of the High land set at Noon which bore N65° west. At the extremity of the low point are some detach'd Rocks; they bore N40°W 5 or 7 miles. Some very high Land appear'd in a NW direction. The Night Foggy & Calm. At 3 AM a Fresh breeze from the SE increas'd the Fog, which gradually dissipated towards Noon. At 6AM we had a Sight of the low Rocky point bearing West. We steerd to the N*orth* and at 10 AM we discover'd the Southern Head in Volcano Bay[5] bearing WNW. At Noon it bore N80°W, Abrupt head Land of yesterday S30° to S35°W, Mountain in Nipon[6] S47°W & the distant Land which last evening bore WNW now bore WSW. Latitude observ'd 41°47′9″N, Account 42°10′N, Long*itu*de 141°17′E, [No] 45 140°56′35″E, Course N*orth*, 37 miles, and we are 27 miles South of our Estimation.

Friday 11 August

The wind soon after Noon shifted to the westward & blew a Fresh breeze till Sunset, when the South point of Volcano Bay bore S37°E 3 or 4 L*eague*s. The Shore from this Head trended S55°W. The Night was variable with a NW breeze at times, & throughout the forenoon it was

[1] Mabechi Kawa.

[2] Possibly Sooty Shearwaters (*Puffinus griseus*), which are sooty brown with white underwings and are found in large numbers in the North Pacific at this time of year.

[3] Shiriya Saki, shown erroneously as Cape Nambu on Broughton's 'A Chart of the N.E. Coast of Asia and Japanese Isles'.

[4] N37°W in Broughton, *Voyage*, p. 387.

[5] Esan Misaki.

[6] The 874-metre-high Asahina Yama in the centre of Shimokita Hanto, 21 miles WSW of Shiriya Saki.

a perfect Calm. At Noon the South point on with the distant Land S15°W, North point[1] N63°W about 4 Leag*ues*, South Volcano[2] S78°W and Land Ex*tremes* N15°E & N80°E. Lat*itude* Observ'd 42°10′16″N, account 42°8′N, Long*itu*de d*itto* 140°56′E, No 45 made the Long*itu*de 140°27′E.

Saturday 12 August

It remain'd Calm till 5 PM when a breeze from the NE carried us into the bay which having enter'd, the Wind drew round to the South & we haul'd round the North Head for the Harbor. At 3 AM we came too in the entrance in 9 fathoms water. In the forenoon we were employ'd seeking a convenient place to wood & water & a situation where the Vessel cou'd be laid on Shore which having found in the SE part of the Harbor, we got under way & ran up having a strong westerly breeze. We moor'd in 3 f*atho*ms water a Cables Length from the Shore being perfectly land-lock'd & Shelter'd from every wind in this Port of which I gave some description in our Voyage last Year. We immediately commenced our necessary operations of refitting the Vessel & preparing for Sea. We had the Winds chiefly from the SE with some Rain & on the 14th it blew strong from that quarter. We were unable to have any observation for the Latitude till the 17th from the constant gloominess of the weather but at intervals throughout the Week we got some Altitudes for the watch. The Natives inhabiting the Harbor were poor and inoffensive living entirely upon Fish & Sea Weed (Fucus sacharinus) and at times we were visited by some Japanese who came as we understood them from the Town of Matzmai to see who we were. On the 18th our old Friends whom we had notic'd last Year for their Civility arriv'd from Matzmai & took up their residence opposite the Ship. From them I acquired a compleat Chart of the Japanese Islands with strong injunctions not to own from whence I procur'd it, as they explain'd it wou'd bring them into Disgrace & punishment. They seem'd anxious to know our time of departure & were desirous we shou'd sail as soon as possible. On Monday the 21st we had compleated our wants and on Tuesday we sail'd out of the Harbor & came too in the entrance as the Wind still continued SE'ly with dark gloomy weather, Hans Oldson Island bearing S5°E 1 mile and we had 9 fathoms water.

The Harbor which the Natives call Endomo is situated in the NW Corner of the Bay in the Latitude of 42°19′24″N and 141°7′36″ East Longitude and bears from the South Head of the opposite Shore, call'd by the Natives Point Esarne, N17°W 11 Leagues. These two points form the entrance into the extensive Bay. You have 50 fathoms in the Center and gradually shoal as you approach either Shore. But the Sketch to which I refer will best explain.[3] We remain'd ten days in the Harbor in the Month of August and most conveniently furnished ourselves with Wood and Water.

The Natives are thinly scatter'd over this part. They appear'd in the bay to have no other employment but fishing for their daily Food while those who liv'd over the Isthmus open to the Sea we always found collecting of the Sea Weed (Fucus sacharinus) which they dried in the Sun and made up in bundles for exportation. Great quantities of this weed drives into the Bay and makes a considerable trade with this Island and Matzmai Town.[4] From thence it is exported to Nipon.

[1] Chikiu Misaki.
[2] The 1,167-metre dormant volcano Yokotsu Take.
[3] Fig. 4 on p. 74.
[4] Present day Matsumae, situated at the SW extremity of the peninsula forming the NW entrance to Tsugaru Kaikyo.

We found the Japanese only Name the parts they more immediately inhabit Matzmai and the Three Towns w[h]ere they chiefly reside are in the Straits. But the Natives term the whole Island Insu which the Japanese acknowledged was the proper Name. We also learn'd by them of the Russians trading to the Harbor of Agodaddy[1] in the Straits of Matzmai[2] and inform'd us how superior it was to Endomo. We saw some trifling plantations of Indian Corn and Millet but no other Signs of cultivation. This I much wonder at as their diet seem'd rather scanty and they have every means to produce Vegetables in which the Japanese establishments abound. We saw no other Animals but Dogs and Bears. A Small Red Fox I kill'd in the Harbor but some of the Natives being cloath'd in Skins made it appear the Country was not deficient in Peltry. Our Researches in the Woods gave us no knowledge of their kinds and except the Sassafras Tree[3] we found no addition to their Timber. Raspberries were plenty.[4]

[1] Present day Hakodate Wan. 'There is another town in the Straits belonging to the Japanese but I could not learn the name'; NMM/RUSI/NM/132(9).

[2] Tsugaru Kaikyo.

[3] A tree, various species of which are found in North America and the Pacific rim, whose bark has medicinal and cosmetic uses.

[4] The preceding three paragraphs taken from UKHO, OD 177.

CHAPTER 5

Passage through Tsugaru Kaikyo and thence to the head of the Gulf of Tartary

Wednesday 23 August
Moderate Breezes from the Eastward with dark gloomy weather. At 9 AM we weigh'd and stood over the Bay towards the Volcano.[1] At Noon we Observ'd in 42°12′7″N when the North point of the Bay [bore] N53°E about 3 Leagues.

Thursday 24 August
Light breezes from the SE quarter. At 3 PM we were close in with the Volcano & tack'd in 7 fathoms which increas'd as we stood off shore. It remain'd after dark Calm till Midnight when a Light breeze sprung up from the SW, which carried us to the South Head of the Bay. This point is very remarkable, the Round Head forming it being bare on the upper part which seems to have proceeded from the lava running down the Sides of it but between these drains it is cover'd with wood. We saw smoke ascending on the NW side of it. Within the Point is a Bay & many Houses scatter'd over it. The Natives call'd this Part Ismare [*in margin*: Esarme[2]]. The Lat*itude* made 41°48′50″N and the Head bore West 2 or 3 miles, South Volcano[3] N58°W, Extremes of Matzmai S69°W and Nipon from S15°E to S34°W & we had no Soundings with 50 fathoms.

Friday 25 August
As we open'd the Straits between Matzmai & Nipon we had a Strong breeze from the Westward & we stood over for Nipon. At 5 PM we tack'd in 70 fathoms water the NE point of Nipon an abrupt Head.[4]* [*in margin*: *Cape Nabo[5]] bearing S40°E 2 Leag*ue*s, Extremes of Nipon[6] W½N and the High Land[7] which is WSW from Abrupt Head S40°W. Between these two Lands the Coast forms an extensive bay of no depth with very low land between. We continued plying to the westward during the Night but to our disappointment in the morning we found we had lost ground. Fine clear pleasant weather. At Noon the Latitude made

[1] The 1,133-metre-high active volcano Komaga Take.
[2] Esarne in Broughton, *Voyage*, p. 274, which is clearly a transcription error.
[3] The 1,167-metre dormant volcano Yokotsu Take.
[4] Shiriya Saki.
[5] C. Nambu on Broughton's 'A Chart of the N.E. Coast of Asia, and Japanese Isles'.
[6] Oma Saki.
[7] Asahina Yama.

41°49′20″N with the same bearings as yesterday, the South Head West[1] 2 Leag*ue*s and we had no Soundings with 80 fathoms of Line. Wind Westerly.

Saturday 26 August

It remain'd Calm till 3 PM when a strong breeze sprung up from the western quarter & we stood over to Cape Nambu[2] [*in margin*: Nabo]. At 7 PM it bore S10°W & the South Head of Volcano bay[3] NW when we tack'd. It was moderate throughout the Night & the wind in the Forenoon veering more to the South-west we were able to fetch up with the South Head which at Noon bore S15°W[4] one mile. Observ'd Lat*itud*e 41°50′27″N. A Strong current still preventing our getting to the westward.[5] We were necessitated to want a change of Wind to ensure our Success. Cape Nambu bears from the South Head S22°E[6] 9 Leagues. No Soundings with 90 fathoms of Line.

Sunday 27 August

Fresh breezes and clear weather from the SW quarter. We kept plying to windward in Esarme Bay which lays round the Point and is well shelter'd from the westerly winds. It affords good Anchorage close in & we found here Many Junks wind bound with cargoes of Sea weed (Fucus facharinus). There are many Houses in the Bay and the Inhabitants appear'd to be chiefly employ'd procuring the Weed above mention'd. Many of them paid us a Visit, the Men only came on board while the Women remain'd in the Boats. A fine Stream of Fresh water is at the back of the Sandy beach which is procur'd with great convenience. The Volcano still continued burning & the West side of the Hill was entirely cover'd with Pumice Stones or Sulphur appearing perfectly white. As the Smoke ascended when to leeward of it we experienc'd a strong Sulphureous Smell. In the Evening we once more stood over for Nipon, the Wind veering to the WNW gave us some hopes as we lay over SW. At Midnight the Wind came to SSW & SW & we tack'd to the westward. In the morning to our surprize we found Cape Nambu bearing South 2 or 3 Leag*ue*s, although our course had been westerly all Night. By being over on the Nipon Shore & the Breeze still continuing Southerly we drew in with the Land & apparently avoided the Current for at Noon Cape Nambu bore S56°E and South Head N5°E, Ex*treme* of Nipon[7] running out to a low point about west. To the westward of the bay between the High Land & Cape Nambu we saw extensive cultivations & pass'd two large Towns. Several Junks were shelterd within the beach of the Eastermost, which is situated on the commencement of that I call Mount Anville[8] or the beginning of High Land. An indif*feren*t Observation made the Latitude 41°31′41″N, Opposite Horison 6 or 7 Leagues dist*an*t.

[1] Oma Saki.

[2] Shiriya Saki.

[3] Esan Misaki.

[4] 'S14°W' in Broughton, *Voyage*, p. 277, but corrected to 'S15°W' in pencil in annotated *Voyage*.

[5] In Tsugaru Kaikyo the flow is the resultant of an east-going branch of the Tsushima Shio and the tidal stream. Except for a few days in the year the current is stronger than the tidal stream and under extreme conditions the resulting flow can run to the east at as much as 7 knots.

[6] 'N22°E' in the MS, which is clearly an error and has been corrected to 'Cape Nambu and Point Esarme bear from each other N22°W and S22°E' in Broughton, *Voyage*, p. 277, and amended erroneously to N22°E and S22°W in the amended *Voyage*.

[7] Oma Saki.

[8] Possibly Asahina Yama

Monday 28 August

Light breezes during the afternoon from the Eastward which enabled us to keep in with the Shore and avoid the Current which we cou'd plainly distinguish without us running with great rapidity. At 7 PM it became suddenly calm & a Small Island[1] laying off the Northern part of Nipon bore S40°W 1 mile. We anchor'd immediately in 12 fathoms stoney bottom. Cape Nambu bearing S66°E & Point Esarme NE, Extremes of Matsmai S75°E and Nipon S56°E. In the morning we had gloomy weather and a breeze from the Eastward with which we got under way, passing close to the rocky point of the Island in 4 f*atho*ms to avoid the Current which when abreast of we met with great force breaking in overfalls against the wind. As the Wind increas'd we run over it going 4 & 5 knots p*e*r hour and at Noon the Island was just in sight bearing S87°E 3 or 4 Leagues, an abrupt Head South[2] and a Head Land making like an Island[3] N17°W. The Hazey weather obscur'd the more distant Lands. We had no Sound*ing*s.

Tuesday 29 August

Soon after noon the Wind died away & we had Light airs from every direction, but chiefly Calm throughout the Night. The current drifted us to the North. At daylight the Head land bore of us N12°E 2 or 3 L*eague*s. It forms the East entrance of the Harbor of Agodaddy which the Japanese inform'd us was frequented by the Russian Vessels & they pointed it out as a much superior Port than Endomo. In the forenoon we had a Fresh breeze from the SW & we stood to the Westward. Towards Noon it was Squally with very heavy rain which entirely obscur'd the Land. Our Soundings had decreas'd from 40 to 23 fath*oms*. [*in margin of* OD 177: Variation; 2°11′, 0°21′, 18′ west]

Wednesday 30 August

The Wind subsiding we had a great deal of heavy rain and towards the evening clear and Calm weather. As we approach'd within two miles of the Shore we had regular soundings to 15 fathoms, with the Harbor of Agodaddy [Hakodate Wan] open to us bearing NE. The West & SW side of the entrance was well wooded & near the Sea we pass'd many Villages situated in the Vallies. A Light air from the Land carried us to the South towards some high broken Land & we were necessitated to bring up in 56 fathoms to avoid the Current. At daylight a breeze sprung up from the Eastward with hazey and gloomy weather and we steer'd to the WSW keeping close in with the Land to avoid the Stream but we had no Soundings. As we pass'd we saw several runs of water precipitated from the Hills into the Sea. These Hills presented a thick Foliage of woods & Clayey cliffs & wore a romantic appearance. The thick weather prevented our distinguishing the opposite side of the Straits. At 10 AM we pass'd an extensive Village with several Junks laying at Anchor off it & several Boats put off to pay us a visit, but a strong breeze prevented their reaching the Vessel. Soon after we open'd two distant Islands[4] with a Rocky point[5] forming the Eastern part of Matzmai

[1] Benten Shima.

[2] Possibly Myojin Saki on the western side of the entrance to Mutsu Wan.

[3] Obana Saki.

[4] O Shima and Ko Shima.

[5] Shirakami Misaki, the NW entrance point of Tsugaru Kaikyo.

bay.[1] At Noon we observ'd in the Bay when the Observation made the Latitude 41°22′N. The Extremes forming the bay bore from E5°N to N65°W, The Center of the Town[2] NNW 2 miles, the Extremes of Nipon from S15°E to S61°E 5 Lea*gue*s.

Thursday 31 August

Fresh breezes and hazey weather the wind at ENE. We sail'd along the Shore close to the Town[3] & the least water was 6 fathoms. As we drew near we observ'd several Horsemen richly dress'd galloping full speed into the Town & the whole of the Inhabitants seemingly collected to observe us. Near the Center at a landing place appear'd a Guard of Soldiers with their Colours in regular Order ready to receive us (I suppose) had we landed. This Town was of very great extent built on the Margin of the Beach & ascending gradually up the Rising grounds Interspers'd with Trees & Gardens. The Houses were of wood, & the roofing of the same universally cover'd with Stones (as describ'd by Thurnburg[4]). The superior Houses were decorated by long pieces of color'd Cloths of various patterns spread length ways which had a pleasing & cheerful effect the ground work being chiefly white. The Towers or public buildings were equally adorned & in many parts were Colors flying. Some Hundred of Junks were laying alongside of each other moor'd close to the Shore & on the Beach were others building & repairing. [*in margin of* RUSI/NM/132 (9): The schooner's force was so insufficient for defence I could not think of putting ourselves into the power of the Japanese government which will account for my not communicating with any part of Japan or the town of Matzmai however desirous I might be of getting information of that celebrated country & its fortress at Insu. Indeed La Perouse with two large frigates did not chuse to risk any intercourse with them. I thought we might have been considered a Pirate as I think we were on the coast of Corea at Chosan harbor.[5]] The bay is about 4 miles in extent the points forming it bearing N70° W and S70°E of each other. At the west extreme is a very small Island[6] connected with the shore by a reef of Rocks within and without it. A Small House is upon it which is a look out in the day and a Light House at Night. The Hills rise to a good height back of the Town but are destitute of Wood & except gardens uncultivated. The Easterly wind brought in to the Bay a great number of Junks & Boats they all kept close in shore to avoid the western current which run at the rate of 3 knots in the Offing in strong riplings. At 3 PM we stood across the Straits in the hopes of making some further remarks. After running Ten miles on a South course from Matzmai, the NW point of Nipon[7] bore S34°E about 3 Lea*gue*s & extended up the Straits to East in which direction we cou'd not see more than 4 Lea*gue*s from the Haze obscuring the Land.[8] This ex*treme* of Nipon projects out from some High bold Rocky Land

[1] Matsumae Wan.

[2] Present day Matsumae.

[3] In annotated *Voyage*, p. 282, inserted in pencil, 'Matsmai Lat 41°29′N, 140°0′E'.

[4] Carl Peter Thunberg, a disciple of Linnaeus, who travelled widely but did not visit Matsumae. His description relates to houses he saw in Nagasaki in August 1775: 'the houses are all built of wood and clay, and covered with tiles'; Charles Peter Thunberg, *Travels in Europe, Africa, and Asia, made between the Years 1770 and 1779*, 4 vols, 2nd edition, London, 1795, III, p. 41.

[5] A later addition as the English edition of La Pérouse's voyage was not published until 1799.

[6] Benten Shima.

[7] Tappi Saki.

[8] Hugging the north side of the strait to avoid the current and the haze prevented Broughton from sighting and charting the extensive inlet of Mutsu Wan.

and makes in Hummocks, at a distance the Extreme being the Highest & falling abruptly to the Sea. The Japanese call this point Cape Sangaroo. It is not more than Six Leagues from Matzmai in a SSE direction. In the evening we were within 3 miles of the Light House Island bearing S54°E & the Matzmai Shore took a NNW direction from the low point & we had 65 fathoms. The Breeze continued Easterly till Midnight when it became Calm. We steer'd NW to that time. At daylight we were within 4 or 5 miles of the Shore but no Sound*ings*. An indifferent observation made the Lat*itude* 41°38′37″N, the low point being the SW Ex*treme* of Matzmai S36°E, Nipon extreme open with it. Ex*treme* North & an Island of[f] the Northern point NW,[1] 1st Island[2] S4°W off which to the west was a small Rock,[3] 2d Island[4] S70°W. These Islands are high but of no extent.

Friday 1 September

Soon after Noon a Strong breeze sprung up from the Eastern quarter and we steer'd North. It was so cloudy over the Land we cou'd see very indistinctly into the Bay we were now crossing but from what we cou'd judge there were no openings. The Land in general High & towards the Sea bare of wood. We pass'd between the N*orth* point & the Island which bore NNW at Noon having a clear passage of 3 or 4 Leagues.[5] It was low at each extreme, of tolerable elevation, about 2 or 3 Leagues in a NE & SW direction. Our Soundings at 8 PM were 65 fathoms 3 or 4 miles off the Matzmai Shore. The Night was Squally & threatening weather from the ENE and at Midnight we bro't too under easy Sail till daylight when the extremes of Matzmai bore from N48°E to S18°E, leaving an extensive Bay between the Points,[6] the Island from S8°W to S18° west 4 or 5 Leag*ue*s nearest shore 5 miles. The Forenoon was variable with calms & variable winds. At last it fix'd in the SE quarter & we steer'd NE, a Junk in company with us pursuing the same Course. At Noon we Observ'd in 42°54′27″N, Matzmai bearing from N52°E to South, nearest Land S10°E 2 or 3 Leag*ue*s and the Island S15° west just in sight. Course North, 75 miles, Accounted Lat*itu*de 42°32′N, Longitude 139°38′E, Current North 22 miles by the Observation.

Saturday 2 September

Strong breezes and very Squally weather which prevented our keeping so close in shore as I wished and the gloomy weather prevented our distinguishing any thing particular. [*in margin of* RUSI/NM/132 (9): According to the Japanese chart, there is a river in the bay which we crossed in the afternoon of some extent.[7]] The Country appear'd highly diversified by Hills & Vales but no Openings promising Shelter. At 7 PM we were within 2 Leagues of the High Land set at Noon[8] from whence the Island trends N60°E from a point[9] which is a continued

[1] Okushiri To

[2] Ko Shima.

[3] Ohiyaku Shima.

[4] O Shima.

[5] Okushiri Kaikyo.

[6] An extensive bay between Motsuta Misaki and Kamui Misaki, 53 miles NNE.

[7] Ishikari Kawa, which flows into the head of Ishikari Wan.

[8] The twin peaks of the 1,255-metre-high Shakotan Take and the 1,298-metre-high Yobetsu Take on Shakotan Hanto.

[9] Shakotan Misaki, the western entrance point of Ishikari Wan.

descent from the High part, a Small Island and Rock appear'd to lay off it.[1] When we had lost sight, it bore N72°E 3 Lea*gue*s, Noon point SE 2 Lea*gue*s. We had moderate and gloomy weather after dark, the Wind gradually veering from the S'East to East & NE which bro't with it rainy weather and a Fresh Gale of Wind. We stood to the NW. At Noon an indifferent Observation made 43°48′N,. Account 43°43′N, Long*itude* d*itt*o 140°3′ East.

Sunday 3 September

Strong breezes & Cloudy weather with an hazey Horizon and a large Swell from the North, wind at NNE. Towards Night it moderated and at 7 we stood to the East but the wind veering more Easterly. At ½ past 10 PM we again tack'd to the NW. AM we had Fresh breezes and hazey weather. Observ'd Latitude 44°15′N, Account 44°9′N, Long*itude* d*itto* 139°31′ East.

Monday 4 September

Moderate breezes & very Cloudy weather, the wind gradually veering to the NW. At 2 PM we tack'd and steer'd to the Eastward to make Matsmai. The Forenoon was Cloudy with light Showers of Rain. At 8 AM we saw an Island of little elevation[2] & soon after another with an high peaked Hill in the Center[3] bearing NbE. The first seen bore East & we haul'd up to pass on the west side of it. At 10 we saw Matsmay extending from East to SE. An indifferent Lat*itu*de made 44°34′29″N, Acc*oun*t 44°33′N, Long*itu*de 140°46′E. Peaked Island Center N8°E 6 or 8 L*eague*s, Flat Islands S70°E to 87½°[E] 2 Lea*gue*s, Matzmai SE to N60°E 6 or 8 Leagues.

Tuesday 5 September

Fresh breezes and hazey weather, wind at NW. We steer'd for the Northern part of Matzmai which we gradually approach'd on a NE Course. At 6 PM having run 7 Leagues the breeze died away & we had 55 fathoms, Sandy bottom, about 4 Leagues off, the extremes bearing from N37°E to S14°E, Flat Islands S12° to S34°W & Peaked Island Center N15°W. It remain'd Calm till midnight when the wind Sprung up from the NE quarter & we stood to the North as it permitted. At 10 AM we were close up with Peaked Island which being unable to weather we pass'd to the South of it ranging the shore at 1½ miles distant in 45 fathoms rocky bottom. The Extreme of Matzmai to the North bears from the North point of this Island N45°E 6 or 8 Leagues & appear'd of very moderate elevation. An Island of considerable extent[4] bears N30°W about 2 Leagues from the SW point of this Island leaving a clear passage between them[5] [in a] NE & SW direction. At Noon we observ'd in 45°6′30″N, the Peaked Island extremes from N10°E to S69°E, from the Center 2 miles N62°E, North Island N2° to 20°W, Matzmai South extreme S37°E 10 or 15 Lea*gue*s. Course N6°W, 37′, Lat*itu*de acc*oun*t 45°12′N, Long*itude* 140°50′ E, No 45 [*blank*].

[1] The 40-metre-high Kamui Iwa.

[2] The islands of Teuri To and Yagishiri To.

[3] Rishiri To, whose summit is a 1,721-metre-high dormant volcano.

[4] Rebun To.

[5] Rebun Suido.

Wednesday 6 September

The afternoon was Light airs & Calms which carried us to the NW. A Canoe paid us a visit, they were in every respect the same with those of Matzmay. They gave us the Names of the Islands & departed. The Northermost they call'd Tee-shee & the other Timo-shee. The latter Island is of an irregular Figure about 6 or 7 Leagues in circumference rising gradually from its base to an uncommon height. Peaked Hill presenting an evident Volcanic Crater much broken into Ragged Points and to appearance Rocky half way down the Mountain exhibiting various colours in the drains, of Earth, Cinders, Pumice Stone or Sulphur. The shore was skirted with Rocks and well cloath'd with Wood till it met the bare part of the Hill which was generally obscur'd by the Clouds. It wore a pleasing aspect from the Sea. The beauty of the Shore; and the Woods were diversified with Fir Trees. As we open'd the passage between the Islands we saw some few Houses. The other Island extends about 4 Leagues in a NbE & SbW direction very moderate in elevation compar'd with Timo-shee, rising abruptly from the Sea in Rocky Cliffs of Horizontal Strata. The Northern part is low and Rocky. Both these Islands are inhabited. During the Night we had light airs from the Eastward & we steer'd in the direction of Tee-shee. In the Forenoon we had very heavy Rain & NE wind & very hazey & Cloudy weather which prevented any Observation. Tee-shee bore from N63°E to S67°E, Peaked Isle or Timo-shee S55°E off shore about 3 leagues. Lat*itu*de account 45°22′N, Long*itude* d*itto* 140°26′E, Course N47°W, 24 miles.

Thursday 7 September

Light breezes and very cloudy weather, Wind at NE. At 5 PM the Wind veering more Northerly we Tack'd in Hopes of weathering Tee-shee & to draw in with [the] Matzmai Shore. At Midnight we had Tee-shee bearing South of us about 2 Leag*ue*s, but we had no Soundings with 80 fathoms. The breeze again fail'd & we had Light airs from the East when we tack'd to the North. At 5 AM we saw a Small Island bearing NbE[1] and the extreme of Matzmay[2] S72°E 7 or 8 L*eague*s. At Noon the Observ'd Latitude gave 45°44′34″N, when Timo-shee Peak bore S8°E 12 L*eague*s, Tee-shee S2°W to 12°W, Extremes of Matzmay just visible S50°E 8 L*eague*s or more, 1st seen Island N13°E, 2d ditto just at Noon visible from the Deck[3] N60°E. Latitude account 45°44′N, Long*itude* d*itt*o 140°48′ East, Variation p*e*r Azimuths 35′ west.[4] [*in margin of* OD 177: Variation 0° 39′ west, 0° 49′ d*itto*, 0°17′ East]

Friday 8 September

Light breezes and cloudy weather with Smooth water, the wind from the SE quarter & we kept our wind for the Eastern Island which prov'd to be the SE extreme of an extensive Island.[5] [*in margin of* RUSI/NM/132 (9): As the Japanese in their chart make an extensive Island to the

[1] Ostrov Moneron. So named by La Pérouse after Paul Mérault de Monneron, *Ingénieur de Chef* on board the *Boussole*; Dunmore, *La Pérouse*, I, pp. xc–xci & II, p. 318.

[2] Soya Misaki, the northern extremity of Hokkaido.

[3] A glimpse of Ostrov Sakhalin.

[4] Here Broughton adds in *Voyage*, p. 290, 'To the South of what is supposed to be the N.W. point of Insoo, between 45°15′N. and 45°N., we could not distinguish the coast. It is therefore probable this may only be an island; and the Japanese chart seems to confirm it. The wind prevented me clearing up this point. These remarks are since our being to the North.'

[5] I.e. Sakhalin. Broughton was off the western entrance of La Pérouse Strait separating Hokkaido from Sakhalin.

N*or*th of Insu I conjectur'd this to be the one.] At ½ past 6 the extremes of it bore from S85°E to N38°E, Center of 1st Island N20°W, Timo-shee S8°W & Tee-shee S20° west about 10 or 12 Leagues, the latter. At Midnight we bro't too till daylight having 61 fathoms fine Sand. At 5 AM we stood in Shore the extremes bearing from S69°E to N20°E, 1st Island NW 3 or 4 Leag*ue*s, distant Land 6 L*eague*s. By Noon we had approach'd within 3 miles of this New Land, the Extremes bearing from N25°E to S20°E, each extreme about 8 or 10 Leag*ue*s distant. The Island bore from S64°W to 70°W 6 or 7 Leag*ue*s. An indifferent Observation made the Latitude 46°27′58″N, Account 46°27′N, Long*itude* d*itt*o 141°14′ East. Cloudy & Hazey weather.

Saturday 9 September

Moderate breezes and cloudy weather. Wind Easterly till 8 AM when it gradually left us. We shap'd our course to the North Keeping from One to two Leagues distant from the Shore, which continued Bold and Cliffy of moderate elevation & swelling into rising grounds well cloath'd with wood and Verdure. The Land continued very strait & little indented and presented the same uniform appearance if we except a Round Hill[1] which form'd the North extreme of yesterday's Noon. We pass'd a few scatter'd Houses in the afternoon. At Noon the extremes bore from N27°E to S2°E each 10 Leag*ue*s apart & we had 20 f*atho*ms 4 miles off, sandy bottom. Lat*itude* 47°29′N, Account 47°28′N, Long*itude* d*itt*o 141°[2]26′E, Course N3°E, 62′, Variation 2°46′W.

Sunday 10 September

The Calm continued till 3 PM when we had a Fresh Wind off the Land at NE, which towards the evening veer'd more Easterly. We kept close haul'd as the wind permitted. At 6 PM the Extremes to the North being low land bore of us N40°E and the round Hill which now form'd the South extreme S26°E and we were 3 or 4 Leagues from the Land. The Night was Squally & the air very cold. Towards morning the wind became South west. At 6 AM we saw the continuation of the same Land bearing N15°E to S15°E, a conspicuous Mountain[3] N37°E and our distance from the Land abreast of us 4 or 5 Leag*ue*s. We haul'd close in till we discern'd low land well wooded filling up some small Opening & afterwards Steer'd North to pass without the extreme,[4] which at Noon bore N20°E 3 or 4 Leag*ue*s, Mountain N51°E, Low extremes S55°E, the more distant Land being[5] obscur'd by the Hazey weather. Lat*itu*de Observ'd 48°34′30″N, Acc*ount* 48°31′N, Long*itude* 141°18′E,. Course North, 65 Miles.

Monday 11 September

Fresh breezes and very thick hazey weather with a Southerly Wind. We steer'd to the N*orth* and NNE as the Coast trended. At 2 PM we pass'd the Point which in the morning made the

[1] Gora Tatarskaya, 350 metres high, situated 1½ miles SSE of present day Port Kholmsk.

[2] '114°' in the MS which is clearly a transcription error.

[3] Gora Izyl'met'yeva, 685 metres high, which from the SW appears pointed and is very prominent, situated 8 miles NNE of Mys Lamanon.

[4] Mys Lamanon.

[5] 'bearing' in the MS, which is clearly a transcription error.

Extreme. It is the only part we have yet seen making a Point & projects out with a gradual descent from the High Land terminating very low. It is situated in the Lat*itude* of 48°46′N & Long*itude* 141°E. I named it [*blank*].[1] The High Round Mountain to the SE of it[2] makes this Coast remarkable. [*in margin of* RUSI/NM/132 (9): I could not reconcile the extent of this land with the Island north of Insoo in the Japanese chart ...] To the South of it the Land is of moderate Elevation but to the North very Mountainous Rising abruptly from the Sea & in Land they appear'd more elevated Cloath'd with wood up to their Summits. The Shore still continued bold & no dangers of any kind seem to obstruct the Navigation of this extensive Coast.[3] At 7 PM we haul'd off for the Night & at 6 AM made the Land where we left it last evening bearing from South to NE, but Hazey & Rainy weather prevented our View being extended beyond 2 or 3 leagues. At Noon & throughout the Morning we had a Fresh gale from the South & the coast preserved the same direction. No Observation. Lat*itu*de acc*oun*t 49°40′N, Long*itu*de d*itt*o 141°57′E, Course N16°E, 68 miles. Cold weather & very Hazey.

Tuesday 12 September

Fresh gales & hazey weather with a large following Swell, the Wind at South and we steer'd NbW. Towards night the Wind moderated and heavy Rain brought on a Calm & in the morning a Northerly Wind with a fine clear Sky which we had not experienced for some time. At 6 PM the Extremes of land bore from N30°E to SE & we were distant 2 Leagues from the Land. Our Sounding 65 fathoms, Muddy bottom. We steerd NW till 9 when we haul'd our wind till daylight. At ½ past 5 the extremes of the Land bore from N22°E to S18°E 4 or 6 Leag*ue*s and we had the pleasure at the same time to see the Coast of Tartary bearing N78°W just visible from the Deck, our Soundings 70 fathoms, muddy bottom. The Forenoon was Calm & very clear weather which enabled us for the first time this Moon to observe distances. A good Observation made our Latitude 50°17′11″N. The North extreme of the land which appear'd to be the terminating Point falling low towards the Sea bore N24°E 5 or 6 Leag*ue*s, other Ex*treme* S14°E, Coast of Tartary from N71° to S83°W 8 or 10 Leagues, a Remarkable Peaked Hill[4] S64°E & a break in the Land S40°E, off the nearest shore 3 or 4 Leag*ue*s. Course N22°W, 40 m*ile*s. Lat*itud*e acc*oun*t 50°21′, Long*itude* d*itt*o 141°34′E. No Current have we yet experienc'd.

Wednesday 13 September

In the afternoon a Light Breeze sprung up from the Southward and we steer'd NNW. At 6 PM the extremes of Western Land bore from N55°W to S80°W 7 or 8 L*eague*s, Noon point N35°E, low land open with it 4°. After running 5 Leagues we bro't too till daylight having

[1] Not named either in Broughton, *Voyage*, p. 204, nor on the chart facing title page. Its latitude agrees with Mys Lamanon, but its description suggests Mys Izyl'met'yeva, which is high and cliffy and prominent from SW.

[2] Gora Izyl'met'yeva is situated SE of Mys Izyl'met'yeva, but NE of Mys Lamanon.

[3] Broughton, in *Voyage*, p. 395, wrote: 'I could not reconcile the extent of this land with the island North of Insoo, in the Japanese chart; neither did it continue the same easterly direction: however, to clear up the matter, we had only to continue our course, keeping well in with the land. The land to the West we could only conclude to be the coast of Tartary; but the extent of the land we had now traced up for 4½ [degrees] of latitude we could not all comprehend. Nor had we any books or charts to clear up our doubts, except Cook's third voyage which only tended to confuse us.'

[4] Gora Kitousi, in 50°12′N, 142°12′E, a conical-shaped 826-metre-high mountain clearly defined among the surrounding hills.

from 55 *fatho*ms to 43 *fatho*ms, Sandy bottom. At daylight we made Sail to the NNE, the Peaked Hill bearing S27°E and we were 4 Lea*gue*s off the Eastern Shore. In the forenoon the Southerly Wind increas'd to a strong breeze & the shores became gradually lower always terminating in very low points. As we saw no double Land we had reason to hope we were near its extremity, but at Noon having Shoal'd our water to 11 fathoms we haul'd over for the Western Land fearing we might from appearances be only in a deep Bay.

A low Sandy point[1] bore N15°E & we saw no Land between it and the Western Shore which bore N30°W, a Round Hill[2] West & the ex*treme* WbS 6 or 8 Lea*gue*s, our distance from the East Shore about 2 or 3 Lea*gues*. ☉ Alt*itude* made 51°25′35″N, Acc*ount* 51°13′N, Course N*orth*, 68 miles, Long*itude* 141°34′E. Variation at ⚓ 3°1′W.[3]

Thursday 14 September

Fresh Breezes and Squally weather from the SW quarter, our Soundings 13 *fatho*ms in crossing over and as we approach'd the west Shore [they] decreas'd to 9 *fatho*ms. In the afternoon the Wind veer'd to WNW and blew Strong & we were necessitated to come too an Anchor 4 Lea*gue*s from the Western Shore, the Round Hill bearing S64°W.[4] The Night was moderate and the Wind coming from SSW we got under way at daylight & stood up the Bay decreasing our water to 3 *fathoms*, when we tack'd & stood off into 4 fathoms, where we came too an Anchor. The Round Hill S58°W, Conic Hill forming the N*orth* Land on the West side N5°W[5] 3 miles, Nearest Land W10°S, No Land from N15°E to 35°E[6] but from thence a continuation of very low Hummocks just appearing above the Horizon. The Western Land was well wooded with Whitish Cliffs rising perpendicularly from the Sea to a good height. Observed Latitude 51°41′21″N, Account 51°43′N, Course N47°W, 24′, Long*itu*de 141°3′E. The Wind continuing Westerly & moderate. The Master was sent to make further Remarks to the North.

Friday 15 September

Fresh breezes & clear weather from the western quarter, the air very Sharp and cool. At Dark the Master return'd who reported having found round the nearest point, which bore N10°E from us, a Bay leading to the Westward[7] at the back of the Conic Hill, 3 or 4 miles deep. In the entrance was 5 *fatho*ms water but the Wind blowing strong out of the Bay they were unable to examine it. It appear'd as they thought to afford Anchorage but open to the Eastward. The South point of the Bay (& the whole of the Land indeed) was compos'd of White & Reddish color'd Rocks so Steep as to prevent their going on Shore. The Woods were universally Firs of a very small growth. It was low at the Head of the Bay. Both going & coming after passing the bank of 3 *fatho*ms they had from 4 to 8 *fatho*ms. For which reason, & the distant High Land to the NE confirm'd it, the Master gave it as his Opinion the passage through seemd pretty

[1] Mys Tyk, a sandy spit, the eastern entrance point of Proliv Nevel'skogo.

[2] Probably the 268-metre-high Gora Davydova, which has a rounded summit, situated close west of Mys Davydova, or possibly the 326-metre-high Gora Tamarna, which has twin peaks, situated 1½ miles farther north.

[3] Since the schooner did not anchor until the following day it seems probable that this statement was inserted here in error instead of the next day.

[4] 'S64°E' in the MS, which is clearly a transcription error.

[5] 'N50°W' in Broughton, *Voyage*, p. 300.

[6] 'N53°E' in Broughton, *Voyage*, p. 300, which is clearly a transcription error.

[7] Bukhta Sushcheva.

evident. At daylight we got under way to clear up all doubts & stood for the South Point of the bay which at Noon bore N6°W 6 or 7 miles, Conic Hill NNW and Round Hill S55°W and our Soundings had been from 3½ to 3 fathoms the last ¾ hour. Observ'd latitude made 51°45′7″N. Fair weather and very Smooth water. [*in margin of* OD 177: Variation 3°36′, 2°38′, 2°47′, 3°50′ west][1]

Saturday 16 September

Fresh Breezes & Clear weather from the Westward we lay up NNW for the Conic Hill, our Soundings still decreasing from 3 f*athoms* to 2 fathoms, in ½ an hour past Noon, when we tack'd. At which time from the Masthead we plainly discern'd very low Land extending over the NE part of the Bay. The nearest part to us bore NEbE 3′ or 4′ & from thence to North within the South point of the Bay was a continuation of Sand Banks partly dry & some ripling only upon the Surface. We kept our Wind to the SSW and were fully satisfied there was no opening to the Sea in the direction the Master suppos'd.[2] At ½ past 2 were we in our old Anchorage having run 6 miles since we tack'd & had only increas'd to 4 f*atho*ms very gradually from 2 f*atho*ms, hard Sandy bottom. We steer'd, as the Wind veer'd to the North of west, SWbS with a fine breeze, Increasing our Water by degrees to 15 f*atho*ms, when Conic Hill point[3] bore N13°E 7 or 8 L*eague*s, Round Hill S70°W, off shore 3 or 4 Leagues. The Night was moderate till 4 AM when a Calm of two hours bro't a Southerly Wind. At 6 AM the Round Hill N72°W, off the western Shore 3 Leagues in 25 fathoms, Clayey bottom. At Noon we observ'd in 51°15′11″N, Extremes of East Land from N86°E to S40°E 10 or 12 Leag*ue*s, Land about our last Anchorage N6°E, Round Hill N53°W, Ex*treme* Land S29°W, off shore 3 or 4 Leagues in 27 fathoms. Accounted Lat*itud*e 51°14′N, Long*itu*de 140°37′E, course S32°W, 31 m*ile*s. The Land projects from the Round Hill to the East & presents a Range of Steep Rocky Cliffs.[4] It wears with little variation much the same appearance both North & South of it, the Shore is considerably indented, but we saw no openings for Shelter.

[1] In Broughton, *Voyage*, p. 300, are the results of observations for longitude by Broughton and Chapman by lunar distances, which are amended in pencil in annotated *Voyage*, page 300, to:

Anchoring Place	Long East per 27 sets of Distances of ⊙ to ☽ East and West of the Moon	141°12′41″·4
	Long. E per Time keeper No 45 Macao rate	141° 0′48″
	Long E per Endermo rate	141°46′48″

[2] The Master, however, was right. There is indeed a passage to the north, but it was not until 1849 that this was finally confirmed. Krusenstern, who carried out surveys at the northern end of Ostrov Sakhalin in 1805, was also convinced that there was no passage, which he felt was confirmed when, 'On my arrival in China I was not a little rejoiced to find Captain Broughton's voyage, which had appeared in print since my departure; and as this is in every one's hands, it will be seen that my ideas upon the junction of Sachalin with Tartary were perfectly well founded. Captain Broughton advanced in a small vessel drawing not more than nine feet, eight miles further north than La Perouse, came into two fathoms water, and at length discovered a bay three or four miles in depth; this he sent a boat to examine and found it closed in on all sides by low sand-hills, nor was there anywhere the smallest trace of a passage. Here then was the extremity of the great gulf of Tartary: and if there still had existed any small channel which escaped the penetrating eye of Captain Broughton, and of Mr. Chapman, the master of his ship, who examined the bay, it must certainly have occasioned some current'; Krusenstern, *Voyage*, II, pp. 181–2. La Pérouse had in fact accepted the insularity of Sakhalin and depicted it as such on his chart, but concluded that the strait was not navigable; see Dunmore, *La Pérouse*, II, p. 313n. See also Ian R. Stone, 'W. R. Broughton and the Insularity of Sakhalin, *The Mariner's Mirror*, 82, 1, February 1996, pp. 76–81.

[3] Mys Sushcheva.

[4] Mys Davydova.

CHAPTER 6

From the head of the Gulf of Tartary along the east coasts of Russia and Korea towards the island of Tsushima

Sunday 17 September

Pleasant weather throughout the afternoon with a Light breeze at SSE. At 4 PM we tack'd within 2 or 3 Leagues of the Round Hill Shore in 27 fathoms. The Land in this part was rather indented and a Small Rocky Islot lay off the Shore. At Sunset it was nearly calm and at dark the wind came from the Eastward. We steer'd SSE to increase our distance from the Shore. The Wind soon increas'd to a Fresh Gale with strong Squally weather and we had a great deal of Lightning. Happily for us the wind Veer'd more Northerly & we made a strait course down the Bay. At 2 AM we bro't too till daylight; our Soundings were 47 fathoms. The Forenoon continued blowing strong from the North and we had constant rain. We cou'd not distinguish the Western Shore, but the Eastern land made very plain & we shap'd our course SW. Before Noon we made the western [shore] which at Noon bore from S60° to N80°W 3 or 4 Leag*ue*s and we had just before lost sight of the Peaked Hill[1] bearing East. Lat*itud*e account 50°8′N, Long*itu*de 140°19′E, Course S10°W, 68 miles. Thick hazey weather with Rain and a considerable Sea.

Monday 18 September

Ditto weather. After running 5 miles the Land bore West and we bro't too the Wind in hopes of our being off the South point of the Bay.[2] Our Soundings were 38 f*atho*ms and the Wind to the westward of North which gave us some hopes of clear weather. At 5 PM we had bro't the Land to bear North and distinguish'd more bearing SSW 4 or 5 Leag*ue*s. We steeer'd for it till dark when we haul'd our Wind under an easy Sail in 30 fathoms 2 Leag*ue*s off shore. The Wind soon after abated and at Midnight it was calm for a short time. A breeze soon after came from the Eastward and we made all sail to clear the Land. Towards Morning it veer'd more Northerly & we stretch'd off to the SE. At ½ past 5 AM the Land bore from North to SW nearest part west 2 or 3 Leagues. We alter'd our Course to SSE and South, opening more Land to the SW. The Forenoon was heavy Squalls with Rain & in general it blew Hard with Hazey weather entirely obscuring the Land. At times we saw the High Land on the Eastern Shore at a great distance. At Noon supposing ourselves near the Southern extremity of the West Shore we bro't too & had 65 f*atho*ms, muddy bottom. Unfortunately we had no Observation. Our account 49°15′N, Long*itude* 140°20′E, Course South, 59 miles. At Noon Fresh Gales &

[1] Gora Kitousi.

[2] I.e. the head of the Gulf of Tartary.

Squally weather, Large Swell. We cou'd barely distinguish the Land bearing WSW 4 or 5 *Leagues*. Cold and very unpleasant weather. Several Albatrosses seen.[1]

Tuesday 19 September

The wind after noon soon increas'd to a Strong Gale with a great deal of Sea & we lay too under a low Sail. Very hazey weather and a Northerly Wind. At 5 PM the haze clear'd away & seeing the Land extending from North to SWbS we bore up and steer'd a South course 4 or 5 Leagues off shore. At 6 PM the extreme South bore S48°W & some High Land abreast of us West about 5 or 6 Leagues. We run till 8 PM in hopes of being on a parallel with the South extreme which we concluded must be the extent of the Land in the S*outh* direction. We had 70 & 80 fathoms during the Night, muddy bottom. The Gale after Sunset gradually abated and at 4 AM it was Calm. A breeze at 6 AM came from the South & we stood in Shore till Nine, the South extreme still bearing S56°W, North extreme NbW, 4 or 5 Leagues off shore. We also discern'd the very High Mountain on the Opposite Shore[2] which was nearly East of us. At Noon the Observation made 48°36′40″N, the Mountain bore N82°E 14 L*eague*s and the West Land from S64°W to NW 6 or 7 Leag*ues*. Lat*itude* acc*oun*t 48°50′N, Long*itude* d*itt*o 140°12′E, Course S7°W, 40 Miles. Moderate and Hazey weather at Noon. Air'd the Vessel with Fires.

Wednesday 20 September

The same mild weather continued throughout the afternoon and we kept plying to the SW. We cou'd observe more distant Land in a WSW direction and the nearest part to us presented the same inhospitable aspect as we notic'd up the Bay. At 6 PM the South extreme S68°W.[3] It terminated in a Hill of Conic Form & we saw more land in the same direction beyond it 7 or 8 L*eague*s. In the Night we had a great deal of Swell from the SE and the wind increas'd from that quarter with very gloomy weather. We kept over for the Eastern Shore and at 6 AM saw the Round Mountain[4] bearing East 8 or 10 Leag*ues*. The wind now veer'd to ENE with Squally and threatening weather. We steer'd SSW to clear the Land. At 8 AM the Western Shore was seen to the NW. Throughout the morning we had Fresh gales & rainy weather & no Sound*ing*s. Lat*itu*de acc*oun*t 48°2′N, Course S4°W, 36 miles, Long*itude* 140°9′ east.

Thursday 21 September

We run till 1 PM and thinking we were sufficiently clear of the Land I thought it most prudent to bring too with our head to the SE. At this time it blew a very hard Gale with thick hazey and Rainy weather. At Sunset it blew with more violence which oblig'd us to close reef the Main Sail & Reef the Bowsprit. We had a great Sea but not much Swell the Wind being from the Land. Every hour if possible seem'd to add force to the wind which

[1] Any one of three North Pacific species of Albatross, possibly the Short-tailed or Steller's Albatross (*Diomedea albatrus*) formerly widespread, now one of the rarest albatrosses.

[2] The 685-metre-high Gora Izyl'met'yeva.

[3] Mys Uspeniya, a spur covered with trees giving it a domed appearance.

[4] Gora Izyl'met'yeva.

continued without any abatement till 2 AM when the Gale broke up. Our little Bark made very good weather Shipping very little water. The Forenoon we had the Wind variable from every direction & the remains of a large Sea. At daylight we had 78 fathoms muddy bottom and we saw the Land bearing from N*orth* to W5°S 5 or 6 L*eagues*. Our observation at Noon made 47°46′N, Account 47°35′N, by which it appears that we drifted more to the West than we suppos'd in the Gale, which the watch confirms. We were employ'd drying our Sails, Cloathing &c this morning and repairing the defective Cordage. Course S30°W, 30 Miles, Long*itude* 139°44′E.

Friday 22 September

Fresh breezes & very cloudy, variable and threatening weather with Showers of Rain and a Southern Swell, the Wind at ENE & we steerd SW. We cou'd not see any thing of the Land but I suppos'd it still trended to the WSW. Having split all our Sails that were bent in the last Gale except the Foretopsail we were busily employd bending our remaining suit. We pursued our Course for 30 Miles at which time we had no Soundings with 90 f*atho*ms. The Wind died away & afterwards came from the South which being directly on Shore we tack'd to the Eastward. By daylight the Wind came round by the West to WNW and blew Fresh. At ½ past 5 we saw the Land from NW to S80°W[1] 6 or 7 Leag*ue*s and by 8 we saw it extending to SWbW. The NW part was high broken Land with bare Rocky Heads the lower parts were of a brownish Hue cover'd with Pine Trees. We had variable weather and little wind. At Noon when the Observation made 46°54′N, Extremes of Land bore from S65°W to N12°W and the Land which bore at 5½h S80°W, bore N50°W about 5 Leagues. To the NE of this Land appear'd a Bay or a Small opening.[2] No Sound*ing*s [with] 70 f*atho*ms. Lat*itud*e account 47°1′N, Long*itude* 139°E, Course S30°W, 60 Miles.

Saturday 23 September

Light breezes and cloudy weather from the Eastern board with a great Swell from the Southward. We steer'd in the direction of the Land SW. Openings appeard to the West of us in the Land but the Haze over it prevented our being convinc'd[3] & we had no weather to induce our approaching the Shore as the Easterly Winds still continued. At 6 PM the South extreme forming a High Hill[4] bore S70°W 6 or 8 Leagues. At 7 we haul'd off South for the Night which was squally & Rainy throughout with a great deal of Lightning. At 2 AM the wind veer'd to South and we were oblig'd to tack to clear the land. Heavy Squalls & a large Sea in the Forenoon Split our Jibb. By Noon the wind veer'd to the West and we had clear weather which gave us a distinct distant view of the Land from West to N35°W 10 or 12 Leag*ue*s. A remarkable Round Top't Hill which I imagin'd to be the same as we set as the extreme last Evening now bore N78°W. Observation made the Latitude 46°22′45″N, Account 46°23′N, Long*itude* 139°13′E, Course S16°E, 32 miles, little wind and a large Swell. By the watch we have been set [*blank*] to [*blank*] these two last days. No Soundings.

[1] Possibly Mys Plitnyak.

[2] There is neither a bay nor a small opening north of Mys Plitnyak, but about 5 miles north of this point a cave in the cliff is very prominent from seaward.

[3] There are no openings along this stretch of coast.

[4] Probably the 495-metre-high Gora Ploskaya, situated 7 miles west of Mys Sosunova.

Sunday 24 September

Light breeze and very clear pleasant weather with the remains of a large swell which prevented our going ahead. In the Night the wind veer'd to NW & blew Fresh. We haul'd up for the Land which at daylight bore from N35°W to 83°W. We had a fine breeze throughout the morning but more westerly which prevented our approach to the Land nearer than 5 Leagues. It made in parts very High & the sides of the Hills afforded a bare prospect. We had no Sound*ing*s with 90 fathoms. Latitude ☉ 45°50′40″N, the Round Topp'd Hill set yesterday N35°W and the extremes from N14°W to S80°W about 6 Leagues. [*in margin of* OD 177: Variation 2°18′, 1°43′, 1°26′ west]

Monday 25 September

Fresh breezes with clear cold weather. The wind veering to the South of west in the afternoon induc'd us to stand in Shore but a violent head Sea retarded our progress. At 5 the wind again came Northerly and we stood to the SW close haul'd, the extremes bearing from WbS to NbW½W. It blew strong from the NW during the Night & moderated towards the Morning with a Chop of a Sea which drove us bodily to leeward. The Land at 6 AM still continued the same direction & the Weather & Sea together wou'd not admit a near approach. At Noon our Latitude 45°4′N, Account 45°9′N, Long*itude* 137°56′E, Course S25°W, 47′. The extremes bore from N25°W to N75°W, High broken Land NW 7 or 8 Leagues.

Tuesday 26 September

Towards the Evening the Sea had entirely subsided and at 3 PM the wind having back'd round to the WSW we stood in for the Land till 7 oclock, when we tack'd off shore. The Wind gradually died away and at midnight it was Calm with very hazey weather. In the Forenoon a breeze came from the NE & we steer'd WbS. We had gloomy weather with some Rain & no Observation. The Land appear'd extending from North (which was the extreme of high broken Land) to N85°W, an opening or Bay[1] N66°W & we were 6 Leagues off shore. Latitude account 44°55′N, Long*itude* acc*oun*t 137°17′E, Course S72° west, 29 Miles. Very Smooth Sea.

Wednesday 27 September

Fresh breezes & dark gloomy weather with a NE wind. We steer'd WSW & SWbW which drew us in with the Land. At 6 the South extreme making in Hummocks bore S85°W between which and a Hilly point bearing N83°W the Land receded making low, but still the whole Coast to us appear'd closely connected particularly the opening notic'd at Noon. It was much broken forming Hills & Vallies along the Shore but very desolate in appearance at l[e]ast as far as our distance enabled us to judge, being 4 or 5 Leag*ue*s [off] & without Soundings. At 11 PM we brought too and the wind veer'd to NW blowing strong in the forenoon. At 4 AM we made Sail upon a wind, the Land extending from N5°E to S19°W about 5 L*eague*s. At Noon the South extreme forming a High Hill,[2] from which some low land projected, bore S84°E 5

[1] Bukhta Golubichnaya in 44°54′N, whose entrance points are high and cliffy and its shores low and sandy, situated to the north of Mys Yegorova.

[2] Gora Piramidal'naya, a prominent pyramid-shaped mountain 635 metres high, situated 6 miles north of Mys Nizmennyy.

or 6 L*eague*s, High Rocky barren Land NWbW 4 L*eague*s, N*orth* Extreme N5°E, No Sound*ing*s. Observ'd Latitude 43°38′N, Account 44′1′N, Long*itude*135°29′E, Course S41°W, 103 miles. Throughout we have pass'd quantities of coarse Grass. Current S20°E, 26 miles, since last Observation.

Thursday 28 September

It became moderate after noon and the wind veer'd round to the SW which brought hazey weather. At ½ past 1 we stood in for the Land and at Sunset being within two Leagues of the Shore we stood off having no Soundings with 90 fathoms of Line. We had run in Shore NWbW 5 Leagues which will make our distance at noon further off than we suppos'd. The Extremes bore from us NNE to WSW. We saw no appearance of any Openings but a Rugged and Rocky shore with some few Scatter'd Trees on the Ridges of the Hills. The interior Land made very High but towards the Sea the elevation was not so considerable forming abruptly at unequal distances. As we saw no Land to the Westward of the extreme point set at Noon, I rather imagine the Coast takes that direction & in its vicinity some River probably empties itself into the Sea. At least I conclude so from the Grass &c before observ'd.[1] This Point for the present we were unable to clear up & the wind continuing SW'ly we kept our Wind to the South and in the Morning the Land was indistinctly seen thro' the Haze to the NW. Observ'd Lat*itude* 42°57′N, Acc*oun*t 43°8′N, Long*itu*de 135°32′E, Course S5°W, 41 [miles].

Friday 29 September

Fresh breezes & hazey wind from the WSW and South. We stood to the SE till 9 PM when we tack'd to the westward. After Sun setting the breeze died away gradually and throughout the forenoon it was very faint with a smooth Sea. During the Night we experienc'd very heavy dews equal to small Rain & the Atmosphere so Hazey we cou'd make no observations for the Long*itude*. Indifferent Lat*itud*e 42°47′N, Account 42°33′N, Longitude d*itt*o 135°2′E, Course S36°W, 37 Miles.

Saturday 30 September

Moderate breezes and Cloudy weather entirely obscuring the Atmosphere. The wind continued steady from the SW quarter. At ½ past 4 PM we discern'd the Land to the NW and at 6 PM the extremes bore from N10°W to 80°W making in detach'd Hills which our distance prevent'd observing the connexion of. The Sun setting clear gave to us a tolerable distinct view of it & we suppos'd its distance 8 or 10 L*eague*s. [At] ½ past [6] we tack'd having no Soundings with 100 fathoms. In the Night we had Showers of Rain and in the morning Foggy weather. At 8AM the wind had veer'd round to North and we steer'd west to make the Land[2] with a fine breeze and Hazey weather. Observ'd Lat*itude* 42°6′17″N, Account 42°8′N, Long*itu*de 134°6′ east, course S63° west, 46 miles. Some small Land Birds & several Gulls seen this day. [*in margin of* OD 177: Variation 5°27′, 4°6′, 3°55′, 5°8′ west][3]

[1] Several minor rivers enter the sea along this stretch of coast.

[2] I.e. parallel to the coast which turns to the west at Mys Povorotnyy.

[3] Signed here 'W. R. Broughton Captain.'

Sunday 1 October

Moderate breezes and very hazey wea*the*r, wind from the NE quarter which towards Night veer'd to the SW and decreas'd to a Calm. In the Forenoon we had Light airs from South. We steer'd to the Westward, and at daylight we distinguish'd the Land bearing NbW.[1] At Noon it extended from N10°E to N22°W 8 or 10 Leagues distant. Very mild pleasant weather enabled us to perform divine Service. Lat*itude* observ'd 42°5′14″N, Lat*itude* account 42°6′N, Long*itude* 133°6′E, Course west, 45 miles. The Land we now found to trend to the westward & we pass'd bunches of Sea Grass (Zostera marina). [*in margin*: Longitude deduced to Noon by No 45 & continued in the Margin throughout. [No] 45 132°55′E]

Monday 2 October

Light winds from the SW quarter with very smooth water carried us to the Land, which at 6 PM extended from N33°E to N23°W 6 or 7 Leagues and we had no Sound*ing*s. The Night was calm and Cloudy and in the forenoon we had a fine breeze from the Eastward. At 6 AM the Land bore from N42°E to N35°W about 6 Leag*ue*s. Low land appear'd near the Sea and very high wooded Mountains at the back. We still pursued the same course and at Noon had no Observation.[2] The extremes bore from N45°E to N10°W and the appearance of the coast to NW, but it was so dark and hazey over the Land we cou'd not well distinguish [it], distance from the nearest Shore 7 or 8 L*eagues*. Course N86°W, 42′, Lat*itude* acc*oun*t 42°8′N, Long*itude* 132°10′E.

Tuesday 3 October

Moderate breezes and hazey weather from the Eastw*ar*d. We pursu'd the same course which I conceiv'd to be parallel with the Coast although the Haze obscur'd it. At 6 PM the only Land we cou'd see bore NNE 8 or 9 Leag*ue*s. The Wind kept veering from East to North with slight Rain and very dark gloomy weather increasing in Squalls. At ½ past 10 having run 7 Leag*ue*s since dark we brought too & Sounded and at 5 AM again made Sail upon a wind. Throughout the morning it blew strong from NWbN with a confused Sea & fair Weather. We pass'd great quantities of Sea grass and were visited by a variety of Land birds & Large flocks of Geese came over us from the NW going to the South. At Noon we discern'd the Land bearing NWbN thro' the haze tho' very faintly & by our seeing it much inclin'd to the South from its former western direction[3]. Our Latitude making 41°44′30″N which is 24 Miles South of our estimation yesterday. Lat*itude* Acc*oun*t 41°52′N, Long*itude* d*itt*o 130° 36′E [*in margin*: No 45 130°9′15″E′]. Several Whales playing about the Vessel. [*in margin of* OD 177: Variation 5°40′, 5°30′ West]

Wednesday 4 October[4]

Fresh breezes and very hazey round the Horizon, the wind at NNW and North. Soon after noon we saw very Mountainous Land to the Westward.[5] At 6 PM we had run in 8 Leagues

[1] The south coast of Russian Siberia in the vicinity of present day Nakhodka.

[2] The schooner was now to the south of present day Vladivostok.

[3] From Mys Gamova the coast trends to the SW and from the mouth of the Tumen River to SSW.

[4] In Broughton, *Voyage*, pp. 319–21, the entries for 4 and 5 October have been transposed.

[5] The mountain range of Hamgyong-sanmaek, culminating in the 2,541-metre-high Kwanmo-bong.

& ⅔ and our distance from the nearest Land was 5 Leag*ues*, the extremes extending from S30°W to N10°E. Between 2 high points bearing from N35°W[1] to S58°W[2] the Land fell Back some little distance[3] and within the 58° point there seem'd to be the appearance of a small opening in the low Land.[4] All this part was bounded by lofty Mountains equal in height or more elevated than Formosa. The Night prevented our making any further remarks. We therfore alter'd our course to the Southward in the suppos'd direction of the Coast. We had fine Moonlight & the weather very mild & pleasant but no Sound*ings*. At 2 PM we haul'd up SW to approach the Land which at 6 AM bore from N10°W[5] to S80°W, a Rocky point[6] N87°W 2 or 3 Leag*ues*. To the South of this point the Shore receded very considerably & presented a very barren surface. The projecting Rocky Land rose very abruptly from the Sea in horizontal Strata of a red & white colour, the higher parts were wooded in Patches. We saw no signs of Inhabitants. At Noon we passed a White Rock much broken into craggy points.[7] It is about 3 Leag*ues* from the Land & bore of us N43°W about 5 or 6 Miles. The Extremes from S70°W to N10°E offshore 5 Leag*ues*, no Soundings. Lat*itude* Observ'd 40°32′36″N, account 40°45′N, Long*itu*de 129°47′W. [*in margin*: [No] 45 129°6′E], Course S28W, 82 Miles. [*in* OD 177: Near this situation the Charts place the I*sland*s of M[o],[8] but we did not see them.] [*in margin of* OD 177: Variation 3°51′, 2°55′, 3°10′ West]

Thursday 5 October

Light Winds & fine weather but uncommonly Hazey to 15° above the Horizon which prevented our observing any objects distinctly on the Land. We steer'd SW as the Coast seem'd to incline to the westw*ar*d and the wind kept gradually veering to the SE. The white Rock appear'd much broken & we saw several small ones detach'd from it.[9] By 6 PM the breeze had veered to SW when we tack'd, the Rocky head set at Noon[10] bearing N30°E and the Southern extreme S75°W. The Land seem'd to form an Opening in a Direction of N15°W.[11] The South point forming double Land was 4 or 5 Leagues distant. We cou'd observe at Sunset very high Mountains still bounded the Coast. The Night was clear & favourable for Lunar Observations which we embrac'd by taking many Distances both East & West of the Moon. At 6 AM the Land was visible thro' the Haze N30°W 12 or 14 L*eagues*, Observed Latitude was 39°40′36″N, Account 39°45′, Long*itude* acc*oun*t 129°48′E, [*in margin*: No 45 129°12′E], course South, 53 miles. Smooth Water & pleasant weather.

[1] Komalsan Dan, situated at the south end of a small promontory, which rises to the 183-metre-high Komal San.

[2] Ŏrang Dan, a rocky treeless point with a pointed summit 119 metres high.

[3] Kyŏngsŏng Man.

[4] Myonggan Chon enters the sea 3½ miles NW of Ŏrang Dan.

[5] 'S10°W' in Broughton, *Voyage*, p. 320, which may be a transcription error.

[6] Musu Dan; a rock 17 metres high stands south of the point.

[7] Nan Do, a white barren rocky islet, situated 9 miles offshore and 14 miles WSW of Musu Dan.

[8] Possibly the island of Mayang Do in 40°00′N; I. Mo is charted in 39°30′N on Roberts's 'A General Chart'.

[9] Several rocks fringe Nan Do.

[10] Musu Dan.

[11] Immyŏng Hae; the land on each side of the entrance to the bay is high. Sŏngjin Hang, one of the most important ports on the NE coast of Korea, is situated in the SW of the bay.

Friday 6 October

Fresh breezes & fair but the same haze still continued. At Noon we stood to the westward with the wind at SSW. At 4 we saw the High Land to the NW. At 6 PM we saw some Flat Land making like two Islands. They bore SW just rising the deck. The extremes of Corea from N*orth* to N75°W 7 or 8 Leag*ues*. At 7 PM we stood off till midnight when we again tack'd. At 8 AM we discern'd the High land of Corea and at Noon we discovered the Islands bearing from S26°W to S36°W and the Coast from N10°W to N40° west 7 Leag*ues*. Latitude observ'd 39°39′15″N, account 39°45′N, Long*itude* 128°46′E, [*in margin*: [No] 45 128°18′E], Course west, 47 Miles. [*in margin of* OD 177: Variation 3°51′, 2°55′, 3°10′ west]

Saturday 7 October

Fresh breezes and very hazey weather from the Southward. As we approach'd the Shore we found the Southern Land which we had before conjectur'd to be Islands to be the continuation of the Corean coast and making the South point of an extensive Bay[1] which we were standing into. At Sun setting we tack'd in the opening of the Bay, the extremes of Land bearing from South to N22°E. We had no Sound*ing*s [with] 50 f*athoms*. The Lofty Mountains seem'd to terminate on the North side of the Bay and in a NW direction from us the Bay receded considerably leaving a small Bay within the great one.[2] The Southern Land rose in Hummocks to a tolerable elevation & our distance from the Head of the Bay we estimated at 7 or 8 Leagues. The wind remain'd steady at South till the morning when it shifted to West and at Noon it was nearly calm with fair & pleasant weather. At 6 AM the Land bore from S25°W to 60° west and we kept our wind in that direction. Observ'd Latitude 38°55′36″N, account 38°56′N, Long*itu*de 128°49′E, [*in margin*: [No] 45 128°20'E], Course S3°E, 43 miles, when the extremes bore from S12°W to N82°W, the Saddle Hill[3] S67°W 5 or 6 Leag*ues*, which is the same Land we set last Night bearing South of our situation.

Sunday 8 October

Light airs and calms with cloudy weather, a Southerly wind & very smooth water. Our situation nearly the same as yesterday, Saddle Hill bearing S82°W 6 Leagues and a Peaked Hill[4] which bore yesterday noon S27°W now bore S43°W. Muster'd the Ships Company and read the articles of War[5] to them. Course S14°E, 7 miles, Latitude account 38°48′N, Long*itude* account 128°51′ East.

Monday 9 October

Light breezes from the South with dark gloomy weather. We kept upon a Wind to the SE and the Peaked Hill at 6 PM bore S48°W 6 or 8 Leag*ues*. Towards Night the wind gradually

[1] Tongjosŏn Man, which is usually defined as lying between Suwŏn Dan and Mayang Do, 80 miles north.

[2] Yŏnghŭng Man. The important port of Wŏnsan Hang is situated on the southern side of Yonghŭng Man.

[3] Possibly Kŭmgang San, which has several peaks, the highest 1,633 metres high, situated 12 miles west of Suwŏn Dan.

[4] The 1,715-metre-high Sŏrak San, situated 36½ miles SSE of Suwŏn Dan. At a distance its pointed summit towers above the other mountains.

[5] The disciplinary code of the Royal Navy.

shifted round to the westward, the Weather clear'd up & we had a fair Night which enabled us to haul in for the Land. At 6 AM the Peaked Hill bore N71°W and the South extreme S12°W, 4 Lea*gue*s offshore. The wind had now increas'd to a strong Gale at North and we steer'd South & SbE till Noon keeping 2 or 3 Lea*gues* off the Land, which afforded a most Sterile prospect. The Hills were much broken with Ridges forming cavities in their Sloping sides towards the Sea. We saw no appearance of any Opening. At Noon the extremes from S15°W to N38°W, the South ex*treme* set at 6AM (N51° west) & we were 5 or 6 miles from the Shore. Fine weather with a large Sea. [An] Indiff*eren*t observation made the Latitude 37°13′N, Acc*oun*t 37°18′N, Long*itude* Acc*oun*t 129°11′E, Course S11°E, 96 Miles.

Tuesday 10 October

Strong gales & clear weather with a large Sea ,Wind at North and we steered SbE keeping 5 miles off Shore. At Sunset the Extremes bore from S18°W to N38°W. At 7 PM we hauled our wind for the Night in hopes of the winds continuance to run down the Coast in the morning & to search for a Harbor which by the Japanese Charts shou'd be near our present situation.[1] After Sunset the Gale moderated but the Heavy swell still remain'd in which we were tumbled about most unmercifully the whole night. At daylight we found ourselves 5 or 6 miles from the Land with nearly the same bearings & in 60 f*athoms*, fine sand. Our Soundings indeed had been regular throughout the Night. Light breezes from the Land admitted our steering along Shore till Noon which presented us the same bare prospect with the variation of a Yellowish Earth mixt with the Heath & Grass that cover'd the Hills. At Noon we Observ'd in the Latitude of 36°24″39 N, Account 36°20′N, Long*itude* 129°25′E, [*in margin*: [No] 45 129°11′40″E], Course S14°E, 49 Miles. The South extreme form'd a low point[2] bearing S14°W, within it to the west the coast fell back some distance & gave us reason to suppose the Corean Port might be there plac'd.[3] North extreme N14°W, off shore 7 or 8 Miles, 60 fath*oms,* fine sand.

Wednesday 11 October

Moderate breezes and very dark gloomy weather with a Southerly wind & the continuation of the large swell. We kept our wind to the SE. At 8 PM we had the wind at west which remaind in the same quarter throughout, blowing Fresh at times with Squalls which entirely prevented our approach to the Land as I wish'd, but we had no recourse. At 6 AM we saw the Land bearing from us N65°W to S65°W about 5 or 6 Leagues. The Northern point we suppos'd to be the same as seen last Night[4] (then forming the South extreme). Just before Noon the Sun appear'd which made the Latitude 35°32′34″N, Account 35°3′N, Long*itude* 129°37′E, [*in margin*: [No] 45 129°30′45″], Course S11°E, 53 miles. Both to day & yesterday we seem to have been affected with a Northerly current.[5] The extremes bore from N47°W to S65°W and a Near point forming in white Cliffs[6] bore west 4 or 5 L*eagues*. [*in margin of* OD 177: Variation 3°12′, 2°0′ west].

[1] Pusan.
[2] Changgigot, the NE point of a promontory, which can be identified in clear weather from almost 20 miles.
[3] Yŏng-Il Man is entered west of Changgigot; the important port of P'ohang is situated at the head of the bay.
[4] Changgigot.
[5] The NE-going Tsushima Current, which in the summer months reaches a maximum rate of about 1 knot.
[6] Probably Hwaam Ch'u, the southern point of a peninsula forming the eastern side of Ulsan Man, but the peninsula now appears ochre in colour.

Thursday 12 October

Fresh breezes with very Squally weather from the SW quarter, which after Midnight veer'd round by the West to NW and it became moderate with pleasant weather. We kept our wind to the West and before Daylight were in sight of Land which prov'd to be the Island of Tzima[1] laying between the Corean Coast and Nipon. At 6 AM it bore from S44°W to west and a Small Island[2] seen only from the Main rigging SSE.[3] We had 70 fathoms and were about 4 Leagues off the Shore. We stood close in & tack'd in 22 fathoms. We soon discovered the Island to be inhabited by their lighting Fires in every part; we also perceivd 4 Japanese Junks working to the westward under the Land. [*in margin of* OD 177: AM Observ'd Distances of ☉ and ☽] At Noon we Observ'd in the Latitude of 34°41′11″N, Account 34°49′N, Long*itude* d*itto* 129°16′E, [*in margin*: [No] 45 129°10′30″E], Course S19W, 54 miles. The North point of the Island[4] bore N88°W, its distance about 3 or 4 Miles, some High Rocks without it.[5] Round them is an extensive Reef and much broken water. They bore N50°W 4′,[6] Extreme [of Island] S20°W. The Coast of Corea appeard very distinctly bearing from N15°W to 83°W 8 or 10 Leagues. This Island was of moderate elevation at each extreme with some High level Land in the Center of it. Some Trees were scattered over the Hills & we saw some cultivation in the Vallies. The Shore in general was very Rocky & the Sea broke upon it with great violence.

[1] Tsushima.

[2] Okino Shima.

[3] 'S24°E' in Broughton, *Voyage*, p. 326.

[4] Maru Saki, the northern extremity of Kami Shima, the northernmost and larger of the two islands which together form Tsushima.

[5] Mitsu Shima, three islets joined by drying reefs.

[6] 'N75°W, five miles' in Broughton, *Voyage*, p. 327.

CHAPTER 7

Off Tsushima and thence to Pusan

Friday 13 October

Moderate breezes & fair weather with a large Swell from the North. The Wind came from[1] the same quarter which enabled us to weather the Rocks & Reef off the North point of the Island having a strong current setting to Windward. The Island we found to trend in the direction of S60°W for 6 or 8 miles & we had great overfalls in this situation having unequal Soundings & a strong Stream setting to windward which occasion'd a great Sea. After passing thro' the Race (it may be call'd) we had the wind to the eastward of the North which carried us over to the Corean Coast by Sunset, having run in the direction of N41°W 22 miles, when the extremes bore from N20°E to S67°W, an Opening in a NW direction[2] about 3 miles, the Island of Tzima South to S23°E & we had 60 fathoms water. The Night was moderate & we plied under an easy Sail. At Daylight we found a strong current had set us to the ENE and we made Sail to the WSW. As we approach'd the Land we perceiv'd several Villages scatter'd along the Shore which appear'd much broken & likely to afford Shelter altho' its external View presented rather an inhospitable prospect. We also saw some Fishing Boats to Seaward. At Noon we had no Observation but our situation nearly the same as last Evening. We had Cloudy & Rainy weather with an Easterly Wind.

Saturday 14 October

Light breeze and Rainy Weather, Wind NE'ly. Having spoke a fishing Boat with whom by signs we comprehended the NW opening (before mentioned) wou'd admit of Anchorage, We bore up at ½ past Noon. Off the North Point of entrance[3] are some stupendous black Rocks detach'd some distance from the Shore.[4] We pass'd within ½ a Cable of them & had no less water than 30 fathoms, muddy bottom. Within we shoal'd to 15 & 10 & 5 [fathoms] regularly. In the entrance, but more on the South side, is a high Rocky Islot.[5] After passing it you open a Sandy Bay to the North when we observ'd cultivated ground and a large Village. To the Westward were several Scatter'd Rocks above water. We therefore turn'd up towards the Bay and came too an Anchor in 4 f*athoms*, Sandy bott*om*. It rain'd incessantly throughout the Evening which prevented our having any Communication with the Shore. In this situation we had the Sea open from S*outh* to S23°E & the Island of Tzima fill'd up the Angles at 10 Leag*ues*

[1] 'to' in the MS, which appears to be a transcription error; 'from' in Broughton, *Voyage*, p. 327.

[2] The entrance to Pusan Hang.

[3] Sŭngdu Mal.

[4] Oryuk To consisting of four islets, one of which is 68 metres high, 5 cables south of Sŭngdu Mal.

[5] Cho Do, situated one mile north of Sangi Mal, the southern entrance point of Pusan Hang.

d*istant*. In the morning we had fair and pleasant weather & our Decks were soon full of Men, Women & Children whose curiosity induc'd them to come off. They were universally cloath'd in Linnen garments made into Jackets and Trowsers, quilted or doubled & some of them [wore] large loose Gowns. The Women wore a Short petticoat over their Trowsers & both Men & Women linnen Boots with Sandals made of Rice Straw. The Hair of the Men was roll'd up in a Knot to the Crown & the Women had theirs plaited & twisted round their Head. Their Features strongly resemble the Chinese particularly their Small Eyes. After breakfast we went in search of water & soon found a convenient run of an excellent sort. This was a great satisfaction to us as we had been so long at Sea we began to be in want. Wood I was sorry to observe appear'd a scarce Article. After taking some Altitudes for the Watch & Observing distances for the Longitude we took a Walk round the Village attended by a numerous party of Inhabitants. We perceiv'd the Harbor to extend to the westward beyond the Rocks from whence it diverg'd to the North & South.[1] Many Villages were scatterd round it, and to the North a large Town seemingly surrounded by a Wall with Battlements. Some Junks lay near it in a Bason surrounded partly by a Mole. Another Mole appear'd to the West of a Similar Form near some White Houses of a Superior construction Enclosed by a thick Wood. All the Villages seem'd to abound in People and the Harbor equally so by the infinity of Boats sailing about. They were similar in figure tho' inferior in Workmanship to the Chinese Boats. They made use of Skulls & Matted Sails. In our walk we remark'd several graves which the Natives pointed out & explain'd to us. They were plac'd in an East & west direction & the Ground elevated over them some height. Trees were planted in a half Circle round some of them & all distinguished by Stone Work. In the afternoon we were visited by some Superior People who were dress'd much better & treated with great deference by the common Sort. They all wore large black Hats with high Crowns & the Brims in Diameter were 3 feet. They tied [them] under the Chin & were manufactured with a strong Gauze not unlike Horse Hair, very Stiff and Strong making good Umbrellas. Each Person carried a Fan with perfume in a fillagree Box attach'd to it, also a Knife handsomely mounted in a Sheath, fasten'd round their Middles. A Boy attended each of them who had charge of their Tobacco Pipes & kept their Dresses Smooth. Most of them wore their Beards long. Their enquiries were to know whence we came. After Dinner we endeavour'd to explain who we were and what we came for, but I fear we comprehended each other very indifferently. They however seem'd satisfied with their entertainment & took their Leave. Afterwards we went on shore to ascend the High Land & take some Bearings. Our View was very extensive but our Angles were useless as we found the Compass strongly attracted by some power in the Mountain which wou'd not let the needle point True, varying eight points from the North & pointing East instead.[2] This Land was Rocky on the Summit but the sides produced coarse Grass on which some Cattle were feeding & in the lower parts were some Paddy Fields. On our return on board we found the Schooner crowded with People nor cou'd we get rid of them till near Dark and then with great difficulty, using almost violence to get them into their Boats. Soon after dark the same Boats having at least 200 people in them wanted to come on board, which not being admitted they came to an Anchor close to us. As we were unacquainted with their intentions we prepared for the worst

[1] Pusan has now been extensively reclaimed. Broughton was unable to ascertain that the coast to the west of the harbour was the 4-mile-long island of Yŏng Do, which separates the north harbour, where he anchored, from the smaller south harbour.

[2] 'A local magnetic anomaly of 10° was reported in 1950 in Pusan Outer Harbour'; *South and East Coasts of Korea, East Coast of Siberia and Sea of Okhotsk Pilot*, 7th edition, Taunton, 2005, p. 165.

but a Boat with some lights coming to them from the Shore, after conversing some time they left us to ourselves. [*in margin*: [No] 45 128°40′33″ at ⚓ in the Harbor of T'shosan[1]]

Sunday 15 October

Fresh breezes and very pleasant weather from the NW quarter. Cool Nights & rather Warm in the Day. We had no Boats come near us till after breakfast when two came with Company Superior to any we had yet seen. Their Habits were made of the same form with the rest only of a finer kind. The Outer Garment of light Blue Gauze or Tiffany & under their Chins, as tying their Hats, with a bow over their right Ear was a String of Beads, some were of Agate others of Amber and one of their Hats tip't with Silver in different parts. It now occur'd to me these three Great Men were those whom we refus'd admittance to last evening, while our suspicions led us to suppose they had some other View to Gratify by coming so late. Their attendants and those in Office around them shew'd the most humble respect, always speaking to them & answering questions in a stooping posture looking on the ground. They had in each Boat some Soldiers who carried Spontoons[2] or Spears with Flags of Blue Sattin, the field & their Arms in yellow Characters. Their Hats were decorated with Peacocks Feathers & with Bamboo Sticks they kept the Boat Men & Populace on shore in Order by making great use of their Power. One of the principal Men who had been on board before, bro't us a present of Rice, Salt fish & Seaweed (Kombu[3] or Fucus sacharinus). After many enquiries respecting us these Great Men express'd a strong desire that we should Sail & having informed them of our wanting Wood & Water they promised to send us some by way of hastening our departure. I also requested they wou'd send us a Bullock as we saw them grazing on Shore which they did not seem inclined to comply with and as Money appear'd of no value we could not procure any, having no other means to induce them. The arrival of these Great Men prevented our having any Visitors on board much to our satisfaction,[4] although we fear'd our Walks on Shore would be equally prevented, which the event prov'd. It did not however prevent our taking the necessary Astronomical Observations we required. Yet the Assemblage of People was always so great as to much impede our operations. In the Afternoon they sent us some water in Jars & Tubs.

Monday 16 October

Fresh breezes & very Squally from the NE quarter with cold and cloudy weather. As the violence of the wind permitted, we were employ'd watering and they also sent us off both wood and water. Each Boat bro't us many Visitors who departed as they delivered their Cargoes. We had very little communication with the Shore owing to the blowing weather.

Tuesday 17 October

The Wind still remain'd in the same quarter but more moderate. In the afternoon we had some deputies from the Great Men to know if we had wood & water enough & when we meant to Sail. I told them in three days but they seem'd very desirous we shou'd make it two

[1] Pusan.
[2] A kind of halberd formerly used by some British infantry officers.
[3] See p. 75, n. 3.
[4] 'This interdiction' omitted to make better sense.

which I wou'd not agree to. After taking some Refreshment & having satisfied their curiosity they took leave. Some of them were very handsomely dress'd, the Colors of their outer Garments being chiefly Sea Green, Light Blue or French Grey, Hightly Glaz'd. These Garments were manufactur'd of Cotton but not Fine. At Noon we observ'd the Latitude on shore opposite the Schooner & we had military attendants after us from the Village who staid with us till we return'd to the Boat. The Night was moderate & fair.

Wednesday 18 October

We had the same wind & Weather but cloudy at times and we were supplied throughout the Day with Water; of wood we had already receiv'd a sufficient Quantity. It was all the same sort and what is call'd the Fox tail Pine. Some of these trees were of a large Growth & the cutting & splitting the large Logs gave us a good deal of trouble. In the afternoon we were compleat with water and our Friends were very desirous we shou'd Sail. I endeavour'd to make them understand we meant to remain two days more to observe the Sun. Towards Night the wind moderated & the Heavy Clouds from the NE produced Rain.

Thursday 19 October

Moderate breezes with constant Rain this Forenoon. The wind still continued from the NE quarter. We had some Visitors in the morning to expedite our departure but I assur'd them we cou'd not sail till one day after the Sun appear'd. They made use of parchment as Covers for their large Hats to keep out the Rain & they had Umbrellas of the same. This rainy weather entirely impeded our progress in the Astronomical Way, as we were anxious to have daily Altitudes to ascertain the Rate of our watch for the short time we remained, and we had been disappointed these two days. The Rain continued without intermission throughout the day & Night. In the afternoon I made an excursion up the Harbor to take some Angles, and our friends soon dispatch'd two Boats after us requesting we wou'd return, which the weather indeed oblig'd me [to do] without their Solicitations, but not till I had taken the necessary bearings.

Friday 20 October

The next day we had Cloudy weather & the appearance of the rain having subsided, but no Sun appear'd. We were necessarily employd preparing for Sea & our Friends again came to know when we intended leaving them. Having convers'd with them respecting my excursion yesterday They explain'd to us that if we landed at the White Houses up the Harbor we shou'd be very ill treated if not put to death & beg'd we wou'd not go away in the Boat any more. Soon after they Landed (as if they still suspected our intention) they sent off four Boats with a Flag & Soldiers in each who anchor'd on each side of us as I would not permit their coming on board. In the Afternoon they left us to ourselves & we only land'd to fill some Water to which they had no objections. This Night was fair & moderate the Clouds gradually dispersing leaving a clear Atmosphere. The wind from the North still continued & a swell came in from the Sea.

Saturday 21 October

Fresh breezes and very clear pleasant weather from the same quarter. Before day light I left the Vessel to compleat a Sketch of the Harbor [see Fig. 9], in which I succeeded as they did not perceive the Boat go away, but finding her absent after daylight threw them into confusion. Fires were lighted upon the Hill & Boats sent in pursuit of us up the Harbor. We however return'd at ½ past 7 AM not having seen any thing of them. The morning being very favorable we took our last Altitudes & hoisted in the Boat. At Breakfast we receiv'd a Visit from one of their Principal Men, who was highly pleas'd at seeing our preparations for Sailing. I presented him with a Telescope and a Pistol, the only two Articles he seem'd desirous of possessing & we parted much pleas'd with each other. We soon after got under way & Made sail out of the Harbor. The entrance of this port is situated in the SE part of the coast of Corea. The South point lies in the Latitude of 35°2′N and 129°7′ East Longitude.

It has a very safe Entrance & no Danger to apprehend. You may keep the Black Rocks close on Board in going in or coming out or indeed border[1] as near as you please on either shore. After passing an abrupt & high head Land 2 Miles West of the Rocks you open a Bay to the North w[h]ere we Anchor'd & remained During our stay. Water is here procured with great convenience which the Chart[2] will explain. Without any pretensions to great accuracy it will answer every purpose to the Navigator and is the best our time & the restrictions we were under admitted us to Make; to that I refer for further particulars. We had very little opportunity to Make any proper remarks upon the People inhabiting this part of Corea, although it is most probable they are the same in their Manners & customs in every other part. We saw several large Villages scatter'd about the Harbor all of them very Popul[o]us. They were generally seated in pleasant situations with Trees interspers'd among them. The Houses are small, all of one Story & thatch'd. After the first Day they gave us no opportunity of seeing them internally & we unfortunately neglected it when in our Power. The Country surrounding the Harbor is much insulated Rising in parts to very high Hills that are bare of Wood & Verdure, but in detach'd places were some scatter'd Pine Trees. The Southern side of the Harbor (which is a Peninsula) is pretty well cover'd with Wood and amongst the Pines were several other Deciderous [*sic*] Trees but of what kind we were unacquainted. We have to regret their Jealousy preventing our acquiring more knowledge respecting the Productions of this Country but in no Instance would they admit our researches. They were well acquainted with fire arms & Great Guns and yet During our Stay we saw no signs of any Offensive weapons whatever amongst them. Nor did they seem any way apprehensive of the small power we possess'd. They only seem'd desirous we should Depart as soon as possible but pressing us at the time in a Civil manner. Indeed this treatment we have been universally accustomed to both amongst the Lieuchieux Isles & Matzmay. It should seem by their behaviour they are by no means desirous of Cultivating an Acquaintance with Strangers and yet many Articles we had very much excited their attention & Curiosity particularly our Woollen Cloathing. As a Commercial People one would suppose they were conversant in Barter but they did not appear anxious to make any exchanges whatever. They have Horses & Black Cattle, Hogs & Poultry. On them I think they set a great Value, as we could not prevail upon them to Part with any. Money indeed they had no Idea of, at least our Coins & Spanish Dollars. Yet they fully Understood the Value of Gold & Silver of which they make use in the workmanship of the Knives &c. Their Grounds were cultivated in the Japaneese Manner in Ridges between the Hills & to their Rice Grounds by Drains they easily convey Water. [*in margin*: Variation 2°48′, 1°8′, 1°11′, 3°34′ west][3]

[1] Possibly a transcription error for 'borrow'.

[2] Fig. 9 on p. 189.

[3] The preceding paragraph is taken from UKHO, OD 177.

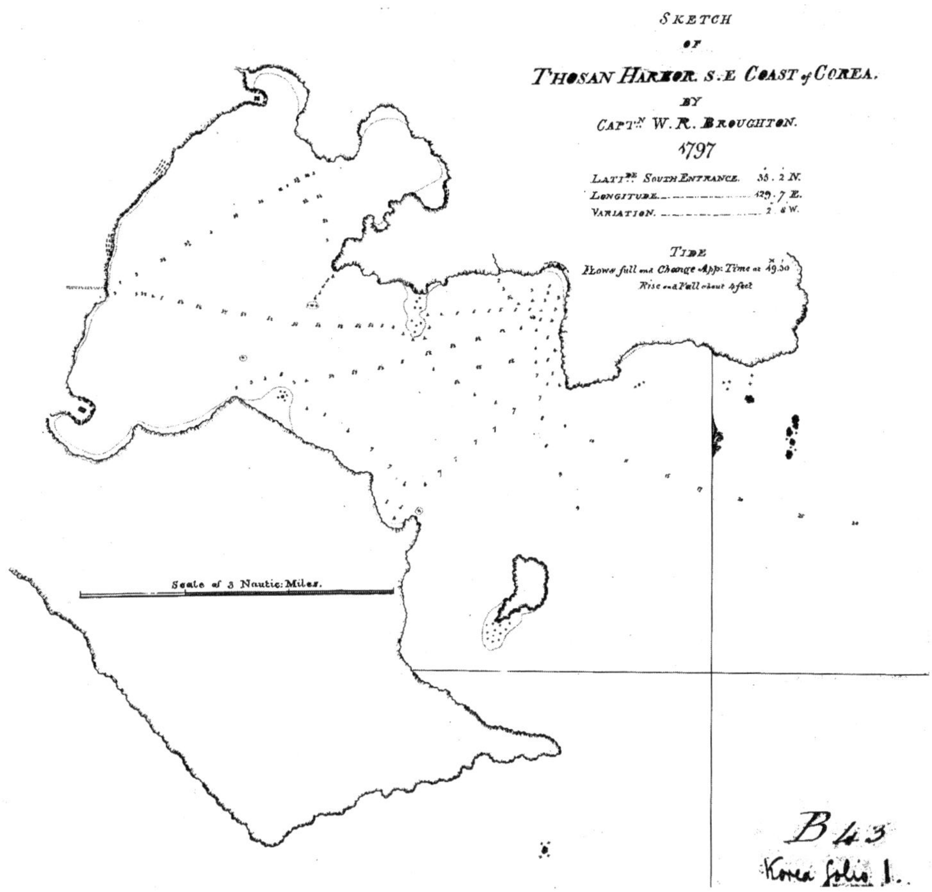

Figure 9. Sketch of Thosan Harbor, S.E. Coast of Corea by Capt[n] W. R. Broughton, 1797. Courtesy of the UK Hydrographic Office.

CHAPTER 8

From Pusan through the Korean Archipelago to Cheju Do and thence to the east coast of China, the south-west coast of Kyushu and thence to Macau and eventual return to England

Saturday 21 October[1]

Fresh breezes & clear pleasant weather. At 9 AM we got under way & made Sail out of the Bay. After passing the Entry points we steer'd SW which course carried us 2 or 3 miles without the Rocks which are scatter'd along shore. These Rocks render'd the approach to the Land intricate from some smaller ones being detach'd from them. They were High & much broken & were at least 2 or 3 Leag*ue*s off the Coast. From the South entrance of the Harbor[2] the Coast falls back to the West & is much indented. I think it probable that some Shelter is to be found were that part more minutely examin'd. As we stood along shore we saw a Junk that appear'd to have come out from there. She stood over for Tzima.[3] At Noon a good observation made the Lat*itude* 34°56′21″N when the South Entry point of the Harbor bore N22°E 8 Miles, Extremes of Corea S60°W, Islands without this Ex*treme* and the Island of Tzima S24°E to 50°E 8 or 10 Leagues.

Sunday 22 October

Fresh breezes & fine clear weather from the NE quarter. We steer'd SW till ½ past 2 PM when seeing some Rocky Islands & Ledges[4] in that direction we steer'd more Southerly to pass without them and the Coast. At 3h the extreme of Corea bore west and we cou'd distinguish its direction no longer for a Cluster of Islands in that quarter with many Rocks between them, some just rising above the water. At 5 PM Tzima bore from S40° to 60°E, a high Rock[5] bore of us N49°W in one with the Island set at Noon[6] without the extreme and we hauled round it WbS. At Sunset the Extremes of Corea N16°E and the Extreme of Rocky Islands N76°W 3 Leagues and we had 40 f*athoms*. muddy bot*tom*. It blew Strong with a large Sea during the Night and we plied in Soundings from 40 to 55 fath*om*s till day light, when the Outer rock we

[1] When Broughton anchored in Pusan he changed from keeping nautical time, which begins and ends at noon, to civil time, which begins and ends at midnight. He reverted to nautical time on sailing at 9 am on 20 October, which was the same date by both civil and nautical reckoning, but with only three hours to run before it became 21 October by nautical reckoning.

[2] Sangi Mal.

[3] Tsushima.

[4] Namhyŏngje Do, which consists of three above-water rocks.

[5] Hong Do, a rugged island and the SE island of the Korean Archipelago.

[6] This island and many of the features mentioned by Broughton in the Korean Archipelago are difficult to identify with any certainty.

had pass'd last evening bore N45°E 4 or 5 miles. The weather was so hazey we cou'd not see more than two Leagues around us and we haul'd up WNW which soon bro't the Rocky Islands in sight altho' very imperfect to our view. At 8ʰ 30′ we had 28 f*atho*ms & soon after we saw a high bare Rock to the west.[1] We pass'd one mile to the South of it in 23 f*atho*ms & before we lost sight of it we saw another[2] which oblig'd us to steer more Southerly. We also perceived extensive Islands to the westward. We sounded every half hour from 23 to 27 f*atho*ms, muddy bottom. At Noon we observ'd in 34°25′N indifferent Lat*itude* when the extensive Islands bore from N44°W to N88°W. The Rock before mention'd N76° west 2 miles & we cou'd see not anything more for the Haze. Course from T'shosan Entrance S53 west, 64 Miles, Latitude account 34°20′, Long*itude* d*itt*o 127°5′E & we had a large Sea.

Monday 23 October

Fresh breezes & uncommon hazey weather round the Horizon, Wind ESE. We steer'd WSW one hour when we discover'd a Rock bearing South[3] which convinc'd me we were surrounded by them. The thick Weather preventd our making any proper remarks for fixing their situation and our Navigation became rather intricate particularly in the Night. We haul'd up NW under shelter of the extensive Islands[4] shoaling our water from 25 to 10 f*atho*ms within two miles of the Shore. These Islands were inhabited & well cultivated. We also saw several Boats a fishing. They were high & in parts well wooded and divided from each other by very narrow passages. Some Ledges of rocks just above water were scatter'd at unequal distances from the land. At Sun set the Ex*treme* of these Islands set at Noon West[5] now bore S48°E & very high Land S50°W[6] and the woody point of the 4th Island having a passage[7] clear between it & more Islands N50°E 2 miles off Shore. We kept plying off this Point[8] during the Night and at daylight it bore N26°W 2 or 3 miles. The same Haze still continued and we stood to the NW till 8AM when we cou'd no longer proceed for Rocks & Islands. We therefore steer'd to the South to pass without the High Land set last Evening[9] S50°W which carried us very near two dangerous rocks lying at some dist*ance* from each other and just above water,[10] but we had no less than 7 f*athoms*. Two large Rocks lay to the SW of this Land.[11] We pass'd between them in 13 f*athom*[12] and to Seaward we cou'd discern thro' the Haze High Rocks surrounding us on all sides & we steer'd more westerly. At 11 AM we lost sight of the high Land[13] N70°E 6 or 7′ & discover'd more Islands ahead; our Soundings had shoal'd to 9 f*atho*ms. At Noon we had Moderate Weather & a bad Obser*vatio*n made our Latitude 34°23′21″. We appear'd to be completely environ'd by Islands, the Clearest passage between them open from N56°W to

[1] Ko Am, a black rock 21 metres in height.
[2] Sejon Do, which has two curiously shaped rocky peaks.
[3] Chak To.
[4] Sori Do
[5] Sori Do.
[6] Oenarodo, the southern island of Naro Yŏlto, which is 393 metres high.
[7] Possibly the passage between northern tip of Sori Do and Kŭmo Yŏlto.
[8] Soryongdan, the southern extremity of Sori Do.
[9] I.e. Oenarodo.
[10] T'anggŏn Yŏ, a conical rock 40 metres high and a smaller rock 3½ metres high close off its south side.
[11] Koktu Yŏ, whose SE islet is 87 metres high.
[12] I.e. between T'anggŏn Yŏ and Koktu Yŏ.
[13] Oenarodo.

N33°W for which we haul'd up with a light wind from the NE. Course S82W,18 miles, Lat*itude* acc*oun*t 34°33′N, Long*itude* d*itt*o 126°44′E, Current of 8′ we found by Latitude this day to the South.

Tuesday 24 October

Light breezes from the NE with the same thick weather. We kept our wind up the passage to the NNW and at 3PM finding a current against us we came too an anchor in 10 fa*thoms,* muddy bottom. This stream afterwards appear'd to be a regular tide setting thro' the Islands from the NW about 1½ miles p*er* Hour. The extremes forming the passage we had come in [through][1] bore from S5°E to S60°E, the latter on with two Islands 2 or 3 miles. Within them were Villages & Cultivation & many Boats passing to & fro. On every other point we were surrounded by Isl*and*s but an opening to Sea appear'd SW of us at least for 2 Leagues, further we cou'd not see. The Night was moderate and Cloudy and in the morning we had a Fresh breeze from the NNW. At 7 AM the Tide having made we got under way & turn'd up to the NW Keeping the Land North of us close on board. The Channel between the High Island south of us[2] and the North shore[3] was about 2 miles wide, Much indented by Bays & many scatter'd Rocky Islands laying in the fair way. We generally carried deep water near them varying from 7 to 15 f*athom*s. The Shores on each side afforded a pleasing view; Many Spots being richly cultivated & interspers'd by Villages with Boats sailing in every direction. Before Noon we came too an anchor in 10 f*athom*s, mud. The Islands which bore from us at our last ⚓ NWbN now bore S74°E 2 or 3 Leag*ue*s and our present situation was between Islands so clos'd with each other as to leave no opening whatever that we could observe might lead to Sea. The Master went to the nearest [island][4] to observe the Latitude and to look out for a passage. He made the Latitude by a quick Silver Horizon[5] 34°30′24″N but the hazy weather prevented his seeing even the Islands at 3 Leagues distance, which prevented our forming any conclusion of their extent to the SW. Lat*itude* account 34°35′N, Longitude d*itt*o 126°25′E. Very mild & pleasant Weather.

Wednesday 25 October

Fresh breezes & hazy weather the wind from the NNW with a clear Atmosphere & clearer round the Horizon. Towards evening it became much more so and we cou'd see 5 or 6 Leagues when the Sun set. After anchoraging we were soon visited from the large Village with questions put to us in written Characters which we were unable to answer. About an hour after a very great Man attended by several Soldiers & other attendants came on board. This Boat was quite gay, the military being well dress'd carrying their silken Flags and one of Red & Purple much larger in the bow. Their motions were accompanied by Trumpets & every Man was arm'd with a Saber. This man of consequence was seated on a handsome Mat with Cushions to rest upon & Pillows. The cushion was cover'd with Leopards Skin. He ask'd many questions but we understood each other little. He was seated upon the Deck his Suite keeping a Most respectful distance & I set down with him upon his Leopard Skin. His principal enquiries however were to know our Numbers and to have them counted before him that he might be

[1] Kŏgŭm Sudo, between the island of Kŏaŭmdo and Kohŭng Pando, part of the Korean mainland to the NE.

[2] Kŏaŭmdo.

[3] Kohung Pando.

[4] Probably Sorokdo, a small island off the SW extremity of Kohŭng Pando.

[5] An instrument better known as an artificial horizon, which can only be set up on shore.

convinced. This was inadmissible on my part which seem'd to give some offence. On my telling him we shou'd stay a day or two he beg'd us to stay several & was desirous we shou'd send our Boat on shore. The same request he made on going away and my objecting to it he seem'd much surprizd. After leaving us we remarkd he went no further than the Island point having sent two Boats away with Messages & leaving two near us at Anchor with his interpreter who was much displeas'd at not being sufferd to count the people. At 3 PM the Haze having clear'd partly we got under way & stood between the Islands to the Westward. We soon observ'd the Great Mans Boat to follow us Hollowing and sounding the trumpets for us to stop, to which we paid no attention. We soon after perceiv'd he landed with the other Boats on the point of the Island we pass'd while we pursued our course with a strong Breeze. What this Man's intentions were I cannot positively say, but to me they appear'd suspicious and I did not wish to see the result of them at the expense of Losing the clear weather. At 4 PM we pass'd an Opening to Sea among rocks and Islands laying off the west end of [a] high Island[1] and at Sunset we had run nearly 12 miles in a WSW direction. The Land to the NNE appear'd very high & broken & connected by Ranges of Hills [to the] back of each other which made us conclude it to be the Continent.[2] At 6 we open'd another opening to Sea & it being dark we came too in 7 f*atho*ms, Muddy Bottom. Our Soundings in the large space we had run over were from 17 to 7 fathoms according to our distance from the Shore. The Island on which we observ'd the Lat*itude* bore N67°E and a deep entrance we conjectur'd to lead to the River NNE.[3] We found the flood Tide to run much stronger than the Ebb & it rose nearly 9 feet by the Lead Line.[4] The morning was moderate and hazy but we cou'd see 4 Leag*ue*s & the Breeze [being] Northerly we therefore weigh'd & stood to the SSW between some Islands. At ½ Past 9 we were close in with what I conceiv'd the continental point from whence it trends to the West but an infinity of small Isles with little Wind preventing our pursuing that Course. At Noon it bore from us N27°E and a High Hummock Isle[5] west 1 mile. We had the Sea open to the SSW & SSE but in every other direction Islands. I counted thirty thro' the Haze. On several of the Large ones were extensive Villages & every accessible part cultivated. Boats were a fishing around us & the Sea was very Smooth having scarcely any Wind. Lat*itude* by land Horizon 34°17′N,[6] Lat*itude* acc*ount* 34°15′N, Long*itude* d*itt*o 126°1′E & we had 10 fathoms.

Thursday 26 October

Light airs and calms carried us in a SW direction which drifted us close in with the Hummock [Island] (or within 30 yards of it). It was very high & we saw some People looking down upon us. I suppose they were there a fishing. We had 12 & 13 fathoms at the same time which gradually deepened as the Tide set us to the SW to 17 and 18 fathoms. We had the wind westerly after dark & we kept plying thro' the SSW opening which seem'd to afford a good passage 4 or 5 miles wide between two large Islands.[7] At 8 PM we came too in 15 fathoms under shelter of the westermost. The Night was moderate and pleasant, wind NW'ly. The current or Tide set

[1] I.e. Kŏaŭmdo.
[2] A correct conclusion.
[3] Actually leading to the head of the important inlet T'ŭngnyang Man.
[4] Clearly by the difference in depths obtained at high and low water by sounding line while at anchor.
[5] Mohwangdo, 115 metres high, situated 2 miles SSE of the SE extremity of Shinjido.
[6] Such an observation would be unreliable unless the horizon to landward was also the true horizon. For instance, if the height of eye of the observer was 20 feet the true horizon would be 5·13 miles.
[7] Ch'ongsando to the east and Taemodo to the west.

to the N'Eastw*ar*d about 1 mile p*er* hour. In the morning we had fine clear weather (for the first time). We saw more Islands & Rocks to the SW and a very extensive Island rising to a great height in the Center bore from S6°W to S25°W.[1] The large Islands forming the passage were well cultivated but no Wood appear'd except round the Villages. Boats we observ'd Fishing but they wou'd not come near us. We got under way & made Sail to the SW deepening our water by degrees to 30 fathoms. By Noon we had just lost sight of the Hummock we pass'd so close to y[esterday] bearing from us N40°E 8 Leagues & about 2′ from a small Flat Island[2] bearing S35E and our observation made the Latitude 33°57′35″N, Lat*itude* acc*oun*t 33°57′, Long*itude* d*itt*o 125°42′E, Course S38°W, 25 miles. The extensive Island which I imagine to be the same the Charts lay Down as Quelpart from S18°E to S25°W 7 or 8 Leagues. We cou'd just discern from the Deck a Rock to the SW.[3] A Group of Rocks[4] also extended from N71°W to S78°W. Sound*ing*s 20 to 30 f*atho*ms. [*in margin*: [No] 45 126°42′0″E]

Friday 27 October

Fresh breezes from the NW quarter which soon after noon brought on the same Hazey weather. We kept our wind to the SW to weather Quelpart and at 4 PM we had pass'd to the NW of the High Rock seen at Noon SW & soon after another[5] nearly the same bearing from it & 2 or 3 miles distant, but we cou'd not see anything of Quelpart for the Haze. At Sunset the two Rocks bore to the Eastw*ar*d of us 5 or 6 miles & the Group to the NNE[6] the same distance nearly. In this situation we plied under an easy Sail till daylight having increas'd our Soundings to 60 f*atho*ms. In the morning the weather having clear'd away partly we saw Quelpart from S5°W to S72°E 4 or 5 L*eague*s. The two Rocks from N70°E to N50°E 3 or 4 miles. During the night we had [been] set towards them 5 or 6 miles more than we expected. The Group of Rocks bore NbE of us and we steer'd South & SSW till Noon, 2 or 3 miles from the Shore, when we were off the west extreme of the Island, the South point[7] bearing S45°E 6 or 7 miles, two Low Islands South of the Point[8] S28°E 8 or 10 miles and the Extreme N27°E about 2 miles off which was a Rocky Island[9] & some Rocks within. The Shores were low & rocky & several patches of white sand we observ'd some distance from the Sea. In many parts were rising Hillocks or Mounts the Sides of the larger ones well cultivated. A Round High Rock upon the Shore[10] bore N85°E & near it the margin of the Coast was cover'd with Habitations & Smoke of Fires in every part. Latitude Obs*erve*d 33°14′10″N, Account 33″23′N, Long*itude* 125°9′ [*in margin*: [No] 45 126°7′E], Course S33°W, 51 m*ile*s.

Saturday 28 October

Fresh breezes and fair weather from the North. At 1 PM the South extreme bore East and we haul'd up for it. After running in 4 miles we observ'd the Flat Island lying South of it to be

[1] Cheju Do.
[2] Changsudo, 72 metres high with a flat wooded summit.
[3] Hwado, which has a distinct flat summit.
[4] Ch'uja Kundo, a group of islands, islets and rocks.
[5] Haeamso, an above-water rock, which is pointed and steep-to, situated 4½ miles SW of Hwado.
[6] Ch'uja Kundo.
[7] Bunamgot.
[8] Marado, 33 metres high, situated 5 miles south of Bunamgot, and the flat island of Kap'a Do, situated midway between Marado and the point.
[9] Chagwido, a grassy island 61 metres high.
[10] Kosan Ak, a rocky peak 148 metres high, standing at the western end of Cheju Do.

connected with it seemingly by Reefs. We also saw breakers to the South of us at no great distance but we had 45 *f*ath*o*ms water, corally bottom. We therefore steer'd out as we came in till having lost sight of them we haul'd up to the SE and at 6 PM the Outer Island bore NE 3 or 4 miles & we haul'd off for the Night plying under easy Sail. We had now entirely lost sight of the Rocky Islands and our Soundings were from 50 to 70 fathoms, Sand & Shells. At 8 AM we were in our former station of last evening and we steered EbN. The Island taking a direction to the ENE we pass'd 2 miles S*outh* of the Reef Isl*an*d[1] in 50 fathoms. The Reef & South Isl*an*d extend between this & the South extreme point in a Northerly direction. At Noon an indifferent Obs*ervatio*n made 33°5′N when the Extreme South of Quelpart bore N55W, Reef Isl*an*d N81°W 3 Leagues, Center of Peaked Hill[2] N12°E and the other extreme N55°E off shore 4 or 5 miles. This Islands greatest extent is in an ENE & WSW direction and the Peak which rises gradually from each extreme is nearly in the Center. By its appearance I imagine it to have been Volcanic & the Rocky shores of the Island seem to confirm it being of a black scorch'd color and in several parts are Rocks of great magnitude & uncommon Form thrown up on the Land, the effect seemingly of some great convulsion. We saw no opening or any Boats or Vessels, but the anchorage though in deep water on the South side was Tolerably Good. The South point lies in the Lat*itude* of 33°11′N & East Longitude 126°20′.[3] [*in margin*: [No] 45 126°22′35″]

Sunday 29 October

Moderate breezes and Hazey weather from the westward. We steerd SE½E 10′ from Noon when the East extreme bore N30°E 5 or 6 L*eague*s & we saw no Islands in any direction. Having satisfied myself in that respect I was now desirous of again making the Corean Coast & we haul'd our wind to the westward with that view. During the time the Moon set & daylight we plied under easy Sail & our Sounding varied from 60 to 56 fathoms, muddy bottom. It was very hazey in the Forenoon which prevented our seeing the Island. We perform'd divine Service & mustered the people &c. Indifferent Lat*itude* made 32°46′N, account 32°43′N, Long*itude* d*itt*o 124°14′E, Course S39°W, 26 miles, Smooth water & wind at SW fluctuating.

Monday 30 October

Light breezes and very cloudy weather till 4 PM when it veer'd round by the west to NW and continued during the Night and forenoon very clear pleasant weather. We tack'd & stood to the westward. Our Soundings had regularly decreasd to 40 fathoms at Noon & we pass'd many large Blubbers[4] or [*blank*] & quantities of Sea weed. At Noon fine clear pleasant weather. Lat*itude* Obs*erve*d 32°31′26″, Account 32°30′, Long*itude* [ditto] 124°27′E, [*in margin*: [No] 45 125°24′30″], Course S69°W, 42 miles. [*in margin of* OD 177: Variation 3°12′, 2°14′ west]

Tuesday 31 October

Moderate breezes & Clear, the afternoon, the Night and morning was very hazey with the wind at NW. We kept our Wind to the westward & at 8 PM we had 25 fathoms which

[1] I.e. Marado.

[2] The 1,924-metre-high Hallasan, situated in the centre of Cheju Do.

[3] The 20′ added in pencil.

[4] A sailor's term for jellyfish or medusa.

continued pretty nearly the same till the Forenoon when we had only 23 fathoms, some sand mix'd with mud. The Sea appeared very white thro' the Haze & we had fair weather. Observed Lat*itude* 32°2′27″N, Acc*oun*t 32°9′N, Long*itu*de [ditto] 123°38′E. [*in margin*: [No] 45 124°37′E] Course S55°W, 50 miles. In the course of the day we saw several Land Birds, a Quail & Dove and among the Number, the former was taken.

Wednesday 1 November

Fresh breezes and very hazey weather. Having run 4 Leagues to 4 PM we had every hour beginning from Noon 22. 20. 20. 21 fathoms, fine black sand, when the water lost its white appearance & became more green. After running two leag*ues* more we had 28 f*atho*ms, sand & mud & our Soundings till Noon never varied to less than 24 and at Noon we had 25, fine grey Sand. The Night was moderate and after midnight the wind was more Northerly and bro't with it clear weather. Latitude only Observation 31°41′33″N, Account 31°35′N, Long*itude* 122°56′, [*in margin*: [No] 45 123°50′E], Course S59°W, 42′.

Thursday 2 November

Light breezes and very clear pleasant weather till midnight when we had a Calm till Noon, which brought the wind from the southern quarter. Upon examining our provisions we found we had not Spirits, Bread, Pease or Callivances[1] to serve the Month. After having well aired the Bread we had the satisfaction to find what remaind was very good. Our Soundings varied from 25 to 19 fathoms, Sandy bottom. Course N80°W, 32 miles, Latitude observed 31°41′33″N, Acc*ount* 31°35′N, Long*itude* d*itt*o 122°19′E [*in margin*: [No] 45 123°17′].

Friday 3 November

Moderate breeze and hazey weather with the wind at South & SW. We stood in for the Coast of China. At midnight we had shoal'd our water to 11 fathoms when we stood to the East under easy Sail. Although we stood out (as we supposd) in the same direction we came in, we had only 8 & 10 fathoms for 8 miles. I rather imagine we must have been affected by a Tide or current that set us differently to our estimation. At daylight we had 15 fathoms & we again made Sail in for the Land. It blew Fresh and we had a confus'd Sea seemingly agitated by a strong Tide. The water was a dirty yellowish color & very thick, which made us conclude we were off the entrance of the Nanking River.[2] As we stood in we perceiv'd several large Junks, some at Anchor and some under way. The Day was fair but hazey round the Horizon. At 10^{h} 45′ we saw Land from the Masthead which on a nearer approach prov'd to be a Ledge of Rocks[3] bearing west of us. At this time we had only 5 fathoms having decreas'd to that in 7 Leag*ue*s run. At ½ past 11 we tack'd having only 4 fathoms, the Ledge S78°W 5 or 6 m*iles*, by which bearing it appear'd the Tide set us to the Northward toward the Junks. From the Mast Head we saw low land to the west & North of the Ledge & we thought we cou'd distinguish the looming of Higher Land but this we

[1] Obsolete form of Calavances, a name for certain varieties of pulse.

[2] The Yangtze or Chang Jiang.

[3] There are no rocks in this position, but drying sandbanks are charted in this vicinity, which is liable to considerable changes.

cou'd not determine. At Noon a good observation made our Latitude 31°58′25″N when the Ledge preserved the same bearing about 8′ just visible from the Deck. This estimation will place it in 31°56′31″N & Long*itude* p*er* Watch 121°44′32″E + 20′, Course N75°W, 43 miles, Lat*itude* acc*oun*t 31°55′N, Long*itude* acc*oun*t 121°29′E [*in margin*: [No] 45 122°13′32″] and we had 4½ f*atho*ms.

Saturday 4 November

Fresh breezes and hazey weather, wind from the SW quarter. We steer'd out East 4½ miles from Noon when having deepen'd our water to 10 fathoms, we bore up NE intending to range the Coast up to the North as our Soundings wou'd admit approach to the Shore. On this course we soon Shoal'd our water to 5 fathoms & we again haul'd out upon a wind & did not deepen our water to 10 fathoms till running 3 Leagues. I suspect this to have been a bank which the tide &c had drifted us upon. We remark'd a Bamboo Buoy upon it and by which we still perceiv'd the tide set us to the North very fast. All the indications of our being in the Mouth of a River seem'd pretty certain & the Nanking River being laid down nearly in this situation I had no doubt upon the subject, but only regretted the weather being so hazey as to obscure Objects at any distance beyond 3 Leagues. As the Wind continu'd favorable for proceeding to the North after deepening our Water at 7 PM to 15 fathoms, we steer'd in that direction till 9 PM when we haul'd our wind to wait for daylight. The Night was very Hazey & towards morning the Wind veer'd round to the NW with almost foggy weather for we cou'd not see 2 Leag*ue*s about us. At 4 AM we made Sail to the North, our Soundings were 19 fathoms. We pass'd some Junks standing different ways. All of them we had yet seen had three Masts and one seen to day of a very large Size had four. They were not painted like those at Canton. Lat*itude* [obs*erve*d] 32°29′N, Account 32°20′N, Long*itude* 122°35′E [*in margin*: [No] 45 123°21′E], Course N61°E, 62 miles. At daylight we had lost the Yellow water, but the Sea was discolour'd in patches.

Sunday 5 November

Fresh breezes with the wind & weather the same. In the afternoon we bore up & pass'd close to a large Junk. Her after Sails were of Cotton very much worn. She was painted on the Forepart & Stern and on her quarters were some Characters in Black. She appeared to have about 20 Men & to sail very heavy. They had Lee Boards but still she kept a bad wind. At Dark we had deepend our water, which continued increasing gradually as we sounded every Hour and we had a Muddy bottom instead of Sand. The Forenoon was very clear and pleasant and we cou'd see a great distance. The Sea had resumed its natural colour and we had 39 fathoms. Observ'd Latitude 33°24′43″N, Account 33°24′N, Longitude d*itt*o 123°37′E [*in margin*: [No] 45 124°20′E], Course N43°E, 75 miles. During the Night and Morning we had a large swell from the NW. Mustered Ships Company.

Monday 6 November

Moderate breezes and very clear pleasant weather. At Sunset it was Calm which continued throughout the Night. At daylight we saw an Island bearing NE of us 8 or 10 Leagues[1] &

[1] Soheugsan Do, the southern island of Huksan Chedo, 569 metres high with a rounded summit, situated 70 miles NW of Cheju Do.

we had 41 f*athoms*. A Light air sprang up from the SW & we steer'd North but it continued very faint. At Noon we observ'd in 33°40′21″N, Account 33°41′, Longitu*de* d*itto*123°52′E [*in margin*: [No] 45 124°44′E], Course N37°E, 20 miles. The Island bore from N48°E to N53°E 7 or 8 Leagues and we had 40 fathoms, muddy bottom. It appear'd of tolerable elevation.

Tuesday 7 November
Light air from the westw*ar*d throughout the afternoon & we pursued our course to the North. At Sunset the Island bore from N62° to 69°E. After midnight we had dark cloudy weather & the wind suddenly shifted to the NW and blew hard in Squalls with heavy Rain. We stood to the NE under a low Sail and at 7 AM we saw the Island to the eastward. The Gale increas'd and at Noon we had a large Sea with Squally weather. Latitude made 33°57′44″N, Account 33°59′N, Longitude d*itt*o 124°13′E, Course NE, 24 miles, & we had 40 fathoms, Mud. The Island upon a close approach appear'd much higher than we had suppos'd. It was very rocky & barren. Off the N*orth* Extreme (which is the Highest part) are several detach'd Rocks at a little distance, one larger than the others. The island bore from N31°E to 52° E 4 or 5 miles.

Wednesday 8 November
Fresh Gales and Squally weather with a large Sea from the North, Wind from the same quarter & tolerably clear, more so to the North & West, but we cou'd see no more Land in that direction. The Wind totally preventing our progress towards Corea excepting running into our former track. I therefore bore away to the SE and at Sunset we saw Quelpart Island bearing indistinctly S55°E and the Island we had left N24°W 12 Leagues. We Sounded in 44 f*athom*s & steer'd South till Midnight, having the Moon very bright. At 12 PM being South of Quelpart we haul'd up SE and at 6 AM the Center Peak[1] N43°E 8 or 10 Leag*ues*. At Noon our Latitude 32°20′7″N when the same Hill bore N51°E 21 Leagues. Lat*itude* acc*ount* 32°25′N, Long*itude* d*itt*o 125°28′E [*in margin*: [No] 45 126°20′E], Course S35°E, 113 miles. The Wind was more moderate and we had fine clear cold weather. T*hermomete*r at 52°. At Noon 45 fathoms.

Thursday 9 November
Fresh breezes and Squally weather, the wind Variable from NE to NNW. We steer'd SEbE. At Sunset we still saw the High Peak of Quelpart bearing due North 24 Leag*ues* & we had 55 f*atho*ms. Having run 60 miles we haul'd up East & we had very clear & pleasant Weather. Our Soundings at Noon had increas'd to 70 fathoms, Sandy bottom. Lat*itude* observ'd 31°43′13″N, Account 31°44′N, Long*itude* 126°49′ [*in margin*: [No] 45 127°37′E], Course S62°E, 77 miles.

Friday 10 November
Moderate breezes & baffling from the NW quarter. In the Forenoon it again increas'd to a Fresh breeze. In the afternoon we saw Rocky Islands[2] to the ENE which at 5 PM bore

[1] Hallasan.

[2] Danjo Gunto, whose three principal islands are Me Shima, the SW and highest, O Shima 3 miles NNE and Yori Shima between them. Tori Shima, an isolated group of above-water rocks, the largest of which is 15 metres high and is situated 18 miles NW of Me Shima.

N65°E[1] 7 or 8 Leagues & we had 75 fathoms. During the Night we run under an easy Sail & at times bro't to the wind but cou'd not reach the bottom with 78 *fatho*ms of line. At 7 AM the Islands bore from N17°W to N33°W 3 or 4 L*eague*s. They were closely connected (in Number five) but the largest were those to the North & South and the Southern Rock was high & Peaked.[2] At 8 AM we haul'd up NE & at 11 we lost sight of them bearing N80°W to N75°W 8 or 10 L*eague*s. At the same time we saw high Land to the ESE[3] and we alter'd our course to East. At Noon we observ'd in 31°56′ North when the High Land bore from S59°E to S69°E 8 or 9 Lea*gue*s. We also saw distant Land to the Eastward from the Masthead.[4] Fair & pleasant weather & no bottom with 78 fathoms of line. Lat*itude* acc*oun*t 31°52′N, Long*itude* 128°20′E [*in margin*: [No] 45 129°6′E], Course N80E, 78 miles.

Saturday 11 November

Fresh breezes & Squally weather. We steer'd SE with the wind at NW. Soon after Noon the Squalls increas'd in Force and a great deal of rain fell which entirely obscur'd the Horizon, at least so as to prevent our seeing the Land at any distance. At ½ past 3 we discern'd to appearance thro' the Haze a Peaked Rock[5] and we alter'd our Course to SbW. The Rock bore SbE & the wind at the same time veer'd to NNW. At ¼ past 5 the Land set at Noon made like an Island and in this point of view the East side presented some Peaked Hills of considerable elevation.[6] It bore from N60°E to N70°E, partly obscur'd by the weather, but the distant Land to the NE we cou'd not perceive. The Rock which form'd very High & Peaked on the N*orth* extreme of a small Island bore S25°E 4 Lea*gue*s and the Island which I conjectur'd to be Meaxima[7] by Mr Thurnburg['s] description of its lofty & peaked Mountains[8] was about 5 L*eague*s. We had Van Keulen's chart on board[9] but cou'd not reconcile the situation of the Lands there plac'd to the point of view we saw them in. For to the Southward of Meaxima as far as Luchieux are no Rocks or Islands laid down. We saw several as this Journal will explain. We haul'd our wind to the westward at ½ past 5 under our lower Sails as it blew very Strong in Squalls and we had a great deal of Sea, but cou'd not get any Soundings with 100 fathoms of line. The Gale continued throughout the Night & morning with frequent Showers. At 8 AM we perceived the Peaked Island bearing N88°E 10 or 12 Lea*gue*s. We steer'd South & SE to Noon & we had the wind from the NE quarter. At Noon an indifferent Observation made the Lat*itude* 30°48′N, Acc*oun*t 30°48′N, Long*itude* d*itt*o 129°4′E, [*in margin*: [No] 45 128°42′E], Course S18°W, 72 m*iles*. Squally weather & large Swell. [*in margin*: NB All the accounted Longit*udes* are to be increas'd by 20′ + since leaving Chosan[10] up to this day].[11]

[1] S65°W in the MS, which is clearly a reciprocal error.

[2] There are three main islands and a number of smaller ones; Me Shima is 105 metres high.

[3] Koshikijima Retto, consisting of three mountainous islands and a number of smaller ones.

[4] Kyushu or possibly the 1,118-metre-high Sakura Shima.

[5] Uji Gunto, a small group of islands, consisting of three main islands and a number of smaller ones.

[6] Koshikijima Retto.

[7] Broughton's identification may be in error. James Horsburgh, in *India Directory, or Directions for Sailing to and from the East Indies*, 2 vols, 3rd edition, London, 1826, II, p. 388, gives the latitude of Meaxima as 31°55′N, which suggests this island is one of the Danjo Gunto islands. However, Findlay, in *Directory*, p. 642 (based on Krusenstern's *Atlas*), gives the position of Meac-sima as 31°40′, 129°42′E, which suggests it is one of the islands of Koshikijima Retto.

[8] 'On the 13th [August 1775], early in the morning, we saw the island of Meaxima, with its lofty and peaked mountain'; Thunberg, *Travels*, III, p. 11.

[9] Possibly Johannes van Keulen's *Nieuwe Plascaart Oost Indien,* 1680, which depicts Maiexima west of Kyushu.

[10] I.e. Pusan.

[11] Shortly before noon Broughton must have passed close to Kusakaki Gunto, consisting of about seventeen bare and precipitous islets and rocks, without sighting them.

Sunday 12 November

Fresh Gales & Squally weather with Showers of Rain. We Steer'd SE to make the Land. We cou'd not haul up more as the wind continued from the NE quarter & the Sea directly upon our Beam. At ½ past 2 PM we saw a very High Round Island right ahead.[1] At 5 PM we saw four more in different directions from S30°E to N60°E, the nearest about 6 *Leagues*.[2] [At] ½ past [5][3] we haul'd our wind & Sounded without any success notwithstanding our proximity to Land. At 1 AM the Moon being well up we wore & Stood in to the Eastward and at daylight we counted 8 Islands extending from S15°E to N43°E,[4] one of these evidently volcanic,[5] the Smoke ascending from it in large Columns. It bore S30°E 7 or 8 *Leagues*. We had the weather more moderate & we made Sail upon a wind to the East*ward* and at 9 the Westermost Island off which is a Pinnacle Rock[6] bore S5°W about 2 Lea*gue*s when we tack'd & plied to the North. As the current seem'd to set us to the SE, which plainly appear'd from our being so near the Islands in the morning when we ought to have been some miles farther from them than our situation of last evening. We stood off till Noon in hopes of increasing our Latitude but our efforts were render'd ineffectual. A good observation made our Latitude 30°2′N, Account 30°22′, Long*itude* 129°13′, Course S9°E, 48 miles, but we found we had been set in addition S38°E 39 miles, which is the only affect of a current we have experienc'd since leaving T'shosan Harbor. It was unfortunate to[o] at this period as I was in hopes to fall in with Ximo's South point which the Wind & Current entirely prevented. The Islands in number 10 extended from S10°E to N45°E.[7] They were of Different Sizes. Three of the largest emitted Smoke at Intervals throughout the Day. In general they presented a Barren Rocky Surface. The volcanic Isles were very high & much broken & the middle one bearing S58°E (the same we first saw Yesterday afternoon) was a Round Elevated Mountain Rising from the Northern part of the Isle with a steep ascent to its Summit.[8] It may easily be seen 20 *League*s or more. We saw it near that distance in July 15th when we pass'd to the East of these Islands after Sailing from Lieuchieux. It was then seen to[o] indistinctly to ascertain its situation with any certainty. The Charts we had place the Island of Tanaesima[9] in the Latitude we observ'd it at Noon but among so many we cou'd not remark to which most properly the Name cou'd be applied. We had no bottom with 100 f*atho*ms of line. [*in margin*: [No] 45 129°23′E]

Monday 13 November

Moderate breezes and very clear pleasant weather from the NE quarter. We bore up at Noon & steerd to the SSW. At Sunset we saw another Island bearing S10°E[10] & the Current seem'd

[1] The 1,935-metre-high Yaku Shima.

[2] The westernmost islands of Osumi Gunto, which includes Tanega Shima and Yaku Shima.

[3] Taken from UKHO OD 177.

[4] Various islands in Osumi Gunto and Tokara Gunto, the next group of islands to the south.

[5] The 978-metre-high Nakano Shima and the 792-metre-high Suwanose Shima in Tokara Gunto are both active volcanoes.

[6] Kuchinoerabu Shima, situated 10 miles NW of Yaku Shima; No Saki, its western extremity, is a pillar-shaped rock, which from the north looks as though it is detached.

[7] Extending from Akuseki Shima to Kuchino Shima.

[8] Nakano Shima.

[9] Tanao Sima on Arrowsmith's 1794, 'Map of the World'; Tanega Shima.

[10] 'S10°W' in the MS, which is clearly a transcription error; S10°E in Broughton, *Voyage*, p. 370. The 290-metre-high Tokara Shima, the southernmost island of Tokara Gunto.

to set us to the East of our Course. The other Islands bore from S49°E to N50°E, the nearest 5 Leag*ues*. At Dark we haul'd away to the west for the Night and the wind gradually died away towards 12 PM. At 6ʰ30′AM it again sprang up from the Eastward & the Islands bore from us nearly the same as last evening about 8 or 10 Leagues distant. At Noon it was nearly Calm with light baffling winds from round the Compass. The Observation made our Latitude 29°36′N, Acc*oun*t 29°34′N, Long*itude* 128°27′E [*in margin*: [No 45] 128°54′E], Course S57°W, 49 miles. The current we experienced was 20 miles to the East of our Account when the Islands bore from S29°E to N66°E 10 or 12 Leag*ues*. No Sound*ings* [with] 100 f*atho*ms [line].

Tuesday 14 November

Light winds & variable continued throughout the afternoon with frequent showers of heavy Rain. At Sunset we saw another Island[1] bearing S10°E and the Southermost of those seen at Noon[2] bore S37°E 10 or 12 Leagues. At 8 PM we had a fine breeze from the NW with clear weather & we kept our wind to the west till 5 AM when we bore away SSW½W to Noon. At Sunrise or 7 AM a small Island[3] bore of us S32°E. At Noon it bore S66°E 6 L*eagues*. Lat*itude* Obs*erve*d 29°0′41″N, Account 28°42′N, Long*itude* 127°33′E,. Course S53°W, 59 m*iles*. The Current appears by the Watch and Latitude to have set us N50°E 28 miles. This seems extraordinary having a strong breeze from the North throughout. The Sights for the watch differ at Noon 7 miles deducing them by the Forenoon & afternoon Altitudes owing to the uncertainty of our Lat*itude* at these times & the Current still affecting us to the Eastward. The Island we suppos'd to be 6 or 7 Leagues. It was of no Size but High & rocky, some smaller ones were near it. Fair & pleasant weather. [*in margin*: [No] 45 128°25′30″]

Wednesday 15 November

Fresh breezes and very clear Weather. The Night was Squally & Cloudy and the Forenoon dark Gloomy weather, the Sun scarcely appear'd. We steer'd SWbS till dark, the wind from the NW, when we haul'd up for the Night under easy Sail. At 6 AM we again resumed our course & made Sail. We had no obser*vatio*n to depend upon but the Alt*itu*de shew'd we were to the South of 27°56′ North, Lat*itude* Acc*oun*t 27°48′N, Long*itude* d*itt*o 126°10′E, Course S48°W, 97 miles. Large swell.

Thursday 16 November

Fresh breezes & dark cloudy weather. We steer'd SW, the wind at NNW. At 5 PM we were by estimation in the Lat*itude* of 27°36′N & Long*itude* [*blank*] but we cou'd not see any Land although our distance did not exceed [*blank*] from the islands laying NW of Lieuchieux. We plied under easy Sail during the Night & at 6 AM made Sail to the WSW. The wind still remain'd to the West of North & blew a Fresh breeze with a large swell. We tried for Soundings during the Night with 100 fathoms of line but had no bottom. Towards Noon the heavy Clouds dispers'd and we had Clear and mild pleasant weather. At Noon we Observ'd in 27°40′53″N, the account 27°21′N, Longitude d*itt*o 124°56′E [*in margin*: [No] 45 126°25′],

[1] Yokate Shima.
[2] Tokara Shima.
[3] Kaminone Shima.

Course S77°W, 67 miles. By the watch the Current has set us N50°E 43 miles since the 14th at Noon and to day we are 20 miles North of our accounted Latitude.

Friday 17 November
Moderate breezes & fair weather from the N'west quarter. We pursued the same Course which carried us about 30 Leagues[1] North of our Track in July last to determine the situation of any Lands we might fall in with or to prove a clear passage in the Parallel of 27°N towards Formosa. During the Night we lay by as usual and at 4 AM had Soundings 62 fath*oms*, Sand & Shells, when we again made Sail to the WSW. In the Morning the wind veerd to NE and blew a gentle breeze to Noon with very clear weather & Smooth Sea & we had 60 fathoms water the same bottom. Latitude observ'd 27°25′11″N, Account 27°13′N, Long*itude* d*itt*o 123°42′E [*in margin*: [No] 45 125°24′E], Course S76°W, 66 miles. Current these 24 hours N39°E 16 miles.

Saturday 18 November
Light breezes and very warm pleasant weather. The wind continued steady at NEbE. During the Night it was very faint & we steer'd to the SW, Sounding every 5 Leagues. At Noon the Observation made the Latitude 26°38′20″N, Account 26°38′N, Long*itude*122°49′E [*in margin*: AM [No] 45 124°39′E], Course SW, 67 miles and we had 75 fathoms, Sandy bottom. Current only 8 miles to the East of account.

Sunday 19 November
Light breezes from the NE quarter & the same continuation of fine weather. At 7 PM our estimated Lat*itude* was 26°22′N when we steer'd west Sounding as before. In the Forenoon the Breeze increas'd and the atmosphere became clouded. At Noon we observ'd in 26°19′43″N, Account 26°16′N, Long*itude* d*itt*o 121°13′E, Course S76°W, 79 miles. After mustering we perform'd Divine Service. By the disagreement of the Forenoon & afternoon Sights we again found the Current had set us 39 miles to the Eastward of our account. Immediately after noon we saw the Peaked Island[2] from the Deck bearing S2°W 40 miles & the small Isle to the North of it.[3] These Islands we pass'd between on the 7th of July last, the day after we left Formosa. At this distance it appear'd like two Islands (by the Saddle between the Peaks). We cou'd not take the bearings of the small Isle. [Sounding] 76 fathoms, sand. [*in margin*: [No] 45 123°30′30″]

Monday 20 November
Moderate breezes continued throughout the afternoon and after dark it suddenly became Calm. The Island was just visible at 5 PM bearing S15°E. In the Forenoon we had cloudy & Squally weather which brought a fine breeze from the NE quarter & we continued our Course to the westward. At noon the Latitude made 26°23′N, Account 26°13′N, Long*itude*

[1] '30 miles' in UKHO 177, which is probably a transcription error.

[2] Uotsuri Shima, the largest island in Senkaku Shoto; see also p. 142, n. 4. Broughton had passed about 45 miles to the north of Sekibi Sho the previous day.

[3] Kobi Sho; see also p. 142, n. 7.

120°15′E [*in margin*: [No] 45 122°37′E], Course N87°W, 52 miles. Soundings were 60 f*athoms*, fine dark Sand. By the watch we found no Current, by the Latitude 10 miles North.

Tuesday 21 November

Strong breezes & Squally dark gloomy weather from the NE. We steer'd WSW with a large Swell. At 6 PM we had 70 fathoms & running 5 Leagues further we had only 44 fath*oms*, sandy bottom. The Night was dark & dismal & we preserv'd our situation by Tacking. Our Soundings continued the same. At daylight we bore away with a Fresh Gale steering SW. At ½ past 10 we saw thro' the Haze the Northern part of Formosa bearing S50°E. At Noon it bore from S65° to S75°E. Latitude observ'd 25°22′4″N, account 25°36′N,[1] Long*itude* 118°53′E [*in margin*: [No] 45 121°8'E], Course S50W, 95 miles & we had 50 fathoms, fine Sand & small Shells. Off shore 4 or 6 Leag*ue*s.

Wednesday 22 November

Fresh gales and very dark gloomy weather from the NE. We steer'd WbS expecting to make the Coast of China by dark, but after running 37 miles we did not see it although we saw it nearly in this situation in July last in a heavy gale of wind, but the Weather prevented our seeing more than 4 or 5 Leagues at this time. Our water had decreas'd from 53 fathoms to 32 f*atho*ms in running over when we bro't too the Wind for the Night with our head to the Eastward. It blew a hard Gale throughout & the Sea was perfectly Luminous every Break[2] appearing like shoal water (with which we were surrounded). At 6 AM our Soundings had increas'd to 42 fathoms, blue mud, & we bore away steering WSW with a large following Sea & very gloomy weather. At Noon we had gradually decreas'd to 27 f*atho*ms & the Lat*itude* by double altitudes made 24°34½′N, Account 24°42′N, Long*itude* 117°38′ [*in margin*: [No] 45 119°46′E], Course S57°W, 81 miles, Very Squally threatening weather.

Thursday 23 November

Fresh Gales & very Squally weather from the NE with a large Sea. Supposing by our soundings we were near the Land, we steer'd SWbW but at 2 PM having 40 fathoms we haul'd up West to make it if possible before Dark, but after running in 9 Leagues we did not see it but had shoal'd our water to 30 fath*oms*, fine sand, when we again brought too in a very hard gale of wind with our head to the NW. The Sea was not quite so luminous as the Night before. We lay too till daylight gradually shoaling our Water to 21 f*athoms*, muddy bottom, when we made Sail to the West. The Gale was moderate and we had fair weather. At ½ past 6 we saw the Land bearing N24°W 3 or 4 leagues & soon after Chapel Island[3] which lies to the South of Amoy.[4] It bore S61°W and at 7 we shap'd our Course to the SW to pass without it and we carried 10 & 12 fathoms. A great many Fishing Boats came out of Amoy & stood off to Sea. They were Small & generally painted white, Green & Red with Eyes to them all. To us they made the Sea look Chearful. The Forenoon we had Strong breezes & hazey weather, but a

[1] '26.36N' in the MS, which is clearly a transcription error; see Broughton, *Voyage*, p. 389.

[2] 'brake' in Broughton, *Voyage*, p. 373, presumably referring to breaking waves.

[3] Dongding Dao.

[4] Xiamen.

clear Sky 10° above the Horizon. At ½ past 10 we lost sight of Chapel Island bearing N33°E 6 or 7 Leagues & pursuing the same course in 12 & 17 fathoms we observ'd in the Latitude of 23°51′39″N, Account made 24°1′N, Long*itude* 116°3′E, Course S67°W, 97′. The High round Hill[1] on the South side of Amoy Harb*o*r bearing N13°E & the Southern Extremes making like Islands S68°W. Some Small Isles abreast us were scatter'd along the Shore from N52°W to 68°W and our distance from them was 2 or 3 Leagues. Very high & barren Land bore NNW from us. The Wind gradually moderating. [*in margin*: [No] 45 118°6′E]

Friday 24 November

Fresh breezes & very hazey weather from the NE quarter. We steer'd SW till 2 PM when seeing two Rocks[2] nearly ahead we haul'd off SSW and at 4 PM having run 13 miles we had pass'd them at two or three miles distance & we resumed our former Course, having no Soundings with 18 & 20 fathoms. At ½ past 5 we saw the Islands of Lamock[3] bearing S70°W 5 or 6 Leagues. At 6 we saw them from the deck and we sounded in 26 fathoms. The Ex*treme* Rock at the same time bore N28°E just visible and the Extremes of the China coast from N5°E to N83° west 6 or 7 Leagues. At 12 PM our estimat'd Lat*itude* was 22°57′N and our Soundings kept regularly the same. At which time we haul'd up WbS & after running 12 miles we shoal'd to 18 fathoms, fine Sand. At daylight or 6 AM we had only 17 fath*om*s and the Coast of China bore from NNE to N25° west 3 or 4 Leagues and we steer'd West to approach the Land. The Sea was cover'd with Fishing boats to a great extent. At Noon we were within 3 miles of the Eastern point[4] of Reyorson Bay[5] (as describ'd in the English Directory by W. Herbert[6]) and the water appearing discolor'd we haul'd out SW but we had no less than 13 fath*om*s. broken Shells. At Noon we made the Latitude 22°36′39″N when the extremes of the Bay extended from N39°E to N70°W from the Small Islands[7] about 2 or 3 Leag*ue*s. Lat*itude* account 22°47′N, Long*itude* ditto 113°6′E [*in margin*: [No] 45 115°51′E], Course S57°W, 140 miles and we found by the Watch we had experienced a set to the WSW 22 miles. Pedro Branco[8] bore from us at Noon SWbW 12 or 13 L*eagues* by [No] 45.

Saturday 25 November

Fresh breezes and very hazey weather so much so as to obscure our discerning any thing in the Coast beyond 2 or 3 Leag*ue*s with any accuracy. We shap'd our course WSW to get sight of Pedro [Branco] before dark, Soundings from 13 to 15 fathoms, very fine sand and Small Shells. At 4 PM we saw the Rock Pedro Branco from the lower Rigging and at ½ past it bore SW from the Deck. At 6 PM having run 32 miles WSW Pedro [Branco] bore South 4 or 5 miles and the watch at the same time by the afternoon sights made our Long*itude* 115°17′E, our estimated Lat*itude* being 22°24′N. The True Lat*itude* of Pedro [Branco] is 22°18′N &

[1] The 128-metre-high Shenan Shan, which is conspicuous.
[2] Xiongdi Yu, two islets, the NW islet 43 metres high and the SE islet 61 metres high.
[3] Nanpeng Liedao.
[4] Tianwei Ji ăo.
[5] Jieshi Wan.
[6] William Herbert, *A New Directory for the East Indies*, possibly the 6th edition, London, 1791.
[7] Xijie Jiao and Dongjie Jiao, situated up to 3 miles SE and SW of Tianwei Jiăo.
[8] Zhenyantou.

115°19′E or 1°47′ east[1] of Macao (by what we made it in the Providence). We run on the same course till midnight having, after we pass'd the Rock, Muddy bottom in 17, 18 & 19 fathoms. At 12 PM we bro't too till 4 AM[2] when we bore up SWbW in 19½ fathoms. After running 10 miles we saw the N*orth* point of the Great Lema[3] bearing west. We haul'd up WNW and after 13 miles sailing on that course it bore South 2 or 3 miles & we took several Altitudes for the Watch, the Mean of these gave the Long*itude* 114°25′E (Poo toy[4] N2°W to 32° west, N*orth* Point G*reat* Lema S23°E 4 or 5′). We steered WbN 14 miles to Noon in 14 & 13 fath*oms* when the mean observ'd Lat*itude* made 22°8′3″N, the North point of the Great Lema then bearing S75°E[5] about 5 Leag*ues* and S*outh* extreme Poo toy N88°E, the Peak of Lantoe[6] N20°W, the Small Island of Atchow[7] N41°W to N76°W about 2′, 8½ *fatho*ms. Latitude account 22°19′N, Long*itude* d*itt*o 111°43′E, Course S69°W, 82 Miles. Current since Yesterday S61°W 22 miles. We make the

Lat*itude* of the Great Lema	22° 4′N, Longitude	114°14′E	}	By the watch's
D*itt*o Pedro Branco	22°18′N	115° 6′E		error in the Typa

Pedro Branco bears N76′E & S76°W from the Great Lema Distance 50 miles.

Sunday 26 November

The afternoon was calm but the Tide being in our favor we drifted between the Islands and Lantoe increasing our water from 10 & 15 to 25 fathoms, the point bearing East 3 or 4 miles. At 8 PM we came too in 12 fathoms, muddy bottom, Lintin[8] Peak NbE, Lantoe Peak N75°E and Macao Town S88°W. It was high water by the Stream at 2 PM. In the morning we had the wind Easterly and the tide still ran down we got under weigh but soon after it fell calm and we were oblig'd to come too again nearly in the same situation. Muster'd the People and perform'd divine Service & enjoin'd them to secrecy respecting our proceedings since leaving Macao and taking from the Officers & Crew all remarks relating to our Voyage. Mean Latitude by 3 Sextants made 22°13′N, Lantoe Peak N80°W, Point of Lantoe on with the center of Great Atchow[9] S61°E, west ext*reme* of Lassammee[10] S5°E. [*in margin*: [No] 45 114°10′E]

Monday 27 November

Moderate breezes from the Northward. It was slack water at 2 PM, the Stream having run down 12 Hours, & we again got under weigh & stood to the westward. As we approach'd Macao we saw a Ship under Sail near the Typa standing towards us. At 4 PM we sent a Boat to her. She prov'd to be a Country Ship that had come out of the Harbor and last from Port

[1] 'West' in the MS, which is clearly a transcription error.
[2] 'PM' in MS, which is clearly a transcription error.
[3] Dangan Dao.
[4] Po Toi Islands.
[5] 'S75°W' in the MS, which is clearly a transcription error.
[6] The peak of Tai Yue Shan, formerly Lantau.
[7] Siu a Chau, the northern of the Soko Islands.
[8] Nei-ling-ting Tao.
[9] Tai a Chau, the southern of the Soko Islands.
[10] Guishan Dao.

Jackson.[1] Having a Pilot on board they were proceeding for Canton River. At Dark we came too in Macao Roads in 3 Fathoms.

During the Night the Wind increasd to a strong breeze and finding our riding most uncomfortable in the Roads we got under weigh & run into the Typa & moor'd near Deadmans Cove[2] for the conveniency of communicating with the Island to refresh our people and regulate the rate of our Time Keeper. We found an English Ship laying here that had lately arrived from the Friendly and Society Islands w[h]ere she had left some settlers.[3] In the morning I went on Shore at Macao to pay my respects to the New Governor, Mons*ieu*r Pinto having been superceded in our Absence by Senor [Don Christopher de Castro].

By observations made in the morning we arrived No 45 shew'd the Long*itu*de 113°43′38″E, the Longitude of the Typa is 113°32′10″E. The error in Longitude therefore since leaving T'shosan is 11′28″ = [in time] 2′54″·5.

W. R. Broughton Captain

The Bearings stand uncorrected for Variation and the daily Longitudes stand uncorrected in the same Manner. Notwithstanding the Error of No 45 was so great in the first part of the Voyage from Macao to Japan.[4]

From Macau to England[5]

I think it unnecessary to expatiate any more on the detail of our voyage, as it was continued to Madras through the Straits of Malacca, and from Madras to Trincomalay; a track of sea well known to the navigator: thence I took the accustomed passage to England, where I arrived in February, 1799, after an absence of four years; and shall consider my exertions as amply rewarded, if this journal in the opinion of the scientific and geographical part of the world (however it may furnish little amusement to the general reader), is considered as adding to the stock of nautical information, and communicating a more extensive knowledge of the globe.

[1] This was the transport *Britannia*, Captain Thomas Dennott, which had sailed from Port Jackson on 2 August; Cumpston, *Shipping Arrivals*, p. 32. She had been waiting in the Typa for six weeks before obtaining permission to proceed to Canton; James Wilson, *A Missionary Voyage to the Southern Pacific Ocean performed in the Years 1796, 1797, 1798 in the Ship Duff*, London, 1799, p. 303.

[2] Not identified, but clearly one of the coves on Ilha de Taipa.

[3] This was the *Duff*, Captain James Wilson, which had taken missionaries to Tahiti and which had anchored in the Typa on 21 November; Wilson, *A Missionary Voyage*, p. 303. In his account Wilson makes no mention of the *Providence Schooner*.

[4] This paragraph from UKHO 177.

[5] The following paragraph from Broughton, *Voyage*, p. 380.

APPENDIX I

Selected Correspondence

I. Broughton to the Admiralty

Letters from Broughton to the Admiralty are to be found in both The National Archives, Kew, and the United Kingdom Hydrographic Office, Taunton, with some duplication between the two. Since those in Kew were clearly received by the Admiralty they are quoted in preference to duplicates held in Taunton, with significant differences between the two noted in footnotes. The collection of letters in Taunton, which are all kept together in OD 177, may possibly be duplicates kept by Broughton and passed to the Hydrographic Office at the completion of his voyage.

1. To Philip Stephens, Secretary of the Admiralty[1]

Recd
Providence in Santa Cruz Bay
at the Island of Teneriffe 7 March 1795

Sir,

I beg leave to acquaint you for the information of their Lordships of the safe arrival of his Maj*esty*'s Ship under my Command at this Island at the 4th Inst., with the Iris Frigate,[2] Reliance[3] & Supply[4] Arm'd Vessels. In company with the latter I propose sailing on the 10th by which time we shall have compleated our Wine & Water & propose touching with the above Ships at Rio Janeiro in the prosecution of my Voyage.

I am Sir
Your Obedi*en*t Ser*van*t
W. R. Broughton
P. Stephens Esq[e]

[1] TNA, ADM 1/1511, Cap Letter B565.

[2] Commanded by William Hargood; she sailed on 8 March bound for Cape Coast.

[3] Commanded by Captain John Hunter, Governor designate of the infant colony of New South Wales, with Henry Waterhouse as second captain. Matthew Flinders was the senior master's mate on board.

[4] Commanded by Lieutenant William Kent. The *Reliance* and *Supply* reached Port Jackson on 7 September 1796.

2. To Philip Stephens, Secretary of the Admiralty[1]

H.M. Ship Providence
at Rio Janeiro 23[d] May 1795
Sir,

I beg leave to acquaint you for the information of their Lordships of the safe arrival of His Majesty's Sloop under my Command in this port on the 4th Instant in Company with His Majesty's Arm'd Vessels Reliance & Supply and of my departure this day for Port Jackson – having compleatly refitted the Sloop & refreshed the Ships Company – Governor Hunter not being ready I did not conceive it necessary to wait but to proceed with all expedition in the prosecution of their Lordships Instructions. And I take this Route to Nootka Sound as more eligible than rounding Cape Horn in the depth of Winter.

I am
Sir
Your Obedient Servant
W. R. Broughton
Philip Stephens Esq[e]

3. To Evan Nepean, Secretary to the Admiralty[2]

Providence in the Harbor of
Port Jackson New South Wales
October 5th 1795
Sir,

I beg leave to acquaint you for their Lordships information of the safe arrival of His Maj*esty's* Sloop under my Command in this Port on the 25th of August. The constant bad Weather we experienc'd in our Run from Rio de Janeiro has rendered it necessary to caulk & Refit the Sloop which will account to their Lordships for my Retention here these last six Weeks.

You will please to inform the Board of the Ships Company under my Command being in perfect Health & of my sailing this day to execute their Lordships Instructions relative to Nootka Sound.

I am Sir
Your Obed*ien*t Serv*an*t
W. R. Broughton
Evan Nepean Esq[e][3]

[1] TNA, ADM 1/1511, Cap Letter B566.
[2] TNA, ADM 1/1511, Cap Letter B567.
[3] Evan Nepean took over as First Secretary to the Admiralty on 3 March 1795.

4. To Evan Nepean, Secretary to the Admiralty[1]

Providence in Monterrey Bay on the coast of California June 15th 1796

Sir,

I beg leave to transmit you for the information of their Lordships the following account of my proceedings after sailing from Port Jackson in New South Wales on the 14th of October 1795.

I steer'd to the north of New Zealand and on the 28th of November I arrived at the Island of Otaheite[2] where I remained twelve days to refresh the Ship's Company. It did not appear that any vessel had touch'd there since his Majesty's Store Ship Daedalus.[3]

We had the pleasure of parting on the most amiable terms with the King Otoo[4] and his Chiefs. I receiv'd from them every kind attention with the most ample supplies of refreshments. I brought from thence at their own consent two Seamen who had been cast away in the Matilda whaler[5] some Years ago and a Native of the Sandwich Islands who wish'd to return to his Country.

On my route to the North I discover'd a small Island. It is situated in the Lat*itud*e of 9°57′ South and 209°35′ East Longitude. I named it the Island of Carolina in compliment to Miss Stephens of the Admiralty.[6] We made the Island of Owhyee[7] January the 1st 1796.but light winds & Strong Currents prevented our anchoring in Karakakooa[8] Bay 'till the 8th. We found here an American Brig call'd the Lady Washington[9] who gave us information of Captain Vancouver having sail'd from the Port of Valparaiso in May 1795 on his return to Europe.

The Ship having sprung a Leak in our run from Otaheite I have tried every means to discover it but all our efforts proved unsuccessful; during this time the Astronomers have settled the Rate of our time Pieces on Shore and on the 1st of February we sail'd. The Natives behaved in the most friendly manner notwithstanding all the Chiefs were absent at the Island

[1] UKHO, OD 177.

[2] Tahiti.

[3] The *Daedalus* was sent out from England with stores for Captain Vancouver. On her arrival in Nootka Sound she was sent by Vancouver to Port Jackson under the command of Lieutenant James Hanson by way of Tahiti, with instructions to rescue survivors from the wrecked whaler *Matilda*.

[4] Tu, also known as Pomare II.

[5] The *Matilda*, Captain Matthew Weatherhead, was wrecked on a reef east of Tahiti on 25 February 1792. The crew returned safely to Tahiti in their boats from where Weatherhead, with two men and two boys, sailed for the North-west coast of America in the *Jenny*, leaving behind twenty-one survivors. On her arrival in Nootka Sound Vancouver arranged for the *Daedalus* to call at Tahiti to rescue the remaining survivors. She anchored in Matavai Bay on 15 February 1793, when six survivors were found on the island of Moorea; John Earnshaw, intro, *A Letter from the South Seas by a Voyager in the 'Daedalus', 1792*, Cremorne, NSW, 1957, unpaginated. In the event only one of the survivors could be prevailed upon to leave the island. A number had already been taken from the island in the *Providence* when Captain Bligh had called there in 1792, shortly after the *Jenny*, leaving behind four or five, who refused to leave; Lamb, *Vancouver's Voyage*, II, p. 748, and Oliver, *Return to Tahiti*, pp. 52–4, 238.

[6] Daughter of Sir Philip Stephens, one of the Lords Commissioners of the Admiralty, whose illegitimate daughter's name was Caroline and not Carolina. The island was not, however, a new discovery, having been sighted by Quiros in 1606 and named by him San Bernardo and possibly discovered by Magellan in 1521; Sharp, *Discovery of Pacific Islands*, pp. 5–6, 61–2. Correctly named Caroline Island on charts.

[7] Hawaii.

[8] Kealakekua.

[9] Commanded by Roger or Robert Simpson; Howay, *A List of Trading Vessels*, p. 32.

of Wohahoo[1] which Island they had lately made a conquest of. After leaving Owhyee I anchored in Whyteetee Bay[2] to make a Sketch of the Harbor[3] in which Mr [William] Brown (who first discovered it) and Mr Gordon the Master of two English Vessels[4] were murdered by the Natives and the Vessels taken.[5] The Inclosed Sketch and copy of the event will best explain the particulars.[6]

It is in this Bay Tai[7] was made King of Owhyee with 16,000 Men [&] has taken up his residence while the Europeans with him are busily employed building a Vessel about 40 Tons[8] to assist him in his intended conquest of the Island of Atooi.[9] The Cultivation of Wohahoo[10] from the destructive war which has taken place, does not afford them sufficient nourishments, they were daily dying & to add to their misery they are universally infected with the Itch in a most virulent degree.[11]

On the 15th of March we arriv'd in Nootka Sound and the following day the principal Chief Maquinna brought me several letters from Lieut. Pearce of the Marines (and the Spanish Officer Brigadier Alava) dated at Nootka March 1795 informing me of the restitution of the Place agreeable to the form and Authority given him by his Grace the Duke of Portland. These letters also gave me authentic information from Capt. Vancouver who in a letter left at Monterrey thus expressed himself 'That having been inform'd another Person was named to negotiate the Business he had determined to quit this Ocean and return to Europe', and actually sail'd from the bay of Monterrey on the 1st of december 1794 with both Vessels in good condition & the Crews in perfect health, since which as before related he had sail'd from Valparaiso in the prosecution of his instructions to return by Cape Horn.

The Leak having gradually increas'd the Ship making from 6 Inches to 12 per hour, I found it absolutely necessary (much as I regretted the occasion) to find the Leak out before I could take any steps for my future proceedings. I accordingly proceeded to the Port of Mawhinnee[12] situated five miles up the Sound, which I found in every respect suitable for the purpose; but the wind and rain much influenced our operations added to this inclemency of the weather. Having lighten'd the Ship and reduc'd her draught of water to ten feet, we haul'd her on shore several times but to no effect as we found the Leak to be much lower down than the water falling would permit us to come at. It therefore became expedient to heave the Ship down and we immediately commenced building a Wharf 100 feet long to which the Ship cou'd lay by a stage of 40 feet as the most expeditious way of unlading and relading.

By the 10th of April we were nearly ready on which day arriv'd the Lady Washington. Her coming in was fortunate for us as it sav'd both time and trouble. On my request the Master gave us every assistance & by her we hove the Ship down. We found 14 feet of the false Keel carried away, and the Leak to have proceeded from an Auger hole through the planking to the

[1] Oahu.

[2] Waikiki Bay, now Mamala Bay.

[3] Honolulu Harbor.

[4] The *Jackal* and *Prince Lee Boo*.

[5] The surviving crews shortly after succeeded in retaking the two vessels.

[6] These do not appear to have survived.

[7] Kamehameha.

[8] The 36-foot schooner *Britannia;* Lamb, *Vancouver's Voyage*, III, pp. 1152–3, 1178–9.

[8] Kauai.

[10] Probably taro, the most widely cultivated food plant in the Hawaiian Islands.

[11] Probably scabies.

[12] Marvinas Bay on the west side of Cook Channel.

Floor timbers which (by neglect) no bolt had ever been driven into. The copper being well preserv'd during her last Voyage had prevented it leaking but now being worn very thin it work'd round the Hole and as it enlarg'd accounted for the gradual increase of Water.

In return for the conveniency afforded us by the Master of the American (Mr Simpson) I employ'd our people in new sheathing & repairing his Vessel & supplied him with an Anchor & Cable, which he was much distress'd for.

By the 16th of May we were ready for Sea but contrary Winds prevented our sailing till the 21st. This Month brought with it mild weather and the Natives generally supplied us with Venison, Fish & Vegetables. We never had any dispute with them, but parted with Maquinna the Chief on most friendly terms. No Vessel of any Nation had been in the Sound since the Spaniards quitted it.

As the ship requir'd caulking & her Mizen Mast to be taken out for fishing, I thought it more convenient to proceed to this Port which lay in our route to the Southward & have the opportunity of acquainting you with my proceedings and future intentions, and, refreshing the Ship's company. On the 5th June we anchor'd in this Bay, and by the 15th compleated our operations & the Ship ready for Sea.

Their Lordships Commands direct me to proceed to Sea and carry into execution that part which relates to Captain Vancouvers Orders of the 8th of March 1791 upon a probability that he would not have been able to undertake the same. But from the above statements respecting Capt. Vancouver & more particularly the Spanish papers giving intelligence from him on the Coast of Chili in May of last Year which leaves no manner of doubt in the opinion of myself and Officers of his not having carried his Orders into execution I have taken upon myself a discretionary power to deviate from their Lordships directions and to proceed to England by that route which I think is most eligible for the further improvements of Geography & Navigation. I beg leave to offer the following reasons for my conduct as I think it will be for his Maj*esty's* service, which I trust will meet the approbation of their Lordships. Mr Crosley, the Astronomer having been sent out expressly to make Observations and to ascertain the situation of unknown Places, has hitherto had no field for complying with his instructions from the Board of Longitude, altho' he has been embark'd for that purpose these two years & a half from our Navigation hitherto having led us to follow the Steps of Capt. Vancouver.

I therefore think (and he is of the same opinion) agreeable to the plan of the late Capt. King in Cooke's 3rd Voyage that if a Survey were made of the Coast of Asia commencing at the Island of Zagaleen[1] situated in the Southern part of the Sea of Ochotzk[2] in 51°½′N° Latitude to the Nanking River[3] in 30° North latitude and the completion of the Japanese Islands left unfinished in Capt. Cookes last Voyage wou'd be very acceptable to Geography for the limits of both the Continents wou'd be Known as far as Navigation was practicable and the Geography of the Northern Pacific wou'd be compleated.

As it is most probable that the above survey cannot be compleated before the middle of 1797 I propose spending my time on that pursuit 'till next Xmas, when it will be necessary to proceed to Canton[4] for Provisions & Stores & continue the survey early in the next Year. This will also give me the opportunity of receiving their Lordships Commands for my future proceedings shou'd any accident have happen'd to Captain Vancouvers Vessels or any other

[1] Ostrov Sakhalin.
[2] Okhotsk.
[3] Chang Jiang, formerly Yangtze.
[4] Now Guangzhou.

Orders they may please to honor me with. I herewith subjoin a weekly account and have the satisfaction to inform you of my having only lost One Man by Sickness and of the Ships company being in perfect health.

I have to request you will please to acquaint their Lordships we have receiv'd every refreshment I requir'd in this Bay, but from some jealousy which renders them more circumspect (than when I was here formerly) they avoided all Society nor wou'd the Governor Señor [Diego de] Borica even give me leave to set up a Tent for the regulation of the Time Pieces Saying his Orders were to that effect as in Capt. Vancouvers time we had made to many Observations at their Settlements:

I am
Sir Your Very Humble Serv*an*t
W. R. Broughton

Evan Nepean Esq[e]

NB This Letter & others were to be sent home by La Vera Cruz by the Commanding Officer at Monterrey Colonel Borica.[1]

5. To Evan Nepean, Secretary of the Admiralty[2]

Providence in the Typa at Macao in China
The 20th of Dec*embe*r 1796
Sir,

I have the pleasure to inform you of the arrival of his Majesty's Sloop under my Command in this Harbor on the 15th Inst. after an absence of five months from the Sandwich[3] Islands, touching at Atooi & Oneehow[4] for Yams and Water, at the latter place I had the misfortune to lose two Marines who were barbarously murder'd by the Natives without any provocation whatever.

From these Islands we traversed the Northern Pacific in the parallel of 28°30′ North Latitude without making any discovery, and early in September we fell in with the Northern part of the Island of Nipon or Japan.[5] From thence we proceeded to explore the Islands of Jesso[6] and the Southern Kurile Islands. On the 16th of September we anchor'd in a Spacious Bay[7] in

[1] The 'others' probably included a letter from Zachary Mudge to John Stuart, first Marquess of Bute, dated 20 June 1796 from Monterey Bay, which is now in the Archivo General de Indias in Seville, the Spanish authorities in Madrid evidently having decided not to forward it to London. When Mudge wrote the letter he was under the impression that Bute was still Ambassador to the Spanish Court; AGI, Estado 39, N.14-8 (5).

[2] UKHO, OD 177.

[3] Hawaiian.

[4] Kauai and Niihau.

[5] Honshu.

[6] Hokkaido. Broughton later used the name Insu instead of Jesso. The confusion between the names Jesso, Matsmaia and Insu for Hokkaido is discussed by Krusenstern; see Krusenstern, *Voyage*, II, pp. 48–50.

[7] Uchiura Wan.

the Island of Matzmai[1] in the Lat*itude* of 43°33′N and Longitude 140°50′E; in the NE'ern corner of this Bay is situated a good Harbor.[2]

We remained ten days compleating our Wood & water having constant intercourse with the Natives who appear'd a most inoffensive race of People answering to the description given of them by the Russian Navigator Spanberg as hairy Men.[3]

The Country is beautifully diversifyed with Hills and Dales producing excellent Timber but little or no Cultivation. The Natives were subject to the Japanese Government to whom they paid great submission; the latter frequently came on board and seem'd jealous of our contracting any intimacy with the Islanders; anxiously wishing us to depart whenever our wants were satisfied. In other respects they were polite and attentive, and on taking leave they presented me with a Map of the Japanese Islands and the Kuriles including the South point of Kamschatka. The writing was in Russian letters and one of the Japanese spoke the Language; we understood he had been at Petersburgh.[4]

After leaving the Bay we continued our route to the North and in the Lat. of 43°50′N and 146°50′ East we came to another Island of few leagues in extent.[5] Seeing no Land to the SE and East but high distant Land[6] to the NE we Shap'd our course to pass thro' a Strait[7] which apparently divided the High Land from Matzmay situated in 44°30′N and 146°25′E; leaving Matzmay to the SW and coasting the west side of the high Land which appear'd to be an Island of some Magnitude. As we proceeded we discover'd more Islands trending to the NE'ward[8] but unfortunately the weather which had hitherto been extremely favorable now became quite the Reverse. As the Gales were chiefly from the NW we were enabled to pass to the West of the Islands 'till we arriv'd off the Northern part of Mareekan,[9] where the Russians (in Capt. Cookes last voyage) are said to have a small Settlement[10] in the Lat. of 47°½′N Lat. I sent an Officer into the Harbor (which is only calculated for small Vessels from a Bar of 2 f*athoms* in the entrance but within there is 5 & 7 fathoms) who reported on his return the place being deserted by the Russians but cou'd not learn from the Natives at what period. There still remain'd the Russian Arms carv'd in wood & painted and Crosses erected in different places.[11]

The Natives of the Island were cloath'd & in other respects were similar to those at Matzmay; it therefore plainly appears all the Islands to the North of Nipon or Japan are

[1] Broughton was mistaken in thinking that the southern part of Hokkaido formed a separate island, perhaps being misled by the map facing the title page of Müller, *Voyages from Asia to America*, in which 'I Matsumay' is of small extent.

[2] Muroran Ko.

[3] Martyn Spanberg or Spangberg, a Danish naval officer in Russian service, took part in both the First and Second Kamchatka Expeditions under Vitus Bering. In 1738 he charted the Kurile Islands as far south as Urup and in 1739 he reached the island of Honshu, which he visited again in 1742. He actually described the inhabitants of Matzmai as having 'pretty long hair all over their bodies'; Müller, *Voyages from Asia to America*, p. 29.

[4] Sankt-Peterburg.

[5] Ostrov Shpanberga.

[6] Ostrov Iturup.

[7] Proliv Yekateriny.

[8] The NE extension of Kuril'skiye Ostrova (the Kurile Islands).

[9] Ostrov Simushir.

[10] This information was given to Cook in Unalaska in 1778 by Grigory Izmaylov who stated 'that on the 12th of May 1771 he sailed from Bolscheretzk, in a Russian Vessel, to one of the Kuril Islands, named Mareekan in the latitude of 47°, where there is a harbour, and a Russian Settlement'; Cook and King, *Voyage to the Pacific Ocean*, II, p. 500.

[11] A bay at the NE end of Ostrov Simushir is now named Bukhta Broughton or Broutona.

inhabited by the same Race of hairy Men. The Northern Isles to Mareekan being subject to the Russians & the Southern ones to the Japanese.

The Winter Season being fast approaching as this was the 17th day of October I had no wish to proceed further to the North, but hop'd to compleat the Survey of the Eastern Coasts of the Islands we had already seen more particularly Ooroop[1] & Nageedsda[2] which by the Russian accounts afford good Harbors. The wind not permitting our passing thro' the passage which divides Mareekan from another Island to the North,[3] we stood to the South and the next day experienc'd a heavy Gale of Wind and Rain, which reduc'd us to a Storm stay-sail in the evening. In the height of the Gale I had the misfortune to fall upon deck & fracture my Right arm above the elbow, an accident to me most unfortunate as my confinement entirely precluded the Survey of the Islands and proceeding as I intended between Japan and the Coasts of Azia. We nevertheless Kept that point in view and the Officers were directed accordingly but the boisterous weather we now met with baffled all our endeavours. A NW wind carried us thro' a passage South of Mareekan in Lat. 46°45′N & Long*itud*e 150°30′ East[4] which soon increasing to a violent Gale we were driven to the Eastward and ever after unable to approach near the Islands. Incessant Gales succeeded each other so rapidly we were seldom able to carry any Sail and whenever we drew near the Land the western Gales with increas'd force drove us from it added to which we had to strive against contrary Currents whose Strength we cou'd seldom detect from little opportunity we had of making Observations.

As my principal pursuit was to examine those parts not seen in Capt. King's Voyage I did not think it necessary to make the coast of Nipon to the north of White point[5] but from thence continued our course to the South exploring the Southern part of Nipon & eastern side of Ximo[6] and the Islands of Likeyo[7] to South of it as well as the weather wou'd admit for seldom cou'd we make the necessary Observations for fixing the points of the Coast.

The prevalence of the westerly Winds prevented our passing thro' Diemen's Strait[8] to the North of Great Likeyo.[9] We therefore ran down on the Eastern side of them and passing the Islands to the East of Formosa[10] we made that Coast in 24°50′N proceeding from there round the South point to China.

Not having entirely recover'd the use of my Arm I am unable to transmit any further account of my proceedings with the necessary Surveys but the Indiamen being on the point of sailing render'd it my duty to convey some information to their Lordships.

I propose carrying into execution early in the Year my Views of compleating the Coast of Tartary[11] and Japanese Islands as express'd in my letter to you from Monterrey in June last.

[1] Ostrov Urup.

[2] A spelling error for Nadeegsda. The identity of this island is uncertain. From the context Broughton appears to consider it to be Ostrov Iturup. For another possible identification see p. 91, n. 6 above.

[3] The passage is Proliv Diany and the island Ostrov Ketoy.

[4] Proliv Bussol'.

[5] 'Witte Hoek, placed by Jansson in latitude 35°24'; Cook and King, *Voyage*, III, p. 402. Taito Saki in 35°18′, whose south side consists of a vertical white chalk cliff and its north side is a white and red bluff; *Japan Pilot*, Vol. II, Taunton, 1979, p. 239.

[6] Kyushu.

[7] Nansei-shoto (Ryukyu Islands).

[8] Osumi Kaikyo.

[9] Okinawa.

[10] Taiwan.

[11] An area covering the east coasts of the Russian province of Primorskiy Kray and the Korean Peninsula.

I have the satisfaction to inform you of the Officers & Ship's Company remaining perfectly healthy.
I am Sir
Your very Humble Ser*van*t
W. R. Broughton
Evan Nepean Esq^{e}

NB This letter sent home by Admiral Rainier in his despatches from China in January 1797 by Capt. Lestock Wilson.[1]

6. To the Commissioners of His Majesty's Navy[2]

Providence in the Typa at Macao
the 29th december 1796

Gentlemen,

Having judged it necessary for the Service on which his Majesty's Sloop under my Command is employ'd to have a smaller Vessel as an Assistant in surveying I have this day purchas'd a Schooner compleatly ready for Sea; she is call'd the Prince William Henry, copper bottom'd and built at Shoreham in Sussex in the year 1788. She is registered at 87 Tons and appears to me every way well calculated for the purpose. Her fix'd price was 1500 £ sterling which I find is her real value in this Country.

I have therefore given three bills of exchange on your board for 1169 £ of the same time & date; & the remaining part of the 1500 £ went to pay the Seamens wages of the Vessel. I beg leave to give you this advice not doubting you honoring the same and Remain
Gentlemen
Your Obedient Servant
W. R. Broughton
To the principal officers
Commissioners of his Majesty's Navy

7. To Evan Nepean, Secretary to the Admiralty[3]

Providence in the Typa at Macao
the 1st January 1797
Sir,

I beg leave to acquaint you I have thought it expedient and for the advantage of the Service on which I have the honor to be employ'd that a Small Vessel shou'd assist his Maj*esty*'s Sloop under my Command in exploring the Coast of Corea & Tartary, with the Japanese Islands, the Nature of Maritime Surveys frequently requiring a more slow investigation than it might

[1] Captain Wilson sailed from Whampoa on 1 January 1797 in the *Exeter* and arrived in the Downs on 24 July; Farrington, *Catalogue*, p. 237.

[2] UKHO, OD 177.

[3] TNA, ADM 1/1517, Cap Letter B512. A duplicate copy of this letter with slightly different wording is held in the UKHO, OD 177.

be proper to Risk in one Vessel only. This consideration with many others (were it necessary I cou'd adduce to convince their Lordships of the propriety my so doing) but I trust their approbation will not require it. I have therefore this day purchas'd a Schooner nearly of 100 Tons every way compleat for his Maj*esty*'s Service, Manning her with the Supernumeraries My Orders directed me to bear for Cap*tai*n Vancouver of the Discovery and have drawn Bills on the Navy Board for the purchase amount to 1,500£ Sterl*ing*.
I have the Honor to be
Your Obed*ien*t serv*an*t
W. R. Broughton
Evan Nepean Esq[e]

Minuted on the flyleaf 'own [indecipherable] and acquaint him that their Lo*rdshi*ps will direct the Navy Board to pay his Bill for the Vessel in question; send copy to N*avy* Bo*ar*d with directions accordingly'

8. To Evan Nepean, Secretary to the Admiralty[1]

His Majesty's Sloop Providence Macao
in China 1st April 1797

Sir,

I beg leave to acquaint you of his Majesty's Sloop under my Command being in every respect fit for Sea, and the Monsoon being nearly favorable for the prosecution of our Voyage in the Northern Pacific Ocean I propose sailing the first Wind to compleat the Survey of the Japanese and Kurile Islands with the coast of Corea &ca intending to return here in the Autumn (shou'd I find sufficient employment in the mean time) to receive any instructions their Lordships may think proper to send me agreeable to my intimation from Monterrey in June last.

By the Cuffnels East Indiaman[2] I send their Lordships a Copy of what we have already done of the Kurile and Liquieux Islands, leaving a duplicate at Canton. Also a Journal in part of my proceedings to elucidate the Chart. I herewith enclose a weekly account of the State and Condition of His Majesty's Sloop under my Command & have the Honor to remain,
Sir
Your most Obedient Servant
W*illia*m Rob*ert* Broughton
Evan Nepean Esq[e]

[1] TNA, ADM 1/1517, Cap Letter B513. A duplicate copy of this letter with slightly different wording is held in the UKHO, OD 177.

[2] The *Cuffnells*, Captain Charles Bowland Cotton, anchored in Long Reach, in the Thames Estuary, on 21 March 1798; Farrington, *Catalogue*, p. 142.

9. To Evan Nepean, Secretary to the Admiralty[1]

His Majesty's Ship Providence Macao
in China 1st April 1797
Sir,

Whereas by their Lordships directions of the 31st of December 1793 I am Ordered to convey Mr Crosley Astronomer appointed by the Board of Longitude to Capt*ai*n Vancouver of His Majesty's Sloop Discovery for the purpose of making Astronomical Observations in a Voyage to Nootka Sound or elsewhere; and on my arrival at Nootka, and afterwards having found Captain Vancouver had returned to England; I beg leave to inform you of my having thought proper to detain Mr Crosley for the purpose of making Astronomical Observations on board His Majesty's Sloop under my Command, in the same manner he was directed to do on board the Discovery, complying with the instructions he has receiv'd from the board of Longitude; as I conceive the Nature of the Service on which I am employ'd renders his abilities equally necessary.

I am
Sir
Your Humble Servant
Will*ia*m R*ober*t Broughton
Evan Nepean Esqe

10. To Evan Nepean, Secretary to the Admiralty[2]

His Majesty's Schooner Providence in the
River of Canton June 10th 1797
Sir,

It is with the greatest concern I have to acquaint You for their Lordships information with the loss of His Majesty's Sloop under my Command.

Having sail'd from Macao early in April we met with constant easterly Winds which impeded our progress to the NE of Formosa. On the 12th of May we pass'd to the North of the Southermost of the Liquieux Islands[3] & examin'd those parts we last Year left unfinished. In completing the Survey of the Western Island (call'd by the Natives Typinsan)[4] the Ship struck upon a Coral Reef distant from the Land five Leagues.[5] This was half an Hour after Dark before which period No appearance of Shoal Water was to be seen from the Mast Head, Nor any other indication to imagine we had any thing to apprehend, being without Soundings & from the formation of the Land upon a Bold Shore. I trust therefore whenever a Court of Enquiry takes place on this unfortunate Event their Lordships will be convinc'd that no blame can be imputable to my Conduct.

[1] TNA, ADM 1/1517, Cap Letter B514.

[2] TNA, ADM 1/1517, Cap Letter B515. A duplicate copy of this letter with slightly different wording is held in the UKHO, OD 177.

[3] Miyako-retto.

[4] Miyako Shima.

[5] Yaebi Se, an extensive area of coral reefs extending 9 miles north of the northern extremity of Miyako Shima.

Every method was ineffectually tried to clear the Reef but the Wind increasing occasion'd the Surf to break with great violence and the Ship striking very hard render'd it doubtful whether we shou'd be able to hoist the Boats out [before the Masts gave way.][1] At the time the Longboat was lower'd into the water. We had 7 feet Water in the Hold and the Carpenter reported the Ship bilged forward; this was near 10 O'clock. The Water soon rose to the upper Deck and we had now only to hope She wou'd remain in the same situation for had She shifted from the Rocks she must have founder'd instantly before we cou'd have got into the Boats.

Finding our efforts of no further effect I found myself under the necessity of ordering the Crew into the Boats for their own preservation the Water had increas'd so Rapidly that few of the Officers cou'd save even their Commissions of which their Lordships will Judge as no private property whatever was Sav'd every Man attending to his duty for the preservation of the Ship till with Reluctance I quitted her at Midnight in the last Boat, the Ship laying nearly on her Beam Ends and the Sea breaking violently over her. We had the good fortune to lose no lives in getting to the Schooner.

At daylight the Boats were sent to the Wreck to see what cou'd be sav'd, but all we cou'd procure was Sails & Spars with some Running Rigging. The Decks were blown up and the Ship almost entirely immers'd in the Water. As our Stock of Provisions and water was very inconsiderable to supply our Numbers (being 110 Souls) I thought it most prudent to find out a place to recruit at and in the Afternoon with the Boats accompanying, we steer'd in between the Islands[2] and the next day with great difficulty (from numerous Shoals) we Anchor'd near an Extensive Village, the Inhabitants humanely bringing us Water, Fire Wood & Millet.[3] Having paid the principals a Visit and Explaining to them our distress'd Situation and Wants, which they appear'd perfectly to Comprehend they most liberally gave us every thing I requir'd & without expecting any Return. These islands are a Group, which they generally named Madgicasimah[4] as the Chart Explains and Typinsan the Westermost off which we were lost in 25°2′0″ North Latitude & 125°25′0″ East Longitude. They are tributary to Great Liquieux and Trade both to China and Japan.[5] Their Manners & Language is Similar to these Nations, but a greater affinity to the Japanese. They were jealous of our seeing their Country, nor cou'd we prevail upon them to admit our making any excursions to acquire further Knowledge respecting its productions and trade. They have Cattle of all kinds and Grain of various sorts in profusion.

During our Stay of four days the Boats were again sent to the Wreck in hopes of procuring some Provisions. They return'd the third day unsuccessful. It had been necessary to go to ⅔ allowance and we now continued it and having present'd these well dispos'd people with some few Trifles,[6] and leaving the Longboat,[7] they seem'd much pleas'd with our attention and we departed from them on the 21st of May with the Cutter and Pinnace in Tow. On the 4th of June (without experiencing most providentially the SW Monsoon) we arrived safe in this River. His Maj*esty*'s Sloop Swift[8] arriv'd the same day with the News of the Spanish War. Captain

[1] Indecipherable; words in square brackets supplied from copy of letter in UKHO.

[2] 'which we knew were inhabit'd' added in the UKHO copy of this letter.

[3] ' with some Hogs' added in the UKHO copy of this letter.

[4] Madjicosemah in Broughton's *Voyage*, p. 210; Sakishima Gunto.

[5] '& Formosa' added in UKHO copy of this letter.

[6] '(all we possess'd)' added in UKHO copy of this letter.

[7] 'compleatly Rigg'd' added in UKHO copy of this letter.

[8] Commanded by Thomas Hayward, who had sailed with Bligh in the *Bounty* and had taken part in the open boat voyage.

Haywood[1] being in want of Men I have compleated his Complement[2] and the Remaining part[3] above the Schooner's Complement I have discharged into the Homeward bound Indiamen who are much distress'd for Men.

The 1st & 2nd Lieutenants with the Lieut*enan*t of Marines[4] I have sent home but the 3rd Lieutenant (who was the Officer of the Watch on the 17th of May) with the Master, Surgeon and some of the Petty Officers & a sufficient Crew[5] for the Schooner Remain with me. I hope their Lordships will approve my Conduct in this respect as I mean to proceed in the Schooner as far as circumstances will admit in the prosecution of my former Plan, and as our deficiency of Nautical Instruments will enable us to act, Returning in the NE Monsoon by this port, and from thence proceed to join Admiral Rainier in India, where I may expect to meet a Court of Enquiry on my Conduct.[6]

Mr Mudge 1st Lieutenant late of the Providence will deliver this letter and the Charts with which he is charg'd Relating to our Voyage and will explain every transaction their Lordships may think necessary to enquire into.

I have great pleasure Sir in communicating that every Officer and Man under my Command behaved with the utmost propriety & with the greatest Exertion possible in the Unhappy Event I have had the Misfortune to relate.

I have the Honor to be
Sir
Your most obedient
humble Servant
W. R. Broughton
Evan Nepean Esqe

11. To Evan Nepean, Secretary to the Admiralty[7]

His Maj*esty*'s Schooner Providence
at Macao June 25th 1797
Sir,

Having consider'd it my duty to alleviate in some degree the distresses of the Officers & Crew of his Maj*esty*'s Sloop Providence late under my Command, I beg leave to acquaint you of my having taken up One thousand Dollars for that purpose Charging the several Sums issued as Money advanced against the respective Names of the Ships Comp*any* in the Ships Books.

[1] Spelled correctly 'Hayward' in the UKHO copy of this letter.

[2] 'the Marines, Four petty Officers and twenty-four Seamen into him for the disposal of R. Admiral Rainier' in the UKHO copy of this letter.

[3] 'fifteen Seamen' in the UKHO copy of this letter.

[4] 'and Astronomer' added in the UKHO copy of this letter.

[5] 'to the number of thirty five' added in the UKHO copy of this letter.,

[6] 'and to deliver the Vessel, Officers & Crew to his disposal' added in the UKHO copy of this letter.

[7] TNA, ADM 1/1517, Cap letter B516. A duplicate copy of this letter with slightly different wording is held in the UKHO, OD 177.

I have to hope their Lordships will approve my Conduct in this Respect as both Officers and Men were entirely destitute of Cloathing (and at the time there was no other way of procuring any). And be pleas'd to give directions to the Navy Board accordingly.[1]
I have the honor
to remain Sir
Your Obedient Servant
W. R. Broughton
Evan Nepean Esq[e]

12. To Evan Nepean, Secretary to the Admiralty[2]

His Majesty's Schooner Providence at Macao
the 25th of June 1797
Sir,

I beg leave to acquaint you that having compleated the Schooners Complement to Thirty Five (Officers included) & Provision'd for Nine Months I propose sailing this day in the prosecution of my former intentions to finish the Survey of the Kurile Islands with the Coast of Corea and the Japanese Islands.

By Lieut*enan*t Mudge their Lordships will be inform'd of my former proceedings to the unhappy period of the loss of His Majesty's Sloop (late under my Command) but in case of accident I leave with M[r] Hall the principal of the Supercargoes a Duplicate of the Charts sent home by *Lieutenan*t Mudge to be forwarded to the Admiralty in case of my not arriving before Xmas.

The remaining part of the Providence's Crew I discharg'd into his Majesty's Sloop Swift to join Admiral Rainier except Fifteen to the Senior Commander of the homeward bound East Indiaman who was much distress'd for Men. The 1st and 2nd Lieut*enant*s the Officer of Marines, Astronomer, Boatswain & Gunner & some of the Petty Officers I have given leave to return to England. The remaining Officers being concern'd in the loss of the Providence I have thought proper to detain them till a Court Martial can take place. I trust the line of Conduct I have pursued may deserve their Lordships approbation As I am actuated by no other motive than a Zeal for the Service and a desire to do my duty to the best of my power.
I have the Honor to be
Sir
Your Obed*ien*t Servant
William Robert Broughton
Evan Nepean Esq[e]

[1] Changed to 'The several Amounts are charg'd in the Ships Books against their respective Names' in the UKHO, OD 177 copy of this letter.

[2] TNA, ADM1/1517, Cap letters B517.

13. To Evan Nepean, Secretary to the Admiralty[1]

Providence Schooner Macao Roads
The 1st of December 1797
Sir,

I beg leave to acquaint you for their Lordships information of my having Sail'd from hence the 26th of June Ult*imo* in the prosecution of my Voyage agreeable to the plan laid down in my former Letters.[2] We proceeded thro the Straits of Formosa[3] examining the western side of the Piscadore Islands[4] & passing several others in our route from the North point of Formosa to Great Liquieux where we Anchor'd in the Entrance of Napachan Harbor[5] situated in the Lat. of 26°11′30″N and 128°13′ East Long. The Inhabitants treated us in the most friendly manner Supplying all our wants liberally without expecting any return, but we had to regret they woud not admit of our Landing which precluded our making any useful observations with respect to their manners & Customs. This Island is very erroneously extended in the Charts, we make its situation N° & South from 26°4′N to 26°53′N & from 128°10′E to 128°45′E. It is of Moderate elevation & well cultivated & abounds in Grain Cattle & poultry. They trade to Japan & Formosa & have some Commerce with the Chinese. On the 16th of July we made the Japanese Coast having experienc'd in our run from Formosa thro' the Liquieux Islands very strong Currents setting to the NE'ward which prevented our passing to the Westward of Nipon as I Intended. We were becalm'd for 10 days in the Lat. of 32°30′N & 132°30′E Long. off the Island of Ximo remaining nearly stationary during that period. This loss of time induced me to give up the Idea of passing thro' the Corean Straits[6] for the present & we pursued our Course to the North along the Coast of Nipon to Volcano Bay[7] seldom losing sight of it & experiencing very fine weather & variable winds which gave me the opportunity of exploring those parts not seen in Cap*tai*n Cooks last voyage, & determining their Situation. It being necessary to compleat our wood & water I prefer'd the Situation of Volcano Bay for that purpose as our operations were not likely to be impeded by the Natives & on the 12th of August [we] Anchor'd in the Harbour of Endomo[8] in the NE[9] corner of the Bay & immediately commenced taking observations for the regulation of our Time piece (which err'd in our passage considerably). We found the Natives the same as last year without the smallest Variation & our Japanese friends soon join'd us from Matzmai[10] to expedite our Departure. As I was more anxious if possible than themselves to shorten our Stay we did not

[1] TNA, ADM 1/1517, Cap Letter B518. A duplicate copy of this letter, dated January 1798, with slightly different wording is held in the UKHO, OD 177, with added 'NB This letter went to England by a Swedish Ship p*er* favor of the Portuguese Minister at Macao.'

[2] 'and more particularly by Mr Mudge' added in copy of this letter in UKHO, OD 177.

[3] Taiwan Strait.

[4] P'eng-hu Ch'un-tao.

[5] Naha Ko.

[6] Korean Strait.

[7] Uchiura Wan.

[8] Muroran Ko.

[9] NW in the original MS, which is clearly an error; NE in copy of letter in UKHO, OD 177.

[10] They were from the Japanese town of Matsumae, situated at the SW extremity of Oshima Hanto, as explained by Müller: 'The Island of *Matmai* is the 15th, and concludes this range [Kuril'skiye Ostrova] ... The Japanese have built a Town on this Island, called by the same Name, *Matmai,* which stands upon the South West Shore, and is inhabited by *Japanese*'; Müller, *Voyages from Asia to America*, p. xxxviii.

give them any offence the short time we remained taking our Departure the 22nd of August. By them we found they had been Visited by a Russian Vessel from Kamschatka in a Harbor call'd Agoddoddy[1] situated in the Straits which divide Matzmai from Nipon.[2] I pass'd through these Straits & found the North part of Japan or Nipon to be in the Lat. of 41°31′N & Long. 140°46′E. They extend from Cape Nambu[3] to Matzmai[4] which we place in Lat. 41°24′N & Long. 139°58′E. There is good anchorage off it. Having clear'd the Straits by the 1st of Sept. We coasted the west side of Matzmai to the Lat. of 45°20′N running nearly in a North Direction from the SW point. Off the North point are two Islands, the North*ermos*t[5] due West from it. One of them is High & peaked wearing every appearance of a Volcano.[6] The Wind was Easterly & we were unable to approach the Northern part of Matzmai & on the following Day we saw more Land to the NE of us, the South part of which in the Lat. 46°N & Long 141°37′E bearing nearly North from Matzmai & of no great distance.[7] By some Charts I had procur'd from the Japanese conjectur'd this might be the large Island laid down off it & as the Wind only admitted our Steering to the North we pass'd on the western Side of it. In this part were some Scatter'd Villages. We found the Coast more Extensive than I expected the Island to be & in the Lat. of 50°20′N & Long. 141°30′E. We discover'd Land Distant to the Westward which we conjectur'd to be the Coast of Tartary & that the Strait we suppos'd ourselves at present in wou'd carry us to the Island of Sagaleen. In the night our Soundings gradually decreas'd & the next Day our Lat. was 51°25′ nearly the Situation of Sagaleen by Cap*tai*n King. The Shores now appear'd lower & more contract'd & we had only 11 fath*oms* 2 or 3 Leagues off Shore. We therefore stood over for the Coast of Tartary & after running 7 or 8 Leagues we Shoal'd to 4 fath*om*s when we came to an Anchor for the Night & Fortunately for us the Wind Shifted to the NW which enabled us to ride with some degree of Safety. In the Morning the Master was dispatch'd in search of an Harbor & to make other remarks. He return'd in the Evening reporting he had seen a Bay open to Easterly winds & passing over a Bank he found deep water within. He was unable to Land but saw no Signs of any Inhabitants. The Country was very barren & only presented a few Fox tail pine Trees of a small growth on the Hills. We in Consequence attempted the next Day to proceed in the Vessel but Shoal water prevented us, & from the Rigging we cou'd see Spits of Sand nearly in our Course & very low Land closing in every direction. I therefore from the unfavourable representation of the Bay by the Master & the evident proofs of their [*sic*] being no River or any opening whatever to admit our proceeding further to the North in this direction. As we were compleatly embay'd in the Lat. of 52°N & 141°E Long. I thought proper to return to the South keeping on with the Tartarian Coast as the Weather permitted. On the 24th of Sept*embe*r after combating some very bad weather we had reached the Lat. of 46°N & 138°E which makes the opening of the Gulf of Tartary[8] as I term it about 60 Leagues wide in the Entrance. The Season being too far advanced to expect our being able to proceed again to the North, towards the Sea of Okotsk and if we succeeded I conceiv'd contrary winds wou'd prevent the Examination

[1] Present-day Hakodate Wan.

[2] Tsugaru Kaikyo.

[3] Shiriya Saki.

[4] 'leaving a clear passage with deep water 5 or 6 Leagues broad the narrowest part' added in the copy of this letter in UKHO, OD 177.

[5] Rebun To.

[6] Rishiri To, the southern and larger island, which rises to a height of 1,721 metres.

[7] Sakhalin.

[8] Tatarskiy Proliv.

of the Corean Coast & Straits I thought it most advisable to give the Latter the preference & Continued our Course accordingly to the Southward. On the 12th of October we made the Island Tzima[1] which lies in the Center of the Straits which divide the Coast of Corea from the Japanese Islands. The Day following we discover'd a Harbor in the Coast call'd T'shosan[2] by the Natives. Here we remain'd a Week our wood & water being nearly expended & our watch requiring regulating. The Coreans treated us with Civility & amply supplied our wants but like our Liquieux friends wou'd not admit of our researches & indeed prevented our having any Communication with the Shore after the first Day placing their Military force in situations to prevent our making any excursions in the Country. The Harbor we found to lay in 35°2′N Lat. & 129°7′E Long. about 10 Leagues NNW from the Island of Tzima which we saw at Anchor very distinctly. The Country here was much broken & insulated seemingly abundant in Grain & Provisions & the Inhabitants very populous all Cloath'd in Linnen Garments & fine in quality according to the Distinction of the person; some of them were elegantly dress'd & Commanded great respect by the power they seem'd to possess over the lower class of People & apparently look'd upon us with great indifference, I suppose from the insignificancy of our small Vessel & our want of Language their comprehending satisfactorily who & what we were which probably made them so solicitous to make us depart from some jealous suspicions or other reason which we coud not divine. They shew'd not the smallest desire for any thing we possess'd except a Telescope. I therefore presented them with one. On the 21st we sail'd & experienc'd a constant succession of thick foggy Weather among the numerous Islands surrounding the Southern part of the Coast which rendered our Navigation intricate & our remarks were useless for want of the necessary observations. I was still in hopes of Continuing the Coast towards the Yellow Sea, and after examining the Island of Quelpart[3] (the Southern part of which is situated in the Latitude of 33°11′N and 126°20′ East Longitude) we proceeded to the NW as the Wind permitted which still continued unfavourable to our Course and we made the coast of China near the Nankin River Having Shoal water and very thick hazy weather. Indeed we had carried Soundings entirely from the Island of Quelpart across the Bay of Nanking.[4] The situation of this part we made in 32°N and 122°10′E and we tack'd and stood to the North with a fair Wind which left us the next day veering to the Northward and entirely stop'd our progress in that direction.

On the 7th of November we made an Island South of the Corean coast in Lat. 34°N and 125°6′E[5] and the Wind still continuing contrary I found it needless to persevere without any hopes of success More particularly as we had only Provisions for the Month at Short allowance (except Salt Meat).

I therefore thought it most advisable to desist in opposing the reigning Winds & stood to the Eastward towards the Islands laying South of Japan situated from 29°N to 31°N & forming Diemens Strait.[6] After exploring them we pursued our Track to the South[7] passing 20 Leag*ue*s to the West of my former Route with a View of ascertaining the position of any Islands that might fall in our way.

[1] Tsushima.

[2] Pusan.

[3] Cheju-do.

[4] Southern part of the Yellow Sea.

[5] Soheugsan Do.

[6] Osumi Kaikyo, separating Tanega Shima from the southern extremity of Kyushu.

[7] SW in the copy of this letter in UKHO, OD 177, which was the course Broughton steered.

On the 21st of November we made the Island of Formosa and experienc'd a continuation of bad weather in passing the Straits, and on the 29th we anchor'd in Macao Roads after the absence of Five Months.
I have the Honor
to Remain Sir
Your Obed*ien*t Servant
W*illia*m Rob*er*t Broughton
Evan Nepean Esqe.

14. To Evan Nepean, Secretary to the Admiralty[1]

Providence's Tender at Macao
the 21st December 1797
Sir,

I beg leave to acquaint you for their Lordships information of my safe arrival here on the 27th of November Ulto after an absence of Five months during which time I have been employ'd Surveying the Coasts of Tartary to 52°N Lat. & the Islands to the North of Formosa towards The Kurile Islands. After leaving a Copy of my proceedings at this port to return them by the India Fleet I propose Sailing to join Admiral Rainier with a view of acquiring a Court Martial upon my Conduct in the loss of His Maj*esty*'s Sloop Providence late under my Command.
I am Sir
Your Obed*ient* Serv*an*t
W. R. Broughton
Evan Nepean Esqe

15. To Evan Nepean, Secretary to the Admiralty[2]

Providence Schooner at Macao in China
the 1st of Jan*uar*y 1798
Sir,

I beg leave to inform you of my arrival here in December after an absence of Five Months during which time I have been employ'd continuing the Survey of the Coast of Tartary &c agreeable to my former Plan which I had the honor to transmit you for their Lordships information in April last. The advanced Monsoon & being short of Provisions prevented my Keeping the Sea [any] longer, but I trust their Lordships will be convinced the existing circumstances in so small a Vessel prevented my researches being so complete as I cou'd have wish'd at least with respect to the Kurile Islands and the Yellow Sea[3] owing to weather & adverse winds.

[1] TNA, ADM 1/1517, Cap Letter B519.

[2] TNA, ADM 1/1520, Cap Letter B448. A duplicate copy of this letter with slightly different wording is held in the UKHO, OD 177, with added 'NB This letter sent home by a Swedish Ship in January and another by a Danish Ship in December just stating our safe arrival.'

[3] Also known as Huang Hai.

Fearing any future accident[1] I thought it prudent to furnish a Copy of our Navigation, which will be transmitted Home by the Chinese Fleet in April next with a general description of my proceedings. At present not having rec'd by the Direct Ships any Orders relative to my future conduct, and conceiving from that circumstance their Lordships have been fully satisfied in the Survey of the SW part of America by Capt Vancouver upon which supposition agreeable to my Letter from California in June 1796; I thought it for the advantage of Geography to return to England by this Route; Nothing further of any consequence remains in my power to execute considering the unfortunate situation I have been thrown into by the Ship wreck of his late Maj*esty*'s Sloop under my Command in May last & the particulars of which event I have related to you by *Lieutenan*t Mudge.I have to request therefore you will acquaint their Lordships that as soon as the Necessary Charts are compleated I propose sailing immediately to join Rear Admiral Rainier with a view of acquiring a Court Martial on my Conduct on the loss of his Maj*esty*'s Sloop Providence and afterwards to return to England, Shou'd the Admiral have no Commands for me to execute. With this intention I have Kept with me in the Tender the necessary Officers & Crew to deliver them to the Fleet on my arrival in India.

It will afford me great satisfaction their Lordships approbation of my Conduct as it has ever been my most anxious Zeal to desire it.

I have the Honor
to be Sir
Your Obed*ien*t Serv*an*t
W. R. Broughton
Evan Nepean Esqe

PS I had the misfortune to bury One Man in my last Cruize.

16. To Evan Nepean, Secretary to the Admiralty[2]

Providence Schooner at Macao in China
15th Jan*uar*y 1798
Sir,

I herewith transmit to their Lordships a Log of my proceedings to elucidate the accompanying Charts which will best explain the course I have pursued in the Survey of the Kurile Islands & the Coast of Tartary, and will serve to clear up all Fabulous accounts relating to those Parts, which has long engaged the attention of the World, from the various accounts given by the Dutch, and Russian Navigators. I trust their Lordships will make the proper allowance for my situation in the Schooner, a Vessel incomplete with the necessary Instruments, Books &c to render the Voyage so compleat as they might desire, but under the

[1] Perhaps referring to the typhoon in which the *Swift* was lost with all hands, including many from the *Providence*'s ship's company, who were returning to England in her.

[2] TNA, ADM 1/1520, Cap Letter B450. A duplicate copy of this letter, dated 21 January 1798, with slightly different wording and without the list of charts, is held in the UKHO, OD 177, with added 'NB This Letter & Charts sent to Capt. Milne of the *Trident* to deliver to Mr Hall, the principal Supercargo Canton, the day we sail'd to be dispatch'd to England by the China Fleet in March next.' The 64-gun ship *Trident* had been sent to Macau to escort the homeward bound China Fleet; Parkinson, *War in the Eastern Seas*, p. 119.

existing circumstances I have some reason to hope they will be convinc'd that every thing was done in my power to explore those Parts which the unfortunate Shipwreck of the Providence prevented being done in a more extensive & compleat manner.

Having receiv'd no directions from their Lordships respecting my future proceedings, I beg leave to inform you I sail this day to join Rear Admiral Rainier as a more expeditious way of acquiring a Court Martial on the conduct of myself, Officers & Crew and to deliver them & the Schooner to his Disposal for his Maj*esty*'s Service. I have the honor to inclose you the State & Condition of the Sloops Crew (late under my Command), and of my having buried one Man by Sickness in my last Cruize.

I am Sir
Your Obedient Servant
W*illia*m R. Broughton
NB The inlosed Charts are as follows.
A General One from 53°N to 21°N including the Tracks of the Providence & her Tender.
Copy of a Manuscript procur'd at the Island of Matzmai, extending from Japan to the Peninsula of Kamschatcka. The Harbors of T'shosan, Endermo & Napachan Roads. I have also to inform you that in April 1797 I had the Honor to transmit to you for their Lordships information an account of my proceedings to that period with an Accompanying Chart by L*ieutenan*t Mudge.

Evan Nepean Esq[e]

17. To Evan Nepean, Secretary to the Admiralty[1]

Madras[2] June the 30th 1798

Sir,

I beg leave to inform you I sail'd from China on the 21st of January ult[o] and agreeable to the tenor of my Letter to you from thence, I proceeded to the East indies to join Rear admiral Rainier, passing through the Straits of Malacca, and arriving in these Roads in March last. Here I remain'd fourteen days victualling and refitting the Schooner for a Voyage to England in case we shou'd not fall in with the Admiral at Trincomalay[3] Harbor; but where I understood he was expected from the Malabar Coast.

In the Month of April we anchor'd in the Harbor and the Admiral not having arriv'd I found it necessary to heave the Schooner down to repair her Copper before our departure for Europe; but before we had compleated her damages the Rear Admiral appear'd in sight[4] and anchor'd on the 25th of the Month.

It was the 19th of May before a Court Martial assembled to inquire into the loss of his Maj*esty*'s late Sloop Providence under my command, the result acquitted me of any blame whatever being attached to my conduct but proving the loss of his Majesty's Sloop to be

[1] TNA, ADM 1/1520, Cap Letter, B451. A duplicate copy of this letter with slightly different wording is held in the UKHO, OD 177, with the addition of 'This Letter & Chart left in charge of Mr Street Purser in his Majesty's Navy expected to leave India by the 1st Ship.'

[2] Now Chennai.

[3] Trincomalee

[4] In the 74-gun ship *Suffolk*.

occasioned by the negligence and disobedience of my Orders by Lieut*enan*t James Giles Vashon, the Officer of the Watch, who was sentenced to be dismissed his Maj*esty*'s Service, but in consequence of his having serv'd with me in the same Capacity (in the Schooner) upon particular professional service the Court submitted his case to the consideration of the Commander in Chief.[1]

On the 24th of May I deliver'd up the command of the Schooner,[2] Officers & Men in obedience to Rear Admiral Rainier's Order, who recommended my proceeding to England by the first convenient opportunity. It was the 15th of June before I cou'd procure a passage to Madras and to the present date before I found any conveyance towards the Cape of Good Hope. I have therefore to request you will please to acquaint their Lordships I am returning to England with all expedition and in consequence of [this] situation I have judged it expedient to draw upon the Navy board for my travelling Expenses, and have to solicit their Lordships will please to give directions for their payment.

I have the honor to remain
Sir
Your very Hum*bl*e Servant
Will*ia*m Robert Broughton
Evan Nepean Esqe

18. To Evan Nepean, Secretary to the Admiralty[3]

London the 6th February 1798 [*sic*]

Sir,

I beg leave to acquaint you with my arrival in England for the information of their Lordships and herewith inclose you an attested Copy of the Court Martial held in my conduct for the loss of his Maj*esty*'s Sloop Providence late under my Command in the East Indies p*er* Order of Rear Adm*ira*l Rainier. I also have to acquaint you I have brought with me for their Lordships information the Journal & Charts of my proceedings since the loss of the said Sloop.

I am Sir
Your very Hum*bl*e Serv*an*t
W. R. Broughton
Evan Nepean Esqe

Minuted '7 Feby Our [recd?] and acquaint him as it appears by the findings of the Court Martial that the Officers who were present only were acquitted desire he will let me know whether he has any charges to prefer [against] any other officers late belonging to the Sloop. Orders for paying her off.'

1 Admiral Rainier clearly did not confirm the sentence as Vashon was appointed almost immediately to the 44-gun frigate *Sybille* in the East Indies fleet.

2 Admiral Rainier then appointed William Hargood in command. She was subsequently commanded by Lieutenant Peter Heywood, who had sailed with Bligh in the *Bounty*; see A. C. F. David, 'Broughton's Schooner and the Bounty Mutineers', *The Mariner's Mirror*, Vol. 63, pp. 207–13.

3 TNA, ADM 1/1520, Cap Letter B449. A duplicate copy of this letter with slightly different wording is held in the UKHO, OD 177, with the addition of 'This Letter & Chart left in charge of Mr Street Purser in his Majesty's Navy expected to leave India by the 1st Ship.'

II. Broughton to the Board of Longitude

To The Commissioners of the Board of Longitude[1]

His Majesty's Sloop
Providence
Macao in China.
1st April, 1797

Gentlemen,

I beg leave to inform You of my having receiv'd Orders from the Lords Commissioners of the Admiralty dated the 31st of December 1793 to convey Mr Crosley Astronomer appointed by your Board to make Astronomical Observations in a Voyage to Nootka Sound or elsewhere; where he might expect to meet His Majesty's Sloop Discovery Captain Vancouver and having certain intelligence of the Discovery's return to England; I have thought proper to inform You (conceiving His Observations to be equally necessary in the Voyage on which I am employ'd) of my having detained him on board His Majesty's Sloop under my Command to act in the same Manner According to your Instructions as he was before directed to do on board of Captain Vancouver.

I have only to regret the want of Instruments preventing his observations being equal to your expectations, but I conceived it highly improper his returning to England while there remained any opportunity of exercising his Abilities.

I am Gentlemen,
Your obedient humble
Servant
Will*ia*m Rob*er*t Broughton

The Commissioners of the
Board of Longitude

III. Crosley to Nevil Maskelyne and the Board of Longitude

1. To Nevil Maskelyne[2]

Rec^{d} Novr 22

H.M.S. Providence
Cawsand Bay, Plymouth
Sound Nov*embe*r 18th 1794

Rev*eren*d Sir,

When I wrote you last the Ship was at St Helens[3] waiting for a fair wind. I observed three Merid*ian* Alt*itude*s ⊙ LL[4] which gave the Latitude 50°47′35″N and two sets of observations

[1] CUL, RGO 14/1, f. 158.

[2] CUL, RGO 14/68, ff. 93–4.

[3] St Helens Roads, a fleet anchorage in the SE approaches to Portsmouth NNE of Bembridge Point, the eastern extremity of the Isle of Wight. Crosley's earlier letter does not appear to have survived.

[4] The sun's lower limb.

on 11th and 17th Oct*obe*r for finding the time,[1] the longitude by the Time-keepers taking a mean of the two days observations is as follows No 248 – 4′8″·0, No 1 – 4′13″·4, No 2 – 4′9″·0, No 56 – 3′50″·5 W in Time,[2] Bembridge point bearing p*er* Compass SWbS distant about 2 Miles allowing the variation 21°W the Latitude of the point is 50°40′38″N and its longitude in degrees No 248 – 1°2′47″, No 1 – 1°4′1″, No 2 – 1°2′55″, No 56 – 0°58′21″W. The latitude is very near that put down in the Requisite Tables, but the longitude differs[3] – I found from the above observations that the Time-keepers had altered their rates from that given me by Mr Bayly[4] at Portsmouth; their daily rates appeared to be No 248 +4″·0, No 1 – 7″·25, No 2 +4″·6, No 56 – 5″·17 as the interval between the observations is only 6 days and I not being used to the motion of the ship suppose there may be some error in my observing, therefore shall allow Mr Bayly['s] rate No 248 +5″·0, No 1 -8″·8, No 2 +5″·0, No 56 -2″·0 p*er* day. The wind coming fair we sailed from St Helens October 21st with the West India Fleet under convoy of H.M. Ships Polyphemus, Saturn, Trusty and Fury Sloop. In the morning the wind suddenly changed to the South when the Fleet bore up for Tor Bay where we anchored about 5 oClock on the 22d Berry Head the southermost point bearing South p*er* compass distant about 2 Miles.[5] I observed some altitudes for the time and took some azimuths but had not an opportunity of making any observations for the Latitude. In a Survey of Tor Bay by Capt*ai*n Knight the Latitude of Berry Head is put down 50°28′0″N and its longitude in time 14′20″W. [I] made use of the above latitude in finding the time and Azimuth when reduced to the Ship the variation of the compass I found 28°52′W which I believe is more than what is generally allowed in the Channel the Azimuths were taken with a common ship's Azimuth compass Capt*ai*n Broughton not having a stand for a better compass made by Adams, shall get a stand made the first opportunity – The Longitude of Berry Head by the Time-keepers is as follows No 248 – 13′13″·9 W, No 1 – 13′49″·1, No 2 – 13′12″·6, No 56 – 12′37″·5 W in Time. They differ from Capt*ai*n Knights[6] longitude all of them making it less.

We sailed from Tor Bay about 1 oClock on 23d[7] October with a fair wind and hazy weather. At 4 in the Evening spoke H.M.S. Artois with the French Frigate La Revolutionaire[8]

[1] I.e. local time.

[2] These were Earnshaw 1, Earnshaw 248, Earnshaw pocket chronometer 2, Earnshaw pocket watch 1522 and Arnold 56. Earnshaw 1 and Arnold 56 were intended for Bodega y Quadra, but as he had died they were delivered to the Governor of Monterey instead. Earnshaw 248 and Earnshaw pocket watch 1522 were Crosley's personal property.

[3] 50°40′15″N, 1°4′45″W in the 2nd (1781) edition of the *Requisite Tables*, but 50°40′59″N, 1°3′26″W in the 3rd (1802) edition, possibly computed from the Grand Trigonometrical Survey.

[4] William Bayly, who served as astronomer in the *Adventure* during Cook's second voyage and in the same capacity in the *Discovery* during Cook's third voyage. He was headmaster of the Royal Naval Academy, Portsmouth, 1785–1807, where there was an observatory. From this it seems likely that the chronometers belonging to the Board of Longitude were sent to Bayly for him to rate them before handing them over to Crosley.

[5] There is no mention of the *Providence* anchoring in Tor Bay in Broughton, *Voyage*, p. 4.

[6] Captain John Knight was a talented and prolific 'amateur' surveyor, whose surveys were published by William Faden. Broughton served under him on two occasions.

[7] Crosley is in error here; they sailed on the 25th.

[8] *La Révolutionaire*, a 44-gun French frigate, commanded by Henri-Alexandre Thévenard, was intercepted by a British squadron on 21 October 1794, about 30 miles west of Ushant. The superior sailing of the 38-gun frigate *Artois*, Captain Edmund Nagle, enabled her singly to bring the French frigate to action. She defended herself with great gallantry, but struck her colours to the *Artois* when the rest of the British squadron closed in; James, *Naval History*, I, p. 211. *La Révolutionaire* was taken into the Royal Navy with the same name and was eventually broken up in 1822.

her prize. We gave them three cheers which was returned by both Ships we then prosecuted our voyage, nothing remarkable happening untill 26th when being about 40 leagues west of Scilly the wind became contrary and blew a hard gale, when the Commodore made the signal to bear up to the nearest port. We accordingly bore up and made this place on 28th at Noon where we anchored, the middle of Drakes Island bearing *per* compass N70°8′E distant about 2 Miles the variation found by the same compass 27°42′W. From the Latitude and longitude of Drakes Island setled by Mr Bayly[1] I call the ships Latitude 50°20′0″N and its longitude in Time 0h 17′13″ West.[2] [I] Should have wrote you before but not being able to give any account of the going of the Time-keepers for altho we have been 21 days here, have never seen the Sun but two mornings. All the rest of the time have had continual rain and fogs. By Observations made on the 7th the longitude *per* Time-keepers No 248 – 16′37″·2 W, No 1 – 17′25″·6, No 2 – 17′21″·7, No 56 – 16′34″·3 and from the observations made on the 13th No 248 – 16′19″·8, No 1 – 17′12″·0, No 2 – 17′23″·7, No 56 – 16′36″·7; their rates from the above observations are No 248 +2″·06, No 1 -11″·05, No 2 +5″·33, No 56 – 3″·63. I believe it is not possible to get an accurate rate of a Time keeper from Observations made on board a Ship without the interval between the observations is 12 or 14 days for when the ship was at Spithead I took observations at intervals of 3 or 4 days but always found different rates. When comparing it with Mr Bayly found it had not altered its rate, but an interval of 12 or 14 days generally gave me very near the true rate. If we [indecipherable] stop here long enough to make any more observations shall take the [indecipherable] they make in the longitude of Drakes Island which divided by the [number] of days since the 6th Oct*obe*r, when I got the time from Mr Bayly, when applied to their Portsmouth rate will probably give a rate nearer what they will keep at Sea than the rate from the preceeding observations – Do not know how long we may remain here at present we are waiting for Admiral Parker[3] who is at Spithead if you have any further instructions to give me if you will please to send them here will be much obliged to you, with an order if the ship has sailed, to be returned to you at Greenwich upon the back of the letter. I have not been the least Sea sick but found great inconvenience from the rolling of the ship in the gale not being able to walk.

I remain with all due respect.

Rev*eren*d Sir,
Your most Obedient
H*um*ble Serv*an*t
John Crosley

[1] 50°21′30″N, 4°13½′W; Cook and King, *Voyage to the Pacific Ocean*, I, p. 13. According to Bayly his observations gave the longitude of Plymouth Sound as 4°33′56″W, but accepted 4°17′ as the true longitude as deduced from former observations; William Bayly, *Original Astronomical Observations*, p. 1.

[2] Or 4°18′W.

[3] Admiral Sir Hyde Parker (1739–1807), who was going to the West Indies to take over as commander-in-chief, Jamaica, 1796–1800. He was commander-in-chief of the British fleet at the first Battle of Copenhagen in 1801 and is chiefly remembered for the signal he sent to Nelson to leave off the action, which Nelson is said to have ignored by lifting his telescope to his blind eye.

2. To Nevil Maskelyne[1]

H.M.S. Providence
Plymouth Sound
December 20 1794

Rev*eren*d Sir,

I received your Letter of the 24th last month and thank you for the problem for finding the Latitude and apparent time which I think may often be very usefully applied when no other sort of observation can be got – Since I wrote you last we have been very near lost upon the Rocks under Mount Edgecumbe. The Commodore having made the Signal for sailing on Nov*embe*r 18th Capt*ai*n Broughton wishing to get further out of Cawsand Bay we weighed, and made sail very near the shore that the next tack we might stand out of the Bay, the ship missed stays when we were obliged to come too with both Bower anchors ahead very near the rocks. On sounding round the ship found a rock very near her stern and the anchors were in bad ground. Cap^{n} B immediately left the ship to get assistance from the Dock Yard to heave the ship further from the rocks – about 5 oClock the next evening the Master Attendant[2] came to our assistance with a Sloop and Launch. The sloop let go an Anchor of 18 Cwt a head, having bent to it three 12½ Inch Cables on an end. We got the end of the third Cable aboard and hove it taught, the Dock Yard Sloop riding to the bight of the Cable between the ship and Anchor to keep the Cable of[f] the rocks – about 11 oClock found the Cable slack hove it in and found it had been cut by the rocks about ½ a Cables length from the ship. In this situation we remained the Night, it blowing a very heavy Gale of wind directly on shore. If either of our own Cables had parted we must certainly have been upon the rocks in a few minutes. As soon as it was daylight we sent the Launch to the sloop to get hold of the end of the Cable we had parted in the night; got it on board and hove it taught, then hove up our two Bower Anchors [and] found both Cables very near cut through. Cut of[f] the wounded part and bent them again. The wind being more moderate in the afternoon of the 20th we hove in one of the Dock Yard Cables and let go our small Bower. About 4 oClock in the Morning it began to blow harder than the preceeding day accompanied with very heavy rain, the Sea breaking entirely over the ship at times, which continued all the morning – about 2 oClock in the afternoon of 21st there came on a very heavy squall of wind and rain which parted our small Bower Cable and the ship Drove before we could let go the Best Bower, but providentially the wind changed suddenly to SSW and almost immediately became calm which relieved us from our apprehension of going a shore. We then hove in the 2d Dock Yard Cable by which means we was so far from the Shore as to be able to make sail, which did the next morning and anchored between Drakes Island and the Main where we remained very quite untill the 16th of this month when some line of Battle Ships going into Hamoaze[3] ordered us out into the Sound where we at present remain.

I carried my Sextant, artificial Horizon and a Pocket watch No 1522 (made by Mr Earnshaw for the purpose of being compared with the Time keepers to take the time on Deck or on shore without moving the Time-keepers which are fixed in the Great Cabin which does not alter its rate above 8 or 10″ p*e*r day) on shore upon Drakes Island w[h[ere I endeavoured to

[1] CUL, RGO 14/68, ff. 95–95v.
[2] Mr Hemmings; Broughton, *Voyage*, p. 5.
[3] The fleet anchorage off Devonport.

find out the place where Mr Bayly observed in 1776. Found the Gunners House he mentioned pulled down and a circular wall built to support a Flag Staff in its place, upon which I set the Quicksilver covered with the roof which was the only place I could find that was level and steady, but it was very much exposed to the wind, and the sun being so low at this time of the Year the observations are not so good as I could wish, but they are the best with my present Instruments I could make. Have inclosed the observations, also those made on board the Ship, have not had any opportunity of finding the Variation of the Compass, or of making Lunar Observations except on [blank]. The weather has been so bad for what little fine weather we have had I applied to the finding the rate of the Time-keepers. Hope when we are clear of the Channel and have quitted the Fleet to send home some observations that may be of consequence. At present do not see any probability of making any in this place, being that time we have only one Boat in use, Cap^{n} B being afraid of losing his men if he sends any of them ashore. Therefore cannot always get a Boat but the wind in general is so high that it is but seldom I can observe to any advantage on shore. On board the Ship cannot make an Observation for we are surrounded by the Fleet.

We have been here nearly 8 weeks waiting for Admiral Parker with the Fleet from Spithead, which has attempted to Sail several times but always being obliged to put back by contrary winds. Yesterday one division of it arrived here which brought the intelligence of his being in Tor Bay and that he would sail from there with the first fair wind and expected the Fleet from this place to join him when he appeared off the Sound ... [remainder of letter wanting]

3. To Nevil Maskelyne[1]

H.M. Sloop Providence
Plymouth Sound,
February 13th 1795

Rev*eren*d Sir

I rec*eive*d your Letter of [*blank*] Ult. on [*indecipherable*] have very little to add to the Observations which I sent you before for altho the weather was very fine during some part of the Frost it was always accompanied with a fresh wind. I made several attempts to observe altitudes of the Sun for finding the time on Drakes island, but always found the wind so high that it shaked the Quicksilver so much that the observations could not be depended on. I attempted to observe on the shore near the ship but found it very difficult to land and the wind shaked the Quicksilver as before – The only observation for the time which I have made was on Drakes Island on 22d Jan*uary* which gave me the following rates of the Watches in an interval of 43 days, No 248 +9″·48, No 1 – 5″·43, No 2 +13″·73 and No 56 – 2″·19 which is but little different from the rates I sent you last except No 248 which is not properly corrected for heat and cold – On comparing it with No 1 I find the same difference since the thaw as it had before the frost when the Thermometer is at the same height. The last height of the Thermometer during the frost at 6 oClock in the morning upon deck was 24°.

I observed the Eclipse of the Moon on February 3d with my Sextant Telescope as follows –

Begining hazy	10h 44′ 55″	App*aren*t Time
End very distinct	13 33 48	

[1] CUL, RGO 14/68, ff. 96–7.

I got the time from the observation of the 22 Jan*uary* allowing the watch No 248 to be gaining 9″·48 p*er* day and to be Jan*uary* 22d at 2h 25′ 16″ +31′ 1″·43 to[o] fast for mean time at Drakes Island. At the beginning of the Eclipse the air was very hazy and remained so till very near the end when after a shower of rain the air was very clear when I observed the end very distinct.

I am afraid you will think me negligent in making so few observations but in our present situation I have made all that lay in my power, our orders being to follow Admiral Parker's motions and expecting to sail every day we have all our Boats stowed for Sea except one which is generally imployed in doing the business of the Ship. On board the Ship I cannot see the horizon being surrounded by the Fleet which now amounts to near eight hundred [ships].

A Frigate came in this afternoon from Lord Howes Fleet, which is in Tor-Bay. Therefore if the wind continuing fair as it is at present we expect to sail tomorrow morning. Will be much obliged to you to send a certificate of my being in the service of the Board of Longitude to Mess[rs] Clementson & Denton Navy Agents, No 14 Clements Inn, who I have appointed my agents to receive what money may be due to me and settle any accounts that may occur during the Voyage. We have remained here so long that the live Stock we bought at Portsmouth, through the inclemency of the weather is greatest Part of it dead so that we shall be obliged to replenish it at the first port we make after leaving this place.

I remain
Rev*eren*d Sir
Your most obedient
Hu*m*ble Ser*van*t
John Crosley

NB The Longitude from the Lunar Observation on Decem[r] 9th should be 17′47″·5 West instead of 17′32″·5 West it being copied wrong in the acc*oun*t I sent you last.

4. To Nevil Maskelyne[1]

H.M.S. Providence
Sydney Cove New
South Wales Oct*obe*r 6th 1795

Rev*eren*d Sir,

I have taken the opportunity of writing by the Ship Young William, Capt*ai*n Mortlake[2] who arrived here yesterday from England after a passage of 4 Months and 10 days and will sail immediately after delivering her cargo to China and from thence to England. When we arrived here there did not appear any chance of sending a Letter home therefore did not write out the Observations, but intended to send them from the Sandwich Islands [as] Capt*ai*n Broughton informed me there might be a probability of sending home from there. We shall drop down the Harbor this morning and if the wind permit sail to morrow. [I] have not had time to write out the observations themselves which I hope you will excuse but have sent the

[1] CUL, RGO 14/68, ff. 98–104.

[2] Captain J. Mortlock, *Historical Records of New South Wales*, 7 vols, Sydney, 1893–1901, II, p. 284. The *Young William* arrived in Sydney on 4 October 1795 and sailed for Canton on 29 October; Cumpston, *Shipping Arrivals*, p. 30.

results. On February 14th the afternoon before we sailed from Plymouth Sound I observed some altitudes of the Sun on board the Ship for finding the time (Drakes Island bearing N82°W distant 8/10 of a Mile) which being reduced to the Island and compared with the Observations made there on January 22d gave the following Rates of the Time-keepers, and how much they were to[o] fast for Mean Time at Drakes Island Feb*ruary* 14th at noon: Viz No 1 +2′20″·50 and losing 5″·28 p*er* day, No 2 +42′4″·55 and gaining 15″·81 p*er* day, No 56 +14′25″·93 and losing 0″·29 p*er* day, No 248 +34′58″·36 and gaining 10″·34 p*er* day. I do not think the above rates very accurate but they were the only ones I could get, intending to find new rates at Teneriffe if we stayed there long enough. We sailed from Plymouth Sound on Sunday February 15th in company of the West India Fleet under the convoy of Admiral Parker and was joined off the Sound by Lord Howes Fleet. On Monday 16th at 6h the Lizard Light bore N27°19′W distant 7 Miles from which I took my departure. On the Afternoon of the 18th Lord Howes Fleet parted company with us and stood in for the Coast of France. We proceeded on our voyage having at intervals heavy squalls of wind and rain. On the 28th Admiral Parker with the West India Fleet parted company and we proceeded with the Ships bound to the coast of Guinea under the convoy of the Iris Frigate. On March 4th at 22h the high land on the Island of Grand Canaria bore S15°8′E and at the same time the Peak of Teneriffe[1] S78°22′W. On March 5 at Noon the high land bore S26°28′E and the Peak S74°52′W. Lat*itude* of the Ship at Noon on March 5th 28°29′22″N. 9 Distances of Spica east of the Moon observed on the 3d and reduced to the 5th at noon p*er* Time-keeper No 1 gave the Long*itude* 15°21′19″W. The Lunar Observations made in Sancta Cruz Bay[2] reduced back to the 5th at noon p*er* No 1 give 15°56′17″W. 17 Distances of the Sun east of the Moon taken on the 14th and reduced back to the 5th at Noon give 15°29′39″W. The mean of these means give the Ships longitude at Noon 15°35′45″W. The difference of Lat*itude* and Long*itude* between the Ship and

High Land	Lat. 0°18′26″S Long. 0°10′28″E
Peak	Lat. 0°11′15″S Long. 0°47′17″W

hence their Lat*itude* and Long*itude*

High Land on Grand Canaria	Lat. 28°19′16″N Long. 15°25′17″ West
Peak of Teneriffe	Lat. 28°18′ 7″N Long. 16°23′ 2″ West

The Longitude March 5 at Noon p*er* Time-keepers No 1 – 15°58′49″W, No 2 – 15°54′19″W, No 56 – 16°9′4″W, No 248 – 15°58′30″W. On the 5th at 15h came to anchor with the best Bower in 35 Fathom water in Sancta Cruz Bay. The Harbor Master came on board next day and shifted our birth about half a Mile to the Northward when the NE point of the Bay bore N62°7′E distant about 7 Miles and the most westerly Church with a very high Tower (called by Capt*ain* Cooke St Francis[3]) N62°18′W distant about 1½ Miles. The evening we came into the Bay I observed 15 Distances of Spica east of the Moon and 9 Distances of Pollux west of

[1] Pico de Teide, 3,715 metres high, which at the time was one of several prime meridians from which longitude was measured.

[2] Bahía Santa Cruz de Tenerife, on the NE side of Tenerife.

[3] 'St Francis's Church, remarkable for its high steeple'; Cook and King, *Voyage to the Pacific Ocean*, I, p. 16.

the Moon. The next evening I observed 7 Distances of Spica east of the Moon which give the Longitude of the Ship as follows –

March 5th Spica east of the Moon	16°18′45″	The mean of the 7 sets give the Longitude of the Ship 16°14′13″W And the Latitude from the mean Of three Merid*ian* Altitudes of the Sun 28°27′31″ North
	16 26 30	
	16 25 0	
Pollux west of the Moon	15 58 45	
	16 7 45	
March 6th Spica east of the Moon	16 9 15	
	16 13 30	

The Time-keepers give the Longitude as below –
No 1 – 16°16′0″W, No 2 – 16°12′45″W, No 56 – 16°29′30″W, No 248 – 16°15′15″W. The fore-going determinations differ from Capt*ain* Cooke who calls the Latitude 28°30′11″ North and the Longitude 16°31′0″W.[1] He lay further to the North than where we anchored for the Peak bore S65°W. We could not see it from our ship. I found the Variation of the Compass from a mean of 6 sets taken by Adams's's[2] compass 16°38′W. Capt*ain* Broughton informed me he did not intend to stop above a day or two therefore did not attempt to find new rates for the Time-keepers which did not appear necessary as they agreed so well with the Lunar Observations and with the Longitude as settled by Chevalier de Borda 16°16′30″ West[3] – The Latitude and Longitude of the NE point of the Bay and St Francis's Church from their bearings and estimated distance will be

NE Point Lat. 28°30′47″N. Long. 16°7′9″W
St Francis Lat. 28°28′13″N. Long 16°15′42″W

We sailed from Sancta Cruz Bay March 12th in company with H.M. Ships Reliance and Supply bound to Port Jackson, having on board His Excellency John Hunter Esq. Governor General of New South Wales. At 4h the NE point of Sancta Cruz Bay bore N13°49′W Distant 6 Leagues. We met with light winds and hazy weather. On March 15 at Noon the Peak bore N3°16′W; at 6h 30′ after having run 17 Miles on a N55°29′W course made good it then bore N24°14′E. Its difference of Latitude and Long*itude* from the Ship on March 15th at Noon will be Lat. 0°36′8″N and Long. 0°2′19″W. The Latitude of the Ship on the 15th from a Merid*ian* Altitude ⊙ L.L. 27°31′30″N carr[y]ing on the observations made at Teneriffe p*er* N 1 to the 15 at Noon give the Longitude 16°32′1″W; 17 Distances of the Sun east of the Moon taken on the 14th and reduced to 15th at Noon give 16°5′23″W. The mean of the two means give the Ships Long*itude* at Noon on 15th 16°18′42″W; The Long*itude* p*er* Time-keepers on the 15th at Noon No 1 – 16°34′33″W, No 2 – 16°31′18″W, No 56 – 17°0′3″W, No 248 – 16°31′33″W. The Lat*itude* and Long*itude* of the Peak from these observations Lat. 28°7′38″N, Long. 16°21′1″W. The Latitude differs considerably from that on March the 5th before we made Sancta Cruz Bay. [I] do not think either of them very accurate for the Ship

[1] Cook and King, *Voyage to the Pacific Ocean*, I, pp. 19–20.

[2] George Adams junior (1750–95).

[3] Cook and King, *Voyage to the Pacific Ocean*, I, p. 20. Captain Jean-Charles Borda (1733–99), a French naval officer in command of the *Boussole*, was sent to Tenerife to test the performance of two chronometers in conjunction with José Varela y Ulloa (1739–94), a Spanish naval officer who had established an observatory there. When Cook called at Santa Cruz in August 1776, they passed the results of their observations to him.

ran only 5 Miles from the 4th at 22h to the 5th at Noon and that upon two courses. From the 15th at noon to 6h 30′ she run 17 miles on three courses; therefore the Peaks distance from the Ship cannot be very accurately determined. Capt*ai*n Cooke determines the Lat. 28°18′0″N and the Longitude 16°30′20″W.[1] On March 15 at 22h saw the extremes of the Isle of Gomera from N20°E to N60°E distant 5 or 6 Leagues, the South end of the Isle of Ferro[2] West, and the Isle of Palma NbW. The weather was so hazy that I could get no observations for the Longitude by the Time-keepers. At noon the weather so hazy that the Islands were not visible therefore [I] could not settle any of them. On March 20th at 22h saw the Island of St Anthony[3] bearing South. On the 21st at Noon the high land[4] bore S25°40′E and after running 25 Miles on a S12°50′W course it then bore S83°40′E at which time I observed two sets of Altitudes for finding the time which gave the Longitude *per* Time-keepers as follows, No 1 – 25°32′15″W, No 2 – 25°27′0″W, No 56 – 26°6′45″W, No 248 – 25°28′0″W. Reducing the Lunar observations made before and after leaving the Island to this Time p*er* Time-keeper No 1 will give the longitude as follows –

31 Observations at Sancta Cruz		25°29′43″W	Before making the Island mean 25 15 35W
17 Do.	on March 15th	25 3 5 W	
19 Do.	on March 26	25 55 49 W	After leaving the Island mean 25°54′36″ West
18 Do.	on March 27	25 38 32 W	
20 Do.	on March 28	25 53 49 W	
14 Do.	on March 29	26 10 15 W	

The mean of both means 25°28′51″W. The difference of Lat*itude* and Longitude between the Island of St Anthony and the Ship at the time the Altitudes were taken is Lat. 0°2′0″S, Long. 0°18′53″E. Consequently the Lat*itude* of the high Land of St Anthony [is] 17°0′46″N and the Long*itude* 25°9′58″W. We proceeded on our Voyage without any thing happening worth notice and crossed the Equator on the 9th of April. The greatest height the Thermometer stood at was 86° [and] we felt very little inconvenience from the heat except two or three days in a Calm. April 22d at Noon the Island of Trinadada[5] bore S43°E dist*ant* 10 or 12 Leagues; the Latitude from Merid*ian* Alt*itude* ⊙ L.L. was 19°57′33″ South and the Longitude p*er* Time-keepers No 1 – 29°34′51″W, No 2 – 28°34′36″W, No 56 – 31°12′6″W, No 248 – 28°49′6″W. Reducing the Lunar Observations made before and after leaving the Island to the 22d at Noon p*er* No 1 the Long*itude* of the Ship on 22d at Noon will be

18	Distances on the	14 April	29°36′16″W	Before making the Island the mean 29°40′18″West
15	Do.	15 April	29 44 21	
11	Do.	22 April	30 3 27	Off the Island
12	Do.	24 April	30 3 26	After leaving the Island the mean 29°54′39″ West
15	Do.	26 April	29 54 59	
20	Do.	27 April	29 48 16	
18	Do.	28 April	29 57 43	
26	Do.	29 April	29 50 11	

[1] Cook did indeed make the latitude of the peak 28°18′N, but his longitude of 16°30′20″W is by lunar observations, while by K1 it is 17°0′30″W; Cook and King, *Voyage to the Pacific Ocean*, pp. 20–21.

[2] Hierro, the SW island of Islas Canarias.

[3] Ilha de Santo Antão, the NW island of Arquipélago de Cabo Verde.

[4] I.e. the summit of Ilha de Santo Antão.

[5] Ilha da Trindade.

The Mean of the three means 29°52′48″W. Supposing the Island distant at Noon 33 Miles the diff*erence* Lat*itude* 0°24′8″S, Long*itude* 0°23′52″E, the Lat*itude* and Long*itude* of Trinadada will be Lat*itude* 20°21′41″ South and Long*itude* 29°28′56″ West. On May 2d at Noon Cape Frio bore N5°15′W distant 8 or 9 Leagues at which time the Lat*itude* was 23°25′34″ South and the Longitude p*er* Time-keepers No 1 – 41°38′35″W, No 2 – 40°20′9″W, No 56 – 43°24′39″W, No 248 – 40°48′56″W. Reducing the Lunar Observations made before we saw the Cape and after leaving it to this time will give the Long*itude* of the Ship on May 22d at Noon viz –

18	Distances on	April 14th	41°40′ 0″W	Before making Cape Frio the mean 41°55′54″ West
15	Do	April 15	41 48 5	
21	Do	April 22	42 7 11	
12	Do	April 24	42 7 10	
41	Do	April 26	42 0 43	
20	Do	April 27	41 52 0	
18	Do	April 28	42 1 27	
20	Do	April 29	41 53 55	
15	Do	on May 3	41 50 16	In sight of the Cape
56	Do	Observed on the Island of Enchados[1]	41 46 5	On the Island of Enchados after leaving the Cape

The mean of the three means 41°50′45″W. Supposing Cape Frio Dist*ant* at noon 24 Miles and the diff*erence* Lat. 0°25′53″N, Long. 0°2′17″W, consequently, the Lat*itude* 22°59′41″S and Long*itude* 41°53′12″W of Cape Frio. On May 4th at 6h we came to Anchor off the Harbor of Rio Janeiro the wind dying away we could not get in, the Sugar Loaf[2] bore N49°W Distant about 10 Miles. On the 5th at 22h weighed and made sail for the Harbor [and] sent the Boats a head to tow. On the 6th at 3h H.M.S. Reliance Saluted the Fort of Sancta Cruz with 13 Guns which was returned with an equal number. At 6h 40′ we came to anchor in 5½ Fath*oms* water. Iimmediately on our anchoring the Guard Boat came along side to know who we were and insisted on leaving a Corporal on board the Ship; which being refused a Lieutenant came on board who said he was ordered by the Vice Roy to put a Guard on board the Ship. Being told by Capt*ai*n Broughton that H.M. Ships never permitted any Soldiers to be put on board, they contented themselves with going round the Ship all Night. Next morning they would not permit any of our Boats to go on shore untill Governor Hunter had visited the Vice Roy, but this was prevented by the Vice Roys secretary and principal officers of the Government going on board the Reliance with a memorial for the Governor to sign stating his reasons for coming into Rio Janeiro with the Force of the Ships and upon what expedition they were going. This being refused by the Governor they returned on shore. In the afternoon they went again on board the Reliance with a paper for Governor Hunter to sign specifying the Names of the Ships and on what account they put into Rio Janeiro; which being agreed to they suffered the Boats to go on shore. When any Officer belonging to the Ships landed they immediately sent a common Soldier after him. This being complained of to the Vice Roy he said it could not be regulated for it was the Order of the Court of Portugal, but to accommodate the Officers they should have an Officer of equal Rank with themselves to attend upon

[1] Ilha das Enxadas.

[2] Pão de Açúcar.

them. After a great many messages between the Vice Roy and the Governor they granted us leave [for us] to pitch a Tent upon a small Island in the Harbor called Enchados, the same I believe where Mr Gooch[1] made his Observations. It is near a mile Distant from the City of St Sebastion and bears due North from it, the Sugar Loaf at the entrance bears S14°2′E distant 4⅓ Miles. May 9 at 20h [I] carried the Time-keepers, hadleys Sextant and artificial horizon on shore to the Tent. From the mean of 4 Meridian Altitudes of the Sun found the Lat*itude* 22°53′17″ South, the Longitude p*er* Time-keepers No 1 – 42°43′15″W, No 2 – 41°14′30″W, No 56 – 44°35′30″W, No 248 – 41°47′0″W. During our stay got equal altitudes of the Sun on 6 different days which gave the Rates of the Time-keepers and how much they differed from mean Time at the Island of Enchados. On May 21st at Noon, viz: No 1 +2h 27′19″·50 and losing 8″·28 p*er* day, No 2 +3h 24′ 3″·07 and gaining 7″·992 p*er* day, No 56 +2h 55′ 15″·66 and losing 1″·24 p*er* day, No 248 +3h 21′ 34″·06 and gaining 9″·227 p*er* day. I observed 56 Distances of the Sun east of the Moon which gave the Longitude 42°50′45″W. Reducing the observ*ations* made before we arrived and those made after we sailed to the Island by the Time-keeper No 1 the Longitude will be as below –

18	Distances on	April 14th	42°44′40″W	185 Observations before making the Island the mean 42°59′55″W
18	Do	April 15	42 52 45 W	
21	Do	April 22	43 11 51 W	
12	Do	April 24	43 11 50 W	
45	Do	April 26	43 2 3 W	
20	Do	April 27	42 56 40 W	
18	Do	April 28	43 6 7 W	
20	Do	April 29	42 58 35 W	
15	Do	May 3	42 54 56 W	
56	Observations made on the Island		42°50′45″W	On the Island of Enchados
21	Observ*ations* on May 27		42°53′8″W	After leaving the Island

The mean of three means 42°57′16″W. The Portuguese Astronomers settled the Longitude 43°18′45″ West from the Eclipses of Jupiters Satellites.[2] The Sugar Loaf at the entrance of the Harbor [was] bearing S14°2′E distant 4⅓ Mile, its lat*itude* 22°57′29″S and Long*itude* 42°50′8″ West. The Variation of the Compass from the mean of 4 Sets taken by Adams compass 4°58′E.

We sailed from Rio Janeiro on Sunday May 24th leaving H.M. Ships Reliance and Supply in the Harbor procuring provisions for the settlement at Port Jackson. We had a pleasant passage untill June 1st when it began to Blow a heavy Gale of wind which continued nearly all the passage to the South Cape[3] of Van Diemen's Land, accompanied with thick cloudy weather and almost continual rain both day and night. We had seldom a sight of the Sun

[1] The astronomer William Gooch, murdered in Oahu on his way to join Vancouver.

[2] Master's Mate Matthew Flinders, serving on board the *Reliance*, also observed a series of eleven altitudes with an artificial horizon and 16 sets of lunar distances on the summit of Ilha das Enxadas, obtaining Latitude 22°53′37″S, longitude 42°55′23″W, compared with the modern value of 22°52′58″S, 43°10′30″W; Ingleton, *Flinders*, p. 23.

[3] Probably South East Cape, which is the southernmost cape of Tasmania, rather than South Cape, situated 5½ miles to the west.

above a few minutes at a time. Capt*ai*n Broughton intending to make Goffs Island,[1] [and] being on June 8th nearly in the Latitude it is laid down in the Charts, we steered in a direction that we might probably see it. In the Afternoon it began to blow a very strong Gale that we were obliged to bring the Ship to the wind under ballanced reefed[2] mizen and storm mizen stay sails. The weather being very hazy, on the 9th we got a very indifferent meridian altitude of the Sun which gave the Latitude 40°19′ South but this cannot be depended upon nearer than 5 or 6 [miles]. The Lat*itude* by account was 40°26′South and the Long*itude* 10°16′W. The weather being exceedingly thick and foggy we altered our course more to the Southward. At 30′ past noon it cleared up and we saw the Island bearing East distant two or three leagues. If this had happened in the night we certainly should have been lost upon it or if the Fogg had not cleared up. It blew so hard we could not have got off. It is laid down in all the Charts and Books we had on board in Lat*itude* 40°3′ South and Long*itude* 2°30′W,[3] so that we supposed ourselves at least 7° from it. We lay to[o] all night under the Lee of the Island with intention of getting some observations next day if the weather would permit, but were not successful, the weather remaining the same as the preceeding day. We then stood away to the ESE. On June 11th at 21h 30′ I observed some Altitudes. The Time keepers gave the Long*itude* as follows No 1 – 4°56′0″W, No 2 – 5°9′30″W, No 56 – 4° 36′15″W, No – 248 4°54′30″W. We had made since we left the Island 4°33′ of east longitude which will give the Longitude of Goffs Island No 1 – 9°29′0″W, No 2 – 9°42′30″W, No 56 – 9°9′15″W, No 248 – 9°27′30″W. The above determinations are not very accurate for at the time when the Altitudes were taken there was a very bad horizon and great motion in the Ship from a very heavy swell, besides the uncertainty of the Longitude from the dead reckoning since we left the Island, but they are accurate enough to prevent the same danger in future.[4] On Sunday Aug*ust* 2d saw Van Diemen's Land bearing NbE p*e*r Compass. We was so far from the Land that we could not distinguish any of the points mentioned by Capt*ai*n Cooke. Next morning we stood in for the land and at Noon observed in Lat*itude* 44°4′50″ South. At this time Swilly Island bore N44°13′W; supposing its latitude 43°55′30″ South as setled by Capt*ai*n Cooke,[5] the difference of longitude between it and the Ship at noon will be 0°12′35″West. Reducing 30 distances of Antares west of the Moon taken the preceeding evening to noon, gives the Longitude of the Ship 147°56′50″E at the time the Altitudes were taken. From which 17′9″being subtracted will give 147°39′41″at noon and 29 distances of the Sun east of the Moon taken on Aug*ust* 9th after leaving the cape and reduced to the same time give 147°24′24″E. The mean of the two means 147°32′7″, consequently the Long*itude* of Swilly Island 147°19′32″E; Capt*ai*n Cooke calls it 147°7′30″E.[6] The Time-keepers give the Long*itude* No 1 – 146°49′16″E, No 2 – 145°35′1″E, No 56 – 148°46′1″E, No 248 – 147°5′46″E; the Var*iation* of the Compass off the South Cape 9°32′E. The wind blowing directly from the land we tacked and stood to the eastward. Aug*ust* 13 at 18h saw the coast of new Holland bearing NWbN. At Noon we were well in with the land about Cape Hawke to the northward of Port Jackson. After beating about with contrary winds on Aug*ust* 18th at

[1] Gough Island, 230 miles SSE of Tristan da Cunha

[2] For an explanation of a balance reef see John Harland, *Seamanship in the Age of Sail*, London, 1984, p. 138.

[3] E.g. Moore, *Practical Navigator*, Table XXIV.

[4] The true situation of Gough Island is 40°19′5″S, 9°56′7″W.

[5] Cook and King, *Voyage to the Pacific*, II, p. 105, except Cook's value is actually 43°55′0″.

[6] Ibid, except Cook's value is 147°6′0″.

2h we came to Anchor in an Harbor called by Capt*ai*n Cooke Port Stephens.[1] As there is no account of more than one Ship the Salamander transport ever being in the Port and they having only made an eye sketch of the Harbor,[2] I assisted Capt*ai*n Broughton to survey it. We found it a very commodious one with plenty of Fresh water and wood on the North side about 1½ mile from the entrance of the Port. I carried my sextant and artificial horizon on shore to the watering place with the time-keeper No 1 and found the Long*itude* per Time-keepers No 1 – 151°55′22″E, No 2 – 150°22′42″E, No 56 – 154°15′51″E, No 248 – 152°15′6″E and the Latitude from two meridian altitudes of the Sun 32°41′33″S. 30 Distances of the Sun west of the Moon give the Long*itude* 152°7′45″E and 30 Distances of α Aquila[3] east of the Moon 152°1′50″E; the mean of both 152°4′47″E. [I] carried one of the common Ships Azimuth compasses on shore which gave the Variation 3°36′E.[4] We met with several parties of the natives who behaved very friendly. Here we met with a very remarkable circumstance; one of the Gentlemen who had the watch upon deck perceived two natives on the Shore making signals to the Ship. The Officer of the Watch manned a Boat to see what they wanted. When the Boat came near them, they called out to the astonishment of the boats crew that they were two Englishmen convicts from Port Jackson and wanted to come on board the Ship [and] that there were two more of them upon an Island at the entrance of the Harbor. They proved to be the 5 Convicts mentioned in Capt*ai*n Hunter's Voyage[5] who stole a Ships boat in Sept*embe*r 1790 and made their escape from Port Jackson. As they had never been heard of before, they were supposed to have been lost. After leaving the entrance of the harbor they coasted along the shore, when meeting with a Gale of wind they were driven on an Island a little to the northward of Port Stephens where they lost their boat and one of them died. They had lived with the natives nearly all the time since they made the Island and had adopted the manners and customs of the Natives (going naked), so much that they appeared to us as much like the Inhabitants and nearly as Black as any of the people we saw. Sent the boat and fetched the two men from the Island. They all seemed glad to leave so miserably [*sic*] a place. We sailed from Port Stephens Aug*ust* 23d and next day about 2h arrived at Sydney cove, Port Jackson. We found here the Endeavour[6] from Bombay [Mumbai] who has brought 130 head of Cattle for the Colony and sailed a few days after we arrived for

[1] 'At four in the afternoon, we passed, at the distance of about a mile, a low rocky point, which I called POINT STEPHENS, on the north side of which is an inlet, which I called PORT STEPHENS: this inlet appeared to me, from the mast head, to be sheltered from all winds'; James Cook in Hawkesworth, *An Account of the Voyages*, III, p. 508.

[2] The convict transport *Salamander*, Master J. [?] Nichol, part of the Third Fleet, reached Sydney on 21 August 1791. On her arrival the *Salamander* was sent to Norfolk Island, calling at Port Stephens during her return to Sydney. Crosley, however, was mistaken in supposing that Nichol's was the only survey of Port Stephens, since in March 1795 Lieutenant-Governor Paterson had sent Charles Grimes, Deputy Surveyor, in the colonial vessel *Francis* to Port Stephens where he spent a week surveying the harbour. For a copy of his survey see *Historical Records of New South Wales*, Vol. 2: *Grose & Paterson 1793–1795*, Sydney, 1893, f.p. 286.

[3] Altair.

[4] Broughton's survey is held in the UK Hydrographic Office, Taunton, C642/2, see Fig. 3 on p. 12.

[5] 'In the night of the 26th of September [1790], five convicts took a punt from Rose-Hill, in which they came down to the look-out, where they exchanged the punt for a four-oared boat, and got off undiscovered. These people certainly meant to go along the coast to the northward, and to attempt getting to some of the Friendly islands; but this project must be almost impossible, and there is every reason to suppose they would perish in a few days'; Captain John Hunter, *An Historical Journal of Events at Sydney and at Sea 1787–1792*, Sydney, 1968, p. 312.

[6] William Bampton, Master, who made a pioneering voyage in 1793 through Torres Strait in the *Shah Hormuzeer* accompanied by Matthew Alt in the *Chesterfield*.

New Zealand to cut Spars.[1] The evening we came in it blew the hardest Gale of wind from the Eastward that they had experienced for many years, accompanied with continual rain which lasted untill 28th. I went on shore and examined the place where Mr Dawes's Observatory was built[2] but found nothing standing but the uprights which supported the roof and the stone Pillar on which he placed his Quadrant. The place was so much exposed to the wind that I could make no observations with the Hadleys Sextant. I fixed a Tent upon the opposite point of the Cove called Cattle Point[3] upon the same place where the Spanish Astronomers[4] had made their Observations about two years before.[5] Mr Dawes's Observatory bore N67°W dist*ant* 1600 feet. The Spaniards made the Latitude from Merid*ian* Altitudes of Stars observed North and South of the Zenith 33°51′42″ South,[6] the Long*itude* from an eclipse of the Sun 15[*blank*].[7] From 42 distances of the Sun and Moon 151[*blank*].[8] By carrying on the Long*itude* of some places in New Zealand as setled by Capt*ai*n Cooke, after a run of 14 days with 4 Time-keepers [*blank*].[9] The mean of the whole the Longitude 151 [*blank*].[10] I found the Latitude from 14 Merid*ian* Altitudes of the ☉ L.L. 33°51′47″ South, the Long*itude* from 45 distances of the Sun East of the Moon 151°25′22″E and from 45 distances of the Sun West of the Moon 150°54′15″E.; the Mean of the two 151°9′48″E. The Time-keepers gave the following Longitudes No 1 – 150°54′0″E, No 2 – 149°9′30″E, No 56 – 153°27′0″E, No 248 – 151°13′30″E. Equal altitudes of the Sun observed on 12 different days give the Rates and the differences of the Time-keepers from mean Time on Cattle Point, Oct*ober* 4th at Noon viz – No 1 slow 10h 46′ 2″·87 and losing 7″·653 p*e*r day, No 2 slow 8h 52′38″·85 and gaining 18″·935 p*e*r day, No 56 slow 9h 17′55″·54 and losing 10″·658 p*e*r day, No 248 slow 9h 11′56″·10 and gaining 12″·80 p*e*r day. H.M. Ships Reliance and Supply arrived here all well on Sep*tembe*r 7th. We found the Colony in a more flourishing situation than we expected, they having great plenty of Grain and Vegetables but animal food very scarce, but there is great prospect of having plenty in a few Years. 4 Cows and a Bull which strayed from the Colony about five years since have been found about 50 Miles up the Country. They have increased their number to upwards of 50 and are very fine cattle.[11] I received information from Capt*ai*n Bampton of the Endeavour, which we found laying here, that two Ships the Duke and Dutchess [*sic*] from the East Indies had fallen in with the coast of New Holland about Nuyts's Land[12] and discovered a Channel through which they sailed

[1] On reaching Dusky Sound Bampton found the *Endeavour* was so leaky that he ran her aground, transferring to a small vessel which had been partly built there a few years earlier by the carpenter of the *Britannia*. After completing this vessel and rigging her as a schooner which Bampton named *Providence,* he sailed for Norfolk Island and from thence to China; Collins, *Account of the English Colony,* p. 384, and A. Charles Begg and Neil C. Begg, *Dusky Bay*, 3rd edition, Christchurch, 1975, pp. 80–84.

[2] On Dawes Point on the west side of Sydney Cove.

[3] Bennelong Point on the east side of Sydney Cove, near the site of Sydney opera house.

[4] During the visit of Spanish expedition commanded by Alejandro Malaspina in March 1793. For an account of their visit see David et al., *The Malaspina Expedition*, III, pp. 67–88.

[5] In March 1793.

[6] Actually 33°51′26″S; David et al., *The Malaspina Expedition*, III, p. 78.

[7] Ibid, 151°2′30″.

[8] Ibid, 151°15′15″.

[9] Ibid, 151°15′25″.

[10] Malaspina does not give a mean of these three observations, but additionally gives the longitude 151°10′15″obtained by the immersion of the second satellite of Jupiter on the night of 24 [25] March; ibid.

[11] The place they were found became known as the Cow Pastures; Collins, *An Account*, I, pp. 365–6, 527n.

[12] Nuyts Archipelago, off the coast of South Australia.

and came out upon the eastern coast to the Southward of Point Hicks,[1] but he could not give any particular account of it.[2] There has allways a passage been supposed to exist, for all Ships sailing near the coast have found near this place a very remarkable current setting them to the Eastward. I hope you will be kind enough to excuse any inaccuracies there may be in this Letter for the shortness of time which I had to write it in, prevented me from taking that care I otherwise should have done and I am with the greatest respect

Rev*eren*d Sir
Your most Obedient
H*um*ble Serv*an*t
John Crosley

PS Will thank you to send to Messrs Clementson & Denton a certificate of my being in the Service of the Board of Longitude on June 7 1795. Capt*ai*n Broughton being on shore has got the Spanish Observations with him which is the reason I have not filled up the Blanks w[h]ere they are mentioned.

5. Extract from Crosley's Astronomical Observations in Nootka Sound[3]

[the left side of the page lost in the spine]

[I can]not account for the great alteration in the rates of Mr Earnshaw's three time-[keep]ers unless they received some damage by unloading the ship in order to heave [her] down. The heavy casks and pigs of iron ballast sometimes struck the deck [and] made it shake violently. The time-keepers being fixed to the bulk head of [the] cabin which communicated with the deck must also have been shaken . [I im]agine this to be the cause of the alteration in their rates, but did not attend to [it at] the time it happened. Mr Arnold's time-keeper was slung on gimbals therefore [the] shaking of the deck would only make it vibrate on the gimbals, which is proved [by] its having gone as well here as at any other place in the voyage. Mr Earnshaw's [time-]keepers not being in gimbals must have been shaken every time the deck was [struck].

6. To the Board of Longitude[4]

The Right Honorable and Honorable the Commissioners of Longitude

My Lords and Gentlemen,

I proceeded according to your appointment bearing date December 7th 1793 on board H.M. Sloop Providence to the North West Coast of America in order to succeed the late Mr

[1] A point on the SE coast of Australia where Captain Cook made his landfall in the *Endeavour* in 1770.

[2] Bampton was mistaken; Hayes did not sail through the Bass Strait. In 6 February 1793 John Hayes of the East India Company sailed from Calcutta in the *Duke of Clarence*, accompanied by the *Duchess*, commanded by William Relf, bound for the north coast of New Guinea where John McCluer had reported finding nutmeg. The early onset of the South-east Trade Wind forced Hayes to round Cape Leeuwin and the south of Tasmania, at that time thought to be part of the mainland of Australia. Hayes carried out some surveys off the SE coast of Tasmania before sailing for New Caledonia and thence to New Guinea; Ida Lee, *Commodore Sir John Hayes: His Voyage and Life*, London, 1912.

[3] CUL, RGO 14/68, f. 115.

[4] CUL, RGO 14/1, ff. 162–164.

William Gooch and make use of the Instruments carried out by him on board the Daedalus and left on board H.M. Sloop Discovery, Capt*ai*n Vancouver. That my voyage from England to the North West Coast of America might be of use to Geography and Navigation I provided myself with Instruments, Books &c and transmitted the observations when opportunity permitted of sending them to the Rev*eren*d Dr Maskelyne, His Majesty's Astronomer Royal.

On the arrival of H.M. Sloop Providence at Monterrey on the Coast of California, the Spanish Governor informed Captain Broughton that Capt*ai*n Vancouver had sailed for England some months before. In consequence of this intelligence I received a letter from Capt*ai*n Broughton desiring my opinion in writing how he could best employ H.M. Sloop under his command in her return to England for the further improvement of Geography and Navigation by completing the survey of any unexplored parts that former Navigators had not been able to complete.

I returned him for answer that it was my opinion if a Survey was made of the Coast of Asia from the sea of Okotstk to the bay of Nanking with the adjacent islands, viz Kurile, Jeso and Japan it would be of more benefit to the Sciences of Geography and Navigation than any other part in those Seas, for the limits of both the Continents of Asia and America would then be known as far as Navigation was practicable and the Geography of the North Pacific Ocean completed.

To effect this purpose [we] sailed from Monterrey June 20th 1796 for the Coast of Japan and arrived there Sep*tembe*r 7th. [We] surveyed one side of a range of islands extending from the north end of Japan to the island of Marikan in Lat*itude* 47°North, intending to complete the other side in our return to the southward, but the weather came on so thick and tempestuous that it was dangerous to venture amongst the Islands. [On] Nov*embe*r 9th made the South east end of the Island of Japan, but were blown off the next day. After beating about several days saw the islands to the Southward of Japan. In our passage from these islands to Macao where we arrived December 11th saw a great many Islands, but when we approached near them had generally thick weather with heavy Gales of wind which afforded but few opportunities of making observations and setling their positions.

I sent to the Rev*erend* Dr Maskelyne by the Cuffnels East India Ship all the Observations made in the voyage from England to Macao and the computations up to August 1st 1796. [I] should have sent them all complete to Macao, but the ship sailing a fortnight sooner than I expected, [I] had not time to copy them into the account.

[We] sailed from Macao April 10th 1797 in company with the Prince William Henry Schooner (Purchased at Macao by Capt*ai*n Broughton) to finish the survey of the Coast of Asia and the adjacent Islands. On May 16th about half past seven in the evening the Providence unfortunately struck on a Reef of Coral Rocks in Latitude 25°2′ North and Longitude 125°40′ East of Greenwich, about 15 Miles distant from an island seen the year before on our passage from Japan. Every effort was made by the Captain, Officers and Ships Company to get the Ship off the Reef, but without success. At 10h 30′ the Carpenter reported the water up to the Orlop Deck and the Ship bilged in her larboard bow. Capt*ai*n Broughton now informed me I had better go below and save what Instruments Books &c that could be got at, he having soon after the Ship struck given an Order that no person should be permitted to look after their things untill it was known whether the Ship could be got off the Reef. It was now almost too late, the water being nearly 2 feet above the Gun Deck and the After Hatchway Ladder washed away: I got from the people on the Gun deck the Time-keeper No 1 belonging to the Navy Board, my Time-keeper, Hadleys Sextant, Journal, Book in which the computations were made during the voyage, and the memorandum books in which the observations were wrote as they were observed on Deck. To preserve the Time-keepers and memorandum books [I] put them into my pockets and got down the stern ladder into the Launch. As soon as I set my feet upon the thwarts of the launch and left go the stern ladder, a sea struck her and drove her under the Ship's Counter which knocked me backwards and set me upon my head in the bow of the launch, by which accident the Time-keepers and memorandum books fell out of my Pockets. The Time-keeper No1 and the books fell into the launch, my own Time-keeper fell overboard and was lost. [I] left my journal, Book in which the computations were made, and Hadleys sextant with the Gunner to lower down into the launch as soon as I was ready to receive them. They were slung and lowering down when a violent sea struck the Ship and she fell on her Beam ends; the Sextant and Books fell overboard . The Sextant was picked up about an hour afterwards by a boat going between the Ship and Schooner, the case was full of water and the Instrument rendered unfit for Astronomical purposes. The Captain, Officers and Ships Company were all safe on board the Schooner by half past twelve o'Clock having saved very little but what they had on their backs when the Ship struck. The boats were sent at day-light to the wreck to see if anything could be got from it. [They] found the ship clear from one end to the other, all the bulk heads being washed away and nothing left but the Guns on the deck. Next day [we] anchored off the island in sight and wooded and watered the Schooner. May 23d sailed for Macao Roads, where we arrived June 3d. [Here we] found the East India Ships laying in Canton River expecting to sail for England in two or three days. June 11th went on board the Crescent Packet, and on the 15th sailed with the India Fleet for England, but meeting with several violent Tuffoons in the China Seas, in which Crescent lost her Bowsprit, Fore Mast, and Main Top Mast, and otherways being much damaged, were obliged to put back to Macao to refit, where we arrived August 1st.

Sailed from Macao Nov*embe*r 20th for England, arrived at St Helena February 27th 1798. May 1st sailed from St Helena in the Marquis of Lansdown, East Indiaman and arrived in the Downs July 7th 1798.

The letter mentioned to have been sent to the Rev*eren*d Dr Maskelyne on board the Cuffnels East India Ship, I received from Capt*ai*n Broughton before the Crescent left Macao. In the passage to England [I] arranged the Observations and entered them into a book, completing them as far as lay in my power. From my letters received from Dr Maskelyne since my arrival in England [I] can finish them to Aug*us*t 1796, but cannot proceed with the remainder untill the arrival of Captain Broughton who has got the Ships Log Book.

My salary from the time of the Shipwreck on May 16th 1797 to my arrival in the Downs on July 7th 1798 [I] shall submit to the consideration of this Honorable Board to allow what they may think proper, but I believe it is the Custom of His Majestys Service in case of Shipwreck, to allow the persons employed their Salaries untill their arrival in England or a Court Martial takes place.

If it should please this Honorable Board to make any allowance for the loss of my Instruments and Books it will be thankfully received by

My Lords and Gentlemen

Your most obedient
Humble Servant
John Crosley

Palatine Houses near
Stoke Newington
Nov*embe*r 30th 1798

Minuted in a different hand 'Sec*retary* to Enquire the expense of his Passage home.'

7. Extract from Crosley's Astronomical Observations regarding the loss of the Providence[1]

May 16th at half past seven in the evening a little before dark the ship struck on a reef of coral rock, her head was then to the north eastward [and] the sails were immediately braced aback. The signal was made to the schooner and the master sent to anchor her as near as possible to heave by her. The ship soon after paid off with her head to the eastward, the main sail was hauled up and the other sails shivered to let her go round about giving her headway. Before this could be affected she again struck fore and aft and remained at last fixed with her head due south. Breakers were then upon each bows. There were from five to fifteen fathoms in the starboard chains and only two and a half fathoms both a head and a stern. The rudder being broken, the masts were struck and the boats hoisted out. At this time the ship did not strike violently and only made 19 inches water. The wind now freshened from the NNW and the sea began to break very heavy, which soon knocked the rudder off. At 9h 30′ began hoisting out the long boat; during this period the ship made water very fast and the violent shocks she received made it doubtful whether the masts would stand. Before the long boat was out, the water had increased so much on the pumps that there was seven feet water in the hold. At this time the schooner anchored near us in 25 fathoms. The ship now suddenly changed her position, swinging now round from South to North by the East and striking more violently than ever. Before hawsers could be carried out [to] the schooner the carpenter reported the water up to the orlop deck and the ship bilged in her larboard bow. The idea of heaving her off was now given up, for if it had succeeded the ship must have foundered in deep water. The Captain and Officers were of opinion that it was impossible to save the ship; on one side there were only six feet of water and on the other three fathoms and a half, the ship nearly on her beam ends, the fore part of her immersed in the sea, and surf breaking over the deck. The people were now employed in collecting arms and ammunition, with armourers and carpenters tools, but from the situation of the ship and the gun deck being full of water with the bulk heads and [chests] washing to and fro they were prevented from saving many. As nothing more could be procured, the ships company were sent into the boats and reached the schooner in safety soon after eleven oclock, both officers and men having lost every thing belonging to them. The pinnace returned for the captain and remaining officers, and at half after twelve they quitted the ship leaving her a perfect wreck. Soon after twelve the wind began to increase, and the schooner having two anchors down in foul ground got one of them up. At 4 oclock parted the other cable, and fortunately casting the right way made sail, and escaped another shipwreck, which must have proved fatal to many.

[1] CUL, RGO 14/68, ff. 133–133v.

8. Crosley's Expenses[1]

The Right Honorable and Honorable the Commissioners of Longitude

D^r^ to John Crosley

	£ s d
To Salary from Decr 7th 1793 to May 16th 1797	1375. 6. 10
By Salary received from Decr 7th 1793 to October 6th 1796	1132. 1. 1
Balance due to John Crosley	£ 243. 5. 9
Expenses for living at Macao from Augt 1st to Novr 20th 1797	60. 15. 0
Expenses for living at St Helena from Feb. 27th to May 1st 1798	66. 11. 6
	£ 370. 11. 9

The above is the statement of my Account to May 16th 1797, the time the Ship was lost, with my expenses since. I have not made any charge of my Passage from China to England the Capt*ain*s of the Indiamen agreed to take what this Honorable Board thought proper to allow.

Instruments lost and damaged on board H.M. Sloop Providence May 16th 1797 belonging to John Crosley[2]

		£ s d
Lost	A Box Time-keeper No 248 made by Earnshaw	68. 5. 0
Lost	A Silver Pocket Watch with Cylindrical Verge and compensation slide for Heat and Cold No 1522 made by Earnshaw	16. 5. 6
Damaged	A Ten inch Brass Sextant by Troughton divided to ten seconds with three extra powers	16. 5. 6
Lost	A stand to the Sextant	3. 11. 6
	A swinging level Horizon	2. 11. 6
	A Roof Horizon with Box and Quicksilver	4. 2. 6
	A Wooden Bottle with Quicksilver	1. 5. 0
	An Ebony Sextant with Brass Arch	6. 6. 0
	A Circular Protractor 6 inches diameter	2. 12. 6
	A Brass Scale divided to 1000th of an inch	3. 3. 0
	A Pair of Beam Compasses with magnifying glasses	3. 13. 6
	A Drawing Board and Square	1. 11. 6
	A Case of Instruments	2. 12. 6
	A Sector 12 Inches radius	0. 7. 6
	An Ivory Plotting Scale	0. 7. 6
	2 Parallel Rule[r]s one 18 Inches the other 2 feet	0. 12. 0
	A Steel drawing pen	0. 3. 6
	A Pocket Compass	1. 11. 6
	A Thermometer	1. 1. 0
	A Chair for observing distances of ☉ and ☽ upon Deck	1. 4. 0
	A Rising Stool	1. 1. 0
		£ 138. 13. 6

[1] CUL, RGO 14/1, f. 164v.

[2] CUL, RGO 14/1, ff. 165–165v.

Books lost on board H.M. Sloop Providence May 16th 1797 belonging to John Crosley

	£ s d
Taylors Logarithims	4. 0.0
De la Land[e]s Astronomy 4 Vols	4. 4.0
Vinces Practical Astronomy	0. 17.6
Zachs Astronomical Tables	1. 1.0
Mayers Tables and Theory of the Moon	0. 10.6
Masons Tables	0. 5.6
Celestial Atlas by Mr Bode of Berlin	1. 5.0
Royal Astronomer [and Navigator] by Heath	0. 12.0
Philosophical Transactions 1st Vol.	0. 16.0
Walker on the Sphere	0. 14.0
Mackenzies Maritim Surveying and Borda on the Reflecting Circle	0. 10.0
Ludlams Astronomical Observations and Clairauts theory of the Moon	1. 1.0
Falconers Marine Dictionary	1. 5.0
Mackay on the Longitude	0. 13.6
Robertsons Navigation 2 Vols	1. 1.0
Nicholsons Navigation	0. 6.0
Garrards Tables	0. 10.6
Requisite Tables	0. 6.0
Gardiners Tables by Calteh	1. 1.0
Tracts from the Nautical Almanacs [not priced]	
Sim[p]sons Euclid	0. 6.0
Crakelts Trigonometry	0. 5.0
Simpsons Geometry and Trigonometry	0. 6.6
Simpsons Algebra	0. 6.0
Leadbetters Astronomy	0. 10.6
Keills Astronomy	0. 6.0
Connoissance des Temps 2 Vols	0. 8.0
Martins Mariners Mirror	0. 5.0
De la Land[e]s Exponition de calcul Astronomique	0. 5.0
Ludlam on Hadleys Quadrant and Vinces Conic Section	0. 5.0
Histoire Abrégé [de la] Mer du Sud 3 Vols and 1 Vol of Charts	3. 3.0
Arrowsmiths Mercator Chart[1]	1. 11.6
	£28. 17.0

9. Payment of Crosley's Account[2]

Gentlemen

We whose names are hereunto subscribed, being part of the Commissioners appointed by Acts of Parliament for the discovery of the longitude at Sea, &c do in pursuance of the power vested in us, hereby desire you will sign bills upon the Treasurer of His Majesty's Navy for

[1] Presumably his 1790 Map of the World on Mercator's projection.

[2] CUL, RGO 14/1, f. 166–166v.

paying to Mr John Crosley (out of any money which may be in his hands unapplied to the uses of the Navy) an allowance after the rate of four hundred pounds per Annum from the seventh day of December 1793, when he received his instructions to proceed to the North-West coast of America to make nautical, astronomical and trigonometrical observations in the room of the late Mr Gooch, to the seventh day of July 1798 when he returned to England and also a further sum of one hundred and sixty seven Pounds ten shillings and sixpence to reimburse him for his Instruments and Books lost when the ship was wrecked on the Coast of China, deducting from the amount of these two figures all such sums of money as may have been already paid to him either by way of imprest at our desire or in consequence of Certificates given to you by the Rev^d Dr Nevil Maskelyne, His Majesty's Astronomer Royal at Greenwich. Given under our hands and seals the [*blank*] Day of [*blank*] 179[-]

To the Principal Officers & Commissioners
of His Majesty's Navy

Rec^d Instructions Decr 7th 1793
Arrived in the Downs July 7 1798
4 years & 212 days

	£	s	d
£ 365 : 400 : : 212:	232	6	7
4 Years	1600		
Instruments & Books	167	10	6
Total	1999	17	1
Already paid according to Mr Crosley's account	1132	1	1
Remains due	867	16	

2 March 1799
Mr Crosley's acc^t

IV. Mudge to the First Marquess of Bute, Ambassador to the Court of Spain

To John Stuart, First Marquess of Bute[1]

Providence Monterey Bay June 20th 1796 (South America)

My Lord,[2]

Since the last I had the honor of writing Your Lordship dated Port Jackson (New South Wales) we stood towards Otahite on our Voyage to Nootka where staying a few days to refresh the crew we proceeded to the Sandwich Islands. We first learnt at this place that Capt. Vancouver had sailed for England near eight Months prior to our arrival, and that he had touched at Valparaiso (Lat. 36°S°) in his way down the coast towards Cape Horn. Your Lordship will easily conceive my chagrin on the occasion having been but three Months out of sixteen Years from actual employment except those two Years I was kept Idleing my time

[1] AGI, Estado 39, N.14-8 (5).
[2] John Stuart, first Marquess of Bute (1744–1814); see the reference to Lady Bute later in the letter.

attending the Admiralty grasping at the faint hope of joining the Chatham. Mr Puget who is gone home in her was a Midshipman and made into the Discovery when I was her first Lieutenant, surely My Lord those Friends to whom I have so long looked up and servd as far as lay in my power will never allow him to be promoted over my head, and I believe My Lord if a certain little hump backed Author, had served eight Years as Midshipman, and eight as Lieutenant (the last five on Discoveries) would alter his tone, and not say what ever is is right. Or rather join with a certain tough, grumbling, Doctor,[1] who asserts a Sailors life is perpetual imprisonment with a good chance of being drownd, or rather the confinement of a Prison without any of its comforts. The middle of February [we] left these Islands and saild for Nootka expecting at least to fulfill part of our mission by seeing the Spaniards quit it altogether, our passage was quick tho' boisterous and got sight of the coast on the sixteenth of April when we Anchord about eight miles to the Southward of the Sound. I set off with one of the Boats next Morning well Arm'd to look into Friendly Cove, and after pulling five hours towards it the Snow falling very thick (Thermometer 28°) I caught a glimpse of the Village of Nootka, but as the Indians were very numerous and the Gale increasing did not land but bore away for the Ship. In this situation we lay the Wind blowing strong off the land for three days and what made us very anxious to get the Ship into safety was her making much Water having sprung a leak between the Sandwich Islands and the Northwest Coast of America. However the Wind having veer'd in from the Sea we Weigh'd, and Anchor'd that same night in the Harbour of Moweena,[2] about five Miles beyond Nootka thinking that Port most eligable for our several occupations. Immediately on our arrival the Indian Chief[3] came on board attended by many Canoes, he brought with him letters, and duplicates of dispatches that were left for Capt. Broughton from Brigadier Alava, and Lieutenant Pearce.[4] In short My Lord we learnt that every thing had been settled relative to Nootka nine Months prior to our arrival. Its impossible for Your Lordship to have the smallest conception of the Astonish'd group who now stood not swallowing the Taylors News,[5] but the Indians, which could only be equalled by the Wild Goose chase on which we were sent. I believe Your Lordship has often heard me say we should loose our season.[6] The Fact is that the business was settled at Nootka when the Ship lay in Plymouth Sound. And we have only to sing the Old Song of the French King marching up the Hill and down again,[7] and so we marchd back again. The climate and situation of Monterey is delightful, but their jealousy will not allow us to enjoy it, indeed they carry it to that length, that we almost believe to be at this moment a War with Spain and what has in some measure confirm'd our suspicion is their putting a miserable Fort of eight Guns

[1] Dr Samuel Johnson, the lexicographer. 'No man will be a sailor who has contrivance enough to get himself into jail; for being in a ship is being in a jail, with the chance of being drowned': James Boswell, *The Life of Samuel Johnson*, ed. J. W. Croker, London, 1848 (first published 1791), p. 117.

[2] Marvinas Bay on the west side of Cook Channel.

[3] Maqinna.

[4] Brigadier General José Manuel de Alava and Lieutenant Thomas Pearce of the Royal Marines, the two Commissioners sent to implement the Third Nootka Convention.

[5] Gossip, rumour, etc. An allusion to 'With open mouth swallowing a taylor's news'; Shakespeare, *King John*, Act IV, Scene ii.

[6] I.e. losing a season's surveying.

[7] Mudge is referring to an old jingle which went:

The King of France went up the hill
With forty thousand men,
The King of France came down the hill
And ne'er went up again.

into some better kind of repair after our arrival. George[1] has been unwell but is now perfectly recovered, indeed we have all experienced a slight Influenza, which I attribute to the change of Climate, Diet, &c, &c as at Nootka the Thermometer general[ly] stood at 30° and here at 70° – at the former we liv'd on Salt Meat, here we had Fresh. He has just brought me a letter addressd to Your Lordship in Hill Street but as Don Boreeca,[2] the Commandant of this place, assures me Your Lordship is at the Capital[3] and at all events its most probable our letters will pass thro that place I therefore think it Eligable to send them for Madrid. We are now Anticipating the Month of September in which we hope to double Cape Horn on our homeward bound passage. In the whole we have been tolerably healthy since our first sailing having lost three men, one by distemper,[4] and the other two were shook off the yards by tempestuous weather.[5] Shall trouble Your Lordship to present my respectful Compliments to Lady Bute, and am My Lord with the greatest respect Your Lordships

Most Obedient, and Much Oblig'd
Humble Servant
Zachary Mudge

[1] Midshipman Lord George Stuart, the fourth son of the Marquess of Bute. He probably accompanied Mudge when both were sent home with leave from China, 10 June 1797. Stuart was promoted rear admiral of the blue on 10 January 1837 and died on 19 February 1841.

[2] Diego de Borica.

[3] The Marquess of Bute was ambassador to Spain in 1795–6.

[4] Samuel Redriffe; died 25 March 1795.

[5] Hugh MacDonald, who fell from the main topsailyard on 16 June 1795; and Patrick Sherry, who fell from the mizen top on 6 June 1796.

APPENDIX II

Crosley's Instructions[1]

By the Commissioners appointed
by Acts of Parliament for the
discovery of the Longitude at Sea
&c &c

Whereas upon the death of Mr William Gooch We have thought fit to appoint You to proceed in his Majesty's Ship the Providence, now fitting out for a Voyage to the North West Coast of America, and to make Nautical Astronomical and Trigonometrical Observations there for fixing the Latitude and Longitude of various points of the Coast and Country and to ascertain their relative position with respect to one another; and also to make Nautical and Astronomical observations during the Voyage out and at home all tending to the improvement of Geography and Navigation: You are hereby required and directed to embark on board the said Ship (the Board of Admiralty having given the necessary Orders for your reception) and to proceed in her on the above mentioned Voyage accordingly: and whereas upon your arrival on the North West Coast of America you may expect to find Capt. Vancouver of his Majesty's Ship the Discovery, you are upon joining him to quit the Providence and to go on board the Discovery (Capt Vancouver having orders to receive you) and then to pursue the operations necessary for Surveying the said Coast and Country, going occasionally on Shore whenever it shall appear conducive to the better carrying on the said Survey; and whereas the several Instruments, Books, Maps, Charts and other things specified in the Schedule herein annexed* [*in margin*: see Schedule annexed to Mr Gooch's Instructions[2]] were furnished by the Astronomer Royal to the late Mr Gooch when he went out in the Daedalus, and are now on board that Ship or the Discovery. You are upon your arrival in America to make application to Capt. Vancouver for the said Instruments, Books, Maps, Charts &ca and take them into your charge and custody, and to make use of them for the several purposes for which they are respectively adapted, taking all possible care of them and upon your return to England to deliver them to Us or our Secretary in the best condition you may be able. In the performance of the above Service You are punctually and faithfully to observe and execute the following Instructions.

I
Observations to be made on Ship board

1st You are every day, if the weather will admit to observe Meridian Altitudes of the Sun for finding the Latitude and also other Altitudes of the Sun both in the Morning and

[1] CUL, RGO 14/9, ff. 70–72v.

[2] For 'List of Properties of the Royal Observatory Signed out by William Gooch', see Greg Dening, *The Death of William Gooch: A History's Anthropology*, Melbourne University Press, 1995, p. 124.

Afternoon at a distance from Noon with the time between measured by a Watch, and the Sun's bearing by the Azimuth Compass at the 1st observation in order to determine both the apparent time of the day and Latitude in case the Sun should be clouded at Noon: – You are moreover to observe distances of the Moon from the Sun and fixed Stars[1] with the Hadley's Sextant, from which you are to compute the Longitude by the Nautical Almanac.

2d You are to wind up the Watches every day as soon after the time of Noon as you can conveniently, and compare them together, and set down the respective times: and you are to note also the times of the Watches when the Sun's morning and afternoon Altitudes, or distances of the Moon from the Sun and fixt Stars are observed: and to compute the Longitude resulting from the comparison of the Watches with the apparent time of the day inferred from the morning and afternoon Altitudes of the Sun.

3d You are to observe, or assist at the Observations of the variation of the Compass, and observe the inclination of the magnetic dipping Needle from time to time.

4th You are to note the height of one or more Thermometers placed in the Air, and in the Shade early in the Morning and about the hottest time of the day, and to observe also the height of the Thermometer within the Vessel near the Watches; and make remarks on the Southern lights when you are far to the South, if any should appear.

5th You are to keep a Ship's journal with the Log worked according to the plain dead reckoning (Lee way and variation only allowed) noting therein the length of the Log line and time of running out of the sand glasses from time to time; and you are to insert therein also another account corrected by the last Celestial Observations, and a third deduced from the Watches.

6th You are to teach such of the officers on board the Ship as may desire it the use of the Hadley's Sextant in taking the Moon's distance from the Sun and fixt Stars and the method of computing the Longitude from these Observations.

7th You are to settle the positions of Head-lands, Islands and Harbors in Latitude and longitude by the Celestial Observations; and also to set down what Longitude the Watches give.

II

Observations to be made on Shore

1st Where ever you land, if time permits, you are to set up the Tent Observatory and Astronomical Clock, either setting the Pendulum to the same length as it was at Greenwich before the Voyage, or noting the difference by the revolutions of the Screw and divisions of the Nut at bottom: and take equal Altitudes of the Sun and fixt Stars, with the universal Theodolite, or with the Hadley's Sextant by reflexion from the Surface of quick silver in a Bason, according to circumstances; for determining the rate of going of the Clock and for finding the time of Noon, and fixing a Meridian for the Transit Instrument which is after that to be made use of.

2d You are to wind up the Watches every day as soon after the time of Noon as you can conveniently; and compare them together, and with the Astronomical clock at the time, and also about the times of equal Altitudes, if any were taken.

[1] I.e. the ten stars whose distances from the Moon are given at three-hourly intervals in the *Nautical Almanac.*

3d You are to observe Meridian Altitudes of the Sun and also fixt Stars,[1] some to the North and some to the South for finding the Latitude with the universal theodolite, or with the Hadley's Sextant by reflexion from the Bason of quicksilver, according to circumstances.

4th You are to observe transits of the Moon and proper fixt Stars over the Meridian with the Transit Instrument, and to make observations of the Eclipses of Jupiter's first satellite, and Occultations of fixt Stars by the Moon, and to take distances of the Moon from the Sun and fixt Stars with the Hadley's Sextant from all of which you are to compute the Longitude of the place by the Nautical Almanac.

5th You are to observe the height of the Tides and the Times of high and low Water, particularly at the full, change, and quarters of the Moon, and note whether there be any differences and what, between the Night and Day tides.

6th You are to observe the variation of the Compass, and the inclination of the dipping Needle.

7th You are to note the height of one or more Thermometers placed in the shade early in the Morning and about the hottest time of the day.

General Directions

You are to take particular care that all your Nautical and Astronomical observations and Trigonometrical observations, whether made on ship board or Shore be kept in a clear, distinct, and regular manner in a Book, or Books, prepared for the purpose, and that they be written therein with all their circumstances immediately after they are made, or as soon after as they can be conveniently transcribed therein from the loose papers, or memorandum books in which they may have been first entered; which book or books are always to be open for the inspection and use of the Commander and Master of the Vessel, and you are to send to Us by every safe conveyance which may offer the results of your several Observations; and also the principal Observations themselves.

For your care, pains and expenses during the time you shall be employed in the above Service, you will be allowed a Salary after the Rate of £400 per Annum, to commence from this day. Given under our hands the 7th day of December 1793.

To
Mr John Crosley
By Command of the
Commissioners of the
Longitude
H. Parker

J. Banks
N. Maskelyne
Thos Hornsby
John Smith
Ao Shepherd
Php Stephens
I. Ibbetson

[1] In this case the reference is to the 60 stars, together with their declinations and right ascensions, listed in Table VII of *Tables Requisite to be used with the Nautical Ephemeris, for finding the Latitude and Longitude at Sea*, first published in 1767, with a second edition in 1781, which was presumably the edition that Crosley would be using.

APPENDIX III

Variant Version of Broughton's Journal from 17 October to 7 December 1796 from UKHO, OD 177

This was the 17th of October and having no wish to proceed North of Marukan[1] I hoped to be able to complete the Eastern coasts of the Isles we had already seen more particularly Ooroop[2] & Nageedsha[3] which by the Russian accounts afford good Harbours. The Wind not permitting our passing thro' the passage which divides Marukan from another island[4] we stood the South & the next day [Tuesday 18 October] experienced a heavy Gale of Wind from the SE with Rain – which reduced us to a Storm Staysail in the Evening. In the Height of the Gale I had the misfortune to fall upon deck and fracture my right Arm above the Elbow. An accident to me most unfortunate as my confinement entirely precluded my carrying on the Survey of the Islands & proceeding as I wish'd between Japan & the Coast of Asia.[5] We nevertheless Kept that point in view and the officers were directed accordingly but the boisterous weather we now met with baffled all our endeavours. The Wind coming to the NW carried us through a passage to the south of Marukan in the Lat*itude* 46°42′N & Long*itude* 150°0′ East.[6] & soon increasing to a Gale we were driven to the Eastward and we were unable afterwards to get near the Islands. Incessant Gales succeeding each other so rapidly we were seldom able to carry any Sail and whenever we approach'd near the Land the Western Gales with increased Force drove us from it. Added to which we had to strive against contrary currents whose Force we cou'd seldom detect from the Little opportunity we had of making Observations.[7] As my principal pursuit was to explore those points not seen in Captain Cookes last Voyage I did not think it necessary to make the coast of Niphon to the north of white point.[8]

[1] Ostrov Simushir.

[2] Ostrov Urup.

[3] A transcription error for Nadeegsda; the identity of this island is uncertain; see p. 91, n. 6 above.

[4] Proliv Diany, which separates Ostrov Simushir from Ostrov Ketoy.

[5] It also appears to have prevented Broughton writing up his journal after the completion of this paragraph as the next dated entry after that is 9 November, although the journal appears to be continuous. This is confirmed in a statement in RUSI/NM/132(8), in which Broughton confirms that he was confined to his cabin until 10 November and that he was totally deprived of the use of his arm for three months, 'tho' the journal was carried on by an Amanuensis daily under my inspection'.

[6] Proliv Bussol', which separates Ostrov Simushir from Ostrova Chernyye Brat'ya. La Pérouse passed through this strait on 30 August 1787, naming it Passage de la Boussole after his ship.

[7] The Oya Shio, which runs south past the Kurile Islands and Honshu until it meets the northern edge of the Kuro Shio in about 36°N. In addition it was not helped by the fact that chronometer No 1 was allowed to run down on 3 November; Broughton, *Voyage*, p. 134.

[8] Taito Saki, a prominent wooded point at the SE extreme of Honshu, the south side of which consists of a vertical white cliff, shown as White P[t] on the chart of Japan in Cook and King, *Voyage to the Pacific Ocean*, III, f.p. 395.

Wednesday 9 November
Moderate breezes and fine Weather. At 2 PM saw the Land of Niphon bearing West. At Sunset the extremes of Niphon bore from S80°W to N40°W, nearest Land N85°W, distant 14 or 16 Miles. We had dark cloudy weather with Rain throughout the forenoon which prevented any meridian Observation. Course S50°W, distance 74 Miles, Lat*itude* Account 34°54′N, Longitude d*itt*o 135°1′ East, ditto from Bearing and distance of Land at 19 Hours, which is suppos'd to be in 140°46′E, 140°54′E. The Error in Longitude by Account being so great I take a new departure from that deduced from the Bearings of the Land.

Thursday 10 November
Fresh Breezes and dark Cloudy weather with small Rain throughout & the wind variable. In the forenoon it fixed in the NW quarter with fine Weather. We plied under easy sail during the night and at daylight the land bore from us N47°W to S85°W and we saw several Fishing boats in shore while we pursued our Course in the direction of the Coast. By our making the land about white point it appears we have experienced Strong Currents setting to the Eastward. Our account & its situation as settled by Captain King differing 6° in Longitude.[1] Five days before making the Land the Time piece having been neglected winding up we now set it a going allowing its Old Rate and taking the Longitude of white point as well settled. At $9^h\ 20^m$ AM white point bore of us N5°E on with a distant Hummock.[2] The shore from thence was of a very moderate elevation with Chalky Cliffs to a small opening[3] where we saw several fishing boats coming out & which bore of us at 10^h30^m AM N52°W, about 3 Leagues Distant. From thence to Noon the Shore took a SW'ly direction very bare of Wood and destitute of Verdure. The fishing boats stood off a Moderate distance from the Shore where lowering down their Sail and Striking their Masts they proceeded to their employment of Fishing without approaching near enough to the Ship to afford any remarks on the Boats or people or an opportunity of speaking to them. Our distance from the Shore never exceeded 2 or 3 Leagues.[4] At Noon we were abreast of a very extensive Town. The Land round it seems to be well cultivated tho' with much apparent labour for the Hills appear very steep and craggy towards the top but the sides are cut into regular plats[5] one above another nearly in the Chinese fashion. In some of the Hollows between the Hills there is a small quantity of Wood but nothing considerable. To the NE'ward of the Town the Land is a little Indented. The Afternoon being Calm many of the Fishing Boats that were in Sight came to the Ship. They supplied us with a large quantity of Fish like those called Balahoes[6] in the West Indies of which each boat seem'd to have a large quantity. They threw them on board & seem'd much pleased to see our people scramble for them having apparently no expectation for any return for them or any idea of Traffic, for when any thing was handed into them they gave no more fish than if

[1] The longitude of White Point taken from the chart 'Part of Japan or Nipon' in Cook and King, *Voyage to the Pacific Ocean*, 3, f.p. 395 is 140°44′E, whereas Broughton's longitude by account on 9 November was 135°01′E; Broughton, *Voyage*, p. 384.

[2] Inubo Saki.

[3] Katsuura Wan, situated 12 miles SSW of Taito Saki.

[4] In annotated *Voyage*, p. 138, inserted in ink, 'At 21h 9′ the SE Point of Japan [Taito Saki] bore N8°E dist. [cropped] miles. Its Latitude 35°13′17″ North, Long*itud*e 140°33′47″E p*er* mean of 31 sets of Lunars.'

[5] Archaic for plot of land.

[6] Not identified.

you had held your hand empty to them. At 6^h30^m AM Extremes from N47°W S85°W, the round Hill N82°W off Shore 6 or 7 Leagues. NB the Round Hill Answered to Captain Kings description situated by him Lat*itude* 35°20′N and Longitude 140°26′ East.[1] At Noon the South extreme or S*outh* Bluff point[2] bore S79°W 10 Miles, Round Hill N58°W, Land about white point or N*orth* Extreme N34°E 8 Leagues, an Island from S64°W to S71°W[3] 6 or 7 Leagues. Off Shore 4 or 5 Miles. Course S50°W, distance 63 Miles, Lat*itude* Account 34°14′N, Lat*itude* Observ'd 34°54′52″N, Longitude Account 139°43′ E, No 1 p*er* the Mean of AM & PM sights 140°25′53″ East. Moderate Breeze and fine clear Weather.

Friday 11 November

PM Moderate breezes and fine weather throughout the afternoon. At 5^h the South extreme same point as set before N30°W 6 Miles, NW extreme[4] N37°W 9 or 10 Miles, a middle point[5] N17°E 5 Miles (NB round this point seems a bay or Harbour out of which many Junks came), Mountain N60°W, two Islands in one[6] S86°W 12 Miles, S*outh* extreme of two Islands S72°W, Hummock Hill[7] S60°W, the outward Island extreme[8] S20°W 10 or 12 Leagues. Between the Mountain and the point bearing N45°W we saw no Land. It appear'd to lead up to the North and seem'd likely to be either an extensive bay or the Entrance of a River or Harbour or probably the entrance towards the City of Jeddo by the description of its situation in the Charts.[9] From the Mountain to the Island bearing S86°W was a continuation of distant high Land & from thence to the Outer Island bearing S20°W where several more Islands [were] seen. We therefore Steer'd to the Southward to pass to the Eastward of them. The Fishing Boats of which there were near two hundred about us are near 36 feet long, 8½ Wide and 2½ Deep, neatly built of Oak Fir & Elm very sharp forward with a rising stern; the sides are continued much further than the Stern which gives them the appearance of being open abaft, the ends of such parts of them as are exposed to the weather are neatly covered with thin sheet copper & they appear very neat, Strong and Ingeniously constructed. About halfway down them is a Deck or false Bottom under which they stow their Fish. Each Boat has a Mast & large oblong sail of Cotton some of which were painted red about 2 or 3 feet from the Yard downwards, others have two long strips of Dark Blue Cotton stitched on to them. There is about 12 Men to each Boat and a Seine about 70 or 80 fathoms long. They are moved by 6 or 8 sculls acting on an outrigger about 2 feet from the Boats side and steer'd by a rudder laying in an oblique direction out astern something in the Chinese manner. The wind was variable with Cloudy weather, Lightning from the Northward throughout the evening & towards morning it blew

[1] Myoken San, a 418-metre high wooded peak, situated 16 miles SW of Taito Saki, the highest hill on Boso Hanto and the first object sighted by a vessel approaching from seaward. Mount King on Broughton's 'A Chart of the N.E. Coast of Asia' and not King's round hill, which was Fuji-san.

[2] Nojima Saki, the SE extremity of Boso Hanto.

[3] O Shima.

[4] Suno Saki, the SW extremity of Boso Hanto and the eastern entrance point of Uraga Suido, leading to Tokyo Wan.

[5] Probably Hachiman-misaki, a narrow cliff-faced promontory, situated 13 miles SSW of Taito Saki.

[6] Presumably the north extremity of O Shima.

[7] To Shima.

[8] Presumably the 853-metre-high Mikura Shima, which would just be visible above the horizon.

[9] Broughton was correct. The extensive bay was Sagami Wan with Uraga Suido, leading to Tokyo Wan, entered at its eastern end, NW of Suno Saki.

Strong from the NNE. We shaped a Course for the Outer Island steering to the South. At daylight we had no Land in Sight having experienced a very strong Current to the Eastward. At 8 AM Saw the Island[1] bearing S59°W and we made Sail for it. At Noon we were in the passage which divides the two Islands[2] & had no soundings. Course S63°W, distance 47 Miles, Lat*itude* Account 34°24′N, Lat*itude* observ'd 34°1′40″N, Longitude account 139°37′0″ East, No1 at Noon 139°58′17″E; bearings, Outer Island[3] bore from S8°W to S23°W 5 Miles, Extremes of second Island[4] from N54°W to N88°W about 3 Miles from the North point, Mountain or round Hill N32°W.

Saturday 12 November

Fresh breezes and fair Weather from the NE quarter. In the forenoon it was moderate and Cloudy. After passing through the passage we open'd to the Southward a Distant Island[5] & a small rock making like a Sail[6] to the SW. We haul'd up for the coast of Niphon to the WNW &West leaving several Islands to the NE of us.[7] At Noon the Land of Niphon bore from N27°E to N53°E 5 or 6 Leagues, Extremes from Deck N45°W, from the Masthead WbN about 12 Leagues, Course N82°W, Distant [*sic*] 78 Miles, Latitude Account 34°13′N, Longitude account 138°4′ East. We sounded throughout but had no bottom.

Sunday 13 November

Moderate Breezes & Squalls with Variable & dark cloudy weather which in the night increas'd to a strong Gale from the SW with heavy rain. We stood off to the SE[8] and at daylight, as the Squalls clear'd up, we saw the rock making like a Sail bearing SE 2 or 3 Leagues. At 9 AM we saw outer Island[9] bearing N74°E and we wore Ship. The wind soon veered to the N*orth* of West & we steer'd NNE with a strong Gale to observe more distinctly the Several Islands. At Noon very squally with Heavy Rain, the Extremes of outer Island bore from S68°E to S76°E, Mountain on with the Center of an Island much broken.[10] Course S64°E, distance 26 Miles, Lat*itude* Account 34°1′N, Latitude observed 33°59′N, Longitude account 138°32′ East.

Monday 14 November

Moderate Breeze & Squally with Rain. We passed to the North of Second Island[11] which evidently appeared to have a Volcanic appearance & from its west extreme are some detached

[1] Presumably Miyake Shima.
[2] I.e. between Miyake Shima and Mikura Shima.
[3] Mikura Shima.
[4] Miyake Shima.
[5] Hachijo Shima.
[6] Inanba Shima.
[7] During the night the *Providence* passed a short distance north of Zeni Su, a group of islets and reefs, the highest 12 metres high, situated 34 miles WSW of Miyake Shima, without sighting them.
[8] During the night the *Providence* again passed a short distance south of Zeni Su without sighting them.
[9] Mikura Shima.
[10] Kozu Shima.
[11] Miyake Shima.

Rocks[1] 3 or 4 Miles off. We pass'd between this Island and another more rugged & broken[2] having a Clear passage of 5 or 6 Leagues. At Sunset we were nearly up with the entrance of the Bay leading to the port of Jeddo. The land forming the SW part bearing from N28°W to N48°W 4 or 5 Leagues, the Mountain N40°W and the outer Island SbE. The wind having now fixed strongly in the NW Quarter I gave up the Idea of making any further progress against it & steer'd to the South to make the Islands called in the Charts Fatecuo.[3] At 7AM we saw them bearing S27°E to SbW and the Weather became Gloomy with Rain. At Noon we had got well up with them & the wind became variable from the East at which time we had the following bearings, Extremes of the Islands of Fatecuo, Small Island N20° to 33° East, Large Island N70°E distant 4 or 5 Leagues, Course S3°E, distance 60 Miles Lat*itude* Account 32°59′N, Longitude Account 138°36′ East.

The Island of Fatecuo bear about NbW and SbE from outer Island[4] distant 13 or 14 Leagues, the Western one is a small high peaked Hill[5] not more than a league in circuit and to appearance inaccessible, notwithstanding which on the north point we saw several Habitations surrounded with patches of cultivated land in ridges as before observed at Niphon. This Island lays about 2 Leagues to the SW'ward of the large Island. Between them appears a clear Channel. After passing it we kept our wind to examine more particularly the large Island but the wind veering to Eastward we were not enabled to approach it within 3 or 4 Leagues. It appears of considerable extent and presented a fertile appearance as far as the Haziness of the weather wou'd permit us to judge. Unfortunately we cou'd not determine its situation exactly not being able to see the Sun at Noon. By our run and the bearings from outer Island we estimate it to lay in Lat*itude* 33°1′N and Longitude 140° East. The unpromising appearance of the Weather gave us no hopes of a change. We however steer'd to the westward in a parallel with it in expectation of having a Meridian altitude next Day; but the weather growing worse and worse, having increased to a Strong Gale from the Eastward with constant Rain, we altered course to WNW to make the Coast of Niphon.

Tuesday 15 November

Strong breezes and dark cloudy weather with incessant Rain which in the fore Noon increased to a strong gale of wind. We steer'd to the westward with the wind from the NE Quarter. Hazy atmosphere prevented our seeing 4 or 5 Mile around. Course N81°W, distance 109 Miles, Latitude account 33°23′N, Longitude 137°51′ East.

Wednesday 16 November

Fresh Gales & thick weather with continual Rain, Wind from the NE quarter. At Dark we brought too the Wind under a Close reef'd Main top Sail and continued so till noon. Course S89°W, distance 39 Miles, Latitude account 33°22′N, Longitude Account 137°5′E.

[1] Onohara Shima, a small group of pinnacle rocks, the highest 117 metres high, situated 5 miles west of the SW extremity of Miyake Shima.

[2] Kozu Shima.

[3] Fatsisio in Broughton, *Voyage*, p. 144, and on Roberts, 'A General Chart ...'; present day Hachijo Shima, with the smaller Ko Shima.

[4] Mikura Shima.

[5] Ko Shima.

Thursday 17 November
Fresh Gales & Hazy. The Wind in the afternoon veer'd to the SW & we steer'd to the NW to make the Land under Courses. At ½ past 3 PM[1] we saw the Land indistinctly through the Haze. We were to the Southward and the wind suddenly flew round to the NW with dark threatening weather. At Sunset the Land extended from NNE to West 4 or 5 Leagues making a low point to the westward.[2] We kept our wind to the west as the Wind admitted, but in the forenoon we had no sight of the Land. Course S51°W, distance 55 Miles, Lat*itude* Acc*oun*t 32°47′N, Lat*itude* Observ'd 32°14′19″N, Longitude Acc*oun*t 136°15′E, No1 135°48′15″ East.

Friday 18 November
Moderate & Cloudy with Calm & Light westerly Winds. Course S51°W, distance 23 Miles, Lat*itude* Account 32°0′N, Lat*itude* Observ'd 31°30′, Longitude Account 135°54′E, Longitude p*er* No1 135°41′20″ East.

Saturday 19 November
Moderate & Cloudy with Calm & Light westerly Winds, Course S36°W, Distance 19 Miles, Latitude Account 31°15′N, Lat*itude* Observ'd 30°48′33″N, Longitude Account 135°41′E, p*er* No1 135°37′ East.

Sunday 20 November
Moderate and pleasant weather with the Wind NE'ly we steer'd WNW to make the Land. Course N71°W, Distance 51 Miles. Lat*itude* Acc*oun*t 31°6′N, Lat*itude* Observ'd 31°3′53″N, Longitude Account 134°45′E, ditto p*er* No1 134°12′10″ East.

Monday 21 November
Squally and dark Cloudy weather. In the Night we hove too not being able to see any distance ahead. With a NE wind we pursued a WNW Course [and] at daylight made Sail. Course N69°W, distance 70 Miles, Latitude Account, 31°29′N, Longitude ditto 133°39′ East.

Tuesday 22 November
Squally weather with heavy showers of Rain. At ½ past 4 we saw the Land bearing WNW. At Sunset it bore from N55°W to S61°W, when we sounded & plied to preserve our situation through the Night. On seeing the Land extending as far as WSW we steer'd for it till dark. It appear'd of moderate Elevation but we cou'd see it but very indistinctly from the Gloominess of the weather. We steer'd a SW course till 8 PM when we haul'd our Wind for the Night in order to examine it next morning. We supposed it to be the south End of the Island of

[1] 'AM' in the MS, which is clearly a transcription error.
[2] Broughton was correct, it was Shiono Misaki, the southernmost point of Honshu.

Ximo[1] which is layd down in Cooks General Chart[2] in Lat*itude* 31°30′N and 133° East Longitude. At daylight in the Morning the weather Calm as it had been most part of the night continued till 22^h when we made sail in for the Land. The Saddle Hill which bore N24°E of us at Noon appear'd like an Island but our distance prevented our determining it to be so.[3] The Land ahead set in the morning bearing WSW we cou'd not see on account of the Haziness of the Horizon in that part. But from WbN to N40°E we saw it appear at intervals which made us conclude it was all continuous. We were fortunate this Day in having Good observations for the Latitude & Longitude which determined our situation very exactly. Course S82°W, distance 36 Miles, Lat*itude* Acc*oun*t 31°22′, Latitude observ'd 32°7′17″, Longitude Account 132°47′E, Longitude No1 132°42′10″ East.

Wednesday 23 November

Fresh Breezes and dark Cloudy weather & Some rain. The wind veer'd from SW to West. At 5 PM we tack'd and stood to the SE. At 8^h[4] the extremes of the Land bore from Saddle Hill N60°E to S50°W, a sloping point of Land S85°W distant about 5 Leagues, a Bluff point[5] N40°W. The Squally and threatening weather increasing with Rain prevented our seeing the Land tho' we must have been within 6 Miles of it. The land to the NE'ward of the Bluff point appear'd to receed considerably having the appearance of an opening, but the indistinct sight we had of it caused by the Badness of the Weather prevented our making any satisfactory observations at Noon.[6] Strong breeze and fine weather with a Large Sea, the High Land bearing N60°W distant 10 or 12 Leagues. Course S17°W, distance 42 M*iles,* Latitude Account 31°27′N, Latitude observ'd 31°34′24″N, Longitude account 132°32′00″ East, Longitude p*er* No 1 133°1′20″ East.

Thursday 24 November

Fresh breezes & Clear Weather. At 4^{h}30^m Extremes of Land in Sight from N55°W to West distant about 10 Leagues. At 5^h Extremes of Land from S70°W to N63°W,[7] flat top Hill[8] N75°W. All those lands we rather supposed to be connected but cou'd not determine as our distance made them but Just visible from the Deck and the wind blowing strong from the NW we were not able to approach them nearer. As far as we can Judge they appear to be of considerable Height in detach'd parts. At 19^{h}30^m we were well up with the extreme of the Land set last evening, the Extremes bearing from S80°W to N12°W offshore 6 Leagues. The starboard point[9] of a deep bay[10] N47°W and distant Island[11] S45°W. At Noon the Extremes set before

[1] Kyushu.

[2] Roberts, 'A General Chart ...'

[3] Broughton appears to have sighted a feature at the southern end of Shikoku.

[4] 3^h in the MS, which appears to be a transcription error.

[5] Possibly Sen Saki on the east coast of Kyushu.

[6] The opening was almost certainly Bungo Suido, the strait separating Kyushu and Shikoku.

[7] 'At 15 h. the land extended from N. 70°W. To N. 63° W.' in Broughton *Voyage*, p. 150, but the log (TNA, ADM 55/147, f. 34v) agrees with the journal.

[8] Not Identified.

[9] Toi Misaki. In annotated *Voyage*, p. 151, inserted in pencil, 'Starboard Point of the Bay Lat. 31°19′4″N, Long. 131°42′53″E p*er* Time k*eepe*r No 1, 131°36′40″E per mean of 31 sets of Lunars.'

[10] Shibushi Wan.

[11] The 1,935-metre high Yaku Shima.

N75°W to N1°W, the starboard point of Bay N33°W 2 or 3 leagues, inner point N59°W, South point of Bay[1] N75°W, Isthmus just open, Volcano[2] N65°W, distant Island S40°W 10 Leagues distance, off nearest Shore 3 or 4 Leagues. This Land is high Rugged and much broken into to ridges, the Hill in general bare of Wood an[d] Rocky. Other parts are again cloathd with wood up to their very summits but we saw not the appearance of Cultivation at the distance we were from it. It however is inhabited as we saw some boats fishing in Shore and a junk standing into the Bay. The Volcano situated in the interior part of the Bay emitted vast Collumns of Smoke which appear'd very evident to every body. Course S60°W, Distance 83 Miles, Lat*itude* Acco*un*t 30°43′N, Lat*itude* observd 31°13′53″, Longitude Account 131°8′E, Longitude No1 131°55′40″ East.

Friday 25 November

Moderate breeze and fine Weather with the wind variable. At 1^h20^m Direction from Barren point[3] to N*orth* extreme of land[4] N1°E. NB Barren point bore N4°W at Noon. At 3^h15^m South point of Bay N65°W 2 or 3 Leagues, North point N6°W about 2 or 3 Leagues, extremes of land[5] from N11°E to S75°W, when we haul'd our wind to the Southward. At 5^h Extremes from N18°E to S75°W, Hummock in the Center of the Bay[6] N43°W about 6 Leagues, South point of Bay N50°W, North point N15°E, Center of the Bay N40°W, low land just visible from the deck[7] S68°W, Distant Island S42°W, connected with which seem'd some Low Land[8] Bearing S25°W, off shore about 6 Miles. This Bay seems to be Capacious and likely to afford good Anchorage but the wind prevented our noticing it more particularly.[9]

At 19^h0^m we saw the low land bearing S67°W which appeard to be in the same direction as that we saw last evening bearing S27°W. The Extremes to the North of the Bay N27°W and the South Extreme set last Evening bearing then S75°W now bore N77°W, Distant Island over the low land S50°W. We stood in for the Land as the wind permitted, our Distance from the low land being[10] 6 or 7 Leagues, it appeared evidently we had experienced a strong Current setting to the Northward.[11] At Noon the Extremes to the North of the bay N15°W, North point of the Bay N23°W, South point of the Bay N42°W, South Extreme N72°W, Extremes of the Low land from S68° to S82°W, Distant Island South Extreme S58°W, other Extreme on with Centre of low Land, off shore 5 or 6 Leagues. Course S11°W, distance 50 Miles, Lat*itude* Account 30°25′N, Latitude [observed] 30°50′15″N, Longitude Account 130°56′ East, Longitude p*er* No1 131°58′45″ East.

[1] Hi Saki.
[2] The 1,118-metre-high Sakura Shima.
[3] Not identified.
[4] 'island' in the MS, which is surely a transcription error.
[5] Ibid.
[6] Possibly Yaku Shima.
[7] The northern extremity of the 301-metre-high Tanega Shima.
[8] The southern extremity of Tanega Shima.
[9] 'There are depths suitable for anchorage [in the bay] ... but the anchorage should be avoided during winds from between E and S. A swell frequently runs into the bay and vessels at anchor usually ride uneasily'; *Japan Pilot*, Vol. 2, 7th edition, Taunton, 1979, p. 163.
[10] 'Bearing' in the MS, which appears to be a transcription error.
[11] Still under the influence of the Kuro Shio.

Saturday 26 November

Moderate breeze and fine Weather. We kept our Wind in for the low Land as far to the South as we coul'd fetch with a fresh Breeze and Smooth water, more low land appearing as we approach'd the Shore. At 5^h0^m we were by estimation within 4 or 5 Miles of the Low Land which extended from S20°W to N75°W, of very moderate elevation, the tops cover'd with scatter'd trees, & towards the Sea white Sandy beaches above which we saw smoke in several places. What was before call'd distant Island now appeared over the low land making very high & at considerable distance bearing S50°W. What was call'd at noon South Extreme now bore N55°W, to the westward of which the Land extended in Hummocks to N68°W 5 or 6 Leagues, the north Extreme near the North point of the Bay N17°E, the South point of the Bay N2°W. The Sun having set we tacked in 45 f*athoms* Water and stood to the Eastward. In our present situation we cou'd not absolutely determine w[h]ether the Low Land join'd to the distant High Land or w[h]ether it was a separate Island (we before supposed it to be an Island) which I rather suppose to be the Case.[1] The opening between the Lands was so much contracted and making in low points from each Shore. I conceiv'd the whole of the Land to be the same Island & that the opening probably terminated in a Deep Bay or harbor imperceptable to us, which point we were for the present obliged to leave undecided.[2] At 14^h0^m the Wind veering more to the SW prevented our standing in shore as we cou'd not have fetchd the Land we had left. Therefore [we] continued our Course to the SE'ward and at Noon found ourselves 21 Miles to the Northward of our Reckoning. Course S47°E, Distance 51 Miles, Lat*itude* Account 30°15′N, Lat*itude* Observ'd 30°36′N, Longitude Account 131°39′ East.

Sunday 27 November

Fresh breezes and Cloudy weather winds from the SW quarter. At Noon moderate breeze and fine weather, Course S42°E, Distance 76 Miles, Latitude Account 29°41′N, Longitude Account 132°36′ East.

Monday 28 November

Moderate breezes and fine weather throughout these 24 hours. In the Latitude and Longitude we were in this day at Noon are several Islands layd down in Arrowsmiths Chart of 1791[3] suppos'd to be discover'd by the Spanyards, but we saw nothing to indicate our being in the vicinity of any. Course S31°E, Distance 77 Miles, Latitude Account 28°34′N, Latitude Observ'd 28°30′N, longitude Account 133°21′East, Longitude p*er* No1 135°4′ East.

Tuesday 29 November

Moderate breezes and fine Weather, winds variable. We steer'd to the SW. At Noon moderate Breeze and cloudy Weather with a swell from the NW. Course S43°W, Distance 67 Miles, Lat*itude* Account 27°41′N, Latitude observ'd 27°43′32″N, Longitude Account 132°30′ East.

[1] Tanega Shima and Yaku Shima are not connected.

[2] Osumi Kaikyo, separating Tanega Shima from the southern extremity of Kyushu, leads towards the major port of Kagoshima.

[3] The non-existent Todos los Santos in 30°45′N. 139°22′E; see Stommel, *Lost Islands*, pp. 108, 132, 133. Possibly a second issue of Aaron Arrowsmith's, *Chart of the World on Mercator's Projection*, on 8 sheets, 1790. Also shown on Arrowsmith's 1794, 'Map of the World', which has Zuyden Island in about 32°N and 138°E.

Wednesday 30 November

Moderate Breezes and cloudy weather, wind from the NW quarter. Towards evening it began to blow fresh with rain. At Noon fresh breezes & Cloudy weather. Course S81°W, distance 118 Miles, Latitude Account 27°26′, Longitude Account 130°20′ East.

Thursday 1 December

Moderate Breezes & dark cloudy weather. We tried for soundings throughout the night and before noon we saw to the westward Land which we supposed to be the Island of Likeo.[1] We pursued the same Course till Noon to distinguish [it] more particularly. At which time the wind veering to the SE with Squally weather, we reef'd the Topsails and kept our Wind to the SW, our Distance from the Shore being 5 or 6 Leagues. From this unfavourable point of View we cou'd discover very little. It appear'd rather low and even, the western [point] falling abruptly [to the sea]. In the Afternoon of the next Day as we steer'd to the Southward it appear'd more broken, the Higher parts being connected by low land. The Southern part presented a pleasing prospect of cultivated Land Diversified with Wood towards the Sea. The Land rose in Clayey Cliffs of a light Yellow with Sandy beaches beneath them. To the Northward on the western Side of the Island it appear'd equally low, the rising Ground loaded with trees. The Islands, that we were not near enough to distinguish them particularly, were of Great Elevation and appear'd to form a Group closely connected with each other. Extremes of the Land indistinctly seen at Noon from S76°W to West, Distant 5 or 6 Leagues. Course S89°W, dist*anc*e 125 M*iles,* Lat*itude* Acco*un*t 27°24′N, Lat*itude* Obs*erve*d 26°47′N, Longitude Account 128°0′ East.

Friday 2 December

Variable weather which in the forenoon increas'd to a Strong Breeze from the N*orth*. During the Evening it was SE'ly. We kept our wind in the SW, as Wind permitted, the land being only visible at times owing to the Squally Weather. At 2^{h} PM Land appearing like a small Island from the Masthead,[2] it bore NWbN about 8 Leagues Distant when the extremes bore from N6°W to S55°W. At 5^{h}16^{m} the Extremes bore from S80°W, about 10 Leagues distant, to N24°W, 7 Leagues, being the same point set at Noon bearing West[3] and which seem'd to be the [North-]Western part of the Island, the Land to the North falling back considerably within it, apparent Island point West, Bluff point N78°W distant 5 or 6 L*eague*s from Shore. The Coast still presented the same level appearance, in General very low. We kept our wind to the SW'ward till midnight when being by estimation in the Latitude of 26°N we plyed to windward under easy sail till daylight, having the wind variable. In the Morning it fixed in the NW quarter with squally and unsettled weather. At 19^{h} the nearest Land bore from us N30°W about 3 Leagues distant, Extremes of low Land N10°E and the Extremes to the Westward N67°W. We saw no land in any other direction tho' some Islands are layd Down in the situation we were now in. At 19^{h}40^{m} the nearest Land on with a point to the Northward N7°W. The Distant point about 5 Leagues off at 21^{h}0 on with a more distant point N26°E. At

[1] Okinawa Shima.

[2] Probably Yoron Shima.

[3] Hedo Misaki, the northern extremity of Okinawa Shima.

the same time the SW point of Likeo on with a low point to the Northward N28°W. NB at $20^{h}0$ Distant Islands were seen bearing WNW. At $21^{h}0$ the outer Island, form'd like a Gunners quoin, bore WbN. At $21^{h}40^{m}$ the SE point of Likeo (the same set N30°W at Daylight) bore N38°E, SW Extreme North Low Land open with it N5°W, making like a small low Island, Extremes of western Island N63°W, East Extreme N65°W, East Extreme of west Island N43°W.

At noon the South part of the Island of Likeo bore from N46°E to N68°E Distant 6 or 7 Leagues, a low Island N*orth* about 8 Leagues, Gunners Quoin Island N34°W 5 Leagues, 2nd Supposed Island N25° & 28°W 6 Leagues, extremes of nearest Island N10° to 18°W nearly 5 Leagues. We cou'd not positively Ascertain the Number of these Islands, but from our distance & the Haziness of the Weather we reckoned 5 of them.[1] Course S48°W, distance 77 Miles, Lat*itude* Acc*oun*t 25°55′N, Lat*itude* Observ'd 25°53′N, Longitude Acc*oun*t 126°57′ East.

Saturday 3 December

Fresh Gales & dark Cloudy weather wind from the NNW. We steer'd a SWbW Course, with as much wind as we cou'd carry d[o]uble reef'd top Sails to, in the direction of some Islands said to lay between Likeo and Formosa. The Gale increas'd with heavy squalls of Wind and Rain & yet the Barometer did not denote any Alteration, the Thermometer continued at 72°. In the forenoon it became more moderate with a very heavy sea from the Northward. At 21^{h} we saw low Land to the Westward,[2] at $22^{h}15^{m}$ the Extremes Bore from N75°W to S79°W, at $23^{h}30^{m}$ the North point on with South Extreme being the Direction of the Shore N52°W, at the Same time South Extreme on with distant low point West. At $21^{h}35^{m}$ Extremes of the Island bore from N85°W[3] to S82°W. The point bearing N85°W is the falling Green point in the Sketch.[4] At $23^{h}40^{m}$ a distant Island seen from the Mast head[5] on with a reef of Rocks off SE point bore north and a hummock which did not appear connected with the Island in one bearing with SE Extreme[6] N37°W, 5 or 6 leagues. Another Island seen from the Mast head appear'd to the NW'ward.[7] At Noon the SE Extreme of the Island bore N9°W 2 or 3 Miles, the other Extreme forming a very low point N85°W, outer part of the reef N22°E about 3 Miles. This reef is composed of Rocks above and under water in an Easterly Direction extending 2 or 3 Miles from the South Extreme. Just within the South Extreme, to the westward, is a small sandy bay above which we discover'd a village interspersed among the trees with some cultivated Ground near it. The Island is Universally Low barren Land & Rocky. On the tops of the rising Grounds [we] saw some Groves of Trees but in general the Country appear'd cover'd with heath and Course grass. We also observ'd several other Villages scatter'd along the Shore and Smoke rising in several places. The Wind still continued blowing so Strong we cou'd not risk having any communication with the Shore, but we ranged the southern side at the distance of 3 or 4 Miles. As we ran to the westward the Shore became more Rocky & much lower with ma[n]y scatter'd Villages on the Most Elevated Grounds. Course S64°W, distance

[1] Kerama Retto, consisting of ten moderately sized islands and a number of smaller ones.

[2] Miyako Shima; Typinsan on Broughton's 'A Chart from the South Point of Formosa to Great Lieuchieux', Broughton, *Voyage*, f.p. 147.

[3] Overwritten and indistinct.

[4] This sketch does not appear to have survived.

[5] Fude Iwa.

[6] But part of the main island.

[7] Possibly Irabu Shima seen over the main island.

118 Miles, Lat*itude* Acc*oun*t 25°1′N, Lat*itude* Observ'd 24°39′20″N, Longitude Acc*oun*t 125°0′ East, Longitude p*er* No1 mean of AM & PM sights 126°0 East.

Sunday 4 December

Moderate Breezes and Variable with Cloudy Weather, we had frequent Soundings. We steer'd WbS keeping three or 4 Miles from the Shore. At $1^h0'$ Extent of the West extreme seem'd to form a small Island[1] bearing from N60°W to N50°W, leaving a passage open with the Larger Island of 5°, between which extended a Reef. At 1^h30^m the Yellow point[2] on with doubtful Island[3] N20°W.[4] At 2^h0^m the west Extreme of Island on with distant Land[5] N11°E, a small reef N45°E,[6] West extremes of low Land[7] N28°W. Between the Island and doubtful Island [is] a very extensive Reef, on the outer part of which we had 18 *fathom*s Water about a League without it. At 3^h10^m West extreme of low Land N12°E, other extreme called doubtful Island N32°E, Extremes of Reef Island[8] from N52° to N76°E 3 or 4 Leagues. Having steer'd a West Course from Noon, at 5^h15^m a very low Island[9] Bore from N65°W to N87°W 5 or 6 Miles, High Land S70°W and Land something lower S85° W.[10] We also saw Distant Land to the Northward[11] Bearing N20°W 5 or 7 Leagues. During the night we ply'd under easy Sail and at 18^h45^m we saw the low Island[12] Bearing N66°W 4 or 5 Leagues. Land we passed Yesterday[13] now Bore NNE to ENE and south Extreme[14] from the Mast head S77°W. At 20^h the low Island from N20°W to N49°W 5 or 6 Miles. This Island is exceedingly low; in the Middle are some trees which give it a small degree of Elevation. Every other part is cover'd with brushwood and surrounded by a Sandy Beach; its direction is ESE & WSW 5 or 6 Miles in extent. The extremes of the High Land, which forms a separate Island, bore from N87°W S70°W, for which we steer'd a WSW Course. At 22^h0^m we lost sight of the low Island which is not visible from the Deck 5 Leagues. At 23 Hours the North Extreme of the High Land N53°W, Bare point (within which is form'd a Bay) N60°W, fall of High Land S80°W, South Extreme S70°W, off Shore 2 or 3 Leagues. At Noon the South Extreme making a low point S54°W, the North Extreme North, Isthmus or Center of Bay N30°W,[15] peak N50°W, fall of High land West. In the same direction, at a considerable Distance, we saw more high land, our distance from the Shore 3 or 4 Miles. The land is of considerable elevation rather high, in some parts much broken, and the Hills connected by low land. It appears in General well cloath'd with

[1] Kurema Shima.

[2] Not identified, but possibly the western extremity of Miyako Shima.

[3] I.e. Kurema Shima.

[4] 'At 1^h40^m Yellow point on with Island N74°W. The other extreme of Island on with Doubtful island N27°W.' omitted as it does not make sense.

[5] Ikema Shima.

[6] 'extremes of distant [*sic*] N3°E to N18°W' omitted as it does not make sense. It may refer to Irabu Shima.

[7] I.e. Irabu Shima.

[8] Probably part of Miyako Shima.

[9] Tarama Shima.

[10] Parts of Ishigaki Shima, Patchusan on Broughton's 'A Chart from the South Point of Formosa to Great Lieuchieux', Broughton, *Voyage*, f.p. 147.

[11] Minna Shima.

[12] Tarama Shima.

[13] Miyako Shima.

[14] Ishigaki Shima.

[15] Miyana Wan.

wood but in other instances the Hills were bare particularly towards the NE. The SE point is very low off which is a small reef and above the beach a very extensive Village. This side of the Island takes a NNE and SSW direction. As we pursued our course we saw smoke ascending in every direction. Course S79°W, distance 64 Miles, Lat*itude* Acco*un*t 24°27′N, Lat*itude* Obs*erve*d 24°23′38″N, Longitude Account 123°51′ East, Longitude p*e*r No1 124°53′55″ East.

Monday 5 December

Moderate breeze and variable throughout the Afternoon. In the fore noon we had strong Breezes with very thick weather from the North. At 1^h15^m the village point bore West 3 Miles. 15 Minutes before, it was on with the distant High Land[1] which now open'd bearing S75°W & which land at 2^h0^m bore due West, Extremes of ditto N73°W leaving a passage open of 13° to the Northward, NE Extreme N17°E, off Shore 4 or 5 Miles. At 2^h30^m the fall of the High land S88°W, SE Extreme N45°E, apparent opening N7°E, two Rocks in the midst of the Breakers N86°W to S75°W, 4 or 5 Miles. Seeing nothing but Breakers extending to the SW'ward and no passage within them, we alter'd our course and at 3^h0^m the extreme of the breakers on with the fall of the High Land bore N85°W. We now perceiv'd a continuation of very low flat Islands to extend entirely from the Village point towards the fall of the High Land and even those insulated spots were not Join'd. They appear'd connected by a reef of rocks on which the Sea broke entirely across, leaving no safe passage between them. At 5^h30^m we approach'd another low Island[2] which bore from S50° to 56°W 2 or 3 Leagues, fall of High Land N65°W, between which and the low Island were some 2 small Hummocks just visible from the Deck[3] bearing N85°W. We were now abreast of the Westermost of the low Islands[4] at about 3 Miles distance, which extended from N6° to 40°W, with a reef joining them N27°W, a second low Island[5] from N18° to 40°E 2 or 3 Leagues, the west end of it at this time on with East extreme of Distant opening and the East end of it on with the Land about Village point. At 18^h30^m the low land Island set last Night in the SW[6] bore of us NNW½W distant 2 or 3 Leagues. At 19^h45^m we saw the 2 small Hummocks bearing N27°W when the West Extreme of the High Land set last Night bearing N65°W now bore North about 5 Leagues, a small opening in the Land N15°E, East Extreme of same Land N34°E, Extreme of low Island from N45°E to N60°E 5 or 6 Miles. At 20^h30^m the two Hummocks North 3 or 4 Leagues, when we lost Sight of the land. This low Island trended nearly NW & SE, its extent about 3 or 4 Miles, leaving a Clear passage between it and the High Land. Ever since daylight the wind increas'd from the NNE with thick Hazy Rainy Weather. No appearance of its altering for the better to permit our having any Communication we therefore pursued our Course to the Westward. Course S77°W, distance 51 M*iles*, Lat*itude* Acco*un*t 24°10′N, Longitude Account 122°54′ East.

[1] Iriomote Shima; Rocho-o-ko-ko on Broughton's 'A Chart from the South Point of Formosa to Great Lieuchieux', Broughton, *Voyage*, f.p. 147.

[2] Hateruma Shima.

[3] Nakanogan Shima.

[4] Aragusku Shima.

[5] Kuro Shima.

[6] I.e. Hateruma Shima.

Tuesday 6 December

Fresh Gales & dark Cloudy weather throughout the night. We brought too under a foresail and in the Morning we made Sail Steering West with a strong Gale, in the hopes of falling in with some more Islands that are lay'd down in this situation, but the Gale increasing in the Afternoon with constant Rain prevented us from seeing more than 3 or 4 leagues round us. At 5^h0^m not thinking it safe to run any longer and lest we might pass any thing during the night we brought too under Fore Sail & Mizen Stay Sail.

In the Morning at Daylight the Gale having moderated a little we bore up under close reef'd Fore top Sail, Fore Sail & close reef'd Main top Sail, and pursued our Course to the Westward. At 2^h0^m the Gale increased with heavy Squalls & Rain which entirely Obscur'd the Horizon. At 21^h30^m the Squall being over we saw the Island of Formosa[1] about 3 Leagues Distant. It appeared to be very Mountainous Land rising abruptly from the Sea to a prodigious Height. An Uncommon fall of water de[s]cending from the Hills nearly in a perpendicular direction pointed out the Land before we saw it thro' the Gloomy Sky. We had frequent squalls & Showers of Rain which prevented our nearer approach to the Land. So we altered our Course along Shore to the Southward with a large following Sea. Course S64°W, Distance 71 Miles, Latitude Acc*oun*t 23°38′ N, Lat*itude* Observ'd 23°45′0″N, Longitude Account 121°44′0 East, Longitude p*er* No1 122°42′12″ East.

Wednesday 7 December

Fresh Gale and Cloudy Weather with the Wind from the NE. The Gale still continued and we steer'd South with a large following Sea, the High Land appearing at Intervals as the Squalls clear'd up. At 4^h0^m the Land bore S65°W. At 8^h30^m Gale continuing and being by estimation in Lat*itude* 22°45′N we brought too for the Night. At 18^h20^m High Land of Formosa bore from West to NNW, a low Island[2] S80°W 4 Leagues. This Island appear'd to be 4 or 5 Leagues from the Main Land. At the Same time the Island of Bottol Tobago Xima[3] SW[4] 10 or 12 Leagues. The weather now became more moderate & at 21^h0^m the low Island N50°W 4 Leagues, Extremes of Bot*ol* Tobago from S23° to S37°W.[5] The Haze now prevented our seeing the Land of Formosa. The [*words to complete the sentence missing*]. At 23^h30^m the East Extreme of Botol Tobago in one bearing with low land S8°E & N8°W. At Noon the Extreme of Botol Tobago from South to S23°E, the South point 8 or 10 Miles, Low Island North on with Land of Formosa, the Extreme point of Formosa in sight to the Southward making in a low point[6] S70°W, soon after Noon Little Tobago on with East Extreme of Botol Tobago S26° East, Some projecting Rocks from the centre of Botol Tobago open 2°. NB this is nearly the Direction of the North Side of the Island. Course S17°W, distance 92 Miles, Lat*itude* Acc*oun*t 22°17′N, Lat*itude* Observ'd 22°15′48″N, Longitude Account 121°15′E, ditto p*er* No1 121°57′45″ East.

[1] Taiwan.
[2] Lü Tao.
[3] Lan Yü.
[4] S6°W in Broughton, *Voyage*, p. 160, but altered in pencil to SWbW in annotated *Voyage*.
[5] S27°W in Broughton, *Voyage*, p. 160.
[6] O-luan Pi.

APPENDIX IV

Remarks on the Japanese & Corean Vessels[1]

The Japanese Junks we had the opportunity of observing were from three hundred to thirty tons burthen and without the smallest variation of the same construction. They have only one Mast, with a Square Sail of Cotton & Notwithstanding they at times Weather'd upon the Schooner we generally sail'd much faster; they were unable to Tack, but always wore in a very short time.

They made use of Iron Creepers[2] for Anchors and their Rudders were of the same construction with the Chinese: and when at anchor hove up by a purchase.

The but[t]-ends[3] and stern were cover'd with Copper to preserve them and the opening in their Sterns is something similar to a Square back'd Vessel; only the sides are carried aft, projecting beyond it some distance.

The Hold was entirely open and they made use of Tanks or Cisterns for carrying water; the Mast was composed of several pieces of Fir rudely made and strongly hooped together. The Floor Timbers, Beams, and main piece of the Rudder were of Oak, the rest of Pine.

The Corean Junks were of a large Size with only one Mast; but we did not see any under sail. Their Boats in the Archipelago of T'Chosan-go were more similar to the Chinese & particularly in their mode of navigating them. They also made use of wooden Anchors, which the Japanese do not; their boats had two Masts and matted Sails, but not extended cross-ways with bamboos, like the Chinese. In the centre of the Vessel they have a Fire place built up with Clay in a square form: they always keep some Fire in it for the conveniency of Smoking.

They made use of no covering like the Chinese Tilts,[4] but cover'd themselves up with coarse dry'd grass Cloaks.

The boats were built of Fir, the Skulls were either Oak or Ash & they appeared to manage them with great dexterity. Some of them would carry from fifty to sixty People, and they sail'd very fast.

[1] TNA, ADM 55/147, ff. 127v–128; reproduced in Broughton, *Voyage*, pp. 378–9, with minor changes in punctuation and capitalization.

[2] Small grapnels.

[3] The extremity of planks when they are united or abut against each other.

[4] Small canopies extending over the stern sheets of a boat, supported by iron or wood work to keep off rain.

APPENDIX V

The Ship's Company

William Robert Broughton, Commander on appointment. Broughton joined the navy on 1 May 1774 and after serving in various ships was promoted lieutenant in 1782. In 1790 he was appointed in command of the *Chatham* to accompany Vancouver to the NW Coast of America, from where he was sent back to England in 1793 with despatches, travelling overland across New Spain [Mexico] from San Blas to Vera Cruz and thence to Cadiz and England. He was promoted commander on 3 October 1793 and sent back to the NW Coast of America in the *Providence* to rejoin Vancouver. For his subsequent career see p. li above.

Zachary Mudge, 1st Lieutenant. Born Plymouth 22 January 1770; served under Captain John Jervis, later Earl St Vincent, in the *Foudroyant.* Sailed with Vancouver in the *Discovery*, returning to England with dispatches in 1792 via China. Sent back to England from Macau, following the loss of the *Providence.* For his subsequent career see p. lv above.

G. J. F. Young, 2nd Lieutenant. Sent home to England from Macau on 11 June 1797. Possibly George Forbes Freeman Young, lieutenant 1790; died 1799.

James G. Vashon, 3rd Lieutenant. Discharged per sentence of Court Martial. For his subsequent career see p. lv above.

William Chapman, Master. Joined the *Providence* in 1793, serving in her until he was discharged to HMS *Trident* on 28 May 1798. For his subsequent career see p. lvi above.

John Floud, Surgeon. Discharged to HMS *Orpheus* on 28 May 1798.

George Young, Lieutenant of Marines. Sent home to England from Macau on 10 June 1797.

John Crosley (1762–1817), Astronomer. Prior to his appointment to the *Providence*, Crosley had spent just under three years as assistant to the Astronomer Royal, Nevil Maskelyne.[1] After surviving the wreck of the *Providence*, Crosley sailed from Macau for England in the packet *Crescent*, but after being struck by a typhoon in the South China Sea was forced to return to Macau for repairs, eventually reaching England on 7 July 1798. For his subsequent career see p. lvii above.

John Cawley, Master's Mate. Cawley gave up a command in the merchant service to join the Navy as a midshipman in 1793, taking part in the Battle of the Glorious First of June in the *Barfleur*. He next joined the *Providence*, transferring to the *Providence Tender* in June 1797. He was discharged to HMS *Suffolk*, 28 March 1798, serving in various ships in the East Indies until 1803, during which time he was promoted lieutenant in 1800. After various other appointments he was on half-pay between 1815 and 1818. In 1821 he was appointed in command of the cutter *Grecian* on the West Indies station, serving in her for three and a half

[1] Derek Howse, *Nevil Maskelyne: the Seaman's Astronomer*, Cambridge, 1989, p. 156.

years at the end of which he was promoted commander in 1825, after which he appears to have been placed on half-pay. He died 29 April 1846.[1]

John J. Haywood, Midshipman. Discharged to HMS *La Sybille* on 28 May 1798. Not in 1807 *Navy List*.

Reginald Bean Hopkins, Midshipman. Discharged to HMS *Victorious* on 28 May 1798; lieutenant 1804; died 13 August 1811.

James Brown Boyd, Midshipman. Sent home on leave on 10 June 1797; lieutenant 1801; retired commander 1832; died 1842.

Lord George Stuart, Midshipman. Sent home on leave on 10 June 1797; lieutenant 1801; commander 1802; captain 1804; CB 1815; rear admiral of the blue 1837; died 19 February 1841.

Honourable Alexander Jones. Sent home on leave on 11 June 1797; midshipman, lieutenant 1800; commander 1806; captain 1811; rear admiral of the blue 1848; rear admiral of the white 1849; retired rear admiral 1851; retired vice admiral 1855; retired admiral 1860; died 8 January 1862. In 1809, when in command of the 3-gun frigate *Amiable*, he captured the French corvette *Iris* off the Texel.

William Mitchell, Boatswain. Sent home on leave on 13 June 1797.
William Forster, Carpenter. Discharged at Trincomalee.
T. Coulston, Carpenter, Surgeon 2nd Mate. Sent home on leave on 10 June 1797.
Stephen Bones, Clerk. Discharged at Trincomalee on 28 May 1798.
John Dining, Master at Arms. Discharged at Macau on 10 June 1797.
George Randal, Carpenter's Mate. Discharged to HMS *Suffolk* on 28 June, 1798.
Edward Miller, Carpenter's Mate, ditto.
Thomas Belcher, Boatswain's Mate, ditto.
Martin Neal, Boatswain's Mate, ditto.
Joseph Mott, Boatswain's Mate, ditto.
Thomas Morrison, Gunner's Mate, ditto.
Robert Miller, Ship's Corporal, ditto.
Charles Bruce, Sailmaker's Mate, ditto.
John Oldroyd, Able Seaman, ditto.
David Henderson, Able Seaman, ditto.
James Steel, Able Seaman, ditto.
John Graham, Able Seaman, ditto.
William Bryan, Able Seaman, ditto.
Andrew Mitchell, Able Seaman, ditto.
John William, Able Seaman, ditto.
John Wilson, Able Seaman, ditto.
Matthew Leonard, Able Seaman, ditto.
Thomas French, Able seaman, ditto.
James Edwards, Able Seaman. Discharged to HM Schooner *Providence* on 28 June 1798.
Thomas Stevens, Able Seaman, ditto.

[1] On 28 May 1834 Cawley submitted to Sir James Graham, First Lord of the Admiralty, a printed 'Statement of the Services of John Cawley, a Commander in H.M. Navy between the Years 1793 and 1834' requesting an appointment to the Ordinary at Plymouth; National Library of Australia.

Nathaniel Ireland, Able Seaman, ditto.
John Hopkins, Able Seaman. Discharged at Trincomalee to return home on 28 May 1798.
James Donald, Able Seaman, ditto.
Samuel Redriff, Able Seaman. Died at sea on 15 March 1795.
Hugh M'Donald, Able Seaman. Killed by falling from the main topsail yard on 16 June 1795.
Patrick Sherry, Landsman. Killed by falling from the mizen top on 6 June 1796.
Daniel Dell, Private. Killed by the natives of Niihau on 30 July 1796.
Jonathan Bird, Private, ditto.
Hans Oldson, Able Seaman. Killed by accident at Hokkaido on 29 September 1796.
Robert Ogilvy, Master's Mate. Lost in HMS *Swift* in June 1797.
John Delahoyde, Master's Mate, ditto.
Robert Mearing, Midshipman, ditto.
Francis Cousins, Surgeon's First Mate, ditto.
William Clerk, Captain's Mate, ditto.
George Cadman, Carpenter's Crew, ditto.
David Starke, Carpenter's Crew, ditto.
George Allen, Gunner's Mate, ditto.
John Thomson, Quartermaster, ditto.
Henry Jones, Quartermaster, ditto.
Peter Swords, Quartermaster, ditto.
William Owen, Quartermaster, ditto.
Robert Gamble, Quartermaster, ditto.
William Coleman, Able Seaman, ditto.
Thomas Capper, Able Seaman, ditto.
Henry Shewman, Able Seaman, ditto.
Abraham Graves, Able Seaman, ditto.
David Stevens, Able Seaman, ditto.
Robert Burn, Able Seaman, ditto.
Alexander Graham, Able Seaman, ditto.
James Clerke, Able Seaman, ditto.
William Dring, Able Seaman, ditto.
Peter Murry, Able Seaman, ditto.
John Davis, Able Seaman, ditto.
Joseph Kennedy, Able Seaman, ditto.
John Moon, Able Seaman, ditto.
Alexander Ducas, Able Seaman, ditto.
James Butcher, Able Seaman, ditto.
Benjamin Braid, Corporal of Marines, ditto.
John Cook, Drummer, ditto.
James Wilkie, Private, ditto.
Joseph Whiles, Private, ditto.
Francis Clerk, Private, ditto.
William Julford, Private, ditto.
John Lloyde, Private, ditto.
William Hertekoal, Private, ditto.
Peter Johnson, Private, ditto.
John Haiton, Private, ditto.

John Bacon, Private, ditto.
William Bacon, Private, ditto.
Thomas Westwood, Private, ditto.
John Pickwick, Private, ditto.
William Thomas, Private, ditto.
Thomas Horn, Private. Died at sea on board the schooner on 14 July 1797.
William Walker, Carpenter's Crew. Killed by accident in China on board the East Indiaman *Glatton.*
Thomas Mullen, Gunner. Killed by accident during his passage from China to England.
Alexander Bishop, Cook. Died in hospital in the Cape of Good Hope in 1798.
John Garland, Quartermaster. Discharged in China to the East Indiaman *True Briton* on 13 June 1797.
John Outridge, Armourer. Discharged to HMS *La Sybille* in Macau on 3 January 1798.
Joseph Grimshire, Cook's Mate. Discharged to the East Indiaman *Carnatic,* in China on 14 June 1797.
James Long, Able Seaman; ditto.
John Martin, Able Seaman; ditto.
Samuel Hudson, Able Seaman. Discharged to the packet *Crescent* on 10 June 1798.
William Anderson, Able Seaman. Discharged to the East Indiaman *Glatton* on 10 June 1797,
William Eastwood, Able Seaman; ditto.
Edward Carpenter, Able Seaman; ditto.
William Smith, Able Seaman; ditto.
William Hurvant, Able Seaman; ditto.
Miles Kimber, Able Seaman; ditto.
Solomon Pollock, Able seaman; ditto.
Samuel Reed, Able Seaman; ditto.
Thomas Allen, Able Seaman; ditto
Robert Shaw, Landsman; ditto.
Joseph Bernard, Landsman; ditto.
Joseph Ashton, Landsman; ditto.
John Lawson, Able Seaman. Run from the schooner at Macau on 21 June 1797.
Thomas Graham, Able Seaman; ditto.

APPENDIX VI

Broughton's Journal of his Journey across New Spain in 1793[1]

The Journal of a Tour across the Continent of New Spain from St Blas in the Northern Pacific Ocean to La Vera Cruz in the Gulph of Mexico by Lieut. W. R. Broughton, in the year 1793. Commander of H.M. Brig Chatham

As it was necessary that Government should have had the earliest information of the misunderstanding that had arisen between the Spanish Commander at Nootka, and Capt*ai*n Vancouver of his Majesty's sloop the Discovery relating to the Articles of the convention agree'd to by the two Courts, it was determin'd that I should cross the continent of New Spain with despatches to England, and on the 12th of January 1793 I left Monterry and embark'd on board a Spanish ~~Brig~~ Ship of War[2] for that purpose with the Spanish Commander Don Juan de la Bodega y Quadra bound to St Blas.[3] The first land we saw was the Island of Guadalupe which is of a considerable height, and visible at the distance of 20 leagues. It lies in the latitude of 28°53S [*sic*], and long*itud*e 241°50 East as settled by Malaspina a Spanish Officer.[4] It affords neither Anchorage or Water, presents a rocky surface, and is generally barren. Two smaller Islands*[5] are plac'd near its Southern Point. The length is about 5 or 6 leagues, the width very inconsiderable; this is the first land the Galleons make in their passage from Manilla.

Cape St Lucas round which we saild is the Southern point of California, and is plac'd by Astronomical observations by Colnet 22.45 – 250°0′E.[6] Vancouver – in Lat*itud*e 22.50S [*sic*] & 250 11 East Long*itud*e from Greenwich.[7] There is a small sandy bay with Anchorage in Isabella island,[8] Lat*itud*e 21°48N, which is distant from St Blas EbS 12 leagues where we

[1] NMM, RUSI/NM/132/2. For a brief description of Broughton's journey across New Mexico see Jim Mockford, 'The Journal of a Tour across New Spain from St. Blas in the North Pacific Ocean to La Vera Cruz in the Gulph of Mexico', *Terrae Incognitae*, 36, 2004, pp. 42–58.

[2] The *Activa*; Cook, *Flood Tide*, p. 394.

[3] Able Seaman John Campbell also embarked on board the *Activa* 'to attend Lieut. Broughton to England'; Lamb, *Vancouver's Voyage*, IV, p. 1645. Broughton makes no mention of Campbell in his journal.

[4] Alejandro Malaspina (1754–1810). For an account of his voyage, see David et al., *The Malaspina Expedition*.

[5] '* Rocks' in margin; they are Isla de Adentro, 227 metres high, and Isla de Afuera, 206 metres high.

[6] James Colnett, *A Voyage to the South Atlantic and round Cape Horn into the Pacific Ocean*, London, 1798, p. 94, in which Colnett gives the longitude as 110°, i.e. west of Greenwich. The fact that Broughton reached England in the summer of 1793, and Colnett did not reach England until November 1794, means that Broughton's journal was written from notes after his return to England.

[7] Vancouver actually placed the cape in 22°52′N, 250°16′18″E; Lamb, *Vancouver's Voyage*, IV, p. 1432. Broughton must have obtained this information from Vancouver on their return to England and before Vancouver made his final adjustments prior to publication of his journal.

[8] Isla Isabel, a small high island 35 miles WNW of San Blas and about 17 miles offshore.

arriv'd on the 1st of February, passing between a white rock,[1] and the point of land which is fortified carrying 4 f*atho*ms water.

The Establishment of St Blas[2] was first settled in 1767 on the low land by the Port, but after a residence of eight years the situation was so unhealthy as to oblige the inhabitants to remove the town to a rocky mountain elevated*[3] 300 feet above the plain, and distant a short mile from the Sea. Towards the Water the Base and Front of this rock are entirely inaccessible, and the ascent, in every part is steep and capable of a long defence. The circumference of this singular hill is nearly a mile, the town is built partly on the summit and on the sides, containing on a moderate computation 4000 Inhabitants, very few of this number are whites, but the majority is compos'd of all colours; The genteeler class of Spaniards consist of the Commissary & Contador,[4] and the families of the Pilots and Merchants, the rest are Soldiers and Seamen belonging to the Arsenal and different Vessels. When the Port is full the Spaniards may compose a tenth part of the Population. The annual expence of the Establishment to the Spanish government is reckon'd at 350,000 Piastres which are brought by land from Mexico[5]. There is no commerce at this Port but what may arise accidentally from the Galleons and other Ships touching here for supplies in their way from China. They find ample refreshment but at an extraordinary price as every article consum'd in this place is brought from a considerable distance, nor is a garden or any kind of cultivation to be seen within eight miles of the town. The entrance to the Port is form'd by a Peninsula elevated 200 feet above the surface of the sea, and entirely commanding the whole passage: ten twenty four pounders are mounted en barbet and six more are ready, ten brass guns are plac'd on the rising ground and if properly situated would render all access very difficult. Close to this rocky point is the deepest water of fourteen and eighteen feet but the carenage which is opposite has no more than ten feet water where the vessels are heav'd down and repair'd abreast of the Arsenal in which is every convenience for Ship-Building: the largest Vessels are oblig'd to be unladen in the road where is good and safe Anchorage in 5 & 6 fathoms water. The entrance into the Port does not exceed 200 yards and is on the Shore to the left. From the Shore to the right a sand bank shoots out into the Sea. It is high water full and change between 12 and 1, and flows 6 feet.*[6] From this point of view the town and neighbouring country make a very noble and picturesque appearance. The governor's house and magazine for Stores are on the summit of the rock and form two sides of a Square but no guns are mounted there. The magazine contains the Stores that are annually sent off to supply the Presidencies [presidios] on the coast of California. 200 men are commonly employ'd at the Arsenal in the manufacture of Cordage from the American Aloe and in working up the Cedar which is of a most excellent kind. This wood is cut on the banks of the river St Iago[7] which falls into the Sea to the north of the Port and is not so convenient for vessels as that at St Blas is from the intricacy of the Channel. There is an arm of the Sea two miles to the South which is not accessible to boats except in the rainy Seasons on account

[1] Inserted in margin: 'NB A Rock Pedro Blanco [Piedra Blanca del Mar] bears from the Island Isabella E½ S 8 Leags and St Blas bears ESE 4 Leags from the Rocks.'

[2] Inserted in margin: 'St Blas Late 21°29′40″No Long. 254.49E.'

[3] Inserted in margin: '*Colnet says 600 feet'; but not given in Colnett, *Voyage for Whaling* nor in Howay, *Journal of Captain James Colnett*; which again suggests that Broughton was in contact with Colnett on the latter's return to England in 1794.

[4] Paymaster.

[5] I.e. Mexico City, here and later in this Appendix. Where ambiguity arises, [City] has been added to the text.

[6] Inserted in margin: '*Colnet says at 10 Oclock flows ten feet'.

[7] Río Grande de Santiago.

of a Bar which runs across within which the water is in depth 9 or 10 fathoms. The country is low, cover'd with wood and entirely intersected by water so that in the rainy Season the town appears to be situated in the midst of an Island. In consequence of this stagnated water the Air is fill'd with Insects especially Mosquitoes. The land and Sea breezes are regular but in the middle of the day the heat is excessive. The rainy season begins in July and continues to November at which time the place is reckon'd unhealthy and generally fatal to Europeans on their first arrival. All who have it in their power leave the place to avoid the contagion which is then prevalent, even the Natives depart also, and there are seldom more than 500 remaining whose circumstances prevent them from going. This Establishment is calld the Burial place of the Spaniards. It has been propos'd by the present Commandant to remove the naval department to Acapulco as more convenient and equally eligible for the Navigators with the North West Coast of America, and the works that have begun for the protection of the Arsenal remain unfinished on that account. It appears probable that this proposition of the Commandant will in time be follow'd as so much Sand is gradually accumulating in the Entrance that the harbour will soon be useless for vessels of any burthen. The Commandant who is a Captain in the Spanish Navy has a Salary of 3000 dollars with other allowances that amount to 2000 more; He has the sole direction of the naval and military departments, and his appointment remains for 5 years. The civil Officer is the Commissary[1] who resides in the Governor's house. Excepting the Europeans the Inhabitants are miserably lodg'd, and so great is the scarcity of Water that it is brought upon Mules from the distance of six Miles. NB I remain'd eight days at St Blas.

On Thursday morning the 14th of February arrived the Viceroy of Mexico's passport for me to proceed to Mexico [City], and the next morning accompanied by Don Joseph Mozino[2] a Gentleman employ'd in botanical researches we pursued our Journey on Horseback. Señor Don Juan de la Bodega y Quadra the Commandante and two other Officers were pleased to shew me the last of their attentions by riding some distance in Company. Too much cannot be said in favor of the universal Politeness and friendly civility received by me from[3] Señor Quadra to whom I acknowledge myself under great obligations.[4]

The first village[5] we came to after leaving St Blas (from which it) is distant eight miles & is situated near the head of this arm of the sea which here terminates in a Lake. The face of the country is wooded, and intersected by several pieces of water. The inhabitants are employ'd in making Salt from the Lake, and in cultivating Gardens for supplying St Blas with vegetables, which are chiefly drawn from this spot and the village of Havastunba[6] seven leagues further from this hamlet remarkable for its bad accommodations where we were oblig'd to sleep at a miserable cabin. The road became very bad as we approach'd Tepic tho' the prospect was varied by several Huts and plantations of the Sugar Cane, Plantain &c.

[1] Francisco Hijosa; Michael E. Thurman, *The Naval Department of San Blas*, Glendale, California, 1967, p. 342.

[2] José Mariano Moziño (1757–1820), who accompanied Broughton as far as Mexico City. For an account of his career, see Iris Higbie Wilson, trans. & ed., *Noticias de Nutka: An Account of Nootka Sound in 1792 by José Mariano Moziño*, Seattle and London, 1970.

[3] 'from me by' in the MS, which is clearly a transcription error.

[4] The above paragraph was originally inserted above the title of this journal signed 'WRB' with 'Tepic' written in the margin.

[5] Possibly Mecatán.

[6] Not identified and possibly now disappeared.

Tepic

The situation of this place forms an agreeable view as the houses are white, and diversified by the intervening verdure of Trees, it is seated at the base of the mountain of San Juan[1] in a plain surrounded by hills on all sides. The streets are regular and intersected at right Angles. The houses generally of one floor are built with unburnt Brick and pav'd with tiling which tends to keep them cool. The population may amount to 5000 Souls, and is compos'd of all Casts. The troops are Europeans and the whole militia is computed at 800 men ready to march to St Blas whenever requir'd, besides there is another company of Europeans who are generally employ'd as Marines. The Commandant of St Blas and the Naval Officers belonging to that Settlement when not employ'd reside here. A small River runs by the Town which would answer many good purposes were the inhabitants inclin'd to make it useful, about three miles below the town it is contracted by its rocky banks to the breadth of eight feet when it throws itself in a beautiful cascade of sixty or seventy feet, thro' a variety of shrubs which render it a very romantic & interesting Object. The mountains have been work'd for minerals particularly a small hill near San Juan but with indifferent success from the greatness of the labour: 24,000 piastres were the Profits which not answering the Expence the working of the Mine was discontinued. The great road to El Rosario[2] leads thro' this town where are the most considerable mines of Silver in New Spain. Tepic has no Trade or Manufacture, but the country to the West is very fruitful in grain and Pasturage which supports an immense number of Cattle. All articles from Europe are of an astonishing Price. The climate is not esteem'd healthy but much superior to that of St Blas; The dress of the women is singular. It is made up by a superfluity of linen and silk, their hair is comb'd back and platted into a long queue reaching below their waist and adorn'd at the top of the head with heavy bunches of Ribbon. Inelegant Ear-rings & Necklaces, Silk Jackets, and Petticoats laden with fringe and gauze adorn their persons which are as naturally devoid of grace as their faces are of beauty. The circumference of Tepic is between 2 and 3 miles.

The road to San Lionel[3] from Tepic passes thro' cultivations of maize, in our way we cross'd the river of Tipsi[4] which in the rainy season is scarcely passable; As the road was continually winding up the rocky mountains we were oblig'd to delay our journey on account of its roughness and to sleep at the wretched village of Ocote[5] scarcely shelterd from the weather by the shatter'd and ruinous mud walls of a most deplorable Venta.[6] Tititan[7] is situated in a fine plain surrounded by an amphitheatre of immense mountains. The Town of Compostella[8] laid 8 leagues to the SW of our route and was formerly the capital of New Galicia till the Presidency was remov'd to Guadalaxara[9] [Guadalajara], since which time it has considerably decreas'd in population. Tititan is a large village containing 6 or 700 inhabitants. The Intendant[10] possesses an extensive Sugar Plantation valued at 170,000 Piastres, 8000 head of Cattle are employ'd in the Estate and 400 horses alone in the manufactory of the Sugar. A small river

[1] The 2,103-metre-high [2,243m on *Times Atlas*] Volcán San Juan.
[2] A small town, about 80 miles NNW of Tepic, on the road to Mazatlán.
[3] Not identified.
[4] Not identified.
[5] Not identified.
[6] Circled and 'Inn' written in margin.
[7] Not identified.
[8] Situated about 20 miles south of Tepic.
[9] 'Gaudalaxara' in the MS, which is presumably a transcription error.
[10] *Intendente* or manager.

runs thro' the village which supplies the mill with water. The road to Ahuatacam [Ahuacatlán] winds round the Volcano of Shirebourne [Volcán Ceboruco] which at times emits Smoke, but it is several years since any Eruptions have happened. The marks of its devastations are scatter'd over the plain in large fragments of Stone, and the S West side is a continuation of its Effects thro' a vast heap of disjointed rocks over which the road has been made with great difficulty for near a mile. Ahuatacam[1] is a large town built as is usual thro' the country, of unburnt brick and contains 1400 inhabitants, from this place to the mountains the cultivation of maize continues, and many Mills are interspers'd to purify the Silver metal that is procur'd in the neighbourhood, which is very trifling, the whole Produce not amounting to 200£ per ann*um*. The pass that leads to Machitiltic[2] is perhaps the most dreadful that can be imagin'd, it is made out of the side of a stupendous rock rising to a great height above, and ending in a precipice beneath wash'd by a torrent and scarcely visible from the immense distance. This road to appearance so dangerous & impracticable is however frequently travell'd by night, here we met a company of Soldiers conveying 200,000 Piastres on Mules from El Rosario to Mexíco. Many mules carried unwrought Silver in pigs, each pig weighing 70 lbs, half a million of dollars was the estimated amount of the Cargo: Those who go in carriages are oblig'd to have them taken to pieces as for 5 leagues the road is impassable for any conveyance of the kind; the village of Madalena [Magdalena] is dirty and miserably poor, situated on the banks of a lake which is seven leagues in circumference, from hence to Tequila we pass'd over barren downs cover'd with Stones; Tequila makes a pretty appearance, interspersed with plantations of the Sugar Cane, Maize, and the Aloe,[3] which plant is in great request for the purpose of making a liquor resembling brandy and which is preserv'd by distillation.[4] The Sugar here is coarse and work'd by Mules, we pass'd by the mine of Butovielo which are work'd partly under the bed of the River St Iago from whence is procur'd the greatest quantity of Silver, thro' the villages of Amenhtlam [Amatitán] and Astellisso[5] till we descri'd the town of Guadalaxara. The country we travell'd over was barren, and the villages wretched, very little provision was to be procur'd, and that consisted of Eggs, and bread made of Maize.

Guadalaxara is seated in the centre of an extensive plain partly cultivated with maize; the rest is Pasture land without any inclosures and there is scarcely a tree to be seen but on the hills and near the villages that are spread in great profusion over the plain. It is the capital of New Galicia, lies 80 leagues to the eastward of St Blas, and 120 westward from Mexico. Its Latitude is 20°45N. The Lake of Chapala [Laguna de Chapala] is 10 leagues from this City and the river St Iago 5. The Streets are regular, and intersect each other at right Angles. The Houses consist of one Story, are flat-roof'd, and have projecting pipes to carry off the water into the Streets. The window glass is very coarse and there is universally an iron casement to secure it from without; The buildings are heavy and clumsy, some are fac'd with stone but generally with unburnt brick and are whitewash'd. The city contains 30 Churches, 15 convents for men, 5 for women, 2 Colleges and the number of inhabitants is computed at 25,000 of which number 3000 are white, the rest of all gradations of Colour. Their manufactures consist of cottons and coarse linens and the abundance of the country renders the articles of living very

[1] 'Ahualatacam' in the MS, which is presumably a transcription error.

[2] Not identified.

[3] The Agave or American Aloe (*Agave americana*), which when tapped yields a copious juice which is fermented into pulque, the traditional native beverage of Mesoamerica.

[4] As tequila.

[5] Not identified.

cheap; but European goods are extravagantly high. The inhabitants are poor as the greater part of the property is divided among the Clergy who are said to enjoy the third part; the Tribunal or Royal Audience[1] is held in this city over which the Governor presides and bears the title of President. Here is also a Cathedral, Bishop and Canons with handsome Salaries,[2] the Bishop's income alone amounts to 80,000 Piastres annually. The Streets are dirty an evident proof of a bad Police,[3] in the Suburbs a large Hospital is building capable of containing 1000 Patients. This city is indifferently supplied with water and from its low situation on a plain sometimes molested with Sand. The manufactory of Tobacco employs 1500 women, and the profit of the duties last year arising from the imports amounted to a million and an half after deducting the expences attending its preparations for Sale.[4] A million is sent to the King payable on the imports of Tobacco.

The road from Guadalaxara to Mexico leads over the hills that encircle the plain, and crosses the river St Iago which is broad but shallow and rocky. It takes its course from the Lake Chapala which is 40 Leagues in circumference. A Bridge has lately been built here as many accidents happen'd by fording the Stream. The country for many leagues was exceedingly barren, on an extensive plain were some scatter'd huts, not the least cultivation appear'd tho' the plain seem'd capable of much improvement by Husbandry. The same stile of Landscape continued over Sunburnt plains destitute of wood & water, some trees were thinly scatter'd on the hills, but we saw but few houses that could give us any idea of population. At Irias[5] the last village in the province of Zapoacam we had a fine view of a distant country bounded by the mountains of Goanaxuato [Guanajuato] celebrated for the richness of their mines and of the province of Mechoacan [Michoacán]. The plains were coverd in some parts by plantations of Chili pepper near the town of Irapuato which is distant 10 leagues from Goanaxuato. It is of some extent, contains 4 Churches, and a convent of Franciscan Monks, the market here is decently supplied, and the population considerable. Salamanca the next town is built amidst fields of maize, wheat, and Mustard. Its population is nearly the same as the town of Irapuato. The market place is handsome, adornd with a Piazza, and encircled by Shops of the Traders. Considerable damage happen'd in this town some years since by the overflowing of the river which destroy'd all the adjacent fields for many miles. Plantations of maize, and corn accompanied us to the town of Passeo,[6] which are intrench'd with small Canals that convey the water to the plantations from wells & which cause no little trouble to the Labourers. Passeo is an extensive Parish with a handsome Church having under its jurisdiction not less than 30,000 Inhabitants of the town and neighbouring Country. Queretaro [Querétaro] is a very handsome town, and has a lively appearance from the profusion of painting on the walls of the houses and Churches. In the centre of the Plaza[7] is a statue of Charles the 4th of Spain, elevated upon a Stone Pillar 50 feet in height. There are supposed to be 12,000 Inhabitants in this town. The manufactures are trifling consisting chiefly of Cotton, this town is under the jurisdiction of the parliament of Mexico; Jesuits of all kinds are here in great plenty, the streets are universally pav'd, and the houses of the higher Order elegantly furnish'd but the Ornaments

[1] Audiencia: the supreme court.
[2] '200,000 Dollars' crossed out.
[3] Public order.
[4] 'In the leaf it is 12 Dollars' crossed out.
[5] Not identified.
[6] Not identified.
[7] Inserted in the margin: 'Square'.

and Pictures in the Churches have no grace or good Taste to recommend them. A regiment of militia is quarterd here, but the town affords no public diversion except that of Cock-Fighting and the general complaint is a want of Society, the few families we convers'd with were polite but seldom associate together. The road from this town went under an Aqueduct rais'd upon Arches 60 feet in height, and carri'd us to the town of St Jean del Rio [San Juan de Río] which is an extensive place, has wide Streets but indifferent houses. The country being Stoney is sparingly cultivated for some time, but we were agreeably surpriz'd by the sight of a plain which fed vast numbers of Cattle, and which was encircled by an immense range of woody hills. The country in the dry season is supplied with water from artificial lakes which is convey'd to the Plantations by Canals. The farm of Annoyo Jasco belongs to the crown and is let at the rate of 8000 Dollars per ann*um*: it formerly was the property of a Jesuit who realiz'd an immense Sum by the production of Corn &c. At this place the Inn is a Stone building but affords little comfort beyond good Stabling tho' it cost the Sum of 83,000 Dollars. The Country now began to improve, the vallies were richly cultivated, and many farmhouses were interspers'd on every side: Tula[1] where we next arrivd is famous in history for the defeat of the Mexicans by Cortez when it is said 20,000 of these unfortunate defenders of their country were destroy'd. It has decreas'd in population since that time, a small river runs by the town into which the waters of the lake of Mexico are emptied by a Canal that is said to have cost 30 millions of Dollars. Our road laid thro' hills that were overrun by the Aloe from which plant a palatable beer is made call'd Putea.[2] From these eminences we had extensive views particularly of the lake of St Christopher of the town of Qapollam [Cuautillán?], and of the hills that encircle Mexico to the North. At a small village on these heights, which is remarkable for paying 9000 dollars annually to its rector tho' the inhabitants are miserably poor, we descended to the valley in which Mexico is situated. We followed the direction of the Aqueduct that goes to Guadalupe[3] which is a small town containing 3 Churches, and a convent of Capucin Friars, and arriv'd on the 8th of March at the celebrated city of Mexico.

Mexico [City]

The road from Guadalupe to Mexico is thro' an Avenue of Poplar trees with a canal on each side, it is two miles in length; we pass'd over several small bridges and enterd the town by the Suburbs which tho' the Streets were clean and well-pav'd presented in the houses a most miserable appearance of Poverty. This city is one of the first in the world for regularity of Streets and buildings, the streets all run at right Angles and some entirely thro' the city; are of a proper breadth, well pav'd and always kept clean and neat. All the new built houses have projecting lamps and the Police[4] is excellently preserv'd by the military; No robbery or Murder has been committed for three years: since the arrival of the present viceroy El Conte de Revillo Gigedo the City has improv'd in that respect as before that time it was as remarkable for its want of government and uncleanliness as it now is for a proper discipline and neatness. The City has many public buildings, but none that deserve particular attention. The Churches and Convents are vast heavy buildings that have no pretensions to Magnificence or beauty of Architecture: The Viceroy's Palace has a plain and handsome front that occupies one side of

[1] Situated about 40 miles NNW of Mexico City.

[2] Pulque; see p. 277, n. 3; 'And which is esteemed very wholesome for digestion' crossed out.

[3] 'Gaudalupe' in the MS, presumably a transcription error; Guadaloupe.

[4] Public order.

the Plaza Major, or great Square. It appears to have been built at different times, its height does not correspond with its extent and the singularity of its entrance is awkward as it is plac'd at one of the corners of the building. The Apartments are furnish'd with damask & pier Glasses[1] plac'd without Taste or Elegance; but the private rooms of the Viceroy are fitted up in a better and more simple stile after the best European manner. His equipage is of English manufacture and very elegant. The Viceroy is a man of Talents and Politeness, and never relaxes in his attention to business. A certain portion of his time is given up to all those who wait upon him either upon business or to pay their compliments. His attentions to me as an English Officer were very flattering, and deserve this small testimony of my grateful remembrance. The etiquette of the Mexican Court will not permit him to dine in company, which is a hard tax to a man who may be fond of Society, but to the Viceroy it is of little consequence; the Salary amounts to 60,000 Dollars annually. The viceroyaltyship must be resign'd at the end of five years. The Square before the palace is irregular tho' well pav'd, in the middle a fountain and an equestrian Statue of the Spanish King are intended to be erected, and the expence of the undertaking is estimated at 80,000 dollars. The Cathedral fills up another side of the Square, and the two remaining are compos'd of houses belonging to the Conquerors descendants and several small Shops which destroy the effect that might arise from an uniformity of building; The houses in this City as in all Spanish America are built after the manner of their convents with cloisters surrounding the Area, and galleries answering above to the Piazza below, the first Story is the one inhabited, the ground floor is for the Servants, Stables &c: The Entrance is thro' a Porte cochere[2] as in Paris: The Furniture is inelegant and tastelessly dispos'd, compos'd chiefly of looking glasses upon white wainscoating, adorn'd at the top with damask, and small Carpets in the middle of the Apartment: Some of the best houses are painted in the inside on the walls, but the floors are generally brick or Stone, and the ceilings stuccoed. The houses are built of a Stone of volcanic origin which has a reddish hue, the Balconies of the windows are made of Iron curiously painted and the outside of the houses being ornamented in the same manner, it gives to the whole City a new and singular appearance: The number of carriages may amount to 600, are drawn by mules, excepting the Viceroy's which is by horses. Walking is not the fashion at Mexico altho' the walks are excellent, the public amusements are the Theatre, Balls both public and private, and Cock-Fighting, the Society is but indifferent when the population of the City is consider'd; This may arise from the little intercourse there is between the natives and the Spanish Families who have generally the most lucrative and honorable places in the government, of course the little Society there is kept up by the families who come from Old Spain as the Creoles generally confine themselves to their own Circles of relations especially the Ladies: The time is here indolently pass'd away, the Spaniards wrap themselves in their morning cloaks and smoke Sigarros. There is little female beauty in this extensive city, and the dress is very unbecoming. The Servants are generally Indians, and the neighbouring villages are very full of the families of the Natives. All Agriculture is perform'd by this class of the Inhabitants indeed it is the only means they have to procure to themselves subsistence as the Spaniards never condescend to undertake manual labour, and generally prefer any little employment as more honorable, or become apprentices to Merchants for a certain time till they are able to take a shop of their own; The Shopkeepers make large fortunes. Every European Article in price exceeds all belief being from 50 to 100 per Cent. It is not reckon'd

[1] Large mirrors.

[2] Gateway and passage for vehicles through the house into the courtyard.

disgraceful in the first Nobility to profit by trading if the King gives them leave, The titled families are here very common and as they are generally purchas'd in Old Spain have little or no Antiquity to boast of.

Mint

24 millions of Dollars are annually coin'd in the mint, and sent to Vera Cruz or to the West Indies to defray the expences of the Spanish Islands. 500 men are employ'd here from morning till evening and the expence of manufacturing the Dollars from the Bullion (which is brought in bars of 1351 lbs weight) is at the rate of 6 to 1000 Dollars. The Intendants Salary amounts to 6000 per annum; We were permitted to see the method of purifying the Silver thro' all its branches and their mode of coining it; There is another extensive building separated from the mint for the purpose of disforming the gold from the Silver as it comes from the mines; The Chymical Process for this purpose is curious, every chymical aid is possess'd by this country except Mercury which is brought from Old Spain, and a mine in Peru; The expence of coining 24 millions of Dollars is 40,000 Dollars the annual waste is computed at 40,000 in the manufacture: The tenth part goes to the King. The principal imports of Mexico are in Tobacco, Pulka,[1] Powder, Cards &c which are sold by Government: Near Mexico at Santa Fe are some well-built Powder mills work'd by water, the expence of building was 400,000 Dollars, the saltpetre here is finely pulveriz'd – 100 men are constantly employ'd in it: This is entirely the property of the King and sold to great profit gaining ⅔% upon each pound, it is 3 leagues from the city upon the declivity of the Eastern Hills from whence is a fine view of the city and neighbouring Country. The Saltpetre when pulveriz'd consists of saltpetre 75, 11 of Sulphur and 14 of Charcoal to 1 lb. of Powder,[2] it is sold from 6 to 10 Rials per pound. The Soil in the neighbourhood abounds in an Alkaline salt proper for the purpose; the collection of Señor Longinus[3] at Santa Fe contains many curious minerals, Petrifactions, Shells &c and ancient ornaments of the Mexicans: About a mile from Mexico to the East is the mineral Spring of Piñon del Baños; the water is nearly 38 Inches of Reaumurs thermometer in heat and is reckon'd salutary for the Rheumatism. The public buildings in Mexico are very much inferior to the exaggerated ideas Europeans are apt to form of them, the Cathedral is a large and heavy building with two plain towers in front, and a Cupola in the centre inferior in height to the towers: This Church is not so preposterously laden with Finery, as the others; but the Tabernacle at the Altar is immensely rich, the figure of the Virgin is of solid gold three feet in height. The candlesticks &c of Silver and the railing round the tabernacle is compos'd of gold Copper, and Brass: The dresses in this Church on particular days are almost incredible for their splendour. The dress of the Archbishop, when he officiates cost 11,000 Dollars. The revenues of the Cathedral are estimated at 100,000 dollars which are divided between the Archbishop, the King, and the Canons. In digging under the foundations of the Cathedral last year a colossal Statue and two ponderous circular Stones rudely carv'd were found, in diameter about 10 feet.

[Then in a different hand]
The Markets are two in number, excellent and guarded by Sentinels to prevent disturbance the booths are built of wood & painted green the gates are regularly shut every evening. The

[1] See Iris H. W. Engstrand, *Spanish Scientists in the New World*, Seattle and London, 1981, pp. 92–3.

[2] I.e. gunpowder.

[3] Perhaps José Longinos Martínez; see Engstrand, *Spanish Scientists*, pp. 20–21.

vegetables are brought in boats which unload in the Centre of the City. Provisions are plentiful & very cheap.

Three leagues to the Westward is a Water Mill for grinding Corn, the erecting of which cost 380,000 Dollars & produces to the Proprietors an annual Revenue of 40,000 Dollars. It is worked with horizonal Stones.

The Hospital of St Hippolete contains 112 Lunatics. This excellent Institution appears to be well regulated & the greatest Attention paid to the unhappy People confined there.

The Inquisition is not to be seen.

The cultivation of tobacco employs 8000 People 3000 of whom are Women – The Gross Amount of the sale of this Article is from 6 to 7 Million Piastres – The expence of labour amounts to about 3000 dollars per week. The net profits are about 3 Millions. The College of St Ildefonso is the best building in the City, it has 400 Scholars.

The Hospitio de Pobres contains between 7 & 800 People &c. The Collegio de Biscayenes 300 Women supported by the Recto of Houses given by the first Founders. The Apartments are extensive & kept very clean. The Buildings are said to have cost upwards of 1 Million of Dollars.

The Revenues of the Province pay 13 Million to the King besides 3 Million to the Governor in Places, Officers Salaries &c.[1]

Mexico [City] is situated in Lat. 19°25′55″N Long. W from Cadiz 93.15.45 in an extensive plain surrounded by lofty Mountains at unequal distances. To the West is the Lake Flecuseo in the deepest part it has only 7 or 8 feet Water. It supplies the City with Fish – and is intersected by Canals, by means of which the City is supplied with Vegetables from a distant part, the soil of Mexico being very unfavourable to the growth of Vegetables on account of its Alkaline Qualities. The Circumference of the City without the Suburbs is about 5 Miles. It is intersected in every direction by Canals after the manner of Holland. The Water not being sufficiently cleand is in some places very offensive. The plain surface of the Country admits but of little variety, the roads are kept in excellent order – Its elevation above the Surface of the Sea 7000 feet or 2333 Yds 1 foot. In consequence of this elevation arises the excellence of its temperatures, the heat being seldom disagreeable. In the morning and evening the Air is very refreshing, but the dust which flies whenever there is the least Wind is very troublesome. The rainy season commences in May or June and ends in October. The rain falling mostly in the Afternoon & Night is soon absorbed by the natural dryness of the Soil & makes the Mornings delightful. The situation tho' entirely surrounded by Marshy grounds is generally admitted to be healthy. Rheumatism and Agues are very uncommon. To the Alkaline Nature of the Soil is attributed the correcting of any bad effects that might arise from its insulated situation. This is the opinion of Mr O. Sulivan reputed [to be] the most able Medical Man in the whole City. [From here in a different hand]

On the 23^d^ March 1793 in the morning accompanied by Senor Salamanca of the Spanish Navy,[2] I took my final leave of Mexico on Horseback. It is from that city 90 leagues to Vera Cruz; my friendly Host came part of the way with us. Our road lay over the Plain towards the Volcanic Mountains[3] to the SE that crosses the Canal that leads to Chalco. 8 leagues brought us to that Village. It is situated on the borders of the Lake and has a constant communication

[1] Here two pages have been torn out.

[2] Teniente de Fragata Secundino Salamanca y Humara (c.1768–1839), who was well known to Broughton, having served in the schooner *Sutil* during the joint Anglo-Spanish survey of Desolation Sound in 1792.

[3] The 5,452-metre Volcán Popocatépetl.

with Mexico by Water, supplying the City daily with Vegetables by Punts navigated by the Indians. After leaving Chalco we ascended the Hills leaving the Volcano Mountains to the South. We here could have enjoyed a delightful view of the Country had not squally Weather with thunder & lightning prevented it. 7 leagues brought us to the Village of Rio Frio where we slept.

In the Morning we pursued our Route thro' the Hills and through forests of Stone Pines, in many parts of which were fires for Cultivation. After 4 or 5 leagues ride we rounded the Volcano Mountains when a fine open Country richly diversified with Corn fields presented itself to our View. 7 Leagues from Rio Frio brought us to the town of St Martin [San Martín] where we dined. This town is of no great extent, it contains one Parish Church and a Convent of Francisco Monks. The Market is well supplied with fruits. In the afternoon through a whole whirlwind of dirt we pursued our route and by 8 at night having rode 7 leagues we reached the City of Puebla, the road chiefly lay through extensive Corn fields and within a league of the City it became barren and rocky – In the evening we paid our respects to the Intendante who received us very politely and in the Morning favor'd us with his Carriage to see the City. This town or City was the first built by the Conquerors 10 years after Mexico. They called it the Citiz[1] de Los Angelos. It is situated about 30 leagues to the Eastward of Mexico upon a Stoney and sandy soil. The Volcano Mountain to the Westward forms a grand and picturesque view at the distance of 10 leagues and to the Eastward is the Mountain of [indecipherable but perhaps Thloisceiles][2] rising gradually to a considerable height. The Country is richly cultivated to its base. This Mountain is not cap'd like the others with snow, but presents a verdant appearance. Near the environs of the town is a mineral spring of a most disagreeable Odour; it possesses some medicinal virtues but has never been properly analys'd. The City is regularly built & paved but very inferior in both respects to Mexico and the Street very offensive in general from the dirt having collected to a great degree. It contains 70,000 inhabitants – 18 Convents 11 of Women and 7 of Men – 12 Churches including the Cathedral, and the Collegio of Saint Espiritu Santo, a Public School. They manufacture here Coarse China and Cloth both Linen and Woollen of a Coarse kind with oil in abundance – The town is well supplied with water by fountains The Cathedral (tho not so large) is more agreeable to the rules of Architecture than that of Mexico and the decorations (tho not so costly) have more taste – It has a Bishop's See and Chapter with handsome Salaries. The Bishop has 80,000 Dollars per Annum. The City possesses no public walks worth mentioning & their Carriages amusements &c are inferior to Mexico in every degree. Indeed the happiness of social intercourse seems universally unknown to the Kingdom of New Spain. The Intendant[e] or Governor is the Chief Civil Officer of the Crown, he superintends the whole Province which extends its jurisdiction to 800,000 Souls. There are also 12 Regidores[3] out of which are chosen 2 Alcaldes[4] for the internal police of the City. The Country is particularly abundant in Corn of all Kinds and its temperature from the vicinity of the Mountains is supposed to exceed all the Provinces of New Spain. The Governor is a Lieut Colonel in the Army with a Salary of 7000 Piastres a year. About 300 Dragoons comprize the military force of this City with their respective officers. The Governors wife is a french woman, Handsome, Polite & Affable. The next day we pursued our route for Vera Cruz thro rocky and Sandy

[1] Partly indecipherable, but perhaps a transcription error for Ciudad = City.

[2] The 4,461-metre La Malinche.

[3] Aldermen.

[4] Mayors.

Country leaving the Mountains of Thloisceiles to the north, Thaslaria[1] to the South the distant hills & part of the Way cloathed with Pine trees; after passing thro' 2 villages we arrived at Pineus 7 Leagues from Puebla[2] w[h]ere we dined and 8 leagues further brought us to Alosso de Agua a miserable Hotel where we slept. The road lay thro a level Country and great cultivation of Maize about the village of Napatua which we passed on our Way.

In the morning we pursued our Journey thro' an extensive plain of pasturage and after riding 6 leagues we breakfasted at the Venta del Soto and 7 leagues further thro a sandy and desert Country with rocky Hills covered with a few palm trees we arrived at the town of Perote. This is a scattered town containing only one Parish Church, the houses chiefly of wood & tild with wooden shingles & is situated on the side of a range of hills & within a ¼ mile is a fort with Barracks for a Regiment of Men intended as a retreat from Vera Cruz. Embrazures are opened in every direction. It is built of Stone & surrounded by a Ditch. It cost 3 Million of Piastres. There are Plantations of Corn &c that thrives very well in the Sandy Soil. At this town the road commences so bad as to oblige every body to leave their Carriages and proceed to Vera Cruz in Litters. Rainy Weather this evening obliged us to lodge in a Cottage about 2 leagues beyond the town and in the Morning the same disagreeable weather continued; 10 leagues from Perote brought us to Xalapa in as many hours thro a very bad road partly paved and very dangerous for Horses. – Very hilly Country Cloathed with Pine trees and as we approached Xalapa it became highly diversified with rising grounds & verdant Vallies agreeably interspersed with Clumps of Woods. On our way we met drives of Mules laden with Quicksilver for Mexico. Xalapa is situated in this beautiful Country, indifferently built with wooden casements to the Windows & very little glass. It contains 4 Churches & being Thursday of the Saint's Week every body was crowding to Church; the town is regular and extensive and paved in the same manner as the road very uneven. This road is paved for 8 or 10 miles at intervals and so badly as to render it of very little advantage. Its use is principally in the rainy season from the Clayness of the Soil rendering it impassable otherwise. By 7 at night we reached the Village of [blank] three leagues further & took up our residence for the night.

At daylight we continued our route for Vera Cruz over the Mountains to Plan del Rio 5 leagues further. A little beyond the Village we had an extensive view of the Country towards Vera Cruz bounded by the Sea. The road in many parts was paved and as we descended the Hills our road lay thro a desert & stoney plain covered with small wood, destitute of Verdure and a strong pathway.

This continued till our arrival at the Village of Antigua on the bank of the River of the same name & during our route of 12 leagues this Morning we have seen no cultivation whatsoever. Some miserable log houses are scattered along the road. Antigua was the ancient Vera Cruz where Cortez first disembarked & built after burning his Vessels. It contains one Church which is going to be repaired as the religious mansion of this Kingdom. It[s] inhabitants do not exceed 3 or 400. Cocoanut trees are dispersed amongst the Cottages tho not in great plenty. – We here crossd the River (which tho wide) is very shallow and pursuing its course for ½ a league brought us to the entrance (which seems to be inaccessible even to Boats), our way from thence to Vera Cruz 5 leagues. We pursued by the Sea Beach and by 8 at night we entered the town by permission of the Governor, who obligingly ordered the gates to be opened. Two forts of no great strength towards the land. It is defended by six Bastions flanking the Walls. The Circuit of the town is about a Mile. It has three gates of entrance which are always closed every Night. The town is indifferently built and paved. Houses of 2 or three

[1] Not identified.

[2] 'Public' in journal which is surely a transcription error for Puebla.

stories with projecting balconies & wooden lattices (the humidity of the Air corrupting Iron). Every house by drains conveys their odour into the street which even remains & occasions a most offensive smell. Its effect in the hot seasons must be very prejudicial to the health of the inhabitants. The Population is estimated at 12,000 Souls, exclusive of the Military, Clergy & Merchants, the greater Number are Blacks of all degrees. It contains 3 Churches & 4 Convents for Men; part of the regiment of Mexico constitute the garrison of the Town. They are of New Spain & this being the centre of Commerce for New Spain the road is continually full of Vessels of all descriptions the Trade being open to the Ports of Spain. Formerly it was confined to Cadiz only & the commerce then was conducted by Pi[l]otos.[1] The Merchants now complain of the Markets being over stocked. No foreign Vessels are admitted & two Brigs of force are employed to prevent the Contraband trade which at present seems very trifling. The Country about the Town is a sandy desert uncultivated, most of the Provisions are brought from Villages at a considerable distance & chiefly from Xalapa the Vegetables. The water in the town is very bad they are not at present employed in conveying water from the River of [blank] by Aqueducts, they partly succeeded once before the swelling of the River destroyed their labour after an expence of 60,000 Dollars. A Handsome pier is well constructed for l[o]ading & unl[o]ading the Vessels with Cranes for the purpose. The landing by it is very good except in the Northerly winds when it is impracticable. The direction of the Road is NW & SE & the prevailing Winds North, SE & SW. The Breezes set in from the Eastward very fresh in the regular seasons which is from [blank] to [blank]. The violence of the North Wind renders the road at times very unsafe, the Men of War are all attached to the Castle rings by many Cables. Not withstanding all the Precautions a line of Battleship was lost here in the year [blank] tho secured by 12 Cables. These accidents are not very common as the Height of the Castle protects them a good deal from the Wind & the Bank of Coral without shelter them from the Sea which breaks with great force over the Pier & against the Sea Wall of the Town. The northerly winds when it blows hard seldom continues more the 24 hours but when moderate it lasts 3 or 4 days preventing Vessels going to Sea. These gales happen chiefly from the Month of October to March & the rainy seasons commence in May & continue to October or November. It seldom rains in the day but generally all the night attended with excessive heat & violent thunders & lightnings. At this time the Merchants return to Xalapa famous for its Salubrity and agreeable temperatures. During my stay of three weeks the Thermometer in the Shade rose from 78 to 81 and the wind daily varying from N° to SE blowing at times very fresh from the North Quarter preventing any communication with the Vessels attended with thick hazy Weather at Sea.

The Governor of Vera Cruz being Inspector General of the Finances at Mexico he resides there during peace and the Governor of the Castle who has the rank of Brigadier in the Army occupies his place at Vera Cruz. The Civil Government remains with him assisted by 12 Regidores &c for the internal police of the Town. On the 1st April sailed the Juno frigate of Spain. This ship had lately arrived with despatches having made his Voyage in 35 days from Cadiz touching at Porto Rico & the Havannah by the Way. I much lamented the Viceroy's not permitting me to go in her but as our Sailing was fixed for the same day when I left Mexico it could not make much difference & we only waited the arrival of Sr Valdes[2] to depart. Some

[1] Spanish naval officers equivalent to masters in the Royal Navy.

[2] Capitán de Fragata Cayetano Valdés y Flores (1767–1835), who was well known to Broughton, having served in command of the schooner *Mexicana* during the joint Anglo-Spanish survey of Desolation Sound in 1792. Valdés was accompanied by José Cardero y Meléndez (1766–c.1810), artist during the *Sutil* and *Mexicana* expedition.

unfortunate delays prevented his arriving with the Viceroy's despatches before the 13th an event I could not foresee or prevent the extraordinary delay it caused in my Voyage to Europe.X[1]

On the 16th April we embarked on board the Minerva Spanish Frigate of 40 Guns & in the morning we sailed from Vera Cruz with the wind at East. The wind the next day veerd more Southerly and we steerd NNE to gain the Parallel of 25°30′N where we might expect to meet Northerly winds. After making the above Lat*itude* we experienced constant breezes from the Eastward and we kept traversing between the Parallels of 24° [and] 26°N for the next fortnight experiencing by our observations very little currents if we except 50′ Southing in one day. We soon after struck soundings on the Tortuga Bank in 120 fathom and kept plying to the East within the bank. – At last a Northerly breeze came without our profiting by it thro' a conviction of its continuance the fallacy of which was pointed out the next day when we were joined by a Brig of War from the Havannah with orders to proceed there, three days after we entered the Port having steered SEbS from the Tortuga Bank in 37 fathoms & Latitude 24°55′N.

Vera Cruz

The only dangers are two Banks in coming in from the Northward and on the Southernmost is 24 feet & the other 19 feet they are avoided by leaving them on your larboard hand coming in. The Mark is the two Centry Boxes in one of the Castle fronting the Anchorage. The Channel to the South East is free of all dangers but what are visible the least water is 23 feet. The Tide is irregular but the perpendicular rise never exceeds 3 feet. 119 Cannon on the Castle chiefly & in general 24 pounders Brass and Cannon. From the look out they are 8 leagues to Sea. Garrison in War 3000 Men at present 120. It is 30 years since its thorough repair & is now in good order. – The Castle lays immediately abreast of the Town distant a short half mile 29 feet alongside the walls of the Castle towards the town it presents a Battery of 30 Guns towards the northern Entrance 40 and nearly the same number in every direction. It is intersected by Canals forming Bastions &c flanking the works in every direction. It is said to have cost 20 Million Sterling building &c. The town of Vera Cruz is situated in the Lat*itude* of 19°12′0″N and Longitude 262°30′E 5 leagues to the Southward of the River Antigua where Hernandez Cortez landed & built the first town. It was afterwards removed here as affording better shelter for vessels under the small Island on which the Castle is built and which constitutes the defence of the Place. The town is encircled by a Wall at the Northern and Southern Extremes.[2]

After remaining three days at the Havannah we sailed for Old Spain in a Spanish Brig of War commanded by a Lieut*enant* & accompanied by Don Cayetano Valdes and Señor Salamanca. The Minerva having on board 3,000,000 of dollars it was thought proper she should not proceed to Europe. On the 10th of ~~April~~ May we saild from the Havannah with the wind easterly. In three days we cleared the straits of Bahama having experienced strong currents making to the North 50′ and 70′ pr day. On the 18th we made the Coast of America in Lat*itude*.35°24′N & Long*itude* by ⊙ and ☽ 76°20′E near Cape Hatteras. In the Evening we stood to the Southward. The next morning with the wind at SE we stood to the NE and on

[1] In the margin 'X Description of the Castle look on the other side'.

[2] Here added in pencil: 'Sr Juan de Ulloa name of Island'.

the 19th in the Evening crossd the Shoal lying off Cape Hatteras in 12 fathom water[1] tho' we could not discern the land; as we increased our distance from the Coast the Wind drew round to the SW blowing a fresh Breeze & continued in that Quarter for some days when it became variable from the Southward round to the northward and at last fixed in the NNW a fresh gale with Cloudy Weather.

On the 22nd May we made the Island of Corvo[2] having experienced a Current to the Northward for these last 4 days. Our estimation right to a league. The greatest extent of this Island does not exceed 4 Miles in the North & South direction. Its North point falls abruptly to the Sea, & the South point is a low extreme. It presents a rocky surface falling perpendicularly to the Sea partly covered with verdure. The summit was so obscured by Clouds as to prevent our perceiving its form. We passed within 2 miles of its N° extreme. Its Lat*itude* 39°42N by an observation which was but indifferent at Noon & we saw it 3½ hours after steering SE 34′ Variation 20°W. In passing we just discerned the Island of Flores bearing from it South 30 West by Compass 4 or 5 leagues.

We continued our course SEbE for Cape St Vincent for 2 days, the Wind remained westerly: 100 leagues to the westward of Corvo we spoke an English Privateer 7 weeks from England called the Bustler she had taken nothing during her Cruize of these Islands.

On the 3d day 25th May the wind came to the NE and blew a fresh gale till we made Cape St Vincent on Sunday Morning. Our reckoning 15 leagues ahead of the ship.

In passing the Cape the Wind came to the northward blowing a moderate Breeze with smooth Water. Saw several Vessels & anchored in the harbour of Cadiz the following day. Meeting as we entered a squadron of Spanish Men of War with a Convoy for the West Indies under the command of Rear Admiral Aristazabal.[3]

[1] The Hatteras Shoals.

[2] With Flores, which lies about 18 miles to the south, the westernmost islands of Arquipélago dos Açores.

[3] Gabriel de Aristizábal (1743–1805) served at Cartagena and later in the Philippines. He then commanded the vessel that took an embassy to Turkey in 1785, and he made a much admired series of charts. He was promoted *Teniente General* in 1793, and in 1795 brought Columbus's remains back to Spain.

WORKS CITED OR CONSULTED

Arrowsmith, Aaron, *Chart of the World on Mercator's Projection,* 8 sheets, London, 1790.

——, 'Map of the World', London, 1794.

Barrington, Daines, *Miscellanies*, London, 1781.

Barrow, John, *The Eventful History of the Mutiny and Piratical Seizure of HMS Bounty*, London, 1831.

Barry, J. Neilson, 'Broughton's Log of a Reconnaissance of the San Juan Islands in 1792', *Washington Historical Quarterly,* 21, 1 (January 1930), pp. 55–60.

Bateson, Charles, *The Convict Ships; 1787–1868,* Glasgow, 1969.

Bayly, William, *The Original Astronomical Observations made in the Course of a Voyage to the Northern Pacific Ocean*, London, 1782.

Beaglehole, J. C., ed., *The Journals of Captain James Cook on His Voyages of Discovery.* III. *The Voyage of the* Resolution *and* Discovery, *1776–1780,* 2 vols, Hakluyt Society, extra ser. 36a–36b, Cambridge, 1967; repr. 1969.

Beals, Herbert K., *For Honor & Country: The Diary of Bruno de Hezeta*, Portland, Oreg. , 1985.

Begg, A. Charles and Neil C. Begg, *Dusky Bay*, 3rd edition, Christchurch, 1975.

Benyowsky, Mauritius Augustus, *Memoirs and Travels of Mauritius Augustus Count de Benyowsky*, London, 1790.

Black, Lydia T., *The Lovtsov Atlas of the North Pacific Ocean*, Kingston, Ontario & Fairbanks, Alaska, 1991.

Blumenthal, Richard W., ed., *With Vancouver in Inland Washington Waters: Journals of 12 Crewmen, April–June 1792,* Jefferson, N.C., 2007.

British Columbia Pilot, Vol. I, 12th edition, Taunton, 2004.

[Broughton, Brian], *A Poem on a Voyage of Discovery, undertaken by a Brother of the Author, with other Sonnets, &c,* London, 1792.

Broughton, William Robert, *A Voyage of Discovery to the North Pacific Ocean*, London, 1804.

Burney, James, *A Chronological History of Voyages and Discoveries in the South Seas*, 5 vols, London, 1803–17.

——, *A Chronological History of North-Eastern Voyages of Discovery*, London, 1819.

Busch, Briton C. and Gough, Barry M., eds, 'The Northwest Fur Trade by the Hon. William Sturgis' (reprinted from *Hunt's Merchants' Magazine*, 14, 1846) in Briton C. Busch and Barry M. Gough, eds, *Fur Traders from New England: The Boston Men in the North Pacific, 1787–1800: The Narratives of William Dane Phelps, William Sturgis and James Gilchrist Swan*, Spokane, Wash., 1997.

The Calcutta Christian Observer, September 1837.

China Sea Pilot, Vol. I, 7th edition, Taunton, 2007.

China Sea Pilot, Vol. III, 4th edition, Taunton, 1982.

The Chinese Repository, Canton, etc., 1832–51.

Colledge, J. J., *Ships of the Royal Navy,* London, 1987.

Collins, David, *An Account of the English Colony in New South Wales,* 2 vols, London, 1798, 1802; repr. Sydney, 1975.

Colnett, James, *A Voyage to the South Atlantic and round Cape Horn into the Pacific Ocean*, London, 1798.

Cook, James and King, James, *A Voyage to the Pacific Ocean ... for making Discoveries in the Northern Hemisphere*, 3 vols, London, 1784.

Cook, Warren L., *Flood Tide of Empire: Spain and the Pacific Northwest, 1543–1819,* New Haven and London, 1973.

Cumpston, J. S., *Shipping Arrivals & Departures Sydney, 1788–1825*, Canberra, 1977.

Dahlgren, E. H., *Were the Hawaiian Islands Visited by the Spaniards before their Discovery by Captain Cook?*, Stockholm, 1916.

David, A. C. F., 'Broughton's Schooner and the Bounty Mutineers', *The Mariner's Mirror*, vol. 63, pp. 207–13.

David, Andrew, 'John Sherriff on the Columbia, 1792: An Account of William Broughton's Exploration of the Columbia River', *Pacific Northwest Quarterly*, 83, 2 (1992), pp. 53–9.

——, 'Vancouver's Survey Methods and Surveys', Robin Fisher and Hugh Johnston, eds, in *From Maps to Metaphors: The Pacific World of George Vancouver*, Vancouver, 1993, pp. 51–69.

——, 'Admiralty Charts and William Faden's Copper Plates', *IMCoS Journal*, Winter 2008, No. 115, pp. 7–11.

David, Andrew, Felipe Fernández-Armesto, Carlos Novi, Glyndwr Williams, eds, *The Malaspina Expedition 1789–1794: The Journal of the Voyage by Alejandro Malaspina*, 3 vols, Hakluyt Society, 3rd ser. 8, 11, 13, London, 2001–4.

Day, Sir Archibald, *The Admiralty Hydrographic Service, 1795–1919*, London, 1967.

Delano, Amasa, *Narrative of Voyages and Travels in the Northern and Southern Hemisphere: Comprising Three Voyages Round the World together with a Voyage of Survey and Discovery in the Pacific Ocean and Oriental Islands*, Boston, 1817.

Dening, Greg, *The Death of William Gooch: A History's Anthropology*, Melbourne, 1995.

Dmytryshyn, Basil, Crownhart-Vaughan, E. A. P., and Vaughan, Thomas, eds and trans, *Russian Penetration of the North Pacific Ocean to Siberia and Russian America, 1700–1797: Three Centuries of Russian Eastward Expansion*, Vol. 2, Portland, Oreg., 1988.

——, ——, ——, *The Russian American Colonies to Siberia and Russian America, 1798–1867*, Vol. 3: *A Documentary Record*, Portland, Oreg., 1989.

Dunmore, John, *French Explorers in the Pacific*, Vol. I: *The Eighteenth Century*, Oxford, 1965.

——, *The Journal of Jean François de Galaup de la Pérouse 1785–1788*, Hakluyt Society, 2nd ser. 179–80, 2 vols, London, 1994.

Du Rietz, Rolf, ed., *Bibliotheca Polynesiana*, Oslo, 1969.

Earnshaw, John, *A Letter from the South Seas by a Voyager in the 'Daedalus', 1792*, Cremorne, NSW, 1957.

Earnshaw, Thomas, *Longitude. An Appeal to the Public*, London, 1808; facsimile edition, British Horological Institute, London, 1986.

Emmons, Louise H., *Neotropical Rainforest Mammals: A Field Guide*, Chicago and London, 1990.

Engstrand, Iris. H. W., trans. & ed., *Noticias de Nutka: An Account of Nootka Sound in 1792 by José Mariano Moziño*, Seattle, 1970.

——, *Spanish Scientists in the New World*, Seattle and London, 1981.

Farrington, Anthony, *Catalogue of East India Company Ship's Journals and Logs 1600–1834*, London, 1999.

Forster, John Reinhold, *History of the Voyages and Discoveries made in the North*, London, 1786.

Findlay, Alexander, *A Directory for the Navigation of the Pacific Ocean*, 2 vols, London, 1851.

Galois, Robert, ed., *A Voyage to the North West Side of America: The Journals of James Colnett, 1786–89*, Vancouver and Toronto, 2004.

The Gentleman's Magazine, Vol. 87 for 1817, part 2.

Golder, Frank A., *Russian Expansion on the Pacific, 1641–1850: An Account of the Earliest and Later Expeditions made by the Russians along the Pacific Coast of Asia and North America; Including some Related Expeditions to the Arctic Regions*, Cleveland, 1914.

——, *Bering's Voyages*, 2 vols, New York, 1925.

Gough, Barry, *The Royal Navy and the Northwest Coast of North America, 1810–1914: A Study of British Maritime Ascendency*, Vancouver, 1971.

——, ed., *To the Pacific and Arctic with Beechey: The Journal of George Peard of H.M.S. Blossom, 1825–1828*, Hakluyt Society, 2nd ser. 143, Cambridge, 1973.

Graham, Gerald S., *Great Britain in the Indian Ocean: A Study of Maritime Enterprise, 1810–1850*, Oxford, 1967.

Grainger, Jon D., *The First Pacific War: Britain and Russia, 1854–1856*, London, 2008.

Hall, Basil, *Account of a Voyage of Discovery to the West Coast of Corea and the Great Loo-Choo Island*, London, 1818.

Harland, John, *Seamanship in the Age of Sail*, London, 1984.

Hattendorf, John, Knight, R. J. B., Pearsall, A. W. H., Rodger, N. A. M. and Hill, Geoffrey, eds, *British Naval Documents 1204–1960*, The Navy Records Society, vol. 131, London, 1993.

Hayes, Derek, *Historical Atlas of the North Pacific Ocean*, London, 2001.

Hawkesworth, John, ed., *An Account of the Voyages undertaken ... for making Discoveries in the Southern Hemisphere*, 3 vols, London, 1773.

Headland, Robert, *Chronological List of Antarctic Expeditions & Related Historical Events*, Cambridge, 1989; 2nd edition, London, 2009.

Herbert, William, *A New Directory for the East Indies*, 6th edition, London, 1791.

Historical Records of New South Wales, 7 vols, Sydney, 1893–1901.

Hoare, James E., 'Captain Broughton, HMS *Providence* (and her tender) and his Voyage to the Pacific Ocean 1794–8', *Asian Studies*, 31, 3 (October 2000), pp. 303–12.

——, 'Captain Broughton, HMS *Providence* (and her tender) and his Voyage to the Pacific Ocean 1794–8', *Papers of the British Association for Korean Studies*, 7 (2000), pp. 49–60.

——, ed., *Britain and Japan: Biographical Portraits*, Vol. 3, Folkestone, Japan Library, 1999, pp. 1–12.

Horsburgh, James, *Directions for Sailing to the East Indies*, 2 parts, London 1809.

——, *India Directory, or Directions for sailing to and from the East Indies*, 2 vols, 3rd edition, London, 1826.

Howay, Frederick William, *The Journal of Captain James Colnett aboard the Argonaut*, Toronto, 1940.

——, *A List of Trading Vessels in the Maritime Fir Trade, 1785–1825*, ed. Richard A. Pierce, Kingston, Ont., 1973.

Howse, Derek, *Nevil Maskelyne: the Seaman's Astronomer*, Cambridge, 1989.

——, 'Britain's Board of Longitude: The Finances, 1714–1828', *The Mariner's Mirror*, 84, 4 (November 1988), pp. 400–417.

Hunter, Captain John, *An Historical Journal of Events at Sydney and at Sea 1787–1792*, Sydney, 1968.

Imray, James F., *North Pacific Pilot, Part I. North America (West Coast)*, 4th edition, London, 1885.

Ingleton, G. C., *Matthew Flinders: Navigator and Chartmaker*, Guildford, 1986.

James, William, *The Naval History of Great Britain*, 6 vols, London, 1837.

Japan Pilot, Vol. I, 8th edition, Taunton, 2006.

Japan Pilot, Vol. II, 7th edition, Taunton, 1979.

King, Robert J., '"The Long Wish'd for object" – Opening the Trade to Japan', *The Northern Mariner/le marin du nord*, 20, 1 (January 2010).

——, 'The Molovsky Expedition and Catherine II's North Pacific Empire', *ASEES*, 21, 1–2 (2007), pp. 97–122.

Krusenstern, Adam J. Von, *Voyage round the World in the Years 1803, 1804, 1805, & 1806 ... on board the Ships Nadeshda and Neva*, 2 vols, London, 1813.

Kruzenshtern, Ivan Fedorovich [A. J. von Krusenstern], *Atlas k puteshestviyu vokrug sveta Kapitana Kruzenshterna ...*, St Petersburg, 1813; facsimile reprint, Amsterdam & New York, 1974.

Lamb, W. Kaye, ed., *George Vancouver, A Voyage of Discovery to the North Pacific Ocean,* 4 vols, Hakluyt Society, 2nd ser. 163–6, London, 1984.

Lee, Ida, *Commodore Sir John Hayes: his Voyage and Life*, London, 1912.

A List of the Company's Civil Servants at their Settlements in the East Indies.

Lind, J., *A Treatise on the Scurvy*, 3rd edition, London, 1772.

London Gazette, No. 13813, 12–14 September, 1795.

Lyon, David, *The Sailing Navy List: All the Ships of the Royal Navy – Built, Purchased and Captured – 1688–1860,* London, 1993.

Lettres édifiantes et curieuses, écrites des missions étrangères, 26 vols, Paris, 1780–83.

Mackay, David, *In the Wake of Cook*, Beckenham, Kent, 1985.

Manning, William Ray, 'The Nootka Sound Controversy', *Annual Report of the American Historical Association for the Year 1904*, pp. 281–478.

Marshall, John, *Royal Naval Biography*, 4 vols with supplements, London, 1823–35.
Marshall, P. J., 'Britain and China in the late Eighteenth Century', in P. J. Marshall, ed., *'A Free Though Conquering People': Eighteenth-Century Britain and its Empire*, Aldershot, 2003.
Meany, Edmond S., *Vancouver's Discovery of Puget Sound*, Portland, Oreg., 1957.
Meares, John, *Voyages made in the Years 1788 and 1789 from China to the North West Coast of America*, London, 1790.
Milet-Mureau, M. L. A., *A Voyage Round the World, Performed in the Years 1785 ... 1788, by the Boussole and Astrolabe, under the Command of J. F. G. de la Pérouse*, 3 vols and atlas, London, 1799.
M'Leod, John, *Voyage of His Majesty's Ship Alceste, along the Coast of Corea to the Island of Lewchew*, 2nd edition, London, 1818.
Mockford, Jim, 'The Journal of a Tour across the Continent of New Spain from St. Blas in the North Pacific Ocean to La Vera Cruz in the Gulph of Mexico', *Terrae Incognitae*, 36 (2004), pp. 42–58.
Moessner, Victoria Joan, ed., *The First Russian Voyage Around the World: The Journal of Hermann Ludwig von Löwenstern, 1803–1806*, Fairbanks, Alaska, 2003.
Moore, John Hamilton, *The New Practical Navigator*, London, 1801.
Morison, Samuel E., *The European Discovery of America: The Southern Voyages A.D. 1492–1616*, New York, 1974.
Müller, Gerhard Friedrich, *Voyages from Asia to America for Completing the Discoveries of the north West Coast of America*, London, 1761.
——, *Voyages et Découvertes faites par les Russes le long des côtes de la Mer Glaciale & sur l'Océan Oriental, tant vers le Japon, que vers l'Amérique*, 2 vols, Amsterdam, 1768.
Oliver, Oliver, *Ancient Tahitian Society*, 3 vols, Honolulu, 1974.
——, *Return to Tahiti: Bligh's Second Breadfruit Voyage*, Melbourne, 1988.
Oxford Dictionary of National Biography, 60 vols, Oxford, 2004.
Pacific Coasts of Central America and United States Pilot, 4th edition, Taunton, 1975.
Pacific Islands Pilot, Vol. III, 10th edition, Taunton, 1982.
Parkinson, C. Northcote, *War in the Eastern Seas, 1793–1815*, London, 1954.
Payne, John, *An Epitome of History*, London, 1794.
Pierce, Richard A., *Russian America: A Biographical Dictionary*, Kingston, Ont. and Fairbanks, Alaska, 1990.
Reid, Alan, 'Broughton's Schooner', *The Mariner's Mirror*, vol. 64 (1978), pp. 241–4.
Regulations passed by Order in Council in January 1806. Reprinted in Brian Lavery, ed., *Shipboard Life and Organization, 1731–1815*, Navy Record Society, vol. 138, 1998, pp. 25–7.
Ritchie, G. S., *The Admiralty Chart: British Naval Hydrography in the Nineteenth Century*, new edition, Bishop Auckland, 1995.
Robbins, Helen H., *Our First Ambassador to China*, London, 1908.
Roberts, Henry, 'A General Chart: Exhibiting the Discoveries made by Captn James Cook in this and his two proceeding Voyages', London, [1784].
Robertson, John, *The Elements of Navigation; Containing the Theory and Practice. With all the Necessary Tables*, 2 vols, London, 1754, with further editions in 1764, 1772, 1780, 1786 and 1796.
Roe, Michael, ed., *The Journal and Letters of Captain Charles Bishop on the North-West Coast of America, in the Pacific and in New South Wales*, Hakluyt Society, 2nd ser. 131, Cambridge, 1967.
Salmond, Anne, *Between Worlds: Early Exchanges between Maori and Europeans 1773–1815*, Auckland, 1997.
Sharp, Andrew, *The Discovery of the Pacific Islands*, Oxford, 1960.
Smyth, W. H., *The Sailor's Word-Book*, London, 1867.
South and East Coasts of Korea, East Coast of Siberia and Sea of Okhotsk Pilot, 7th edition, Taunton, 2005.
Staunton, Sir George Leonard, *An Authentic Account of an Embassy from the King of Great Britain to the Emperor of China*, 2 vols, London, 1797.
Stephan, John J., 'The Crimean War in the Far East', *Modern Asian Studies*, 3 (1969), pp. 257–77.
——, *Sakhalin: A History*, Oxford, 1971.
Stommel, Henry, *Lost Islands: The Story of Islands that have vanished from Nautical Charts*, Vancouver, 1984.

Stone, Ian R., 'W. R. Broughton and the Insularity of Sakhalin, *The Mariner's Mirror*, 82, 1 (February 1996), pp. 76–81.

Sturgis, William, 'The Northwest Fur Trade by the Hon. William Sturgis', *Hunt's Merchants' Magazine*, 14 (1846), pp. 532–9.

Syrett, David and DiNardo, R. L., *The Commissioned Sea Officers of the Royal Navy 1660–1815*, The Navy Records Society, 1994.

Tables Requisite to be used with the Nautical Ephemeris for finding the Longitude at Sea, 2nd edition, London, 1781.

The Times Concise Atlas of the World, 8th edition, London, 2000.

Thunberg, Charles Peter, *Travels in Europe, Africa, and Asia, made between the Years 1770 and 1779*, 4 vols, 2nd edition, London, 1795.

Thurman, Michael E., *The Naval Department of San Blas*, Glendale, Calif., 1967.

Tovell, Freeman M., *At the Far Reaches of Empire: The Life of Juan Francisco de la Bodega y Quadra*, Vancouver, 2008.

Tracy, Nicholas, *Who's Who in Nelson's Navy: 200 Naval Heroes*, London, 2006.

Tuck, Gerald, *A Guide to Seabirds on the Ocean Routes,* London, 1980.

Vale, Brian, 'The Conquest of Scurvy in the Royal Navy 1793–1800: A Challenge to Current Orthodoxy', *The Mariner's Mirror*, 94, 2 (May 2008), pp. 160–75.

Wagner, Henry R., *Cartography of the Northwest Coast of America to 1800*, 2 vols, Berkeley, 1937.

Walbran, John T., *British Columbia Coast Names 1592–1906*, reprint edition, Vancouver, 1971.

Waxell, Sven, *The American Expedition*, London, Edinburgh and Glasgow, 1952.

Wilson, Iris Higbie, trans. & ed., *Noticias de Nutka: An Account of Nootka Sound in 1792 by José Mariano Moziño*, Seattle and London, 1970.

Wilson, James, *A Missionary voyage to the Southern Pacific Ocean performed in the Years 1796, 1797, 1798 in the Ship Duff*, London, 1799.

Wrangell, Ferdinand, *Narrative of an Expedition to the Polar Sea in the Years 1820, 1821, 1822, & 1823*, Fairfield, Wash., 1981.

INDEX